Mechanical Vibrations
Theory and Applications

FRANCIS S. TSE
University of Cincinnati

IVAN E. MORSE
University of Cincinnati

ROLLAND T. HINKLE
Michigan State University

Mechanical Vibrations
Theory and Applications

SECOND EDITION

CBS PUBLISHERS & DISTRIBUTORS PVT. LTD.
New Delhi • Bengaluru • Chennai • Kochi • Mumbai • Pune

ISBN : 81-239-0846-6

First Indian Edition : 1983
Reprint : 1998, 2002, 2004

This edition has been published in India by arrangement with
Allyl & Bacon Inc., London

Published by Satish Kumar Jain and produced by V.K. Jain for
CBS Publishers & Distributors Pvt. Ltd.,
CBS Plaza, 4819/XI Prahlad Street, 24 Ansari Road, Daryaganj,
New Delhi - 110002, India. • Website: www.cbspd.com
e-mail: delhi@cbspd.com, cbspubs@airtelmail.in
Ph.: 23289259, 23266861, 23266867 • Fax: 011-23243014

Branches:

- ***Bengaluru:*** Seema House, 2975, 17th Cross, K.R. Road, Bansankari 2nd Stage, Bengaluru - 560070
  • Ph.: +91-80-26771678/79 • Fax: +91-80-26771680
  • E-mail: cbsbng@gmail.com, bangalore@cbspd.com
- ***Pune:*** Bhuruk Prestige, Sr. No. 52/12/2+1+3/2, Narhe, Haveli (Near Katraj-Dehu Road By-pass), Pune - 411041
  • Ph.: +91-20-64704058/59, 020-32392277 • E-mail: pune@cbspd.com
- ***Kochi:*** 36/14, Kalluvilakam, Lissie Hospital Road, Kochi - 682018, Kerala • Ph.: +91-484-4059061-65
  • Fax: +91-484-4059065 • E-mail: cochin@cbspd.com
- ***Chennai:*** 20, West Park Road, Shenoy Nagar, Chennai - 600030
  Ph.: +91-44-26260666, 26208620 • Fax: +91-44-42032115
  • E-mail: chennai@cbspd.com
- ***Mumbai:*** 83-C, Ist Floor, Dr. E. Moses Road, Worli, Mumbai-400 018, Maharashtra Ph.: +91-9833017933, 022-24902340/24902341
  • E-mail: mumbai@cbspd.com

Printed at :
J.S. Offset Printers, Delhi

Contents

Preface

Vibration is the study of oscillatory motions. The ultimate goals of this study are to determine the effect of vibration on the performance and safety of systems, and to control its effects. With the advent of high performance machines and environmental control, this study has become a part of most engineering curricula.

This text presents the fundamentals and applications of vibration theory. It is intended for students taking either a first course or a one-year sequence in the subject at the junior or senior level. The student is assumed to have an elementary knowledge of dynamics, strength of materials, and differential equations, although summaries of several topics are included in the appendices for review purposes. The format of its predecessor is retained, but the text material has been substantially rewritten. In view of the widespread adoption of the International System of Units (SI) by the industrial world, SI units are used in the problems.

The objectives of the text are first, to establish a sense of engineering reality, second, to provide adequate basic theory, and finally, to generalize these concepts for wider applications. The primary focus of the text is on the engineering significance of the physical quantities, with the mathematical structure providing a supporting role. Throughout the text, examples of applications are given before the generalization to give the student a frame of reference, and to avoid the pitfall of overgeneralization. To further enhance engineering reality, detailed digital computations for discrete systems are presented so that the student can solve meaningful numerical problems.

The first three chapters examine systems with one degree of freedom. General concepts of vibration are described in Chapter 1. The theory of time and frequency domain analysis is introduced in Chapter 2 through the study of a generalized model, consisting of the mass, spring, damper, and excitation elements. This provides the basis for modal analyses in subsequent chapters. The applications in Chapter 3 demonstrate that the elements of the model are, in effect, equivalent quantities. Although the same theory is used, the appearance of a system in an engineering problem may differ greatly from that of the model. The emphasis of Chapter 3 is on problem formulation. Through the generalization and classification of problems in the chapter, a new encounter will not appear as a stranger.

Discrete systems are introduced in Chapter 4 using systems with two degrees of freedom. Coordinate coupling is treated in detail. Common methods of finding natural frequencies are described in Chapter 5. The material in these chapters is further developed in Chapter 6 using matrix techniques and relating the matrices to energy quantities. Thus, the student would not feel the artificiality in the numerous coordinate transformations in the study.

The one-dimensional wave equation and beam equation of continuous systems are discussed in Chapter 7. The material is organized to show the similarities between continuous and discrete systems. Chapter 8, on nonlinear systems, explains certain common phenomena that cannot be predicted by linear theory. The chapter consists of two main parts, conforming to the geometric and analytical approaches to nonlinear studies.

The digital computation in Chapter 9 is organized to follow the sequence of topics presented in the prior chapters and can be assigned concurrently with the text material. The programs listed in Table 9-1 are sufficient for the computation and plotting of results for either damped or undamped discrete systems. Detailed explanations are given to aid the student in executing the programs. The programs are almost conversational and only a minimal knowledge of FORTRAN is necessary for their execution.

The first five chapters constitute the core of an elementary, one-quarter terminal course at the junior level. Depending on the purpose of the particular course, parts of Section 3-5 can be used as assigned reading. Sections 3-6 through 3-8, Section 4-9, and Sections 5-4 through 5-6 may be omitted without loss of continuity.

For a one-semester senior or dual-level course, the instructor may wish to use Chapters 1 through 4, Chapter 6, and portions of Chapter 7 or 8. Some topics, such as equivalent viscous damping, may be omitted.

Alternatively, the text has sufficient material for a one-year sequence at the junior or senior level. Generally, the first course in mechanical vibrations is required and the second is an elective. The material covered will give the student a good background for more advanced studies.

We would like to acknowledge our indebtedness to many friends, students, and colleagues for their suggestions, to the numerous writers who contributed to this field of study, and to the authors listed in the references. We are especially grateful to Dr. James L. Klemm for his suggestions in Chapter 9, and to K. G. Mani for his contribution of the subroutine $PLOTF in Appendix C.

Francis S. Tse
Ivan E. Morse
Rolland T. Hinkle

1

Introduction

1-1 PRIMARY OBJECTIVE

The subject of vibration deals with the oscillatory motion of *dynamic systems.* A dynamic system is a combination of matter which possesses mass and whose parts are capable of relative motion. All bodies possessing mass and elasticity are capable of vibration. The mass is inherent of the body, and the elasticity is due to the *relative* motion of the parts of the body. The system considered may be very simple or complex. It may be in the form of a structure, a machine or its components, or a group of machines. The oscillatory motion of the system may be objectionable, trivial, or necessary for performing a task.

The objective of the designer is to control the vibration when it is objectionable and to enhance the vibration when it is useful, although vibrations in general are undesirable. Objectionable vibrations in a machine may cause the loosening of parts, its malfunctioning, or its eventual failure. On the other hand, shakers in foundries and vibrators in testing machines require vibration. The performance of many instruments depends on the proper control of the vibrational characteristics of the devices.

The primary objective of our study is to analyze the oscillatory motion of dynamic systems and the forces associated with the motion. The ultimate goal in the study of vibration is to determine its effect on the performance and safety of the system under consideration. The analysis of the oscillatory motion is an important step towards this goal.

Our study begins with the description of the elements in a vibratory system, the introduction of some terminology and concepts, and the discussion of simple harmonic motion. These will be used throughout the text. Other concepts and terminology will be introduced in the appropriate places as needed.

1-2 ELEMENTS OF A VIBRATORY SYSTEM

The elements that constitute a vibratory system are illustrated in Fig. 1-1. They are idealized and called (1) the *mass*, (2) the *spring*, (3) the *damper*, and (4) the *excitation*. The first three elements describe the physical system. For example, it can be said that a given system consists of a mass, a spring, and a damper arranged as shown in the figure. Energy may be stored in the mass and the spring and dissipated in the damper in the form of heat. Energy enters the system through the application of an excitation. As shown in Fig. 1-1, an excitation force is applied to the mass m of the system.

The *mass* m is assumed to be a rigid body. It executes the vibrations and can gain or lose kinetic energy in accordance with the velocity change of the body. From *Newton's law of motion*, the product of the mass and its acceleration is equal to the force applied to the mass, and the acceleration takes place in the direction in which the force acts. *Work* is force times displacement in the direction of the force. The work is transformed into the *kinetic energy* of the mass. The kinetic energy increases if work is positive and decreases if work is negative.

The *spring* k possesses elasticity and is assumed to be of negligible mass. A spring force exists if the spring is deformed, such as the extension or the compression of a coil spring. Therefore the spring force exists *only* if there is a *relative displacement* between the two ends of the spring. The work done in deforming a spring is transformed into *potential energy*, that is, the strain energy stored in the spring. A *linear spring* is one that obeys *Hooke's law*, that is, the spring force is proportional to the spring deformation. The constant of proportionality, measured in force per unit deformation, is called the stiffness, or the *spring constant* k.

The *damper* c has neither mass nor elasticity. Damping force exists *only* if there is *relative motion* between the two ends of the damper. The work or the energy input to a damper is converted into heat. Hence the damping element is *nonconservative*. *Viscous damping*, in which the damping force is proportional to the velocity, is called *linear damping*. Viscous damping, or its equivalent, is generally assumed in engineering.

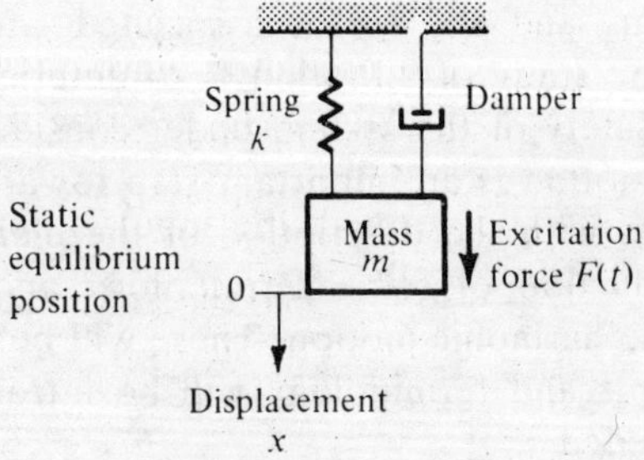

FIG. 1-1. *Elements of a vibratory system.*

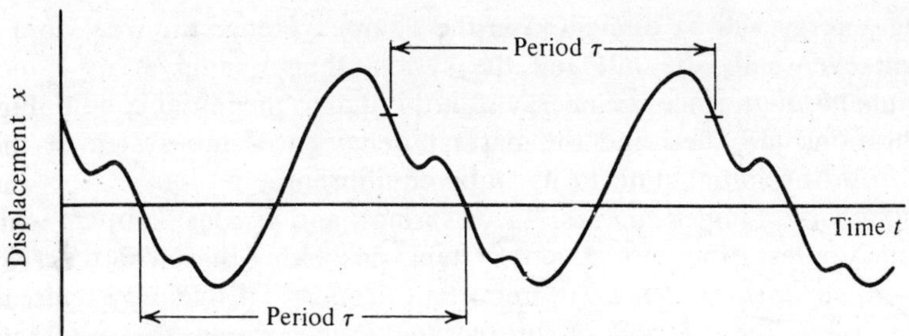

FIG. 1-2. *A periodic motion.*

The *viscous damping coefficient* c is measured in force per unit velocity. Many types of nonlinear damping are commonly encountered. For example, the frictional drag of a body moving in a fluid is approximately proportional to the velocity squared, but the exact value of the exponent is dependent on many variables.

Energy enters a system through the application of an *excitation.* An excitation force may be applied to the mass and/or an excitation motion applied to the spring and the damper. An excitation force $F(t)$ applied to the mass m is illustrated in Fig. 1-1. The excitation varies in accordance with a prescribed function of time. Hence the excitation is always *known* at a given time. Alternatively, if the system is suspended from a support, excitation may be applied to the system through imparting a prescribed motion to the support. In machinery, excitation often arises from the unbalance of the moving components. The vibrations of dynamic systems under the influence of an excitation is called *forced vibrations.* Forced vibrations, however, are often defined as the vibrations that are caused and maintained by a periodic excitation.

If the vibratory motion is *periodic,* the system repeats its motion at equal time intervals as shown in Fig. 1-2. The minimum time required for the system to repeat its motion is called a *period* τ, which is the time to complete one *cycle* of motion. *Frequency* f is the number of times that the motion repeats itself per unit time. A motion that does not repeat itself at equal time intervals is called an *aperiodic* motion.

A dynamic system can be set into motion by some *initial conditions,* or disturbances at time equal to zero. If no disturbance or excitation is applied after the zero time, the oscillatory motions of the system are called *free vibrations.* Hence free vibrations describe the natural behavior or the *natural modes* of vibration of a system. The initial condition is an energy input. If a spring is deformed, the input is potential energy. If a mass is given an initial velocity, the input is kinetic energy. Hence initial conditions are due to the energy initially stored in the system.

If the system does not possess damping, there is no energy dissipation. Initial conditions would cause the system to vibrate and the free vibration of an undamped system will not diminish with time. If a system possesses damping, energy will be dissipated in the damper. Hence the free vibrations will eventually die out and the system then remain at its *static equilibrium position.* Since the energy stored is due to the initial conditions, free vibrations also describe the natural behavior of the system as it relaxes from the initial state to its static equilibrium.

For simplicity, lumped masses, linear springs, and viscous dampers will be assumed unless otherwise stated. Systems possessing these characteristics are called *linear systems.* An important property of linear systems is that they follow the *principle of superposition.* For example, the resultant motion of the system due to the simultaneous application of two excitations is a linear combination of the motions due to each of the excitations acting separately. The values of m, c, and k of the elements in Fig. 1-1 are often referred to as the system *parameters.* For a given problem, these values are assumed *time invariant.* Hence the coefficients or the parameters in the equations are constants. The equation of motion of the system becomes a linear ordinary differential equation with constant coefficients, which can be solved readily.

Note that the idealized elements in Fig. 1-1 form a *model* of a vibratory system which in reality can be quite complex. For example, a coil spring possesses both mass and elasticity. In order to consider it as an idealized spring, either its mass is assumed negligible or an appropriate portion of its mass is lumped together with the other masses of the system. The resultant model is a *lumped-parameter,* or *discrete, system.* For example, a beam has its mass and elasticity inseparably distributed along its length. The vibrational characteristics of a beam, or more generally of an elastic body or a *continuous system,* can be studied by this approach if the continuous system is approximated by a finite number of lumped parameters. This method is a practical approach to the study of some very complicated structures, such as an aircraft.

In spite of the limitations, the lumped-parameter approach to the study of vibration problems is well justified for the following reasons. (1) Many physical systems are essentially discrete systems. (2) The concepts can be extended to analyze the vibration of continuous systems. (3) Many physical systems are too complex to be investigated analytically as elastic bodies. These are often studied through the use of their equivalent discrete systems. (4) The assumption of lumped parameters is not to replace the basic understanding of a problem, but it simplifies the analytical effort and renders a technique for the computer solution.

So far, we have discussed only systems with rectilinear motion. For systems with *rotational motions,* the elements are (1) the mass moment of inertia of the body J, (2) the torsional spring with spring constant k_t, and (3) the torsional damper with torsional damping coefficient c_t. An angular displacement θ is analogous to a rectilinear displacement x, and an excitation torque $T(t)$ is analogous to an excitation force $F(t)$. The two types of systems are compared as shown in Table 1-1. The comparison is

TABLE 1-1. *Comparison of Rectilinear and Rotational Systems*

RECTILINEAR	ROTATIONAL
Spring force $= kx$	Spring torque $= k_t\theta$
Damping force $= c\dfrac{dx}{dt}$	Damping torque $= c_t\dfrac{d\theta}{dt}$
Inertia force $= m\dfrac{d^2x}{dt^2}$	Inertia torque $= J\dfrac{d^2\theta}{dt^2}$

shown in greater detail in Tables 2-2 and 2-3. It is apparent from the comparison that the concept of rectilinear systems can be extended easily to rotational systems.

1-3 EXAMPLES OF VIBRATORY MOTIONS

To illustrate different types of vibratory motion, let us choose various combinations of the four elements shown in Fig. 1-1 to form simple dynamic systems.

The spring-mass system of Fig. 1-3(a) serves to illustrate the case of *undamped free vibration.* The mass m is initially at rest at its static equilibrium position. It is acted upon by two equal and opposite forces, namely, the spring force, which is equal to the product of the spring constant k and the static deflection δ_{st} of the spring, and the gravitational force mg due to the weight of the mass m. Now assume that the mass is displaced from equilibrium by an amount x_0 and then released with zero initial velocity. As shown in the free-body sketch, at the time the mass is released, the spring force is equal to $k(x_0+\delta_{st})$. This is greater than the gravitational force on the mass by the amount kx_0. Upon being released, the mass will move toward the equilibrium position.

Since the spring is initially deformed by x_0 from equilibrium, the corresponding potential energy is stored in the spring. The system is conservative because there is no damper to dissipate the energy. When the mass moves upward and passes through equilibrium, the potential energy of the system is zero. Thus, the potential energy is transformed to become the kinetic energy of the mass. As the mass moves above the equilibrium position, the spring is compressed and thereby gaining potential energy from the kinetic energy of the mass. When the mass is at its uppermost position, its velocity is zero. All the kinetic energy of the mass has been transformed to become potential energy. Through the exchange of potential and kinetic energies between the spring and the mass, the system oscillates periodically at its *natural frequency* about its static

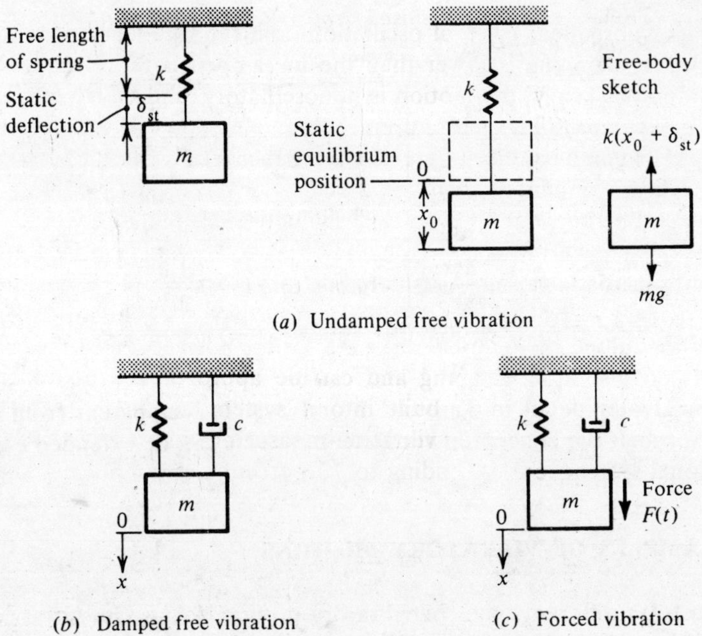

(a) Undamped free vibration

(b) Damped free vibration

(c) Forced vibration

FIG. 1-3. *Simple vibratory systems.*

equilibrium position. Hence natural frequency describes the rate of energy exchange between two types of energy storage elements, namely, the mass and the spring.

It will be shown in Chap. 2 that this periodic motion is sinusoidal or *simple harmonic.* Since the system is conservative, the maximum displacement of the mass from equilibrium, or the *amplitude* of vibration, will not diminish from cycle to cycle. It is implicit in this discussion that the natural frequency is a property of the system, depending on the values of m and k. It is independent of the initial conditions or the amplitude of the oscillation.

A mass-spring system with damping is shown in Fig. 1-3(b). The mass at rest is under the influence of the spring force and the gravitational force, since the damping force is proportional to velocity. Now, if the mass is displaced by an amount x_0 from its static equilibrium position and then released with zero initial velocity, the spring force will tend to restore the mass to equilibrium as before. In addition to the spring force, however, the mass is also acted upon by the damping force which opposes its motion. The resultant motion depends on the amount of damping in the system. If the damping is light, the system is said to be *underdamped* and the motion is oscillatory. The presence of damping will cause (1) the eventual dying out of the oscillation and (2) the system to oscillate more slowly than without damping. In other words, the amplitude decreases with each subsequent cycle of oscillation, and the frequency of vibration with viscous damping is lower than the undamped natural frequency. If the damping is heavy, the motion is nonoscillatory, and the system is said to be *overdamped*. The mass, upon being released, will simply tend to return to its static equilibrium position. The system is said to be *critically damped* if the amount of damping is such that the resultant motion is on the border line between the two cases enumerated. The free vibrations of the systems shown in Figs. 1-3(a) and (b) are illustrated in Fig. 1-4.

All physical systems possess damping to a greater or a lesser degree. When there is very little damping in a system, such as a steel structure or a simple pendulum, the damping may be negligibly small. Most mechanical systems possess little damping and can be approximated as undamped systems. Damping is often built into a system to obtain the desired performance. For example, vibration-measuring instruments are often built with damping corresponding to 70 percent of the critically damped value.

If an excitation force is applied to the mass of the system as shown in Fig. 1-3(c), the resultant motion depends on the initial conditions as well as the excitation. In other words, the motion depends on the manner by which the energy is applied to the system. Let us assume that the excitation is sinusoidal for this discussion. Once the system is set into motion, it will tend to vibrate at its natural frequency as well as to follow the frequency of the excitation. If the system possesses damping, the part of the motion not sustained by the sinusoidal excitation will eventually die out. This is the *transient motion*, which is at the natural frequency of the system, that is, the oscillation under free vibrations.

The motion sustained by the sinusoidal excitation is called the *steady-state vibration* or the *steady-state response*. Hence the steady-state response must be at the excitation frequency regardless of the initial

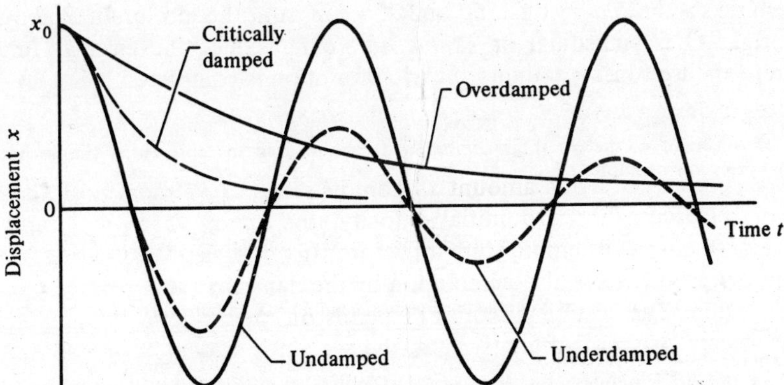

FIG. 1-4. *Free vibration of systems shown in Figs.* 1-3(a) *and* (b). *Initial displacement* = x_o; *initial velocity* = 0.

conditions or the natural frequency of the system. It will be shown in Chap. 2 that the steady-state response is described by the particular integral and the transient motion by the complementary function of the differential equation of the system.

Resonance occurs when the excitation frequency is equal to the natural frequency of the system. No energy input is needed to maintain the vibrations of an undamped system at its natural frequency. Thus, any energy input will be used to build up the amplitude of the vibration, and the amplitude at resonance of an undamped system will increase without limit. In a system with damping, the energy input is dissipated in the damper. Under steady-state condition, the net energy-input per cycle is equal to the energy dissipation per cycle. Hence the amplitude of vibration at resonance for systems with damping is finite, and it is determined by the amount of damping in the system.

1-4 SIMPLE HARMONIC MOTION

Simple harmonic motion is the simplest form of periodic motion. It will be shown in later chapters that (1) harmonic motion is also the basis for more complex analysis using Fourier technique, and (2) steady-state analysis can be greatly simplified using vectors to represent harmonic motions. We shall discuss simple harmonic motions and the manipulation of vectors in some detail in this section.

A *simple harmonic motion* is a *reciprocating motion*. It can be represented by the circular functions, sine or cosine. Consider the motion of the point P on the horizontal axis of Fig. 1-5. If the distance OP is

$$OP = x(t) = X \cos \omega t \tag{1-1}$$

where t = time, ω = constant, and X = constant, the motion of P about the origin O is sinusoidal or *simple harmonic.** Since the circular function repeats itself in 2π radians, a cycle of motion is completed when $\omega\tau = 2\pi$,

* A sine, a cosine, or their combination can be used to represent a simple harmonic motion, For example, let

$$x(t) = X_1 \sin \omega t + X_2 \cos \omega t = X\left(\frac{X_1}{X} \sin \omega t + \frac{X_2}{X} \cos \omega t\right)$$

$$= X(\sin \omega t \cos \alpha + \cos \omega t \sin \alpha) = X \sin(\omega t + \alpha)$$

where $X = \sqrt{X_1^2 + X_2^2}$ and $\alpha = \tan^{-1}(X_2/X_1)$. It is apparent that the motion $x(t)$ is sinusoidal and, therefore, simple harmonic. For simplicity, we shall confine our discussion to a cosine function.

In Eq. (1-1), $x(t)$ indicates that x is a function of time t. Since this is implicit in the equation, we shall omit (t) in all subsequent equations.

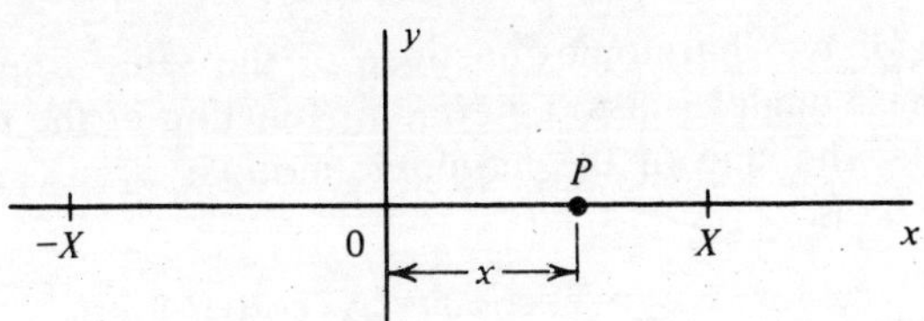

FIG. 1-5. *Simple harmonic motion:* $x(t) = X \cos \omega t$.

that is,

$$\text{Period } \tau = \frac{2\pi}{\omega} \text{ s/cycle} \tag{1-2}$$

$$\text{Frequency } f = \frac{1}{\tau} = \frac{\omega}{2\pi} \text{ cycle/s, or Hz*} \tag{1-3}$$

ω is called the circular frequency measured in rad/s.

If $x(t)$ represents the displacement of a mass in a vibratory system, the velocity and the acceleration are the first and the second time derivatives of the displacement,† that is,

$$\text{Displacement } x = X \cos \omega t \tag{1-4}$$

$$\text{Velocity } \dot{x} = -\omega X \sin \omega t = \omega X \cos(\omega t + 90°) \tag{1-5}$$

$$\text{Acceleration } \ddot{x} = -\omega^2 X \cos \omega t = \omega^2 X \cos(\omega t + 180°) \tag{1-6}$$

These equations indicate that the velocity and acceleration of a harmonic displacement are also harmonic of the same frequency. Each differentiation changes the amplitude of the motion by a factor of ω and the *phase angle* of the circular function by 90°. The phase angle of the velocity is 90° leading the displacement and the acceleration is 180° leading the displacement.

Simple harmonic motion can be defined by combining Eqs. (1-4) and (1-6).

$$\ddot{x} = -\omega^2 x \tag{1-7}$$

where ω^2 is a constant. When the acceleration of a particle with rectilinear motion is always proportional to its displacement from a fixed point on the path and is directed towards the fixed point, the particle is said to have simple harmonic motion. It can be shown that the solution of Eq. (1-7) has the form of a sine and a cosine function with circular frequency equal to ω.

* In 1965, the Institute of Electrical and Electronics Engineers, Inc. (IEEE) adopted new standards for symbols and abbreviation (IEEE Standard No. 260). The unit hertz (Hz) replaces cycles/sec (cps) for frequency. Hz is now commonly used in vibration studies.

† The symbols $\dot{x}$ and $\ddot{x}$ represent the first and second time derivatives of the function $x(t)$, respectively. This notation is used throughout the text unless ambiguity may arise.

The sum of two harmonic functions of the same frequency but with different phase angles is also a harmonic function of the same frequency. For example, the sum of the harmonic motions $x_1 = X_1 \cos \omega t$ and $x_2 = X_2 \cos(\omega t + \alpha)$ is

$$\begin{aligned} x = x_1 + x_2 &= X_1 \cos \omega t + X_2 \cos(\omega t + \alpha) \\ &= X_1 \cos \omega t + X_2(\cos \omega t \cos \alpha - \sin \omega t \sin \alpha) \\ &= (X_1 + X_2 \cos \alpha)\cos \omega t - X_2 \sin \alpha \sin \omega t \\ &= X(\cos \beta \cos \omega t - \sin \beta \sin \omega t) \\ &= X \cos(\omega t + \beta) \end{aligned}$$

where $X = \sqrt{(X_1 + X_2 \cos \alpha)^2 + (X_2 \sin \alpha)^2}$ is the amplitude of the resultant harmonic motion and $\beta = \tan^{-1}(X_2 \sin \alpha)/(X_1 + X_2 \cos \alpha)$ is its phase angle.

The sum of two harmonic motions of different frequencies is not harmonic. A special case of interest is when the frequencies are slightly different. Let the sum of the motions x_1 and x_2 be

$$\begin{aligned} x = x_1 + x_2 &= X \cos \omega t + X \cos(\omega + \varepsilon)t \\ &= X[\cos \omega t + \cos(\omega + \varepsilon)t] \\ &= 2X \cos \frac{\varepsilon}{2} t \cos\left(\omega + \frac{\varepsilon}{2}\right)t \end{aligned}$$

where $\varepsilon \ll \omega$. The resultant motion $x(t)$ may be considered as a cosine wave with the circular frequency $(\omega + \varepsilon/2)$, which is approximately equal to ω, and with a varying amplitude $[2X \cos(\varepsilon/2)t]$. The resultant motion is illustrated in Fig. 1-6. Every time the amplitude reaches a maximum, there is said to be a *beat*. The beat frequency f_b, as determined by two

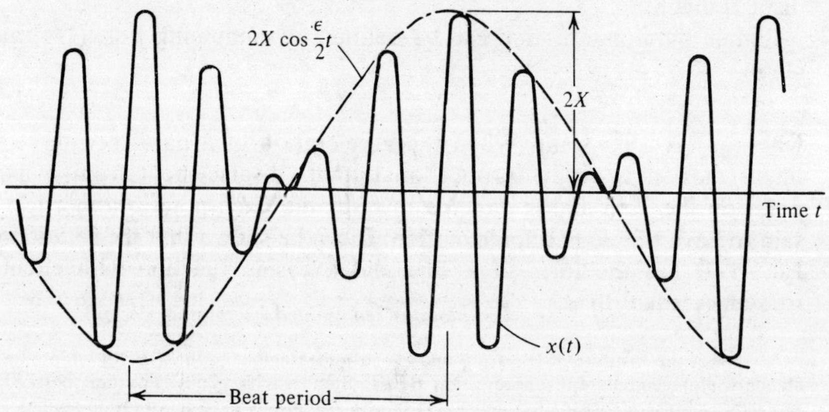

Fig. 1-6. *Graphical representation of beats.*

consecutive maximum amplitudes, is

$$f_b = f_2 - f_1 = \frac{\omega + \varepsilon}{2\pi} - \frac{\omega}{2\pi} = \frac{\varepsilon}{2\pi} \tag{1-8}$$

where f_1 and f_2 are the frequencies of the constituting motions. The more general case, for which the amplitudes of x_1 and x_2 are unequal, is left as an exercise.

The phenomenon of beats is common in engineering. Evidently beating can be a useful technique in frequency measurement in which an unknown frequency is compared with a standard frequency.

1-5 VECTORIAL REPRESENTATION OF HARMONIC MOTIONS

It is convenient to represent a harmonic motion by means of a rotating *vector* **X** of constant *magnitude** X at a constant angular velocity ω. In Fig. 1-7, the displacement of P from the center O along the x axis

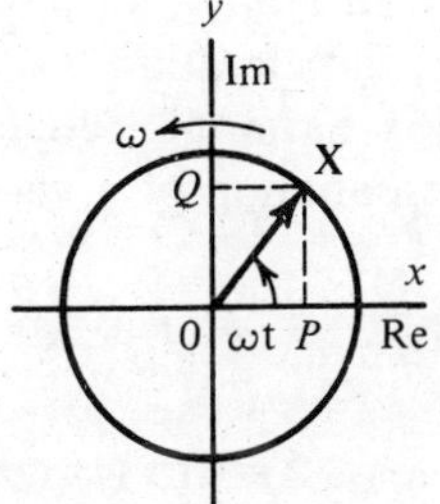

(*a*) Vectorial representation

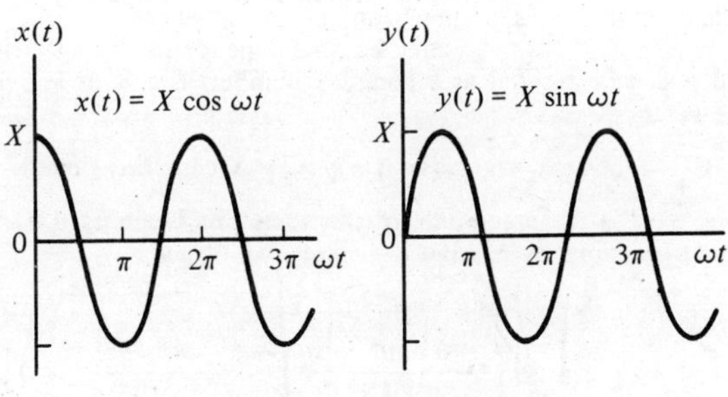

(*b*) Harmonic motions

FIG. 1-7. *Harmonic motions represented by rotating vector.*

* In complex variables, the length of a vector is called the *absolute value* or *modulus*, and the phase angle is called the *argument* or *amplitude*. The length of the vector in this discussion is the amplitude of the harmonic motion. To avoid confusion, we shall use *magnitude* to denote the length of the vector.

is $OP = x(t) = X \cos \omega t$. This is the projection of the rotating vector **X** on the diameter along the x axis. Similarly, the projection of **X** on the y axis is $OQ = y(t) = X \sin \omega t$. Naming the x axis as the "real" axis and the y axis as the "imaginary" one, the rotating vector **X** is represented by the equation*

$$\mathbf{X} = X \cos \omega t + jX \sin \omega t = Xe^{j\omega t} \qquad \textbf{(1-9)}$$

where X is the length of the vector or its magnitude and $j = \sqrt{-1}$ is called the imaginary unit.

If a harmonic function is given as $x(t) = X \cos \omega t$, it can be expressed as $x(t) = \text{Re}[Xe^{j\omega t}]$, where the symbol Re denotes the real part of the function $Xe^{j\omega t}$. Similarly, the function $y(t) = X \sin \omega t$ can be expressed as $y(t) = \text{Im}[Xe^{j\omega t}]$, where the symbol Im denotes the imaginary part of $Xe^{j\omega t}$. It should be remembered that a harmonic motion is a reciprocating motion. Its representation by means of a rotating vector is only a convenience. This enables the exponential function $e^{j\omega t}$ to be used in equations involving harmonic functions. The use of complex functions and complex numbers greatly simplifies the mathematical manipulations of this type of equations. In reality, all physical quantities, whether they are displacement, velocity, acceleration, or force, must be real quantities.

The differentiation of a harmonic function can be carried out in its vectorial form. The differentiation of a vector **X** is

$$\begin{aligned} \frac{d}{dt}\mathbf{X} &= \frac{d}{dt}(Xe^{j\omega t}) = j\omega Xe^{j\omega t} = j\omega \mathbf{X} \\ \frac{d^2}{dt^2}\mathbf{X} &= \frac{d}{dt}(j\omega Xe^{j\omega t}) = (j\omega)^2 Xe^{j\omega t} = -\omega^2 \mathbf{X} \end{aligned} \qquad \textbf{(1-10)}$$

* A complex number z is of the form $z = x + jy$, where x is the real part and y the imaginary part of z. Both x and y may be time dependent. For a specific time, x and y are numbers and z can be treated as a complex number. Let **X** in Fig. 1-7(a) be a complex number. The vector **X** is

$$\mathbf{X} = x + jy = X(x/X + jy/X) = X(\cos \omega t + j \sin \omega t)$$

where $X = \sqrt{x^2 + y^2}$ is the magnitude of the vector **X**. Defining $\theta = \omega t$ and expanding the sine and cosine functions by Maclaurin's series, we obtain

$$\begin{aligned} \mathbf{X} &= X(\cos \theta + j \sin \theta) \\ &= X\left[\left(1 - \frac{\theta^2}{2!} + \frac{\theta^4}{4!} - \cdots\right) + j\left(\theta - \frac{\theta^3}{3!} + \frac{\theta^5}{5!} - \cdots\right)\right] \\ &= X\left(1 + j\theta - \frac{\theta^2}{2!} - j\frac{\theta^3}{3!} + \frac{\theta^4}{4!} + j\frac{\theta^5}{5!} - \cdots\right) \\ &= X\left[1 + \frac{(j\theta)}{1!} + \frac{(j\theta)^2}{2!} + \frac{(j\theta)^3}{3!} + \frac{(j\theta)^4}{4!} + \frac{(j\theta)^5}{5!} + \cdots\right] \\ &= Xe^{j\theta} = Xe^{j\omega t} \end{aligned}$$

The equation $e^{\pm j\theta} = \cos \theta \pm j \sin \theta$ is called Euler's formula.

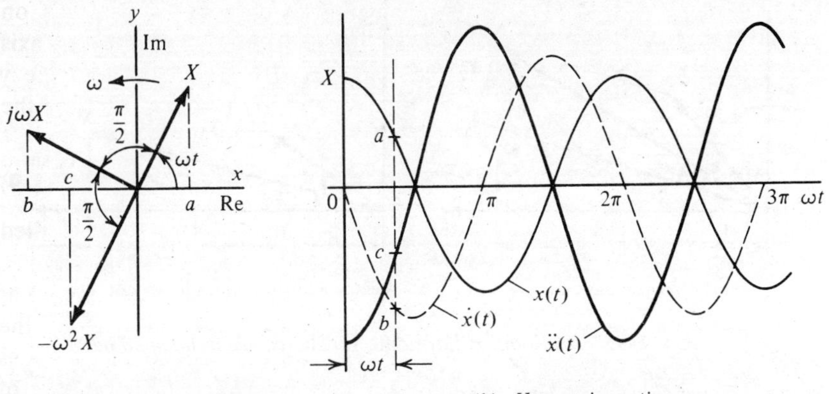

(a) Vectorial representation (b) Harmonic motions

FIG. 1-8. *Displacement, velocity, and acceleration vectors.*

Thus, each differentiation is equivalent to the multiplication of the vector by $j\omega$. Since X is the magnitude of the vector $\mathbf{X}$, ω is real, and $|j| = 1$, each differentiation changes the magnitude by a factor of ω. Since the multiplication of a vector by j is equivalent to advancing it by a phase angle of 90°, each differentiation also advances a vector by 90°.

If a given harmonic displacement is $x(t) = X\cos\omega t$, the relations between the displacement and its velocity and acceleration are

$$\begin{aligned}
\text{Displacement } x &= \text{Re}[Xe^{j\omega t}] = X\cos\omega t \\
\text{Velocity } \dot{x} &= \text{Re}[j\omega Xe^{j\omega t}] = -\omega X\sin\omega t \\
&= \omega X\cos(\omega t + 90°) \\
\text{Acceleration } \ddot{x} &= \text{Re}[(j\omega)^2 Xe^{j\omega t}] = -\omega^2 X\cos\omega t \\
&= \omega^2 X\cos(\omega t + 180°)
\end{aligned} \tag{1-11}$$

These relations are identical to those shown in Eqs. (1-4) to (1-6). The representation of displacement, velocity, and acceleration by rotating vectors is illustrated in Fig. 1-8. Since the given displacement $x(t)$ is a cosine function, or along the real axis, the velocity and acceleration must be along the real axis. Hence the real parts of the respective vectors give the physical quantities at the given time t.

Harmonic functions can be added graphically be means of vector addition. The vectors $\mathbf{X}_1$ and $\mathbf{X}_2$ representing the motions $X_1\cos\omega t$ and $X_2\cos(\omega t + \alpha)$, respectively, are added graphically as shown in Fig. 1-9(*a*). The resultant vector $\mathbf{X}$ has a magnitude

$$X = \sqrt{(X_1 + X_2\cos\alpha)^2 + (X_2\sin\alpha)^2}$$

and a phase angle

$$\beta = \tan^{-1}\frac{X_2\sin\alpha}{X_1 + X_2\cos\alpha}$$

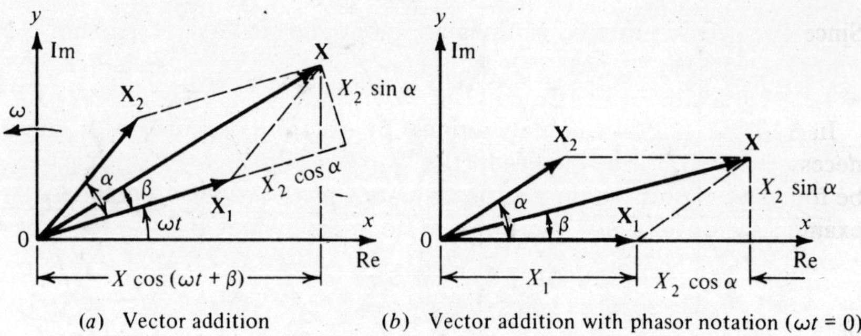

(*a*) Vector addition (*b*) Vector addition with phasor notation ($\omega t = 0$)

FIG. 1-9. *Addition of harmonic functions: vectorial method.*

with respect to $\mathbf{X}_1$. Since the original motions are given along the real axis, the sum of the harmonic motions is $\text{Re}[\mathbf{X}] = X\cos(\omega t + \beta)$. The addition operation can readily be extended to include the subtraction operation.

Since both $\mathbf{X}_1$ and $\mathbf{X}_2$ are rotating with the same angular velocity ω, only the relative phase angle of the vectors is of interest. It is convenient to assign arbitrarily $\omega t = 0$ as a datum of measurement of phase angles. The vector $\mathbf{X}_1$, $\mathbf{X}_2$, and their sum $\mathbf{X}$ are plotted in this manner in Fig. 1-9(*b*). Note that the vector $\mathbf{X}_2$ can be expressed as

$$\mathbf{X}_2 = X_2 e^{j(\omega t + \alpha)} = (X_2 e^{j\alpha})e^{j\omega t} = (X_2 \cos\alpha + jX_2 \sin\alpha)e^{j\omega t}$$

or

$$\mathbf{X}_2 = \bar{X}_2 e^{j\omega t} \tag{1-12}$$

The quantity $\bar{X}_2 = X_2 e^{j\alpha}$ is a complex number and is called the *complex amplitude* or *phasor* of the vector $\mathbf{X}_2$. Similarly, $\bar{X} = Xe^{j\beta}$ in Fig. 1-9(*b*) is the phasor of the vector $\mathbf{X}$.

Harmonic functions can be added algebraically by means of vector addition. Using the same functions $x_1 = X_1 \cos\omega t$ and $x_2 = X_2 \cos(\omega t + \alpha)$, their vector sum is

$$\begin{aligned}\mathbf{X} = \mathbf{X}_1 + \mathbf{X}_2 &= X_1 e^{j\omega t} + X_2 e^{j(\omega t + \alpha)} = (X_1 + X_2 e^{j\alpha})e^{j\omega t}\\ &= (X_1 + X_2 \cos\alpha + jX_2 \sin\alpha)e^{j\omega t}\\ &= Xe^{j\beta}e^{j\omega t} = Xe^{j(\omega t + \beta)}\end{aligned}$$

where

$$X = \sqrt{(X_1 + X_2 \cos\alpha)^2 + (X_2 \sin\alpha)^2}$$

and

$$\beta = \tan^{-1} \frac{X_2 \sin\alpha}{X_1 + X_2 \cos\alpha}$$

Since the given harmonic motions are along the real axis, their sum is

$$x = \text{Re}[\mathbf{X}] = \text{Re}[Xe^{j(\omega t+\beta)}] = X\cos(\omega t+\beta)$$

In representing harmonic motions by rotating vectors, it is often necessary to determine the product of complex numbers. The product can be found by expressing the complex numbers in the exponential form. For example, the product of the complex numbers $\bar{A}$ and $\bar{B}$ is

$$\bar{A}\bar{B} = (a_1+ja_2)(b_1+jb_2)$$

or

$$\bar{A}\bar{B} = (Ae^{j\alpha})(Be^{j\beta}) = ABe^{j(\alpha+\beta)} \qquad \textbf{(1-13)}$$

where $A = \sqrt{a_1^2+a_2^2}$ and $B = \sqrt{b_1^2+b_2^2}$ are the magnitudes of the numbers and $\alpha = \tan^{-1} a_2/a_1$ and $\beta = \tan^{-1} b_2/b_1$ are their phase angles. Equation (1-13) indicates that

$$\text{Magnitude of } \bar{A}\bar{B} = (\text{magnitude of } \bar{A})(\text{magnitude of } \bar{B}) \qquad \textbf{(1-14)}$$

$$\text{Phase of } \bar{A}\bar{B} = (\text{phase of } \bar{A}) + (\text{phase of } \bar{B}) \qquad \textbf{(1-15)}$$

Obviously, the multiplication operation can be generalized to include the division operation.

Example 1. Manipulation of Complex Numbers

(a) $\bar{A} = 1+j\sqrt{3} = \sqrt{1+3}(\cos 60° + j\sin 60°) = 2e^{j\pi/3} = 2\underline{/60°}$
The symbol $2\underline{/60°}$ is a convenient way of writing $2e^{j\pi/3}$. It represents a vector of magnitude of two units and a phase angle of 60° or $\pi/3$ rad *counter-clockwise relative to the reference x axis.*

(b) $\bar{A} = \bar{A}_1 + \bar{A}_2 = (1+j2)+(4+j3) = 5+j5 = 5\sqrt{2}\underline{/45°}$

(c) $\bar{A} = \bar{A}_1\bar{A}_2 = (1+j\sqrt{3})(4+j3) = (2e^{j\pi/3})(5e^{j0.642})$

$= 10e^{j(\pi/3+0.642)} = 10\underline{/60°+36.8°} = 10\underline{/96.8°}$

(d) $\bar{A} = \dfrac{\bar{A}_1}{\bar{A}_2} = \dfrac{1+j\sqrt{3}}{4+j3} = \dfrac{2\underline{/60°}}{5\underline{/36.8°}} = \dfrac{2}{5}\underline{/60°-36.8°} = 0.4\underline{/23.2°}$

(e) $\bar{A} = 2j = 0+j2 = 2\left(\cos\dfrac{\pi}{2} + j\sin\dfrac{\pi}{2}\right) = 2e^{j\pi/2} = 2\underline{/90°}$

(f) $\bar{A} = j(1+j\sqrt{3}) = (e^{j\pi/2})(2e^{j\pi/3}) = 2\underline{/90°+60°} = 2\underline{/150°}$

(g) $\bar{A} = \dfrac{1+j\sqrt{3}}{j} = \dfrac{2\underline{/60°}}{1\underline{/90°}} = 2\underline{/60°-90°} = 2\underline{/-30°}$

The last two examples indicate that the multiplication of a vector by j advances the vector counter-clockwise by a phase angle of 90° and a division by j retards it by 90°.

1-6 UNITS

Since there will be a change from the English engineering (customary) to the International System of Units (SI), the two systems of units will co-exist for some years. The student and the practicing engineer will need to know both systems. We shall briefly discuss the English units and then the SI units in some detail in this section.

Newton's law of motion may be expressed as

$$\text{Force} = (\text{mass})(\text{acceleration}) \tag{1-16}$$

Dimensional homogeneity of the equation is obtained when the force is in pounds lb_f, the mass in slugs, and the acceleration in ft/sec^2. This is the English ft-lb_f-sec system in which the mass has the unit of lb_f-sec^2/ft. A body falling under the influence of gravitation has an acceleration of g ft/sec^2, where $g \simeq 32.2$ is the gravitational acceleration. Hence one pound-mass lb_m exerts one pound-force lb_f under the gravitational pull of the earth. In other words, 1 lb_m weighs one pound on a spring scale. If a mass is given in pounds, lb_m or weight, it must be divided by g to obtain dimensional homogeneity in Eq. (1-16).

The in.-lb_f-sec system is generally used in the study of vibrations. The gravitational acceleration is $(32.2)(12) = 386$ in./sec^2. Hence the weight is divided by 386 in order to obtain dimensional homogeneity in Eq. (1-16). We assume that the gravitational acceleration is constant unless otherwise stated. In the derivation of equations, the mass m is assumed to have the proper units.

The International System of Units (SI) is the modernized version of the metric system.* SI consists of (1) seven well-defined base units, (2) derived units, and (3) supplementary units.

The *base units* are regarded as dimensionally independent. Those of interest in this study are the meter, the kilogram, and the second. The *meter* m is the unit of length. It is defined in terms of the wave-length of a krypton-85 lamp as "the length equal to 1 650 763.73 wave-lengths in vacuum of the radiation corresponding to the transition between the levels $2p_{10}$ and $5d_5$ of the krypton-86 atom." The *kilogram* kg is the unit

* A description of SI and a brief bibliography can be found in the literature. A kit, containing the above and several other publications, is obtainable from the American Society of Engineering Education:

The International System of Units (SI), Edited by C. H. Page and P. Vigoureux, US National Bureau of Standards, Special Publication 330, Revised 1974.

ASME Guide SI-l, ASME Orientation and Guide of SI (Metric) *Units*, 6th ed., ASME, United Engineering Center, 345 E. 47th St., New York, N.Y. 10017, 1975.

Some References on Metric Information, US National Bureau of Standards, Special Publication 389, Revised 1974.

ASEE Metric (SI) Resource Kit Project, One Dupont Circle, Suite 400, Washington, DC 20036.

TABLE 1-2(*a*). *Examples of SI Derived Units*

QUANTITY	SI UNIT NAME	SYMBOL	IN TERMS OF SI BASE UNITS	IN TERMS OF OTHER UNITS
Area	square meter	m^2		
Volume	cubic meter	m^3		
Speed, velocity	meter per second	m/s		
Acceleration	meter per second squared	m/s^2		
Density, mass density	kilogram per cubic meter	kg/m^3		
Specific volume	cubic meter per kilogram	m^3/kg		
Frequency	hertz	Hz	s^{-1}	
Force	newton	N	$m \cdot kg \cdot s^{-2}$	
Pressure, stress	pascal	Pa	$m^{-1} \cdot kg \cdot s^{-2}$	N/m^2
Energy, work	joule	J	$m^2 \cdot kg \cdot s^{-2}$	$N \cdot m$
Power	watt	W	$m^2 \cdot kg \cdot s^{-3}$	J/s
Moment of force	meter newton	$N \cdot m$	$m^2 \cdot kg \cdot s^{-2}$	

TABLE 1-2(*b*). *SI Supplementary Units*

QUANTITY	SI UNIT NAME	SYMBOL
Plane angle	radian	rad
Solid angle	steradian	sr

of mass. The standard is a cylinder of platinum-iridium, called the International Standard, kept in a vault at Sévres, France. The *second* s is the unit of time. It is defined in terms of the frequency of atomic resonators. "The second is the duration of 9 192 631 770 periods of the radiation corresponding to the transition between the two hyperfine levels of the ground state of the cesium-133 atom."*

The *derived units* are formed from the base units according to the algebraic relations linking the corresponding quantities. Several derived units are given special names and symbols. The *supplementary units* form a third class of SI units. Examples of derived units and the supplementary units are shown in Tables 1-2(*a*) and (*b*), respectively.

* Page and Vigoureux, *op. cit.*, p. 3.

TABLE 1-3. *Prefixes for Multiples and Submultiples of SI Units*

MULTIPLE	PREFIX	SYMBOL	SUBMULTIPLE	PREFIX	SYMBOL
10^{12}	tera	T	10^{-1}	deci	d
10^{9}	giga	G	10^{-2}	centi	c
10^{6}	mega	M	10^{-3}	milli	m
10^{3}	kilo	k	10^{-6}	micro	μ
10^{2}	hecto	h	10^{-9}	nano	n
10	deca	dc	10^{-12}	pico	p
			10^{-15}	femto	f
			10^{-18}	atto	a

The common prefixes for multiples and submultiples of SI units are shown in Table 1-3. Examples of conversion from the English to the SI units are given in Table 1-4. Note that a common error in conversion is to become ensnared in too many decimal places. The result of a computation cannot have any more significant numbers than that in the original data.

For uniformity in the use of SI units, the recommendations* are: (1) In *numbers*, a period (dot) is used only to separate the integral part of numbers from the decimal part. Numbers are divided into groups of three to facilitate reading. For example, in defining the meter above, we have "...equal to 1 650 763.73 wave-lengths..." (2) The *type* used for symbols is illustrated in Table 1-2. The lower case roman type is generally used. If the symbol is derived from a proper name, capital roman type is

TABLE 1-4. *Examples of English Units to SI Conversion*[a]

TO CONVERT FROM	TO	MULTIPLY BY
Inch (in.)	meter (m)	2.540 000 E−02
Pound-mass (lb_m)	kilogram (kg)	4.535 924 E−01
Pound-mass/inch^3 (lb_m/in.^{3})	kilogram/meter^3 (kg/m^3)	2.767 990 E+04
Slug	kilogram (kg)	1.459 390 E+01
Pound-force (lb_f)	newton (N)	4.448 222 E+00
Pound-force-inch (lb_f-in.)	newton-meter (N · m)	1.129 848 E−01
Pound-force/inch (lb_f/in.)	newton/meter (N/m)	1.751 268 E+02
Pound-force/inch^2 (lb_f/in.^{2})	pascal (Pa)	6.894 757 E+03
Horsepower (550 ft-lb_f/sec)	watt (W)	7.456 999 E+02

[a] The table gives the conversion from the in.-lb_f-sec units to the SI units. The second and radian are commonly used in both systems and no conversions are needed. For example, the damping coefficient c from Table 2-2 has the units of lb_f-sec/in. The value of c is multiplied by 175.1 to obtain the value of c in N · s/m.

*Page and Vigoureux, *op. cit.*, p. 10.

used for the first letter. The symbols are not followed by a period. (3) The *product of units* is denoted by a dot, such as N · m shown in Table 1-2. The dot may be omitted if there is no risk of confusion with another unit symbol, such as N m but not mN. (4) The *division of units* may be indicated by a solidus (/), a horizontal line, or a negative power. For example, velocity in Table 1-2 can be expressed as m/s, $\frac{\text{m}}{\text{s}}$, or $\text{m} \cdot \text{s}^{-1}$. The solidus must not be repeated on the same line unless ambiguity is avoided by parentheses. For example, acceleration may be expressed as m/s^2 or $\text{m} \cdot \text{s}^{-2}$ but not m/s/s. (5) The *prefix symbols* illustrated in Table 1-3 are used without spacing between the prefix symbol and the unit symbol, such as in mm. Compound prefixes formed by the use of two or more SI prefixes are not used.

1-7 SUMMARY

Some basic concepts and terminology commonly used in vibration are described in this chapter.

The idealized *model* of a simple vibratory system in Fig. 1-1 consists of (1) a rigid mass, (2) a linear spring, (3) a viscous damper, and (4) an excitation. The inertia force is equal to the product of the mass and its acceleration as defined by Newton's law of motion. The spring force is proportional to the spring deformation, that is, the relative displacement between the two ends of the spring. The damping force is proportional to the relative velocity between the two ends of the damper. An excitation may be applied to the mass and/or other parts of the system.

If a system is unforced, the energy stored due to the initial conditions will cause it to vibrate about its static equilibrium position. If damping is zero, the system will oscillate at its natural frequency without diminishing in amplitude. If the system is underdamped, the amplitude of oscillation will diminish with each cycle and the frequency is lower than that without damping.

If a periodic excitation is applied to a system, the vibration consists of (1) a steady-state response and (2) a transient motion. The former is being sustained by the excitation and is therefore at the excitation frequency. The latter is due to the initial energy stored in the system and is at its damped natural frequency. Resonance occurs when the system is excited at its natural frequency. The amplitude at resonance is limited only by the damping in the system.

A simple harmonic motion is a reciprocating motion as shown in Fig. 1-5. Alternatively, it can be represented by means of a sinusoidal wave or a rotating vector as shown in Figs. 1-7 and 1-8. These representations are artifices for the convenience of visualization and manipulation only. Using these representations, it can be shown that the velocity leads the displacement by 90° and the acceleration leads the velocity by 90°. A complex amplitude, shown in Fig. 1-9, is called a *phasor*. It has a magnitude and a phase angle relative to the reference vector.

A complex number has magnitude and direction. It can be added (subtracted) by adding (subtracting) the real and imaginary parts separately. The product (quotient) of complex numbers is determined by Eqs. (1-14) and (1-15).

$$\text{Magnitude of } \bar{A}\bar{B} = (\text{magnitude of } \bar{A})(\text{magnitude of } \bar{B})$$
$$\text{Phase of } \bar{A}\bar{B} = (\text{phase of } \bar{A}) + (\text{phase of } \bar{B})$$

The in.-lb_f-sec system is generally used in vibration. The gravitational constant is $g \simeq 386\ \text{in./sec}^2$. There will be a change to the International Systems of Units (SI). The gravitational constant in SI units is $g \simeq 9.81\ \text{m/s}^2$. Examples of SI units and the conversion from the English to the SI units are given in Tables 1-2 and 1-4, respectively.

PROBLEMS

1-1 Describe, with the aid of a sketch when necessary, each of the following:

(a) Spring force, damping force, inertia force, excitation.

(b) Kinetic energy, potential energy.

(c) Free vibration, forced vibration, a conservative system.

(d) Steady-state response, transient motion.

(e) Discrete system, continuous system.

(f) Natural frequency, resonance.

(g) Initial conditions, static equilibrium position.

(h) Rectilinear motion, rotational motion.

(i) Periodic motion, frequency, period, beat frequency.

(j) Superposition.

(k) Underdamped system, critically damped system.

(l) Amplitude, phasor, phase angle.

1-2 A harmonic displacement is $x(t) = 10 \sin(30t - \pi/3)$ mm, where t is in seconds and the phase angle in radians. Find **(a)** the frequency and the period of the motion, **(b)** the maximum displacement, velocity, and acceleration, and **(c)** the displacement, velocity, and acceleration at $t = 0$ s. Repeat part **(c)** for $t = 1.2$ s.

1-3 Repeat Prob. 1-2 if the harmonic velocity is $\dot{x}(t) = 150 \cos(17t + \pi/2)$ mm/s.

1-4 An accelerometer indicates that the acceleration of a body is sinusoidal at a frequency of 40 Hz. If the maximum acceleration is 100 m/s^2, find the amplitudes of the displacement and the velocity.

1-5 Repeat Prob. 1-4 if the acceleration lags the excitation by 15°. What is the excitation frequency?

1-6 A harmonic motion is described as $x(t) = X \cos(100t + \psi)$ mm. The initial conditions are $x(0) = 4.0$ mm and $\dot{x}(0) = 1.0$ m/s.

(a) Find the constants X and ψ.

(b) Expressing $x(t)$ in the form

$$x = A \cos \omega t + B \sin \omega t$$

and find the constants A and B.

1-7 Given $x(t) = X \cos(100t + \psi) = A \cos 100t + B \sin 100t$, find A, B, X, and ψ for each set of the following conditions:

(a) $x(0.1) = -8.796$ mm and $x(0.2) = 10.762$ mm

(b) $x(0.1) = -8.796$ mm and $\dot{x}(0.1) = -621.5$ mm/s

(c) $x(0.1) = -8.796$ mm and $\ddot{x}(0.2) = -10.76 \times 10^4$ mm/s^2

(d) $x(0) = 4.0$ mm and $\ddot{x}(0.2) = -10.76 \times 10^4$ mm/s^2

1-8 A table has a vertical sinusoidal motion with constant frequency. What is the largest amplitude that the table can have, if an object on the table is to remain in contact?

1-9 Find the algebraic sum of the harmonic motions x_1 and x_2.

$$x = x_1 + x_2 = 2 \sin(\omega t + \pi/3) + 3 \sin(\omega t + 2\pi/3)$$
$$= X \sin(\omega t + \alpha)$$

Find X and α. Check the addition graphically.

1-10 The motion of a particle is described as $x = 4 \sin(\omega t + \pi/6)$. If the motion has two components, one of which is $x_1 = 2 \sin(\omega t - \pi/3)$, determine the other harmonic component.

1-11 In a sketch of x versus t for $0 \le t \le 0.4$ s, plot the motions described by each of the equations: $x_1 = 5 \sin 10\pi t$; $x_2 = 4 \sin(10\pi t + \pi/4)$; $x_3 = 3 \sin(10\pi t - \pi/4)$.

1-12 A periodic motion is described by the equation

$$x = 5 \sin 2\pi t + 3 \sin 4\pi t$$

In a plot of x versus t, sketch the motion for $0 \le t \le 1.5$ s.

1-13 Repeat Prob. 1-12 if

(a) $x = 5 \sin(2\pi t + 30°) + 3 \sin(4\pi t + 60°)$

(b) $x = 5 \sin(2\pi t + 90°) + 3 \sin(4\pi t + 180°)$

1-14 Is the motion $x(t) = \cos 10t + 3\cos(10+\pi)t$ periodic?

1-15 Find the period of the functions

(a) $x = 3\sin 3t + 5\sin 4t$

(b) $x = 7\cos^2 3t$

1-16 Determine the sum of the harmonic motions $x_1 = X_1 \cos \omega t$ and $x_2 = X_2 \cos(\omega + \varepsilon)t$, where $\varepsilon \ll \omega$. If beating should occur, find the amplitude and the beat frequency.

1-17 Sketch the motion described by each of the following equations:

(a) $x = 5e^{-2t}\sin(10\pi t + \pi/4)$

(b) $x = 5e^{-2t}\sin(10\pi t + \pi/4) + 7\sin 4\pi t$

for $0 \le t \le 1.0$ s.

1-18 Express the following complex numbers in the exponential form $Ae^{j\theta}$.

(a) $1 + j\sqrt{3}$ **(e)** $3/(\sqrt{3} - j)^2$

(b) -2 **(f)** $(\sqrt{3} + j)(3 + 4j)$

(c) $3/(\sqrt{3} - j)$ **(g)** $(\sqrt{3} - j)/(3 - 4j)$

(d) $5j$ **(h)** $[(2j)^2 + 3j + 8]$

1-19 The motion of a particle vibrating in a plane has two perpendicular harmonic components: $x_1 = 2\sin(\omega t + \pi/6)$ and $x_2 = 3\sin \omega t$. Determine the motion of the particle graphically.

1-20 Repeat Prob. 1-19 using $x_1 = 2\sin(2\omega t + \pi/6)$ and $x_2 = 3\sin \omega t$.

2

Systems with One Degree of Freedom—Theory

2-1 INTRODUCTION

The one-degree-of-freedom system is the keystone for more advanced studies in vibrations. The system is represented by means of a generalized model shown in Fig. 1-1. The common techniques for the analysis are discussed in this chapter.

Examples of one-degree-of-freedom systems are shown in Fig. 2-1. Though such systems differ in appearance, they all can be represented by the same generalized model in Fig. 1-1. The model serves (1) to unify a class of problems commonly encountered, and (2) to bring into focus the concepts of vibration. The applications to different types of problem will be discussed in the next chapter.

Four mathematical techniques are examined. These are (1) the energy method, (2) Newton's law of motion, (3) the frequency response method, and (4) the superposition theorem. Our emphasis is on concepts rather than on mathematical manipulations.

Since vibration is an energy exchange phenomenon, the simple energy method is first presented. In applying Newton's second law, the system is described by a second-order differential equation of motion. If the excitation is an analytical expression, the equation can be solved readily by the "classical" method. If the excitation is an arbitrary function the motion can be found using the superposition theorem. The frequency response method assumes that the excitation is sinusoidal and examines the system behavior over a frequency range of interest.

Note that a system will vibrate in its own way regardless of the method of analysis. The purpose of different techniques is to find the most convenient method to characterize the system and to describe its behavior. We treat Newton's second law and the superposition theorem as

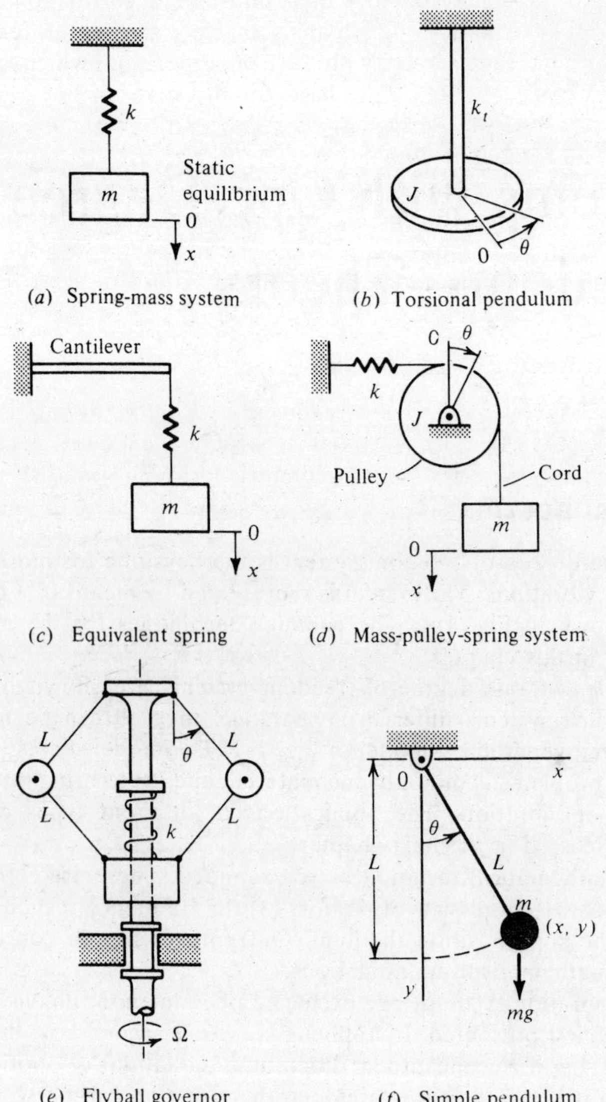

FIG. 2-1. *Examples of systems with one degree of freedom.*

time domain analysis, since the motion of the mass is a time function, such as the solution of a differential equation with time as the independent variable. The frequency response method assumes that both the excitation and the system response are sinusoidal and of the same frequency. Hence it is a *frequency domain analysis.* Note that time response is intuitive but it is more convenient to describe a system in the frequency domain.

It should be remarked that there must be a correlation between the time and the frequency domain analyses, since they are different methods to consider the same problem. In fact, superposition, which is treated as a time domain technique, is the basis for the study of linear systems. The convolution integral derived from superposition can be applied in the time or the frequency domain. We are presenting only one aspect of this very important theorem and shall not discuss methods of correlation. The mathematical correlation of the time and frequency analyses is not new. Its implementation, however, was not practical until the advent of computers, instrumentation, and testing techniques in recent years.

2-2 DEGREES OF FREEDOM

The number of *degrees of freedom* of a vibratory system is the number of *independent spatial coordinates* necessary to define its configuration. A configuration is defined as the geometric location of all the masses of the system. If the inter-relationship of the masses is such that only one spatial coordinate is required to define the configuration, the system is said to possess *one degree of freedom.*

A rigid body in space requires six coordinates for its complete identification, namely, three coordinates to define the rectilinear positions and three to define the angular rotations. Ordinarily, however, the masses in a system are constrained to move only in a certain manner. Thus, the *constraints* limit the degrees of freedom to a much smaller number.

Alternatively, the number of degrees of freedom of a system can be defined as the number of spatial coordinates required to specify its configuration minus the number of *equations of constraint.** We shall illustrate these definitions with a number of examples.

The one-degree-of-freedom systems shown in Fig. 2-1 are briefly discussed as follows:

1. The spring-mass system in Fig. 2-1(a) has a mass m suspended from a coil spring with a spring constant k. If m is constrained to move only in the vertical direction about its static equilibrium position 0, only one spatial coordinate $x(t)$ is required to define its configuration. Hence it is said to possess one degree of freedom.
2. The torsional pendulum in Fig. 2-1(b) consists of a heavy disk J and a shaft of negligible mass with a torsional spring constant k_t. If the system is constrained to oscillate about the longitudinal axis of the shaft, the configuration of the system can be specified by a single coordinate $\theta(t)$.

* Such a system is called a holonomic system; it is the only type of system considered in this text. For a discussion on holonomic and nonholonomic systems, see, for example, H. Goldstein, *Classical Mechanics*, Addison-Wesley Publishing Company, Inc., Reading, Mass, 1957, pp. 11–14.

3. The mass-spring-cantilever system in Fig. 2-1(c) has one degree of freedom if the cantilever is of negligible mass and the mass m is constrained to move vertically. By neglecting the inertial effect of the cantilever and considering only its elasticity, the cantilever becomes a spring element. Hence a simple spring-mass system is obtained from the given mass m and an equivalent spring, constructed from the combination of the spring k and the cantilever.

4. The mass-pulley-spring system in Fig. 2-1(d) has one degree of freedom if it is assumed that there is no slippage between the cord and the pulley J and the cord is inextensible. Although the system possesses two mass elements m and J, the linear displacement $x(t)$ of m and the angular rotation $\theta(t)$ of J are not independent. Thus, either $x(t)$ or $\theta(t)$ can be used to specify the configuration of the system.

5. A simple spring-loaded flyball governor rotating with constant angular velocity Ω is shown in Fig. 2-1(e). If a disturbance is applied to the governor, its vibratory motion can be expressed in terms of the angular coordinate $\theta(t)$.

6. The simple pendulum in Fig. 2-1(f) is constrained to move in the xy plane. Its configuration can be defined either by the rectangular Cartesian coordinates $x(t)$ and $y(t)$ or by the angular rotation $\theta(t)$. The (x,y) coordinates, however, are not independent. They are related by the *equation of constraint*

$$x^2 + y^2 = L^2 \tag{2-1}$$

where the length L of the pendulum is assumed constant. Thus, if $x(t)$ is chosen arbitrarily, $y(t)$ is determined from Eq. (2-1).

Several systems with *two degrees of freedom* are shown in Fig. 2-2.

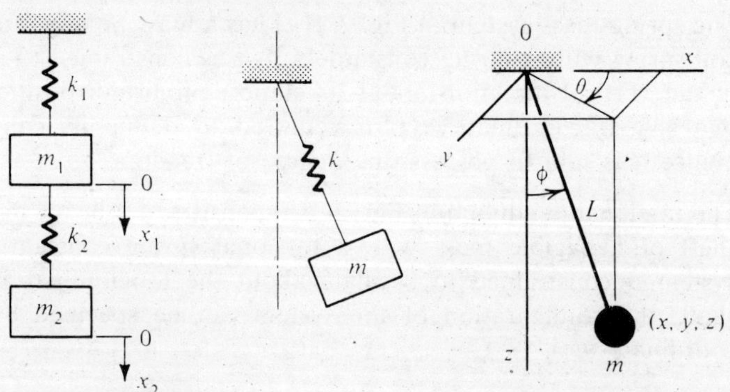

(a) 2-mass-2-spring system (b) Spring-mass system (c) Spherical pendulum

FIG. 2-2. *Two-degree-of-freedom systems.*

1. The two-spring-two-mass system of Fig. 2-2(*a*) possesses two degrees of freedom if the masses are constrained to move only in the vertical direction. The two spatial coordinates defining the configuration are $x_1(t)$ and $x_2(t)$.

2. The spring-mass system shown in Fig. 2-2(*b*) was described previously as a one-degree-of-freedom system. If the mass m is allowed to oscillate along the axis of the spring as well as to swing from side to side, the system possesses two degrees of freedom.

3. The pendulum in space in Fig. 2-2(*c*) can be described by the $\theta(t)$ and $\phi(t)$ coordinates as well as by the $x(t)$, $y(t)$, and $z(t)$ coordinates. The latter are related by the equation of constraint $x^2+y^2+z^2=L^2$. Thus, this pendulum has only two degrees of freedom.

2-3 EQUATION OF MOTION—ENERGY METHOD

The equation of motion of a *conservative* system can be established from energy considerations. If a conservative system in Fig. 2-3 is set into motion, its total mechanical energy is the sum of the kinetic energy and the potential energy. The kinetic energy T is due to the velocity of the mass, and the potential energy U is due to the strain energy of the spring by virtue of its deformation. Since the system is conservative, the total mechanical energy is constant and its time derivative must be zero. This can be expressed as

$$T+U=(\text{total mechanical energy})=\text{constant} \tag{2-2}$$

$$\frac{d}{dt}(T+U)=0 \tag{2-3}$$

To derive the equation of motion for the spring-mass system of Fig. 2-3, assume that the displacement $x(t)$ of the mass m is measured from its

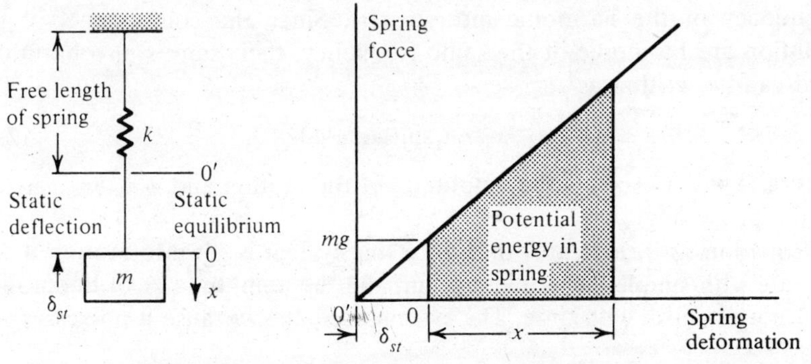

FIG. 2-3. *Potential energy in spring.*

static equilibrium position. Let $x(t)$ be positive in the downward direction. Since the spring element is of negligible mass, the kinetic energy T of the system is

$$T = \tfrac{1}{2}m\dot{x}^2 \tag{2-4}$$

The corresponding potential energy of the entire system is the algebraic sum of (1) the strain energy of the spring and (2) that due to the change in elevation of the mass. The *net* potential energy of the system about the static equilibrium is

$$\begin{aligned} U &= \int_0^x (\text{total spring force})\,dx - mgx \\ &= \int_0^x (mg + kx)\,dx - mgx \\ &= \tfrac{1}{2}kx^2 \end{aligned} \tag{2-5}$$

Substituting Eqs. (2-4) and (2-5) into Eq. (2-3) gives

$$\frac{d}{dt}\left(\tfrac{1}{2}m\dot{x}^2 + \tfrac{1}{2}kx^2\right) = (m\ddot{x} + kx)\dot{x} = 0$$

Since the velocity $\dot{x}(t)$ in the equation cannot be zero for all values of time, clearly

$$m\ddot{x} + kx = 0 \tag{2-6}$$

$$\ddot{x} + \omega_n^2 x = 0 \tag{2-7}$$

where $\omega_n^2 = k/m$. The equation of motion of the system can be expressed as shown in Eq. (2-6) or (2-7).

It can be shown that the solution of Eq. (2-7) is of the form

$$x = A_1 \cos \omega_n t + A_2 \sin \omega_n t \tag{2-8}$$

where A_1 and A_2 are arbitrary constants to be evaluated by the *initial conditions* $x(0)$ and $\dot{x}(0)$.* It is apparent that ω_n in Eq. (2-7) is the circular frequency of the harmonic motion $x(t)$. Since the components of the solution are harmonic of the same frequency, their sum is also harmonic and can be written as

$$x = A \sin(\omega_n t + \psi) \tag{2-9}$$

where $A = \sqrt{A_1^2 + A_2^2}$ is the amplitude of the motion and $\psi = \tan^{-1} A_1/A_2$ is the phase angle.

Equation (2-9) indicates that once this system is set into motion it will vibrate with simple harmonic motion, and the amplitude A of the motion will not diminish with time. The system oscillates because it possesses two

* The arbitrary constants A_1 and A_2 can be evaluated by conditions specified other than at $t = 0$. It is customary and convenient, however, to use initial conditions.

types of energy storage elements, namely, the mass and the spring. The rate of energy interchange between these elements is the *natural frequency* f_n of the system.*

$$f_n = \frac{\omega_n}{2\pi} = \frac{1}{2\pi}\sqrt{\frac{k}{m}} \text{ Hz} \tag{2-10}$$

Note that the natural frequency is a property of the system. It is a function of the values of m and k and is independent of the amplitude of oscillation or the manner by which the system is set into motion. Evidently, only the amplitude A and the phase angle ψ are dependent on the initial conditions.

Example 1

Determine the equation of motion of the simple pendulum shown in Fig. 2-1(f).

Solution:

Assume (1) the size of the bob is small as compared with the length L of the pendulum and (2) the rod connecting the bob to the hinge point 0 is of negligible mass. The mass moment of inertia of the bob of mass m about 0 is

$$J_0 = (J_{cg} + mL^2) \simeq mL^2$$

where J_{cg} is the mass moment of inertia of m about its mass center. If the bob is sufficiently small in size, then $J_{cg} \ll mL^2$.

The angular displacement $\theta(t)$ is measured from the static equilibrium position of the pendulum. The kinetic energy of the system is $T = \frac{1}{2}J_0\dot{\theta}^2 \simeq \frac{1}{2}mL^2\dot{\theta}^2$. The corresponding potential energy is $U = mgL(1-\cos\theta)$, where $L(1-\cos\theta)$ is the change in elevation of the pendulum bob. Substituting these energy quantities in Eq. (2-3) gives

$$mL^2\ddot{\theta} + mgL\sin\theta = 0 \tag{2-11}$$

$$\ddot{\theta} + \frac{g}{L}\sin\theta = 0 \tag{2-12}$$

The equation of motion of the simple pendulum is as shown in Eq. (2-11) or (2-12). If it is further assumed that the amplitude of oscillation is small, then $\sin\theta \simeq \theta$ and Eq. (2-12) becomes

$$\ddot{\theta} + \frac{g}{L}\theta \simeq 0 \tag{2-13}$$

This is of the same form as Eq. (2-7) and the solution follows. The frequency of oscillation of a simple pendulum is $\omega_n = \sqrt{g/L}$.

* It is convenient to call ω_n the natural frequency instead of the natural circular frequency. In the subsequent sections of the text, natural frequency will refer to f_n or ω_n unless ambiguity arises. Similarly, frequency may refer to f or ω.

Note that, if small oscillations are not assumed, Eq. (2-11) is a nonlinear differential equation and elliptical integrals are used for the problem solution. The dependent variable $\theta(t)$ and the independent variable t are related by*

$$t = \int_{\theta_0}^{\theta} \frac{d\theta}{\sqrt{\dot{\theta}_0^2 + \dfrac{2mgL}{J_0}(\cos\theta - \cos\theta_0)}} \tag{2-14}$$

where θ_0 and $\dot{\theta}_0$ are the initial conditions at $t = 0$. It is conceivable that, if the pendulum is given a sufficient large initial velocity, the pendulum may continue to rotate about the hinge point. Thus, $\theta(t)$ will increase with time and the motion will not be periodic.

Small oscillations will be assumed throughout this text unless otherwise stated. This assumption greatly simplifies the effort necessary to obtain the solution. Furthermore, the answers will be relevant for most problems, such as in predicting the onset of resonance in a vibratory system.

Example 2

Figure 2-4 shows a cylinder of mass m and radius R_1 rolling without slippage on a curved surface of radius R. Derive the equation of motion of the system by the energy method.

Solution:

The kinetic energy of the cylinder is due to its translational and rotational motions. The translational velocity of the mass center of the cylinder is $(R - R_1)\dot{\theta}$. The angular velocity of the cylinder is $(\dot{\theta}_1 - \dot{\theta})$. Since the cylinder

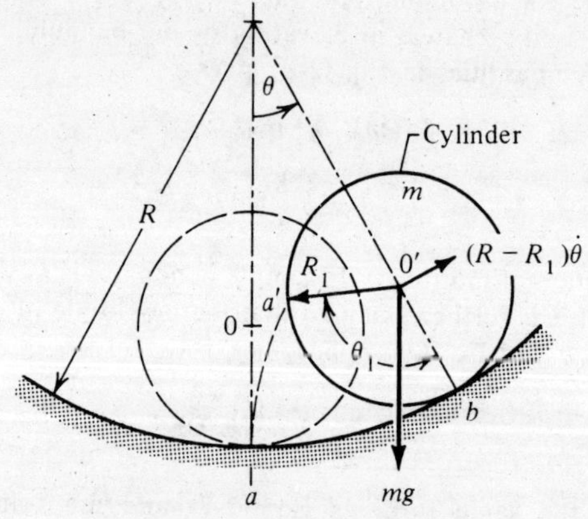

FIG. 2-4. *Cylinder on curved surface.*

* T. von Karman and M. A. Biot, *Mathematical Methods in Engineering*, McGraw-Hill Book Co., New York, 1940, pp. 115–119.

rolls without slippage, the arc $\widehat{ab} = \text{arc}\,\widehat{a'b}$ and $R\theta = R_1\theta_1$. Hence the angular velocity can be written as $(R/R_1 - 1)\dot{\theta}$. The total kinetic energy T of the cylinder is

$$T = \tfrac{1}{2}m[(R - R_1)\dot{\theta}]^2 + \tfrac{1}{2}J_0[(R/R_1 - 1)\dot{\theta}]^2$$

where $J_0 = \frac{1}{2}mR_1^2$ is the moment of inertia of the cylinder about its longitudinal axis. The potential energy U is due to the change in elevation of the mass center of the cylinder with respect to its static equilibrium position, that is,

$$U = mg(R - R_1)(1 - \cos\theta)$$

Substituting the T and U expressions into Eq. (2-3) gives

$$\left[\frac{3}{2}m(R - R_1)^2\ddot{\theta} + mg(R - R_1)\sin\theta\right]\dot{\theta} = 0$$

$$\ddot{\theta} + \frac{2g}{3(R - R_1)}\theta \simeq 0$$

where $\theta \simeq \sin\theta$ for small oscillations. Comparing this with Eq. (2-7), the natural frequency ω_n of the system is equal to $\sqrt{2g/3(R - R_1)}$.

The natural frequency of a *conservative system* can be deduced by *Rayleigh's method.* Natural frequency is the rate of energy interchange between the kinetic and the potential energies of a system during its cyclic motion. As the mass passes through the static equilibrium position, the potential energy is zero. Hence the kinetic energy is maximum and is equal to the total mechanical energy of the system. When the mass is at a position of maximum displacement, it is on the verge of changing direction and its velocity is zero. Correspondingly, its kinetic energy is zero. Thus, the potential energy is maximum and is equal to the total mechanical energy of the system. As indicated in Eq. (2-9), the motion is harmonic when the system is vibrating at its natural frequency. The maximum displacement, or amplitude, is A and the maximum velocity is $\omega_n A$. Equating the maximum kinetic and potential energies, we have

$$T_{\max} = U_{\max} = \text{total energy of the system} \tag{2-15}$$

or

$$\tfrac{1}{2}m(\omega_n A)^2 = \tfrac{1}{2}kA^2$$

$$\omega_n = \sqrt{k/m}$$

Example 3. Equivalent mass of spring: Rayleigh method

The mass of the spring shown in Fig. 2-5 is not negligible. Determine the natural frequency of the system by Rayleigh's method.

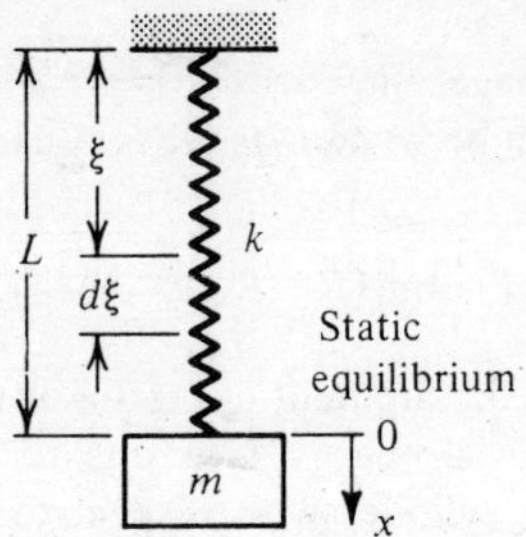

FIG. 2-5. *Equivalent mass of spring: Rayleigh method.*

Solution:

Let L be the length of the spring k when the system is at its static equilibrium position. Assume that, when the end of the spring has a displacement $x(t)$, an intermediate point ξ of the spring has a displacement equal to $\frac{\xi}{L}x(t)$. Thus, $x(t)$ defines the configuration and the system has only one degree of freedom.

The kinetic energy of the system is due to the rigid mass m and the mass of the spring k. The kinetic energy of an element of the spring of length $d\xi$ is $\frac{1}{2}(\rho d\xi)\left(\frac{\xi}{L}\dot{x}\right)^2$, where ρ = mass/length of the spring. Let $x = A \sin \omega_n t$. Hence the maximum kinetic energy of the system is

$$T_{max} = \tfrac{1}{2}m\dot{x}_{max}^2 + \int_0^L \tfrac{1}{2}\rho\left(\frac{\xi}{L}\dot{x}_{max}\right)^2 d\xi$$

$$T_{max} = \tfrac{1}{2}\left(m + \frac{\rho L}{3}\right)\dot{x}_{max}^2$$

$$= \tfrac{1}{2}(m + \rho L/3)(\omega_n A)^2$$

From Eq. (2-5), the maximum potential energy of the system is

$$U_{max} = \tfrac{1}{2}kx_{max}^2 = \tfrac{1}{2}kA^2$$

The natural frequency is obtained by equating the maximum kinetic and potential energies, that is,

$$\tfrac{1}{2}(m + \rho L/3)(\omega_n A)^2 = \tfrac{1}{2}kA^2$$

$$f_n = \frac{\omega_n}{2\pi} = \frac{1}{2\pi}\sqrt{\frac{k}{m + \rho L/3}}$$

This equation shows that the inertial effect of the spring can be accounted for by adding one-third of the mass of the spring to the rigid mass m. The natural frequency can then be calculated as if the system were to consist of a massless spring and an equivalent rigid mass of $(m + \rho L/3)$.

The approximate method above indicates that the natural frequency is independent of the mass ratio $\rho L/m$, that is, the mass of the spring to that of the rigid mass. For a heavy spring with a light mass, a larger fraction of the spring-mass would have to be used for the frequency calculation. The error, however, is less than one percent, as compared with the exact value, when the spring-mass is equal to the rigid mass.*

2-4 EQUATION OF MOTION—NEWTON'S LAW OF MOTION

Newton's law of motion is used to establish the differential equation of motion of one-degree-of-freedom systems in this section. The emphasis is on concepts of vibration rather than the technique for solving the equation.

The generalized model representing this class of problems is shown in Fig. 2-6. The displacement $x(t)$ of the mass m is measured from the static equilibrium position. Displacement is positive in the downward direction, and so are the velocity $\dot{x}(t)$ and the acceleration $\ddot{x}(t)$. A positive force on the mass m will produce a positive acceleration of the mass and vice versa. Referring to the free-body sketch, the forces acting on the mass m are (1) the gravitational force mg which is constant, (2) the spring force $k(x+\delta_{st})$ which always opposes the displacement, (3) the damping force $c\dot{x}$ which always opposses the velocity, and (4) the excitation force which is assumed to equal to $F \sin \omega t$.

Newton's law of motion (second law) states that the rate of change of momentum is proportional to the impressed force and takes place in the direction of the straight line in which the force acts. If the mass is constant, the rate of change of momentum is equal to the mass times its acceleration. From the free-body sketch in Fig. 2-6, the equation of motion of the system is

$$m\ddot{x} = \sum(\text{forces})_x \tag{2-16}$$

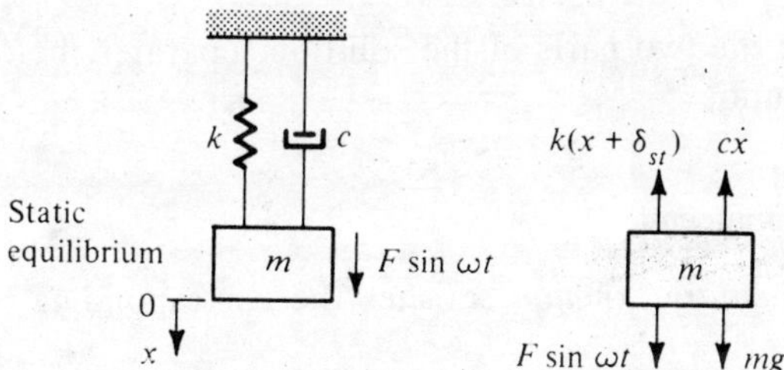

FIG. 2-6. *Model of systems with one degree of freedom.*

* S. Timoshenko and D. Young, *Vibration Problems in Engineering*, 3d ed., D. Van Nostrand Co., Inc., New York, 1954, pp. 306–314.

or

$$m\frac{d^2}{dt^2}(x+\delta_{st}) = -k(x+\delta_{st}) - c\frac{d}{dt}(x+\delta_{st}) + mg + F\sin\omega t \tag{2-17}$$

$$m\ddot{x} + c\dot{x} + kx = F\sin\omega t \tag{2-18}$$

Note that the gravitational force mg is equal to the static spring force $k\delta_{st}$. This is obvious in a simple problem. The implications are (1) that the static forces must cancel in a vibratory system, and (2) that only the dynamic forces need be considered. This concept may be helpful for more complex problems.

The equation above can be derived whether the general position of the mass is considered above or below the static equilibrium position or whether the mass is moving upward or downward. Thus, the equation is true for all time and for all positions of the mass. The verification of this statement is left as an exercise.

Using *d'Alembert's principle*, Eq. (2-16) can be expressed as

$$\sum(\text{forces})_x - m\ddot{x} = 0 \tag{2-19}$$

The quantity $-m\ddot{x}$ is called the *inertia force*. In other words, introducing the appropriate inertia force, we can say that the impressed force on the mass is in equilibrium with the inertia force. Thus, the dynamic problem is reduced to an "equivalent" problem of statics.

2-5 GENERAL SOLUTION

The equation of motion for the model in Fig. 2-6 is a second-order linear ordinary differential equation with constant coefficients, Eq. (2-18). The general solution $x(t)$ is the sum of the *complementary function* $x_c(t)$ and the *particular integral* $x_p(t)$ as shown in Eq. (D-13), App. D.

$$x = x_c + x_p \tag{2-20}$$

Let us consider the two parts of the solution separately before discussing the general solution.

Complementary Function

The *complementary function* satisfies the corresponding homogeneous equation

$$m\ddot{x} + c\dot{x} + kx = 0 \tag{2-21}$$

The solution is of the form

$$x_c = Be^{st} \tag{2-22}$$

where B and s are constants. Substituting Eq. (2-22) into Eq. (2-21) gives

$$(ms^2+cs+k)Be^{st}=0 \tag{2-23}$$

Since the quantity Be^{st} cannot be zero for all values of t, we deduce that

$$ms^2+cs+k=0 \tag{2-24}$$

This is called the auxiliary or the *characteristic equation* of the system. The roots of the characteristic equation are

$$s_{1,2}=\frac{1}{2m}(-c\pm\sqrt{c^2-4mk}) \tag{2-25}$$

Since there are two roots, the complementary function is

$$x_c=B_1e^{s_1t}+B_2e^{s_2t} \tag{2-26}$$

where B_1 and B_2 are arbitrary constants to be evaluated by the initial conditions.

Let us rewrite the equations above in a more convenient form by defining

$$\frac{k}{m}=\omega_n^2 \quad \text{and} \quad \frac{c}{m}=2\zeta\omega_n \quad \text{or} \quad \zeta=\frac{c}{2\sqrt{km}} \tag{2-27}$$

where ω_n is the *natural circular frequency* of the system and ζ is called the *damping factor*. Since m, c, and k are positive, ζ is a positive number. Using the definitions of ζ and ω_n, Eqs. (2-21), (2-24), and (2-25) become

$$\ddot{x}+2\zeta\omega_n\dot{x}+\omega_n^2x=0 \tag{2-28}$$

$$s^2+2\zeta\omega_ns+\omega_n^2=0 \tag{2-29}$$

$$s_{1,2}=-\zeta\omega_n\pm\sqrt{\zeta^2-1}\,\omega_n \tag{2-30}$$

When $\zeta>1$, Eq. (2-30) shows that the roots are real, distinct, and negative, since $\sqrt{\zeta^2-1}<\zeta$. Thus, no oscillation can be expected from the complementary function in Eq. (2-26) regardless of the initial conditions. Since both the roots are negative, the motion diminishes with increasing time and is aperiodic.

When $\zeta=1$, Eq. (2-30) shows that both the roots are equal to $-\omega_n$. Thus, the complementary function is of the form

$$x_c=(B_3+B_4t)e^{-\omega_nt}$$

where B_3 and B_4 are constants. The motion is again aperiodic. Since the $\lim_{t\to\infty}e^{-\omega_nt}=\lim_{t\to\infty}te^{-\omega_nt}=0$, the motion will eventually diminish to zero.

When $\zeta<1$, the roots are complex conjugates.

$$s_{1,2}=-\zeta\omega_n\pm j\sqrt{1-\zeta^2}\,\omega_n \tag{2-31}$$

where $j=\sqrt{-1}$. Defining

$$\omega_d=\sqrt{1-\zeta^2}\,\omega_n \tag{2-32}$$

and using Euler's formula $e^{\pm j\theta}=\cos\theta\pm j\sin\theta$, the complementary function x_c in Eq. (2-26) becomes

$$x_c=e^{-\zeta\omega_n t}(B_1e^{j\omega_d t}+B_2e^{-j\omega_d t})$$

or

$$x_c=e^{-\zeta\omega_n t}[(B_1+B_2)\cos\omega_d t+j(B_1-B_2)\sin\omega_d t] \tag{2-33}$$

Since the displacement $x_c(t)$ is a real physical quantity, the coefficients, (B_1+B_2) and $j(B_1-B_2)$, in Eq. (2-33) must also be real. This requires that B_1 and B_2 be complex conjugates. Hence Eq. (2-33) can be rewritten as

$$x_c=e^{-\zeta\omega_n t}(A_1\cos\omega_d t+A_2\sin\omega_d t) \tag{2-34}$$

or

$$x_c=Ae^{-\zeta\omega_n t}\sin(\omega_d t+\psi) \tag{2-35}$$

where A_1 and A_2 are real constants to be evaluated by the initial conditions. The harmonic functions in Eq. (2-34) are combined to give Eq. (2-35), where $A=\sqrt{A_1^2+A_2^2}$ and $\psi=\tan^{-1}(A_1/A_2)$. The motion described by Eq. (2-35) consists of a harmonic motion of frequency ω_d and an amplitude $Ae^{-\zeta\omega_n t}$, which decreases exponentially with time.

For the three cases enumerated, the type of motion described by $x_c(t)$ depends on the value of ζ. The system is said to be *overdamped* when $\zeta>1$, *critically damped* when $\zeta=1$, and *underdamped* when $\zeta<1$. This was explained intuitively in Chap. 1. Note that (1) $x_c(t)$ is vibratory only if the system is underdamped; (2) the frequency of oscillation ω_d is lower than the natural frequency ω_n of the system; and (3) in all cases, $x_c(t)$ will eventually die out, regardless of the initial conditions or the excitation. Hence the complementary function gives the *transient motion* of the system. As a limiting case, if the system has no damping, the amplitude of $x_c(t)$ will not diminish with time. Furthermore, Eq. (2-35) shows that the frequency ω_d and the rate of the exponential decay in amplitude are independent of the arbitrary constants of the equation. In other words, they are properties of the system, independent of the initial conditions or the manner by which the system is set into motion.

The *critical damping coefficient* c_c is the amount of damping necessary for a system to be critically damped, that is, $\zeta=1$. From Eq. (2-27), when $\zeta=1$, we have

$$c_c=2\sqrt{km} \tag{2-36}$$

Hence the *damping factor* ζ can be defined as

$$\zeta = \frac{c}{c_c} \tag{2-37}$$

It is a measure of the existing damping c as compared with that necessary for a system to be critically damped.

Example 4. Damped-Free Vibration

A machine of 20 kg mass (44 lb_m) is mounted on springs and dampers as shown in Fig. 2-7. The total stiffness of the springs is 8 kN/m (45.7 lb_f/in.)

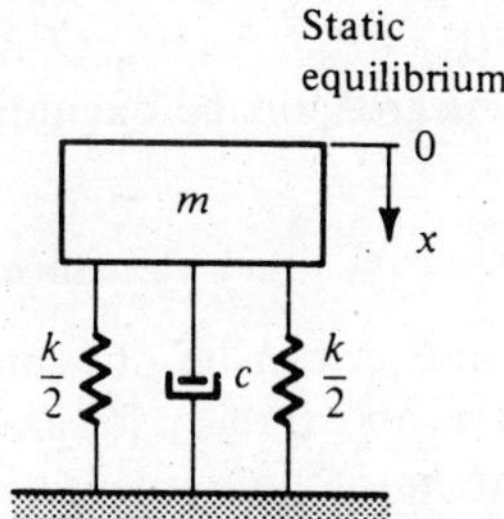

FIG. 2-7. *Damped-free vibration.*

and the total damping is 130 N·s/m (0.74 lb_f-sec/in.). If the system is initially at rest and a velocity of 100 mm/s (3.9 in./sec) is imparted to the mass, determine (**a**) the displacement and velocity of the mass as a time function, and (**b**) the displacement at $t = 1.0$ s.

Solution:

The displacement $x(t)$ is obtained by the direct application of Eq. (2-34). The parameters of the equation are

$$\omega_n = \sqrt{k/m} = \sqrt{8000/20} = 20 \text{ rad/s}$$
$$\zeta = c/(2m\omega_n) = 130/[2(20)(20)] = 0.1625$$
$$\omega_d = \sqrt{1-\zeta^2}\,\omega_n = 20\sqrt{1-0.1625^2} = 19.7 \text{ rad/s}$$

(**a**) Substituting these values in Eq. (2-34), we obtain

$$x = e^{-3.25t}(A_1 \cos 19.7t + A_2 \sin 19.7t)$$
$$\dot{x} = -3.25e^{-3.25t}(A_1 \cos 19.7t + A_2 \sin 19.7t) + 19.7e^{-3.25t}(-A_1 \sin 19.7t + A_2 \cos 19.7t)$$

Applying the initial conditions gives

$$x(0)=0 \therefore A_1=0$$

$$\dot{x}(0)=100 \text{ mm/s} \therefore A_2=100/19.7=5.07$$

$$\therefore x=5.07e^{-3.25t} \sin 19.7t \text{ mm}$$

$$\dot{x}=e^{-3.25t}(-16.47 \sin 19.7t+100 \cos 19.7t)$$

$$=101.3e^{-3.25t} \cos(19.7t+9.5°) \text{ mm/s}$$

(b) The displacement at $t=1$ s is

$$x(t=1)=5.07e^{-3.25} \sin 19.7=0.162 \text{ mm}$$

Particular Integral

The *particular integral* for the excitation $F(t)=F \sin \omega t$ in Eq. (2-18) is of the form

$$x_p=X \sin(\omega t-\phi) \tag{2-38}$$

The values of X and ϕ can be obtained by substituting Eq. (2-38) into (2-18). This is left as an exercise. It can be shown that the amplitude X of the steady-state or *harmonic response* is

$$X=\frac{F}{\sqrt{(k-\omega^2 m)^2+(\omega c)^2}} \tag{2-39}$$

$$X=\frac{F/k}{\sqrt{(1-\omega^2 m/k)^2+(\omega c/k)^2}} \tag{2-40}$$

and

$$\phi=\tan^{-1}\frac{\omega c}{k-\omega^2 m} \quad \text{or} \quad \phi=\tan^{-1}\frac{\omega c/k}{1-\omega^2 m/k} \tag{2-41}$$

X is the amplitude of the steady-state response and $-\phi$ is the phase angle of $x_p(t)$ relative to the excitation $F \sin \omega t$, that is, the displacement lags the excitation by ϕ rad. For convenience, the last two equations are often expressed in nondimensional form. Substituting the relations $k/m=\omega_n^2$ and $\omega c/k=2\zeta\omega/\omega_n$ and defining $r=\omega/\omega_n$, these equations become

$$\frac{X}{F/k}=\frac{1}{\sqrt{(1-r^2)^2+(2\zeta r)^2}}=R \tag{2-42}$$

and

$$\phi=\tan^{-1}\frac{2\zeta r}{1-r^2} \tag{2-43}$$

where R is called the *magnification factor* and r the *frequency ratio* of the excitation frequency to the natural frequency of the system. Equations (2-42) and (2-43) are plotted in Figs. 2-8 and 2-9 with ζ as a parameter.

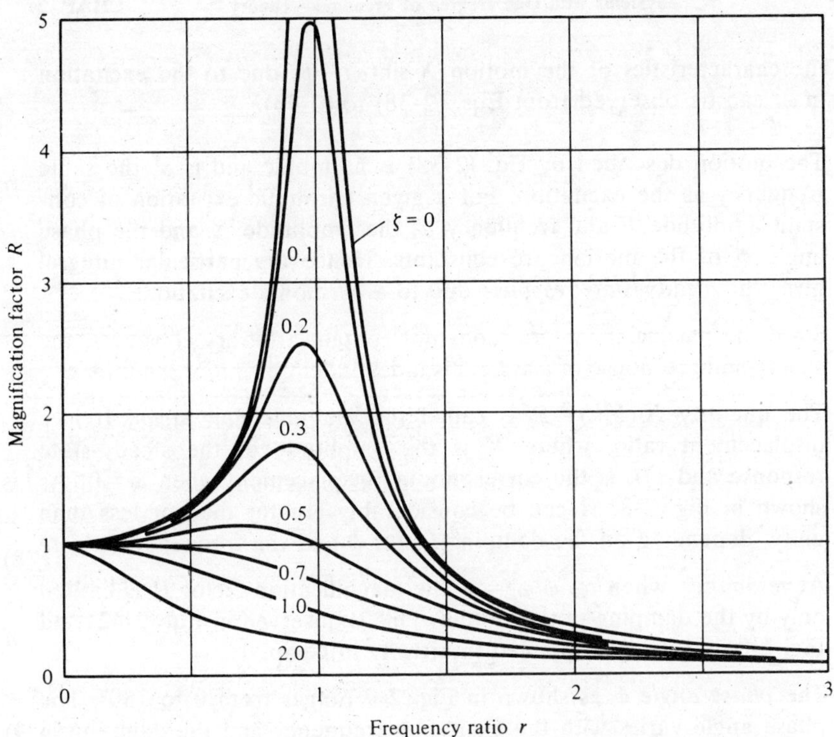

FIG. 2-8. *Magnification factor-versus-frequency ratio; system shown in Fig. 2-6.*

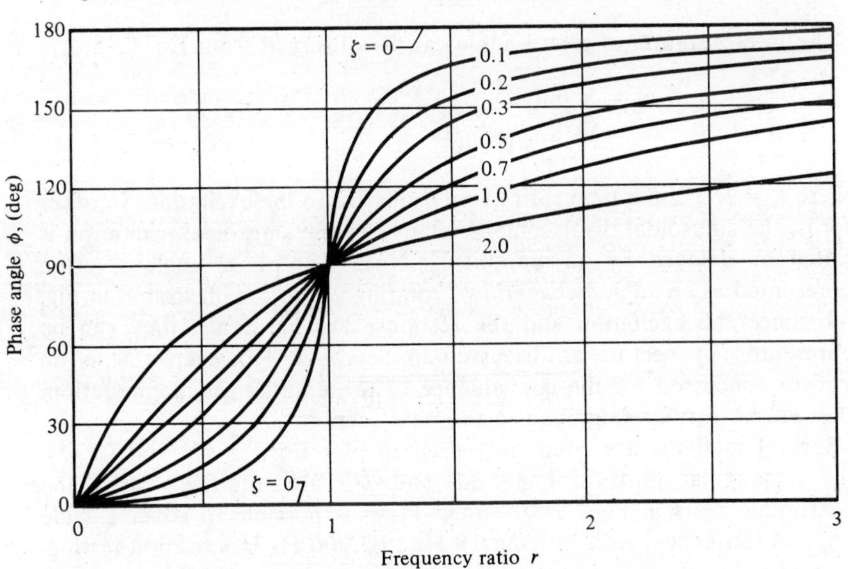

FIG. 2-9. *Phase angle-versus-frequency ratio; system shown in Fig. 2-6.*

The characteristics of the motion $X\sin(\omega t-\phi)$ due to the excitation $F\sin\omega t$ can be observed from Eqs. (2-38) to (2-43).

1. The motion described by Eq. (2-38) is harmonic and is of the same frequency as the excitation. For a given harmonic excitation of constant amplitude F and frequency ω, the amplitude X and the phase angle ϕ of the motion are constants. Hence the particular integral gives the *steady-state response* due to a harmonic excitation.
2. Since the particular integral does not contain arbitrary constants, the steady-state response of a system is independent of *initial conditions.*
3. The quantity $X/(F/k)\triangleq R$ is called the *magnification factor.* It is a displacement ratio, where X is the amplitude of the steady-state response and F/k is the corresponding displacement when $\omega=0$. As shown in Fig. 2-8, R can be considerably greater than or less than unity, depending on the damping factor ζ and the frequency ratio r.
4. At *resonance,* when $r=\omega/\omega_n=1$, the magnification factor R is limited only by the damping in the system. This is observed in Eq. (2-42) and Fig. 2-8 and it was explained intuitively in Chap. 1.
5. The *phase angle* ϕ as shown in Fig. 2-9 ranges from 0 to 180°. The phase angle varies with the excitation frequency and the damping in the system. Without damping, the phase angle can only be either 0 or 180°. At resonance, when $r=1$, the phase angle is always 90°.

The interpretation of phase angle can be observed from Eq. (2-38).

$$\begin{aligned} x_p &= X\sin(\omega t-\phi)=X\sin\omega(t-\phi/\omega) \\ &= X\sin\omega(t-t_\phi) \end{aligned}$$

where $t_\phi=\phi/\omega$ is the time shift of $x_p(t)$ relative to the excitation. In other words, the sinusoidal displacement relative to the sinusoidal excitation is shifted or *delayed* by an amount t_ϕ. Note that phase angle is often represented as an angle between two rotating vectors as illustrated in Fig. 1-9. Since the excitation and the response are harmonic, they can be represented by vectors as discussed in Sec. 1-5. However, this is an artifice, concocted for the convenience of presentation or manipulation. This will be further examined in the next section.

Several methods are commonly used to plot Eqs. (2-42) and (2-43). The rectangular plots in Figs. 2-8 and 2-9 are intuitive. Using the logarithmic plots in Figs. 2-10 and 2-11, it is possible to cover a wide range of frequency, such as from 10 Hz to 3,000 Hz in vibration testing. Correspondingly, the range of the magnification factor R, called the *dynamic range,* can be presented conveniently. The logarithmic plots also greatly facilitate the data interpretation in vibration testing.

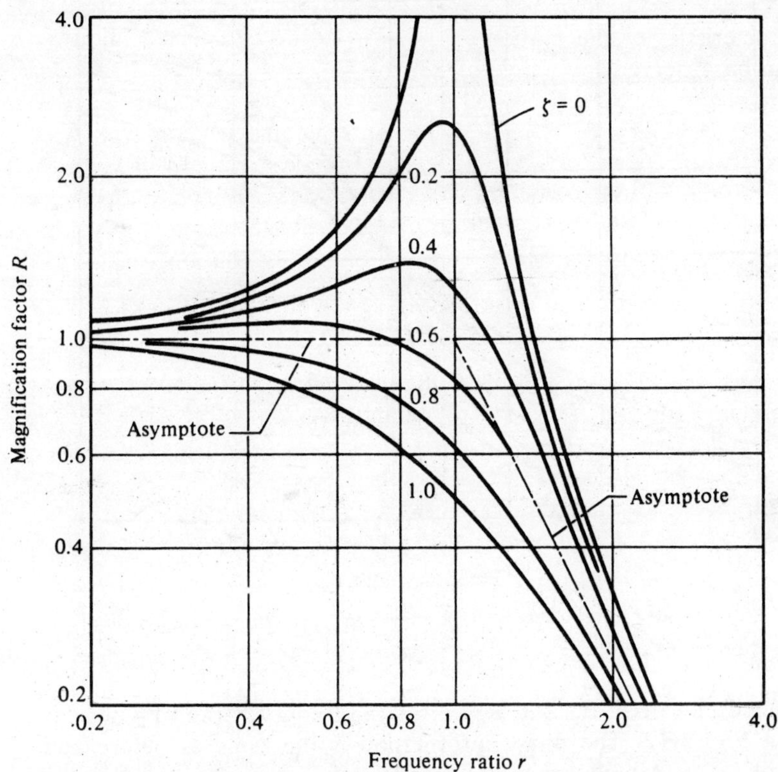

FIG. 2-10. *Magnification factor-versus-frequency ratio; system shown in Fig. 2-6.*

Another convenient method to present the steady-state response data is shown in Fig. 2-12. This format is often used to present the performance characteristics of instruments for vibration measurement. The magnification factor R as defined in Eq. (2-42) is a displacement ratio.

$$R = \frac{\text{Amplitude of steady-state displacement}}{\text{Amplitude of static displacement}} = \frac{X}{F/k} \tag{2-44}$$

Similarly, the velocity ratio R_v and the acceleration ratio R_a can be defined. Since the steady-state response is $x(t) = X\sin(\omega t - \phi)$, the steady-state velocity amplitude is ωX and the acceleration amplitude is $\omega^2 X$. Dividing these quantities by ω_n and ω_n^2, respectively, the velocity ratio R_v and the acceleration ratio R_a are

$$R_v = \frac{\omega X}{\omega_n F/k} = \frac{\omega}{\omega_n} R \tag{2-45}$$

$$R_a = \frac{\omega^2 X}{\omega_n^2 F/k} = \left(\frac{\omega}{\omega_n}\right)^2 R \tag{2-46}$$

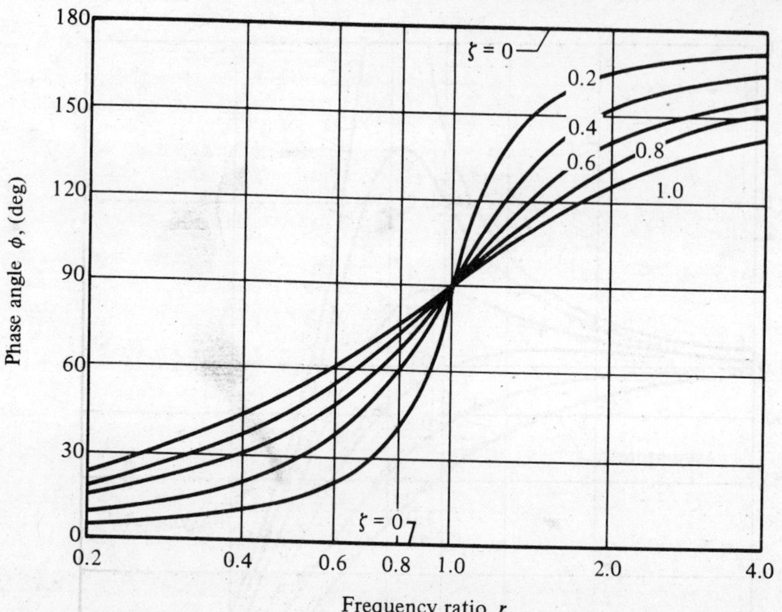

FIG. 2-11. *Phase angle-versus-frequency ratio; system shown in Fig. 2-6.*

A combined plot of R, R_v, and R_a versus the frequency ratio $r = \omega/\omega_n$ is shown in Fig. 2-12. The phase information is the same as before and needs not be presented again. At steady state, the velocity leads the displacement by 90° and the acceleration leads the velocity by 90°. The steady-state response data can be presented in other convenient forms but we shall not pursue the subject further.

General Solution

The general solution of the equation of motion in Eq. (2-18) represents the system response to a harmonic excitation and the given initial conditions. Assume that the system is underdamped, which is often encountered in vibration. Substituting Eqs. (2-35) and (2-38) into (2-20), the general solution due to a harmonic excitation is

$$x = x_c + x_p$$

$$x = Ae^{-\zeta\omega_n t}\sin(\omega_d t + \psi) + X\sin(\omega t - \phi) \qquad \textbf{(2-47)}$$

where X and ϕ are calculated from Eqs. (2-42) and (2-43), respectively. Note that only the constants A and ψ are arbitrary. They are evaluated by applying the initial conditions to the general solution in Eq. (2-47).

The physical interpretation of this equation was explained in Chap. 1. As the harmonic excitation and the initial conditions are applied to the

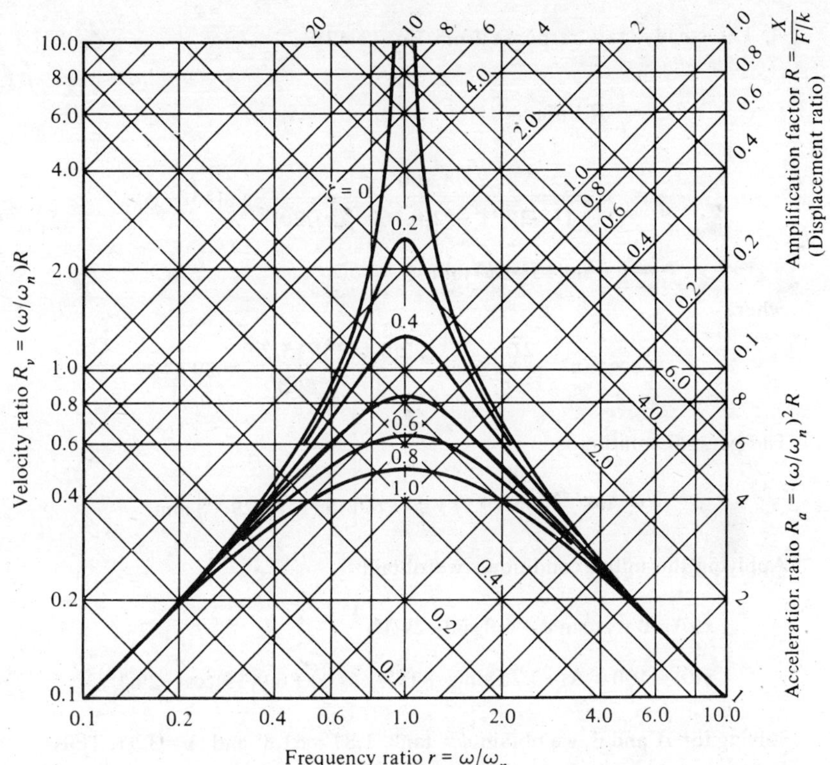

FIG. 2-12. *Magnification factor, velocity ratio, and acceleration ratio-versus-frequency ratio r for various damping factor ξ; system shown in Fig. 2-6.*

system, it tends to follow the excitation and to vibrate at its own natural frequency. Since x_p is sustained by the excitation, it must be at the excitation frequency. On the other hand, x_c is not sustained by the excitation and it is the transient motion. The frequency ω_d of the transient motion is that of the free vibration of the system.

Example 5

Find the steady-state response and the transient motion of the system in Example 4, if an excitation force of 24 sin 15t N (5.4 sin 15t lb_f) is applied to the mass in addition to the given initial conditions.

Solution:

The displacement of the mass m is obtained by the direct application of Eq. (2-47). The system parameters are identical to those calculated in Example

4. The steady-state response from Eq. (2-42) is

$$x_p = \frac{F}{k} R \sin(\omega t - \phi)$$

$$= \frac{24/8{,}000}{\sqrt{[1-(15/20)^2]^2 + [2(0.1625)(15/20)]^2}} \sin(15t - \phi)$$

$$= 6.0 \sin(15t - \phi) \text{ mm}$$

where

$$\phi = \tan^{-1}\frac{2\zeta r}{1-r^2} = \tan^{-1}\frac{2(0.1625)(15/20)}{1-(15/20)^2} = 29.1^\circ$$

The general solution is

$$x = Ae^{-3.25t}\sin(19.7t + \psi) + 6.0\sin(15t - 29.1^\circ) \text{ mm}$$

Applying the initial conditions, we obtain

$$x(0) = 0 = A\sin\psi + 6.0\sin(-29.1^\circ)$$

$$\dot{x}(0) = 100 = A(-3.25\sin\psi + 19.7\cos\psi) + 6.0(15)\cos(-29.1^\circ)$$

Solving for A and ψ, we obtain $\psi = \tan^{-1} 1.87 = 61.8^\circ$ and $A = 3.31$. Thus,

$$x = 3.31e^{-3.25t}\sin(19.7t + 61.8^\circ) + 6.0\sin(15t - 29.1^\circ) \text{ mm}$$

The equation is plotted in Fig. 2-13.

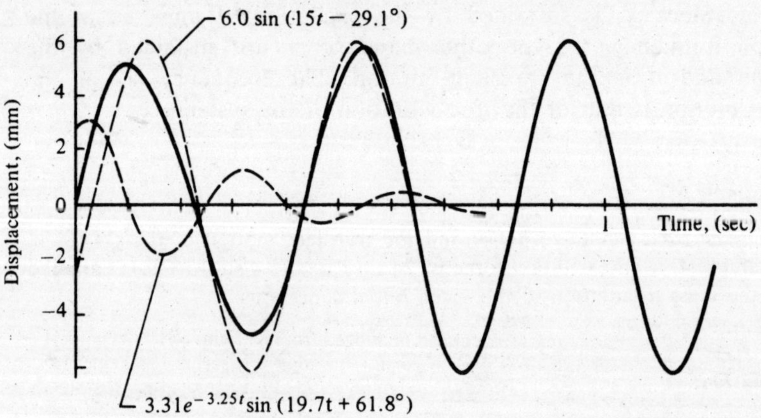

FIG. 2-13. *Displacement vs time; Example 5.*

2-6 FREQUENCY RESPONSE METHOD

Frequency response method is a harmonic analysis. A sinusoidal excitation is applied to a system and its steady-state response is examined over a frequency range of interest. For a linear system, both the excitation and the system response are sinusoidal of the same frequency. This can be verified from the theory of differential equations.

The method is generally used for vibration measurement. The implication is that it is more convenient to describe a system by its Fourier spectrum (see Chap. 3). With the advent of instrumentation and computers, the pulse technique has become a popular test procedure. The results from pulse testing, however, are generally expressed as frequency response data. A vast amount of vibration measuring has been done in the past decade or two. This field of study will gain further prominence in the future.*

We shall discuss mechanical impedance method and sinusoidal transfer function in this section.

Impedance Method

Mechanical impedance method is a harmonic analysis. It represents the sinusoidal functions in the equation of motion by means of rotating vectors as discussed in Sec. 1-5. We shall first represent the forces in a system by means of rotating vectors and then derive the mechanical impedance of the system and its components.

The equation of motion of the one-degree-of-freedom system in Fig. 2-6 and its steady-state response from Eq. (2-38) are

$$\begin{aligned} m\ddot{x} + c\dot{x} + kx &= F \sin \omega t \\ x &= X \sin(\omega t - \phi) \end{aligned} \qquad \textbf{(2-48)}$$

Using the vectorial representation of harmonic motions, the equations above can be expressed as

$$m\ddot{x} + c\dot{x} + kx = \bar{F}e^{j\omega t} \qquad \textbf{(2-49)}$$

$$x = \bar{X}e^{j\omega t} \qquad \textbf{(2-50)}$$

* The importance of vibration measurement can be judged from a quotation: "Mechanical Maintenance is one of the largest industries on earth. Vibration measuring has helped to lower costs by as much as 50%." M. P. Blake, Monograph #721111, Lovejoy, Inc., Downers Grove, Ill., Dec. 1972.

For discussion of vibration measurement, see, for example, M. P. Blake and Wm. S. Mitchell, *Vibration and Acoustic Measurement Handbook*, Spartan Books, New York, 1972.

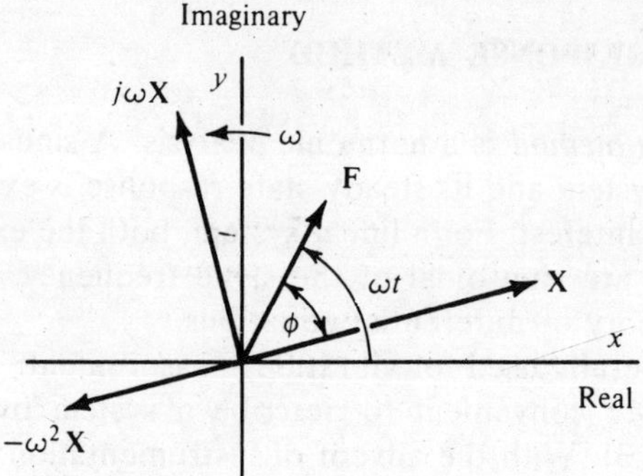

FIG. 2-14. *Displacement, velocity and acceleration vectors.*

The force vector is $\mathbf{F} = \bar{F}e^{j\omega t}$ and the displacement vector is $\mathbf{X} = \bar{X}e^{j\omega t}$. $\bar{F}$ and $\bar{X}$ are the phasors of $\mathbf{F}$ and $\mathbf{X}$, respectively.*

The force vector $\mathbf{F}$ and the displacement vector $\mathbf{X}$ are shown in Fig. 2-14. The corresponding velocity and acceleration vectors are obtained from $\mathbf{X}$ by differentiation as shown in Eq. (1-10). The velocity vector is $j\omega\mathbf{X}$ and the acceleration vector is $-\omega^2\mathbf{X}$. The relative positions of the vectors are illustrated in the figure.

The harmonic forces are obtained by multiplying each of these vectors by the appropriate constants. The spring force always resists the displacement. Hence the spring force vector is $-k\mathbf{X}$. Similarly, the damping and the inertia-force vectors are $-j\omega c\mathbf{X}$ and $\omega^2 m\mathbf{X}$, respectively. These vectors are shown in Fig. 2-15(a). Although these are rotating vectors, their relative positions or phase angles are constant. For dynamic equilibrium, the vectorial sum of the forces due to the spring, the damper, and the mass is equal and opposite to the applied force as indicated in Eq. (2-49). Hence the force vectors form a closed polygon as shown in Fig. 2-15(b).

Figure 2-16 shows the relation of these vectors for an excitation force of constant magnitude but for frequency ratio $r \gtrless 1$, where $r = \omega/\omega_n$. The

* The phasor notation is often a source of confusion for some students. A phasor is a time independent complex coefficient, which together with the factor $e^{j\omega t}$ gives a complex time function.

From Sec. 1-5, a phasor is a complex amplitude or a complex number. It denotes the magnitude and phase angle of a vector relative to the reference vector. In this case, the force is the reference vector and its phase angle is zero. Thus, $\bar{F} = F$ or $\bar{F} = Fe^{j\alpha}$ where $\alpha = 0$. If given $F(t) = F\sin\omega t$ and $x = X\sin(\omega t - \phi)$, the displacement vector is $\mathbf{X} = Xe^{j(\omega t - \phi)} = \bar{X}e^{j\omega t}$. Hence the phasor of $\mathbf{X}$ is $Xe^{-j\phi}$.

More generally, if given $F(t) = F\sin(\omega t + \beta)$, the steady-state response is $x = X\sin(\omega t + \beta - \phi)$. In phasor notation, the force vector is $\mathbf{F} = Fe^{j(\omega t + \beta)} = \bar{F}e^{j\omega t}$ and the displacement vector is $\mathbf{X} = Xe^{j(\omega t + \beta - \phi)} = \bar{X}e^{j\omega t}$. Hence the phasor of $\mathbf{F}$ is $Fe^{j\beta}$ and that of $\mathbf{X}$ is $Xe^{j(\beta - \phi)}$. The relative amplitude and the phase angle between the force and the displacement remain unchanged.

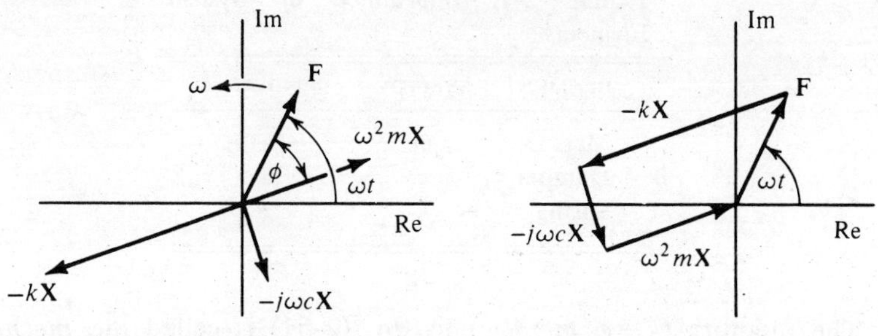

(*a*) Force vectors (*b*) Polygon of force vectors

FIG. 2-15. *Vectors representing harmonic spring, damping, inertial, and excitation forces.*

magnitudes and the phase angles indicated may be compared with those shown in Figs. 2-8 and 2-9. Since the interest is in the relative amplitudes and phase angles of the vectors, the vectors are rotated clockwise by an amount $(\omega t - \phi)$. This is equivalent to choosing $(\omega t - \phi)$ as the datum of measurement.

The displacement $x(t)$ is obtained by substituting Eq. (2-50) in (2-49). Factoring out the $e^{j\omega t}$ term, we get

$$[(j\omega)^2 m + (j\omega)c + k]\bar{X} = \bar{F} \tag{2-51}$$

or

$$\bar{X} = \frac{\bar{F}}{k - \omega^2 m + j\omega c} = Xe^{-j\phi} \tag{2-52}$$

where

$$X = |\bar{X}| = \frac{F}{\sqrt{(k - \omega^2 m)^2 + (\omega c)^2}} = \frac{F}{k} R \tag{2-53}$$

$$-\phi = \angle \bar{X} = -\tan^{-1} \frac{\omega c}{k - \omega^2 m} \tag{2-54}$$

where R is the magnification factor defined in Eq. (2-42).

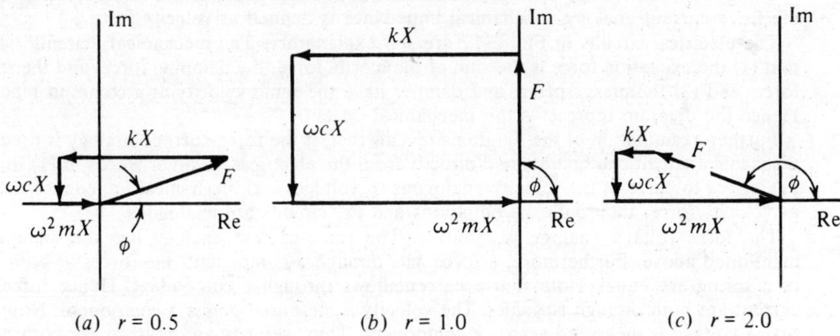

(*a*) $r = 0.5$ (*b*) $r = 1.0$ (*c*) $r = 2.0$

FIG. 2-16. *Polygon of force vectors for frequency ratio $r \gtrless 1$: $F =$ constant and $\zeta = 0.25$.*

TABLE 2-1. *Impedance of System Elements*

ELEMENT	SYMBOL	IMPEDANCE
Mass	m	$-\omega^2 m$
Damper	c	$j\omega c$
Spring	k	k

The quantity $(k-\omega^2 m+j\omega c)$ in Eq. (2-51) is called the *mechanical impedance* of the system in Fig. 2-6. It has the dimension of *force per unit displacement.* The definition follows conveniently from Newton's law of motion in Eq. (2-51). Similarly, for the elements m, c, and k, the corresponding mechanical impedances are defined as $-\omega^2 m$, $j\omega c$, and k, respectively. These are tabulated in Table 2-1. In the literature, mechanical impedance is also defined as force per unit velocity, although this definition is by no means standardized.*

* Mechanical impedance was defined by analogy from Ohm's law. Let us briefly examine the analogy.

Electrical impedance was defined from the generalization of Ohm's law $V=RI$, where V is the voltage drop *across* the resistor R and I the current flow *through* R. The generalization is $V=ZI$, where Z is the impedance of a component or a network.

Let us rewrite Eq. (2-49) and compare it with the RLC circuit in series and in parallel shown in Fig. 2-17.

$$m\frac{d\dot{x}}{dt}+c\dot{x}+k\int \dot{x}\,dt=F(t)$$

$$L\frac{di}{dt}+Ri+\frac{1}{C}\int i\,dt=v(t)$$

$$C\frac{dv}{dt}+\frac{1}{R}v+\frac{1}{L}\int v\,dt=i(t)$$

where $F(t)$, $v(t)$, and $i(t)$ are the sources of force, voltage, and current, respectively. Since all the equations above are of the same form, either the *force-voltage* analogy or the *force-current* analogy can be used to define mechanical impedance.

Using the force-voltage analogy, mechanical impedance is defined as force/velocity. Using the force-current analogy, mechanical impedance is defined as velocity/force.

The electrical circuits in Fig. 2-17 are self-explanatory. The mechanical "circuit" shows that (1) the excitation force is the sum of the inertia force, the damping force, and the spring force, and (2) the mass, spring, and damper have the same velocity at a common junction. Hence the diagram represents the mechanical system.

Further examination of the diagrams reveals that, if the force-current analogy is used, the mechanical circuit can be obtained directly from the electrical. In other words, if (1) force is analogous to current, (2) velocity analogous to voltage, and mechanical impedance defined as velocity/force, then *both* the equations and the circuits are analogous.

The force-voltage analogy is intuitive. The force-current analogy has the advantages mentioned above. Furthermore, a force acts *through* a component; the forces at both ends of a spring are equal. Note that a current flows through a component. Hence force and current are both *through variables.* The velocity is measured *across* a component. Note also that a voltage is measured across a component. Thus, velocity and voltage are both *across variables.* Using this concept and the force-current analogy, the electrical and mechanical circuits should be alike.

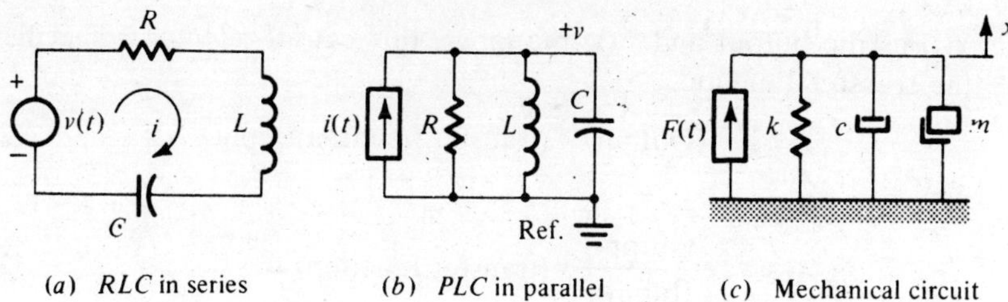

(a) RLC in series (b) PLC in parallel (c) Mechanical circuit

FIG. 2-17. *Comparison of electrical and mechanical circuits.*

Example 6

Use the impedance method to find the steady-state response of the system described in Example 5.

Solution:

The impedance of the system is

$$(k-\omega^2 m)+j\omega c=[8000-15^2(20)]+j15(130)$$
$$=3500+j1950=4007\underline{/29.1°}$$

From Eq. (2-52) we have

$$Xe^{-j\phi}=\frac{24}{4007\underline{/29.1°}}=0.006\underline{/-29.1°}$$

which is the phasor or complex amplitude of the displacement vector **X**

$$\mathbf{X}=Xe^{j(\omega t-\phi)}=0.006e^{j(15t-29.1°)}$$

Since the given excitation force is $F\sin\omega t$, which is equal to $\text{Im}(Fe^{j\omega t})$, the displacement $x(t)$ is $\text{Im}[Xe^{j(\omega t-\phi)}]$, which is $X\sin(\omega t-\phi)$.

$$x=6.0\sin(15t-29.1°)\text{ mm}$$

Transfer Function

The *transfer function* is a mathematical model defining the input-output relation of a physical system. If the system has a single input and a single output, it can be represented by means of a block diagram shown in Fig. 2-18. The system response $x(t)$ is caused by an excitation $F(t)$. Naming

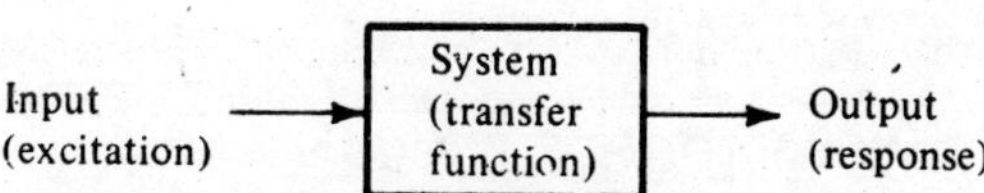

FIG. 2-18. *Block diagram of linear systems.*

$x(t)$ as the output and $F(t)$ the input, this *causal relation* is specified by the transfer function.

$$\text{(Output)} = \text{(transfer function)(input)} \tag{2-55}$$

or

$$\frac{\text{(Output)}}{\text{(Input)}} = \text{(transfer function)} \tag{2-56}$$

Consider the system in Fig. 2-6 as an example. The equation of motion is

$$m\ddot{x} + c\dot{x} + kx = F(t)$$

Using the impedance method and substituting $j\omega$ for the time derivative in the equation, we obtain

$$\frac{x}{F}(j\omega) = \frac{1}{(k - \omega^2 m) + j\omega c} = G(j\omega) \tag{2-57}$$

The symbol $G(j\omega)$ indicates that G is a function of ω. Similarly, x/F is a function of ω, that is, the symbol does *not* indicate a product of (x/F) and $(j\omega)$. $G(j\omega)$ is the sinusoidal transfer function of the system.

Comparing Eqs. (2-57) and (2-52), it is evident that the transfer function is another technique to present the frequency response data of a system. Moreover, the data in Figs. 2-8 to 2-12 are the nondimensional plots of the sinusoidal transfer function. The transfer function of a complex system can be obtained from test data. Thus, a system can be identified by data from its frequency response test.

Note that the transfer function defined in Eq. (2-55) is an operator. It operates on the input to yield an output. It is often called a ratio of output per unit input. This is not a ratio in the normal sense of the word as a ratio of two numbers. As illustrated in Eq. (2-57), the transfer function is a complex number. Furthermore, it is dimensional, such as displacement per unit force. It is more appropriate to think of a transfer function simply as an operator.

Example 7

Determine the frequency response of the system described in Example 5 by means of its transfer function.

Solution:

From Eq. (2-57) the transfer function is

$$G(j\omega) = \frac{1}{(k - \omega^2 m) + j\omega c}$$

$$= \frac{1}{[8000 - 15^2(20)] + j15(130)} = 0.25 \times 10^{-3} \underline{/-29.1^\circ}$$

Hence $\frac{x}{F}(j\omega) = 0.25 \times 10^{-3}\underline{/29.1^\circ}$. Since the excitation is 24 sin 15t, the magnitude of the displacement is

$$X = 24(0.25 \times 10^{-3}) = 0.006 \text{ m}$$

or

$$x = 6.0 \sin(15t - 29.1^\circ) \text{ mm}$$

Resonance, Damping, and Bandwidth

It is observed in Fig. 2-8 that the height of the resonance peak is a function of the damping in the system. One of these frequency response curves is reproduced in Fig. 2-19. It can be shown that the peak of the resonance curves occur at $r = \sqrt{1 - 2\zeta^2}$. If $\zeta \leq 0.1$, the peaks occur at $r \simeq 1$. Thus, from Eq. (2-42) the value of the maximum amplification factor is

$$R_{max} \simeq \frac{1}{2\zeta} \qquad \textbf{(2-58)}$$

The damping in a system is indicated by the sharpness of its response curve near resonance and can be measured by the *bandwidth*. The bandwidth is $(r_2 - r_1)$ as shown in Fig. 2-19, where $r = \omega/\omega_n$ is a frequency ratio and r_1 and r_2 are the frequency ratios at the *half-power points*.* The amplification factor R at r_1 and r_2 is $R = R_{max}/\sqrt{2}$. Substituting this in Eq. (2-42) and letting $R_{max} = 1/2\zeta$ shown in Eq. (2-58), we obtain

$$\left(\frac{1}{2\zeta}\right)\left(\frac{1}{\sqrt{2}}\right) = \frac{1}{\sqrt{(1-r^2)^2 + (2\zeta r)^2}}$$

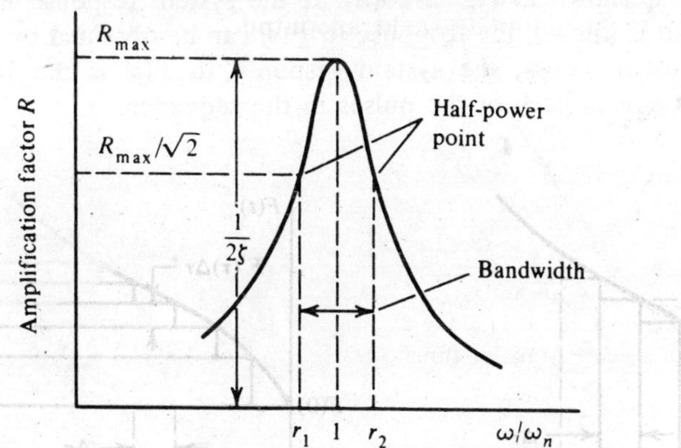

FIG. 2-19. *Harmonic response curve showing bandwidth and half-power points.*

* This terminology is commonly used in electrical engineering. The power dissipation P in a resistor R is $P_1 = I_1^2 R$. At half power, $P_2 = P_1/2 = I_2^2 R$. Thus, $I_2 = I_1/\sqrt{2}$.

Assuming $\zeta \leq 0.1$ and solving for r, we get

$$r_{1,2} \simeq 1 \pm \zeta \tag{2-59}$$

$$\text{Bandwidth} = r_2 - r_1 = \frac{\omega_2}{\omega_n} - \frac{\omega_1}{\omega_n} = 2\zeta \tag{2-60}$$

A factor Q is also used to define the bandwidth and damping.

$$Q = \frac{1}{\text{bandwidth}} = \frac{1}{2\zeta} \tag{2-61}$$

Q is used to measure the *quality* of a resonance circuit in electrical engineering. It is also useful for determining the equivalent viscous damping in a mechanical system.

2-7 TRANSIENT VIBRATION

We shall show that the transient vibration due to an arbitrary excitation $F(t)$ can be obtained by means of superposition. Although the method is not convenient for hand calculations, it can be implemented readily using computers.

The equation of motion of the model in Fig. 2-6 for systems with one degree of freedom and an excitation $F(t)$ is

$$m\ddot{x} + c\dot{x} + kx = F(t) \tag{2-62}$$

One method to solve the equation is to approximate $F(t)$ by a sequence of pulses as shown in Fig. 2-20(*a*). If the system response to a typical pulse input is known, the response to $F(t)$ can be obtained by superposition. In other words, the system response to $F(t)$ is the sum of the responses due to each of the pulses in the sequence.

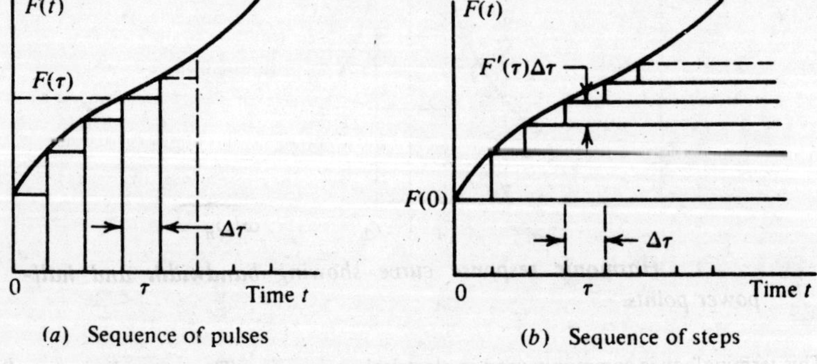

(*a*) Sequence of pulses (*b*) Sequence of steps

FIG. 2-20. *F(t) approximated by pulses and steps.*

Impulse Response

The system response due to a unit impulse input with zero initial conditions is called its *impulse response.* A rectangular pulse of duration or width T_0 and height $1/T_0$ is shown in Fig. 2-21(a). The area of this pulse is unity. To obtain a *unit impulse,* let the pulse width T_0 approach zero while maintaining the pulse area at unity. In the limit, we have a unit impulse $\delta(t)$ as defined by the relations

$$\begin{aligned} \delta(t) &= 0 \qquad \text{for} \qquad t \neq 0 \\ \int_{-\infty}^{\infty} \delta(t)\, dt &= 1 \end{aligned} \tag{2-63}$$

This impulse occurs at $t=0$ as shown in Fig. 2-21(b). If a unit impulse occurs at $t=\tau$, it is defined by the relations

$$\begin{aligned} \delta(t-\tau) &= 0 \qquad \text{for} \qquad t \neq \tau \\ \int_{-\infty}^{\infty} \delta(t-\tau)\, dt &= 1 \end{aligned} \tag{2-64}$$

Note that $\delta(t-\tau)$ is a unit impulse translated along the positive time axis by an amount τ.

Any function, not necessarily a rectangular pulse, having the properties above can be used as a unit impulse and is called the *Dirac delta function.* Mathematically, a unit impulse must have zero pulse width, unit area, and infinite height. It seems that an impulse cannot be realized in applications. In pulse testing of real systems, however, an excitation can be considered as an impulse if its duration is very short compared with the natural period ($=1/f_n$) of the system.

From Eq. (2-62), the equation of motion with an excitation $F(t)=\delta(t)$ is

$$m\ddot{x} + c\dot{x} + kx = \delta(t) \tag{2-65}$$

Assume that the system is at rest before the unit impulse $\delta(t)$ is applied,

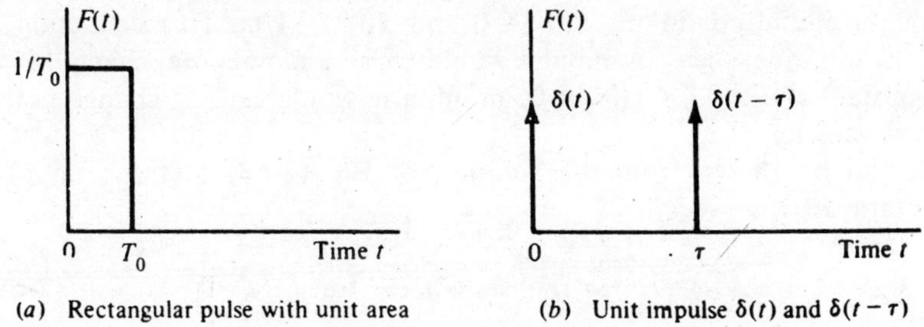

(a) Rectangular pulse with unit area (b) Unit impulse $\delta(t)$ and $\delta(t-\tau)$

FIG. 2-21. *Rectangular pulse and unit impulse.*

that is,

$$x(0^-) = \dot{x}(0^-) = 0 \qquad \textbf{(2-66)}$$

Since $\delta(t)$ is applied at $t = 0$, it is over with at $t \geq 0^+$. Thus, (1) the system becomes unforced for $t \geq 0^+$, and (2) the energy input due to $\delta(t)$ becomes the initial conditions at $t = 0^+$.

To find the initial conditions at $t = 0^+$, we integrate Eq. (2-65) twice for $0^- \leq t \leq 0^+$. Thus,

$$m[x(0^+) - x(0^-)] + \int_{0^-}^{0^+} cx\, dt + \iint_{0^-}^{0^+} kx\; dt\,dt = \iint_{0^-}^{0^+} \delta(t)\; dt\,dt$$

From Eq. (2-63), the first integration of $\delta(t)$ gives a constant and the second integration for $0^- \leq t \leq 0^+$ is zero. Hence the right side of the equation is zero. If $x(t)$ does not become infinite, its integration over this infinitesimal interval is also zero. Thus,

$$m[x(0^+) - x(0^-)] + 0 + 0 = 0$$

If $x(0^-) = 0$ as indicated in Eq. (2-66), we have $x(0^+) = 0$.

Now, integrating Eq. (2-65) once for $0^- \leq t \leq 0^+$ gives

$$m[\dot{x}(0^+) - \dot{x}(0^-)] + c[x(0^+) - x(0^-)] + \int_{0^-}^{0^+} kx\; dt = \int_{0^-}^{0^+} \delta(t)\; dt$$

From Eq. (2-63), the right side of this equation is unity. The third term on the left side is zero if $x(t)$ does not become infinite. The second term is zero as explained above. Thus,

$$m[\dot{x}(0^+) - \dot{x}(0^-)] + 0 + 0 = 1$$

For $x(0^-) = \dot{x}(0^-) = 0$ in Eq. (2-66), the initial conditions at $t = 0^+$ due to a unit impulse at $t = 0$ are

$$x(0^+) = 0 \qquad \text{and} \qquad \dot{x}(0^+) = 1/m \qquad \textbf{(2-67)}$$

The homogeneous equation equivalent to Eq. (2-65) is

$$m\ddot{x} + c\dot{x} + kx = 0 \qquad \textbf{(2-68)}$$

with the initial conditions $x(0^+) = 0$ and $\dot{x}(0^+) = 1/m$. This deduction is almost intuitive, since an impulse would cause a momentum change. If m is constant and $x(0^-) = \dot{x}(0^-) = 0$, an impulse would cause a change in the initial velocity.

It can be shown from the solution of Eq. (2-68) that the impulse response $h(t)$ is

$$h(t) = \frac{1}{\omega_d m} e^{-\zeta\omega_n t} \sin \omega_d t, \qquad \text{for} \qquad t > 0 \qquad \textbf{(2-69)}$$

If a unit impulse $\delta(t - \tau)$ occurs at $t = \tau$ as shown in Fig. 2-21(b), the response is delayed by an amount τ, that is,

$$h(t-\tau)=\frac{1}{\omega_d m}\,e^{-\zeta\omega_n(t-\tau)}\sin\omega_d(t-\tau)\qquad\text{for } t>\tau \tag{2-70}$$

where $h(t-\tau)=0$ for $t<\tau$.

Convolution Integral

Let an excitation $F(t)$ be approximated by a sequence of pulses as shown in Fig. 2-20(a). The *strength* of a pulse is defined by the pulse area. The strength of a typical pulse in the sequence at time τ is the area $F(\tau)\Delta\tau$. The system response to a typical pulse is the product of its unit impulse response and the pulse strength, that is, $h(t-\tau)[F(\tau)\Delta\tau]$. By superposition, we sum the responses due to each of the pulses in the sequence and obtain

$$x(t)=\sum h(t-\tau)F(\tau)\Delta\tau$$

As $\Delta\tau$ approaches zero, the summation becomes the *convolution integral*

$$x(t)=\int_0^t F(\tau)h(t-\tau)\,d\tau=\frac{1}{\omega_d m}\int_0^t F(\tau)e^{-\zeta\omega_n(t-\tau)}\sin\omega_d(t-\tau)\,d\tau \tag{2-71}$$

This is the system response for the input $F(t)$ with zero initial conditions. An alternative form of the integral is

$$x(t)=\int_0^t F(t-\tau)h(\tau)\,d\tau \tag{2-72}$$

In other words, the response of a linear system to an arbitrary excitation is the convolution of its impulse response and the excitation. This statement is known as Borel's theorem.

If the initial conditions are not zero, the complete solution is obtained by the superposition of the particular solution due to the excitation and the complementary solution due to the initial conditions. Substituting the initial conditions $x(0)=x_0$ and $\dot{x}(0)=\dot{x}_0$ into Eq. (2-34) gives the *complementary solution*

$$x=e^{-\zeta\omega_n t}\left(x_0\cos\omega_d t+\frac{\dot{x}_0+\zeta\omega_n x_0}{\omega_d}\sin\omega_d t\right) \tag{2-73}$$

The *particular solution* is shown in Eq. (2-71). By superposition, the *complete solution* is

$$x=e^{-\zeta\omega_n t}\left(x_0\cos\omega_d t+\frac{\dot{x}_0+\zeta\omega_n x_0}{\omega_d}\sin\omega_d t\right)+\int_0^t F(\tau)h(t-\tau)\,d\tau \tag{2-74}$$

The convolution integral is a powerful tool in the study of linear systems. Although Eq. (2–71) cannot be conveniently applied by hand calculations, it can be implemented readily using computers. The example to follow is not indicative of the amount of algebraic computation involved in applying the convolution integral by hand calculations.

Example 8*

A box shown in Fig. 2-22 is dropped through a height H. Find the maximum force transmitted to the body m when the box strikes the floor. Assume there is sufficient clearance between m and the box to avoid contact.

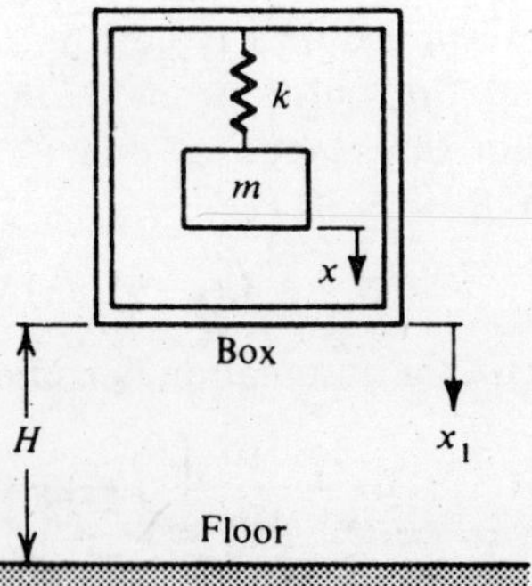

FIG. 2-22. *Drop test.*

Solution:

Let $x(t)$ be the relative distance between m and the box and $t_0 = \sqrt{2H/g}$ the time for the box to strike the floor. Assume that on striking the floor the box remains in contact with the floor. Let us consider the time interval for the free fall and that after striking the floor separately.

During the free fall, the absolute displacement of m is $(x + x_1)$. Hence the equation of motion of m is

$$m(\ddot{x} + \ddot{x}_1) = -kx \qquad \text{or} \qquad m\ddot{x} + kx = -m\ddot{x}_1$$

where

$$x_1 = \tfrac{1}{2}gt^2 \qquad \text{or} \qquad \ddot{x}_1 = g$$

Hence the equation of motion becomes

$$\ddot{x} + \omega_n^2 x = -g$$

If the box is initially at rest before the fall, we have zero initial conditions. Applying Eq. (2-74) where $F(t) = -g$ we get

$$x = 0 + \int_0^t (-g)h(t-\tau)\, d\tau$$

* R. D. Mindlin, "Dynamics of Package Cushioning," *Bell Syst. Tech. Jour.*, 24, (July 1945) pp. 353–461.

The expression for $h(t-\tau)$ is obtained from Eq. (2-70). Since the system is undamped, the equation above becomes

$$x = -\frac{g}{\omega_n}\int_0^t \sin \omega_n(t-\tau)\, d\tau$$

$$= -\frac{g}{\omega_n^2}(1-\cos \omega_n t)$$

From the time that the box strikes the floor at $t=t_0$, the system becomes unforced. Redefining the time from the instant of impact, the initial conditions are

$$x(0) = x(t_0) = -\frac{g}{\omega_n^2}(1-\cos \omega_n t_0)$$

$$\dot{x}(0) = \dot{x}(t_0) + gt_0 = gt_0 - \frac{g}{\omega_n}\sin \omega_n t_0$$

where $x(t_0)$ and $\dot{x}(t_0)$ are obtained from the $x(t)$ above and gt_0 is the velocity of the box assembly at $t=t_0$. Applying Eq. (2-74) with $F(t)=0$ gives

$$x = -\frac{g}{\omega_n^2}(1-\cos \omega_n t_0)\cos \omega_n t + \left(\frac{gt_0}{\omega_n} - \frac{g}{\omega_n^2}\sin \omega_n t_0\right)\sin \omega_n t$$

The maximum force transmitted to m is $m(\ddot{x}_{max}) = kX$, where X is the amplitude of $x(t)$. Thus, the maximum force is

$$\text{Force} = \frac{kg}{\omega_n^2}\sqrt{(1-\cos \omega_n t_0)^2 + (\omega_n t_0 - \sin \omega_n t_0)^2}$$

Indicial Response

The system response due to a unit step input with zero initial conditions is called the *indicial response.* A *unit step function* $u(t)$ shown in Fig. 2-23(a) has the property

$$u(t) = \begin{cases} 1 & \text{for } t>0 \\ 0 & \text{for } t<0 \end{cases} \tag{2-75}$$

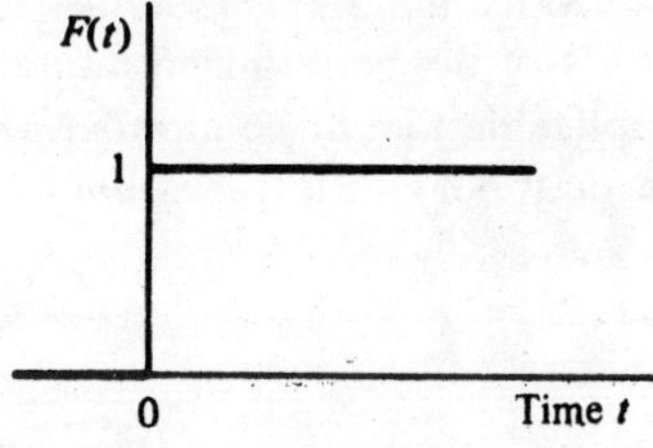

(a) Unit step function $u(t)$

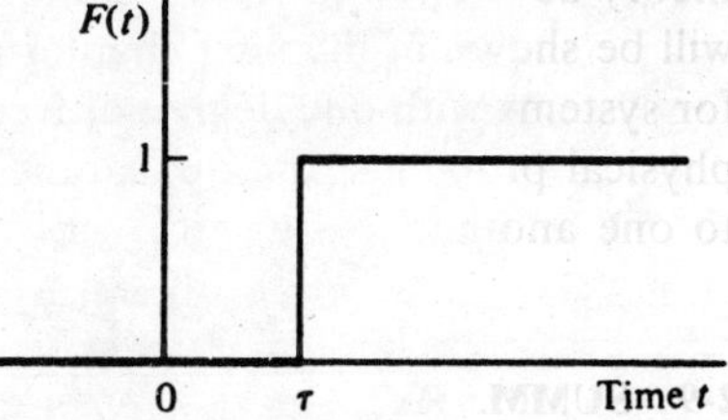

(b) Unit step function $u(t-\tau)$

FIG. 2-23. *Unit step functions.*

A unit step translated along the positive time axis by an amount τ to become $u(t-\tau)$ is shown in Fig. 2-22(b).

$$u(t-\tau)=\begin{cases}1 & \text{for } t>\tau \\ 0 & \text{for } t<\tau\end{cases} \tag{2-76}$$

An arbitrary function $F(t)$ can alternatively be approximated by a sequence of steps as illustrated in Fig. 2-20(b). Following the steps enumerated for the impulse response, it can be shown that the indicial response $x_u(t)$ is

$$x_u(t)=\frac{1}{k}\left[1-\frac{1}{\sqrt{1-\zeta^2}}\, e^{-\zeta\omega_n t}\sin(\omega_d t+\phi)\right] \tag{2-77}$$

where $\phi=\sin^{-1}\sqrt{1-\zeta^2}$. The system response to an arbitrary input $F(t)$ is the superposition of the responses due to each of the individual steps. Thus,

$$x(t)=F(0)x_u(t)+\int_0^t F'(\tau)x_u(t-\tau)\,d\tau \tag{2-78}$$

where $F'(\tau)$ is the time derivative of $F(t)$ estimated at $t=\tau$. The term $F(0)x_u(t)$ is to account for the step at $t=0$, since the slope does not take into account $F(0)$. Equation (2-78) is referred to as *Duhamel's integral* or the *superposition integral.*

2-8 COMPARISON OF RECTILINEAR AND ROTATIONAL SYSTEMS

The discussions in the previous sections centered on systems with rectilinear motion. The theory and the interpretations given are equally applicable to systems with rotational motion. The analogy between the two types of motion and the units normally employed are tabulated in Table 2-2. The responses of the two types of systems are compared in Table 2-3.

Extending this analogy concept, it may be said that systems are analogous if they are described by equations of the same form. The theory developed for one system is applicable to its analogous systems. It will be shown in the next chapter that our study of the generalized model for systems with one degree of freedom is applicable to a large number of physical problems, the appearance of which may bear little resemblance to one another.

2-9 SUMMARY

The number of degrees of freedom of a system is the number of spatial coordinates required to specify its configuration minus the number of

TABLE 2-2. *Analogy between Rectilinear and Rotational Systems*

QUANTITY	RECTILINEAR SYSTEM			ROTATIONAL SYSTEM		
		UNIT			UNIT	
	SYMBOL	ENGLISH	SI UNIT	SYMBOL	ENGLISH	SI UNIT
Time	t	sec	s	t	sec	s
Displacement	x	in.	m	θ	rad	rad
Velocity	$\dot{x}$	in./sec	m/s	$\dot{\theta}$	rad/sec	rad/s
Acceleration	$\ddot{x}$	in./sec^2	m/s^2	$\ddot{\theta}$	rad/sec^2	rad/s^2
Mass, moment of inertia	m	lb$_f$-sec^2/in.	kg	J	in.-lb$_f$-sec^2	m^2 · kg
Damping factor	c	lb$_f$-sec/in.	s · N/m	c_t	in.lb$_f$-sec/rad	m · s · N/rad
Spring constant	k	lb$_f$/in.	N/m	k_t	in.-lb$_f$/rad	m · N/rad
Force, torque	$F = m\ddot{x}$	lb$_f$	N = m · kg/s^2	$T = J\ddot{\theta}$	in.-lb$_f$	m · N = m^2 · kg/s^2
Momentum	$m\dot{x}$	lb$_f$-sec	s · N = m · kg/s	$J\dot{\theta}$	in.-lb$_f$-sec	m^2 · kg · rad/s
Impulse	Ft	lb$_f$-sec	s · N	Tt	in.-lb$_f$-sec	m^2 · kg · rad/s
Kinetic energy	$T = \frac{1}{2}m\dot{x}^2$	in.-lb$_f$	J	$T = \frac{1}{2}J\dot{\theta}^2$	in.-lb$_f$	J
Potential energy	$U = \frac{1}{2}kx^2$	in.-lb$_f$	J	$U = \frac{1}{2}k_t\theta^2$	in.-lb$_f$	J
Work	$\int F\,dx$	in.-lb$_f$	J = m · N = m^2 · kg/s^2	$\int T\,d\theta$	in.-lb$_f$	J = m · N = m^2 · kg/s^2
Natural frequency	$\omega_n = \sqrt{k/m}$	rad/sec	rad/s	$\omega_n = \sqrt{k_t/J}$	rad/sec	rad/s
	$f_n = \dfrac{\omega_n}{2\pi}$	Hz	Hz	$f_n = \dfrac{\omega_n}{2\pi}$	Hz	Hz

TABLE 2-3. *Response of Rectilinear and Rotational Systems*

ITEM	RECTILINEAR SYSTEM	ROTATIONAL SYSTEM
System		
Equation of motion	$m\ddot{x}+c\dot{x}+kx=F(t)$	$J\ddot{\theta}+c_t\dot{\theta}+k_t\theta=T(t)$
System response	$x=x_c+x_p$	$\theta=\theta_c+\theta_p$
Initial conditions	$x(0)=x_0,\ \dot{x}(0)=\dot{x}_0$	$\theta(0)=\theta_0,\ \dot{\theta}(0)=\dot{\theta}_0$
Complementary function, $x_c(t)$	$x_c=Ae^{-\zeta\omega_n t}\sin(\omega_d t+\psi)$ $\omega_n^2=k/m,\ \zeta=\frac{1}{2}c/\sqrt{km},\ \omega_d=\sqrt{1-\zeta^2}\,\omega_n$	$\theta_c=Ae^{-\zeta\omega_n t}\sin(\omega_d t+\psi)$ $\omega_n^2=k_t/J,\ \zeta=\frac{1}{2}c_t/\sqrt{k_tJ},\ \omega_d=\sqrt{1-\zeta^2}\,\omega_n$
Particular integral, $x_p(t)$		
(1) $F(t)=F\sin\omega t$	$x_p=X\sin(\omega t-\phi)$ $X=\dfrac{F}{k}R=\dfrac{F/k}{\sqrt{(1-r^2)^2+(2\zeta r)^2}}$ $r=\omega/\omega_n,\ \phi=\tan^{-1}2\zeta r/(1-r^2)$	$\theta_p=\theta\sin(\omega t-\phi)$ $\Theta=\dfrac{T}{k_t}R=\dfrac{T/k_t}{\sqrt{(1-r^2)^2+(2\zeta r)^2}}$ $r=\omega/\omega_n,\ \phi=\tan^{-1}2\zeta r/(1-r^2)$
(2) $F(t)=\delta(t)$	$h(t)=\dfrac{1}{\omega_d m}e^{-\zeta\omega_n t}\sin\omega_d t$	$h(t)=\dfrac{1}{\omega_d J}e^{-\zeta\omega_n t}\sin\omega_d t$
(3) $F(t)$ arbitrary	$x_p=\int_0^t F(\tau)h(t-\tau)\,d\tau$	$\theta_p=\int_0^t T(\tau)h(t-\tau)\,d\tau$

equations of constraint. Many practical problems can be represented by systems with one degree of freedom, the model of which is shown in Fig. 2-6.

The methods of study in this chapter can be broadly classified as the time and the frequency response methods.

The energy method treats the free vibration of a conservative system. The total mechanical energy is the sum of the kinetic energy T and the potential energy U. Since the total energy $(T+U)$ is constant, the equation of motion is derived from

$$\frac{d}{dt}(T+U)=0 \tag{2-3}$$

Rayleigh's method assumes that (1) the motion is sinusoidal, and (2) the maximum kinetic energy is equal to the maximum potential energy. Thus, if the mass of a spring is not negligible, an equivalent mass m_{eq} of the spring can be calculated from its kinetic energy. Hence the natural frequency of the system is $\omega_n = \sqrt{k/(m+m_{eq})}$.

The equation of motion from Newton's second law is

$$m\ddot{x} = \sum (\text{forces})_x \tag{2-16}$$

All the static forces can be neglected if $x(t)$ is measured from the static equilibrium position of the system. The general solution by the "classical" method is

$$x = x_c + x_p \tag{2-20}$$

where $x_c(t)$ is the complementary function representing the transient motion and $x_p(t)$ the particular integral due to the excitation. The roots of the characteristic equation from Eq. (2-24) or (2-29) dictate the form of $x_c(t)$. This gives the "natural" behavior of the system. If the system is underdamped, $x_c(t)$ is sinusoidal with exponentially decreasing amplitude as shown in Eq. (2-35).

If the excitation is harmonic, $x_p(t)$ is harmonic and at the excitation frequency. The harmonic response, described by Eq. (2-38), is shown graphically in Figs. 2-8 to 2-12. Many schemes can be used for this graphical presentation.

The general solution in Eq. (2-47) shows that the arbitrary constants A and ψ occur only in $x_c(t)$. They are evaluated by applying the initial conditions to the general solution, since it is the entire solution that must satisfy the initial conditions.

In the frequency response method, vectors are used to represent the sinusoidal functions in the equation of motion. Denoting the excitation by the vector $\bar{F}e^{j\omega t}$ and substituting $j\omega$ for d/dt in the equation of motion, the amplitude X and the phase angle ϕ of the steady-state response

$x(t) = \sin(\omega t - \phi)$ are as shown in Eqs. (2-53) and (2-54). The mechanical impedance and the sinusoidal transfer function methods are variations of this technique, although they are very important in vibration measurement.

An arbitrary excitation $F(t)$ can be approximated by a sequence of pulses as shown in Fig. 2-20(a). The system response due to $F(t)$ is the sum of the responses due to the individual pulses. In other words, if the impluse response in Eq. (2-69) of a system is known, its response due to $F(t)$ can be obtained by superposition. This gives the convolution integral in Eq. (2-71).

Zero initial conditions are generally assumed when applying the convolution integral. If the initial conditions are not zero, the complementary solution in Eq. (2-73) due to the initial conditions is added directly to yield the complete solution in Eq. (2-74). In contrast, the classical method first obtains the general solution of the nonhomogeneous equation and then evaluates the constants of integration by applying the initial conditions to the general solution.

The theory discussed in this chapter is equally applicable to systems with rectilinear and rotational motions. The two types of systems are compared in Tables 2-2 and 2-3.

PROBLEMS

Assume all the systems in the figures to follow are shown in their static equilibrium positions.

2-1 Use the energy method to determine the equations of motion and the natural frequencies of the systems shown in the following figures:

(a) Figure 2-1(b). Assume the mass of the torsional bar k_t is negligible.

(b) Figure 2-1(d). Assume there is no slippage between the cord and the pulley.

(c) Figure 2-1(f). Consider the mass of the uniform rod L.

(d) Figure P2-1(a). Assume there is no slippage between the roller and the surface.

(e) Figure P2-1(b). Assume there is no slippage between the roller and the surface. Neglect the springs k_2 and let the springs k_1 be under initial tension.

(f) Repeat part **e**, including springs k_1 and k_2. Assume all the springs are under initial compression.

(g) Figure P2-1(c). Assume there is no slippage between the pulley and the cord.

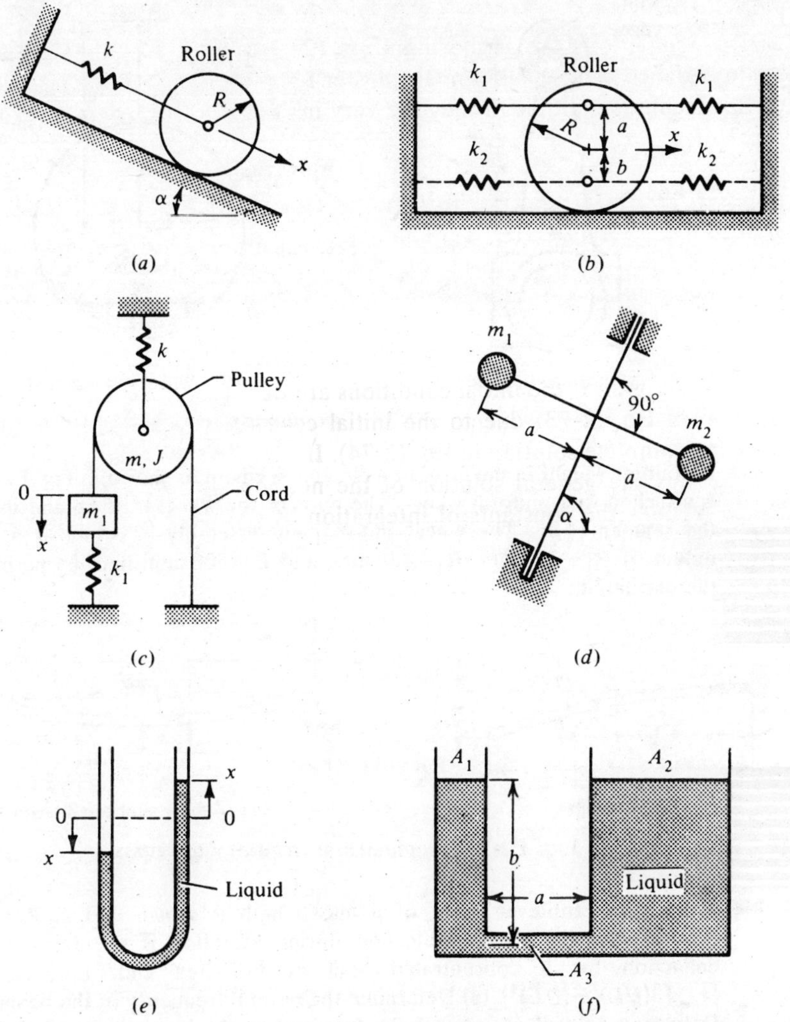

FIG. P2-1. *Vibratory systems.*

(h) Figure P2-1(*d*). Assume $m_2 > m_1$.

(i) Figure P2-1(*e*). The U tube is of uniform cross section.

(j) Figure P2-1(*f*). The cross sectional areas are as indicated.

2-2 A connecting rod of 2.0 kg mass is suspended on a knife edge as shown in Fig. P2-2(*a*). If the period of oscillation is 1.2 s, find the mass moment of inertia J_{cg} of the rod about its mass center *cg*.

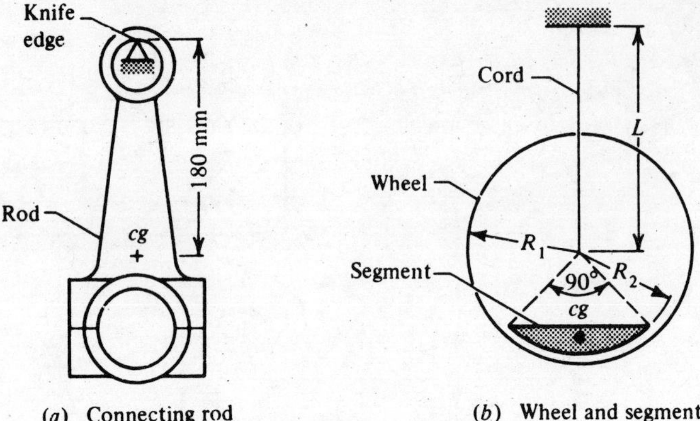

(a) Connecting rod (b) Wheel and segment

FIG. P2-2. *Pendulums.*

2-3 A counter weight in the form of a circular segment as shown in Fig. P2-2(*b*) is attached to a uniform wheel. The mass of the wheel is 45 kg and that of the segment 4 kg. The wheel-and-segment assembly is swung as a pendulum. If $R_1 = 250$ mm, $R_2 = 230$ mm, and $L = 500$ mm, find the period of the oscillation.

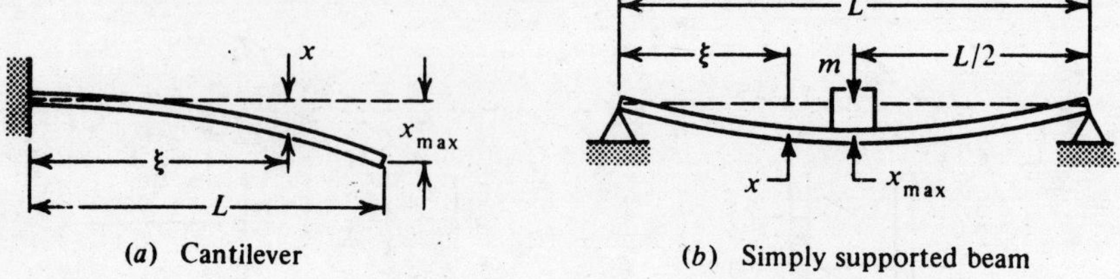

(a) Cantilever (b) Simply supported beam

FIG. P2-3. *Fundamental frequency of beams.*

2-4 A uniform cantilever beam of ρ mass/length is shown in Fig. P2-3(*a*). Assume that the beam deflection during vibration is the same as its deflection for a concentrated load at the free end, that is, $x = \frac{1}{2}x_{max}[3(\xi/L)^2 - (\xi/L)^3]$. **(a)** Determine the natural frequency of the beam. **(b)** Define an equivalent mass at the free end of the beam for this mode of vibration.

2-5 Repeat Prob. 2-4 if the deflection curve is assumed as $x/x_{max} = \xi/L$. What is the percentage error in the natural frequency as compared with Prob. 2-4? Note that the assumed deflection curve does not satisfy the boundary condition at the fixed end, since the slope at the fixed end must be zero.

2-6 Repeat Prob. 2-4 if a mass m is attached to the free end of the cantilever.

2-7 A simply supported uniform beam with a mass m attached at midspan is shown in Fig. P2-3(*b*). The mass of the beam is ρ mass/length. Assume that the deflection during vibration is the same as the static deflection for a concentrated load at midspan, that is, $x = x_{max}[3(\xi/L) - 4(\xi/L)^3]$ for $0 \leq \xi \leq L/2$. **(a)** Find the fundamental frequency of the system. **(b)** What is the equivalent mass of the beam at $L/2$?

2-8 Repeat Prob. 2-7 if the deflection curve is assumed $x/x_{max} = \xi/(L/2)$ for $0 \le \xi \le L/2$.

2-9 A uniform bar of ρ mass/length with an attached rigid mass m is shown in Fig. P2-4(a). Assume the elongation of the bar is linear, that is, $x/x_{max} = \xi/L$. Find the frequency for the longitudinal vibration of the bar.

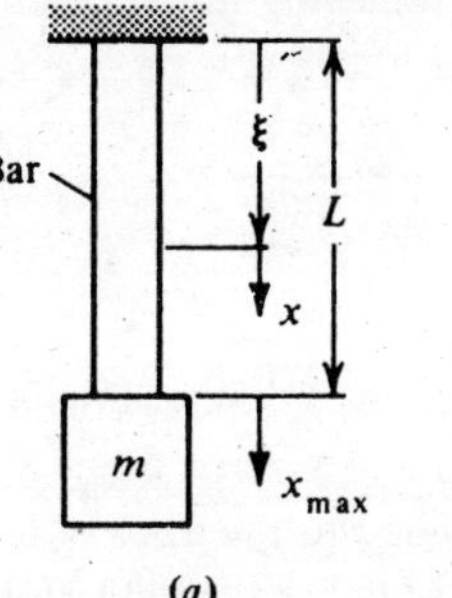

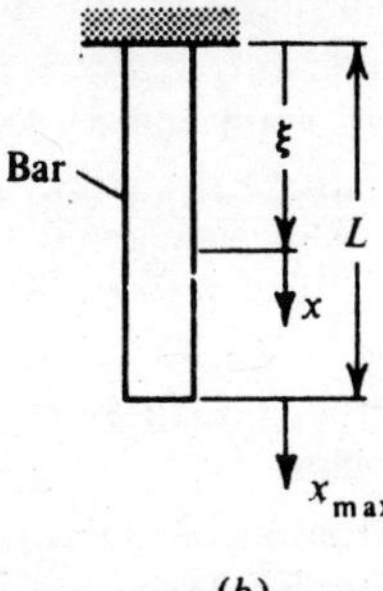

FIG. P2-4. *Fundamental frequency of bars.*

2-10 A uniform bar of ρ mass/length is shown in Fig. P2-4(b). Assume the maximum deflection of the bar is due to its own weight. Find the fundamental frequency for the longitudinal vibration.

2-11 Repeat parts (**a**) to (**j**) of Prob. 2-1, using Newton's law of motion.

2-12 Referring to Fig. 2-4, let the cylinder be of mass $m = 45$ kg, $R_1 = 120$ mm, and $R = 500$ mm. (**a**) Derive the equation of motion by means of Newton's second law. (**b**) Find the natural frequency.

2-13 Find the solutions of the homogeneous equation $\ddot{x} + 9x = 0$ for the following initial conditions:

(**a**) $x(0) = 1$ and $\dot{x}(0) = 0$

(**b**) $x(0) = 0$ and $\dot{x}(0) = 2$

2-14 Find the solutions of the homogeneous equation $\ddot{x} + 4\dot{x} + 13x = 0$ for the following initial conditions:

(**a**) $x(0) = 1$ and $\dot{x}(0) = 0$

(**b**) $x(0) = 0$ and $\dot{x}(0) = 2$

(**c**) $x(0) = 1$ and $\dot{x}(0) = 2$

2-15 Assuming the initial conditions $x(0) = \dot{x}(0) = 0$, find the solution of each of the following nonhomogeneous equations:

(**a**) $\ddot{x} + 4\dot{x} + 13x = 5e^{-t}$

(**b**) $\ddot{x} + 4\dot{x} + 13x = 5\sin(4t + \pi/3)$

(**c**) $\ddot{x} + 4\dot{x} + 13x = 5e^{-t}\sin 4t$

2-16 The equations of motion are given as (**a**) $m\ddot{x}+c\dot{x}+kx=F\sin(\omega t+\alpha)$, and (**b**) $m\ddot{x}+c\dot{x}+kx=F\cos(\omega t+\beta)$. Derive the steady-state response of each of these equations by the method of undetermined coefficients.

2-17 A machine of 20 kg mass is mounted as shown schematically in Fig. 2-7. If the total stiffness of the springs is 17 kN/m and the total damping is 300 N · s/m, find the motion $x(t)$ for the following initial conditions:

(**a**) $x(0)=25$ mm and $\dot{x}(0)=0$

(**b**) $x(0)=25$ mm and $\dot{x}(0)=300$ mm/s

(**c**) $x(0)=0$ and $\dot{x}(0)=300$ mm/s

2-18 Repeat Prob. 2-17 if an excitation force $80\cos 35t$ N is applied to the mass of the system.

2-19 An excitation of $20\sin(10t-30°)$ N is applied to the mass of a mass-spring system with $m=18$ kg and $k=7$ kN/m. (**a**) Find the motion $x(t)$ of the mass for the initial conditions $x(0)=\dot{x}(0)=0$. (**b**) Repeat part *a* if the damping in the system is $c=200$ N · s/m.

2-20 The equation of motion of the system in Fig. 2-6 is $m\ddot{x}+c\dot{x}+kx=F\sin\omega t$. Represent the forces by rotating vectors and indicate the positions of the vectors for the following conditions:

(**a**) m is moving downward and it is below the equilibrium 0.

(**b**) m is moving upward and it is below the equilibrium 0.

(**c**) m is moving upward and it is above the equilibrium 0.

(**d**) m is moving downward and it is above the equilibrium 0.

2-21 The equations of motion are given as (**a**) $m\ddot{x}+c\dot{x}+kx=F\sin(\omega t+\alpha)$, and (**b**) $m\ddot{x}+c\dot{x}+kx=F\cos(\omega t+\beta)$. Find the steady-state response for each of the equations by the method of mechanical impedance.

2-22 Find the steady-state response of the system described in Prob. 2-14 if an excitation force $F=20\cos 35t$ N is applied to the mass of the system.

2-23 A mass-spring system with damping has $m=2$ kg, $c=35$ N · s/m, $k=4$ kN/m, and an excitation $F=30\cos\omega t$ N applied to the mass. Use the mechanical impedance method to find the steady-state amplitude X and the phase angle ϕ for each of the following excitation frequencies:

(**a**) $\omega=6$ rad/s

(**b**) $\omega=\omega_n$

(**c**) $\omega=120$ rad/s

2-24 Derive the equations of motion for each of the systems shown in Fig. P2-5. Derive expressions for the steady-state response of the systems by the mechanical impedance method.

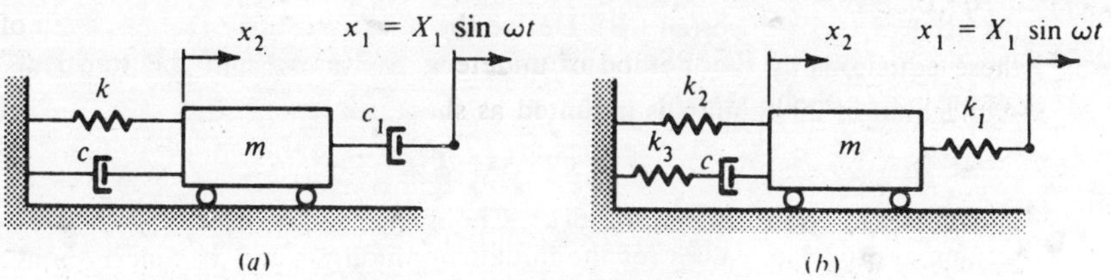

FIG. P2-5. *Vibratory systems.*

2-25 A constant force is applied to an underdamped mass-spring system at $t=0$. Assuming zero initial conditions, (**a**) derive the equation for the response $x(t)$, (**b**) find the time at which the first peak of $x(t)$ occurs, and (**c**) derive an equation relating this peak response and the damping factor ζ of the system.

2-26 Consider the equation $\tau\dot{x}+x=C$, where τ and C are constants. If $x(0)=0$, find the solution (**a**) by the method of undetermined coefficients, and (**b**) by the convolution integral in Eq. (2-72). (**c**) Repeat the problem for $x(0)=x_0$.

2-27 Repeat Prob. 2-26 for the equation $\tau\dot{x}+x=Ct$.

2-28 Given the equation of motion of an undamped system

$$m\ddot{x}+kx=F(t) \qquad \text{or} \qquad \ddot{x}+\omega_n^2 x=F(t)/m$$

derive the equation for the transient response $x(t)$ shown in Eq. (2-74) by (1) multiplying the equation above by $\sin \omega_n(t-\tau)$, and (2) integrating by parts for $0\le\tau\le t$, that is,

$$x(t)=x_0 \cos \omega_n t+\frac{\dot{x}_0}{\omega_n}\sin \omega_n t+\frac{1}{m\omega_n}\int_0^t F(\tau)\sin \omega_n(t-\tau)\,d\tau$$

2-29 Given the equation of motion of an underdamped system

$$m\ddot{x}+c\dot{x}+kx=F(t) \qquad \text{or} \qquad \ddot{x}+2\zeta\omega_n\dot{x}+\omega_n^2 x=F(t)/m$$

derive the equation for the transient response $x(t)$ shown in Eq. (2-74) by (1) multiplying the equation above by $e^{-\zeta\omega_n(t-\tau)}\sin \omega_d(t-\tau)$ and (2) integrating by parts for $0\le\tau\le t$, that is,

$$x(t)=e^{-\zeta\omega_n t}\left(x_0 \cos \omega_d t+\frac{\dot{x}_0+\zeta\omega_n x_0}{\omega_d}\sin \omega_d t\right)$$
$$+\frac{1}{m\omega_d}\int_0^t F(\tau)e^{-\zeta\omega_n(t-\tau)}\sin \omega_d(t-\tau)\,d\tau$$

2-30 Given the system equation and the initial conditions

$$\ddot{x}+10\dot{x}+100x=60 \qquad \text{and} \qquad x(0)=1,\ \dot{x}(0)=2$$

find the transient response $x(t)$ by means of Eq. (2-74). Check the answer by means of the classical method.

2-31 Assuming zero initial conditions, find the transient response $x(t)$ of a system described by the equation

$$\ddot{x}+2\zeta\omega_n\dot{x}+\omega_n^2 x=At$$

by means of Eq. (2-74), where $A=$ constant. Use the classical method to check the answer.

Computer problems:

2-32 Use the program TRESPSUB listed in Fig. 9-2(a) to find the transient response $x(t)$ of the system

$$m\ddot{x} + c\dot{x} + kx = F(t)$$

Let $F(t)$ be as shown in Fig. P2-6(a). Choose values for m, c, k, F, and T. Assume appropriate values for the initial conditions x_0 and $\dot{x}_0$. Select about two cycles for the duration of the run and approximately twenty data points per cycle.

Consider the problem in three parts as follows:

(**a**) $F(t) = 0$, $x(0) = x_0$ and $\dot{x}(0) = \dot{x}_0$

(**b**) $F(t) \neq 0$, $x(0) = 0$ and $\dot{x}(0) = 0$

(**c**) $F(t) \neq 0$, $x(0) = x_0$ and $\dot{x}(0) = \dot{x}_0$

Verify from the computer print-out that $x(t)$ in part **c** is the sum of the parts **a** and **b**. In other words, this is to demonstrate Eq. (2-74) in which the response due to the initial conditions and the excitation can be considered separately.

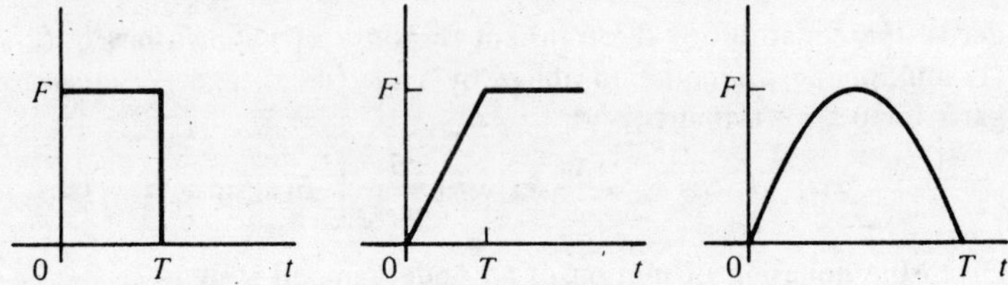

(a) Rectangular pulse (b) Step input with rise time (c) A half sine pulse

FIG. P2-6. *Excitation forces.*

2-33 Repeat Prob. 2-32 for the excitation $F(t)$ shown in Fig. P2-6(b).

2-34 Repeat Prob. 2-32 for the excitation $F(t)$ shown in Fig. P2-6(c).

2-35 Select any transient excitation $F(t)$ and repeat Prob. 2-32.

2-36 It was shown in the pendulum problem in Example 1 that the equation of motion is nonlinear for large amplitudes of vibration. Consider a variation of the pendulum problem in Eq. (2-11).

$$mL^2\ddot{\theta} + c_t\dot{\theta} + mgL \sin\theta = \text{torque}\,(t)$$

or

$$\ddot{\theta} + 2\zeta\omega_n\dot{\theta} + \omega_n^2 \sin\theta = T(t)/mL^2$$

where c_t is a viscous damping factor and $T(t)$ a constant torque applied to the system. Select values for ζ, ω_n, $T(t)$, and the initial conditions $\theta(0)$ and $\dot{\theta}(0)$. Using the fourth-order Runge-Kutta method as illustrated in Fig. 9-1(a), write a program to implement the equation above.

2-37 (**a**) Repeat Prob. 2-36 but modify the program for plotting, as illustrated in Fig. 9-4(a). (**b**) Plot the results using PLOTFILE listed in Fig. 9-5(a).

3

Systems with One Degree of Freedom—Applications

3-1 INTRODUCTION

This chapter is devoted to the application of the theory developed in Chap. 2 to a large class of problems, the appearance of which may differ appreciably from that of the generalized model in Fig. 2-6. The emphasis is on problem formulation and the generalization of each type of system. The approach is to reduce the equation of the system to the form of a one-degree-of-freedom system shown in Eq. (2-18), that is,

$$m_{eq}\ddot{x} + c_{eq}\dot{x} + k_{eq}x = F_{eq}(t) \tag{3-1}$$

where m_{eq}, c_{eq}, k_{eq}, and $F_{eq}(t)$ are the equivalent mass, damping coefficient, spring constant, and excitation force, respectively.* Once the equation is developed, the interpretation follows the general theory discussed in the last chapter.

The equivalent quantities in Eq. (3-1) may be self-evident for simple problems. For example, the equivalent spring force $k_{eq}x$ is that which tends to restore the mass to its equilibrium position. The restoring force can be due to a spring, gravitation, the buoyancy of a liquid, a centrifugal field, or their combinations. Alternatively, from energy considerations, k_{eq} is a quantity in the total potential energy $\frac{1}{2}k_{eq}x^2$ of the system due to a displacement in the x direction. Similarly, m_{eq} is a quantity in the total kinetic energy $\frac{1}{2}m_{eq}\dot{x}^2$ of the system due to $\dot{x}$. The c_{eq} accounts for all the energy dissipation associated with $\dot{x}$. $F_{eq}(t)$ could be due to a force

* For convenience of writing, the subscript (eq) is omitted from subsequent equations unless ambiguity arises.

and/or a motion applied to the system or an unbalance in the machine. The product of F_{eq} and the displacement x has the unit of work.*

The generalized model shown in Fig. 2-6 consists of four elements, namely, the mass, the damper, the spring, and the excitation. The systems considered in this chapter are grouped according to the elements involved. If a system does not possess one of these elements, such as a damper, it is simply omitted from the equation of motion in Eq. (3-1). We shall begin with the simple mass-spring system.

3-2 UNDAMPED FREE VIBRATION

The simplest vibratory system is one that consists of a mass and a spring element. If a system is lightly damped, it can be approximated by a simple spring-mass system. Neglecting the damper and the excitation, Eq. (3-1) becomes

$$m\ddot{x} + kx = 0 \tag{3-2}$$

From Eq. (2-8), the solution of Eq. (3-2) is

$$x = A_1 \cos \omega_n t + A_2 \sin \omega_n t$$

where A_1 and A_2 are constants. Substituting the initial conditions $x(0) = x_0$ and $\dot{x}(0) = \dot{x}_0$ gives

$$x = x_0 \cos \omega_n t + \frac{\dot{x}_0}{\omega_n} \sin \omega_n t \tag{3-3}$$

or

$$x = A \sin(\omega_n t + \psi) \tag{3-4}$$

where

$$A = \sqrt{x_0^2 + (\dot{x}_0/\omega_n)^2} \qquad \text{and} \qquad \psi = \tan^{-1} \frac{x_0}{\dot{x}_0/\omega_n} \tag{3-5}$$

Example 1. Equivalent Mass

A machine component at its static equilibrium position is represented by a uniform bar of mass m and length L and a spring k in Fig. 3-1(a). Derive the equivalent system shown in Fig. 3-1(b).

Solution:

The equivalent mass m_{eq} is obtained by considering the kinetic energy T of the system as illustrated in Example 3, Chap. 2. Assuming the spring is of

* Although the concept of equivalent quantities may not be fully utilized in this chapter, they are introduced early in the text because (1) the one-degree-of-freedom system is basic in vibration, and (2) the concept of equivalent or generalized quantities is essential for more advanced studies in later chapters.

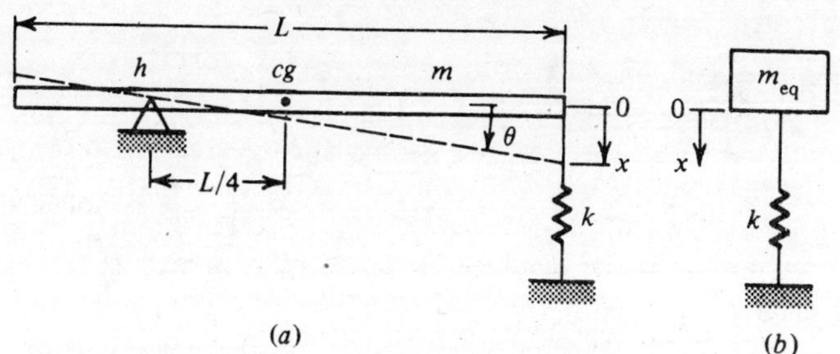

FIG. 3-1. *Equivalent mass m_{eq}.*

negligible mass, we get

$$T = \tfrac{1}{2}J_h\dot{\theta}^2 = \tfrac{1}{2}m_{eq}\dot{x}^2$$

where

$$J_h = J_{cg} + m\left(\frac{L}{4}\right)^2 = mL^2(\tfrac{1}{12} + \tfrac{1}{16}) = \tfrac{7}{48}mL^2$$

Substituting $\dot{x} \simeq (3/4)L\dot{\theta}$ in the kinetic energy equation yields

$$m_{eq} = \tfrac{7}{27}m$$

Example 2. Equivalent Mass Moment of Inertia

A pinion-and-gear assembly is shown in Fig. 3-2. It is often convenient to refer the mass moment of inertia J of the assembly to a common shaft. Find the J_{eq} of the assembly referring to the motor shaft.

Solution:

Let N_1 be the number of teeth on the pinion and N_2 that of the gear. The gear ratio is $n = N_1/N_2$. Let θ_1 and θ_2 be the angular rotations of the pinion J_1 and the gear J_2, respectively. The kinetic energy T of the assembly referred to the motor shaft is

$$T = \tfrac{1}{2}J_1\dot{\theta}_1^2 + \tfrac{1}{2}J_2\dot{\theta}_2^2 = \tfrac{1}{2}J_{eq}\dot{\theta}_1^2$$

Substituting $\dot{\theta}_2 = n\dot{\theta}_1$ in T, we get

$$J_{eq} = J_1 + n^2J_2$$

Hence the equivalent mass moment of inertia of J_2 referring to the motor shaft 1 is n^2J_2.

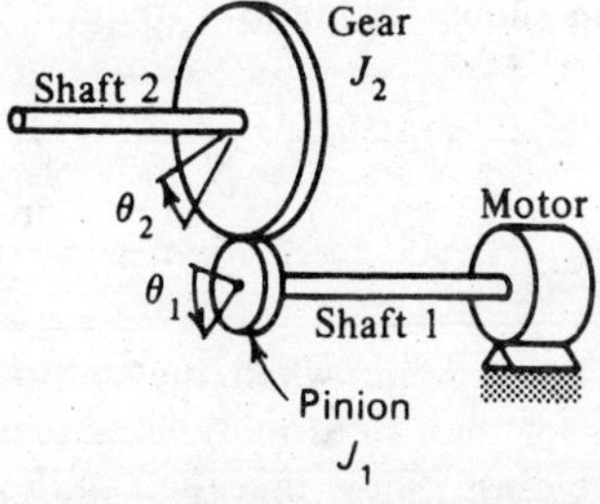

FIG. 3-2. *Equivalent mass moment of inertia J_{eq}.*

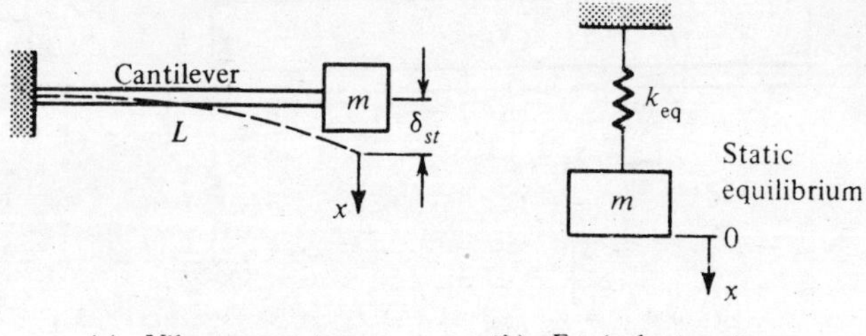

(a) Vibratory system (b) Equivalent system

FIG. 3-3. *Equivalent spring.*

Similarly, equivalent springs can be calculated. Let k_{t2} be the torsional stiffness of shaft 2. It can be shown by equating potential energies that the equivalent stiffness of shaft 2 referring to shaft 1 is n^2k_{t2}. The equivalent spring in a system can assume various forms. We shall illustrate the equivalent spring with the examples to follow.

Example 3. Equivalent Spring

Figure 3-3 shows that the static deflection δ_{st} of a cantilever beam is due to the mass m attached to its free end. Find the natural frequency of the system.

Solution:

The equivalent system is as shown in Fig. 3-3(b) if (1) the cantilever is of negligible mass and (2) m is small in size compared with L. The static deflection δ_{st} due to the concentrated force mg at the free end of a beam of length L is

$$\delta_{st} = \frac{mgL^3}{3EI}$$

where EI is the flexural stiffness of the beam. The equivalent spring constant k_{eq} is defined as force per unit deflection.

$$k_{eq} = \frac{mg}{\delta_{st}} = \frac{3EI}{L^3}$$

From the equivalent system, the natural frequency is

$$f_n = \frac{1}{2\pi}\sqrt{\frac{k_{eq}}{m}} = \frac{1}{2\pi}\sqrt{\frac{3EI}{mL^3}} = \frac{1}{2\pi}\sqrt{\frac{g}{\delta_{st}}}$$

Example 4. Springs in Series

Springs are said to be *in series* when the deformation of the equivalent spring k_{eq} is the sum of their deformations. Assume the cantilever in Fig. 3-4(a) is of negligible mass. Show that the cantilever and the spring k_2 are in series.

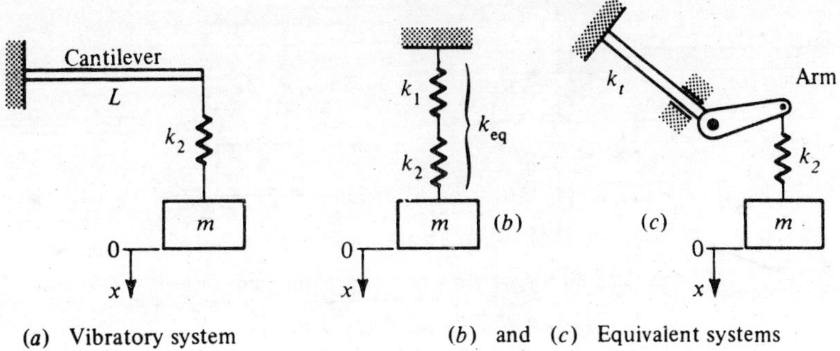

(a) Vibratory system (b) and (c) Equivalent systems

FIG. 3-4. *Springs in series.*

Solution:

The cantilever can be replaced by an equivalent spring k_1 of spring constant $3EI/L^3$ as in Example 3. The equivalent system is as shown in Fig. 3-4(*b*). A unit static force applied at m in the x direction will cause the spring k_1 and k_2 to elongate by $1/k_1$ and $1/k_2$, respectively. The corresponding elongation of the equivalent spring is $1/k_{eq}$. Thus,*

$$\frac{1}{k_{eq}} = \frac{1}{k_1} + \frac{1}{k_2} \tag{3-6}$$

or

$$k_{eq} = \frac{3EIk_2}{3EI + k_2L^3}$$

The system in Fig. 3-4(*c*) consists of a torsional shaft with an extended arm and a spring k_2 in series. If the mass of the shaft and its arm are negligible, this system reduces to that of Fig. 3-4(*b*).

Example 5. Springs in Parallel

Springs are said to be *in parallel* when (1) the equivalent spring force is the sum of the forces of the individual springs, and (2) the springs have the same deformation. A disk J is connected to two shafts shown in Fig. 3-5(*a*).

(a) Show that the shafts are equivalent to springs in parallel.

(b) Determine the natural frequency of the system for torsional vibrations.

* Students often wonder why the equation for springs in series is like electrical impedance Z in parallel, where $Z = V/I$, V = voltage, and I = current. Referring to the discussion on analogy in Sec. 2-6 and using the *current-force analogy*, we define the impedance of a spring $Z_k = \delta/F = 1/k$, where δ = deformation, F = force, and k = spring constant. It can be shown in a mechanical "circuit" that the impedance of the springs in Fig. 3-4 are in series. Thus,

$$Z_{k_{eq}} = Z_{k_1} + Z_{k_2} \quad \text{or} \quad \frac{1}{k_{eq}} = \frac{1}{k_1} + \frac{1}{k_2}$$

Equation (3-7) for springs in parallel can be explained in the same manner.

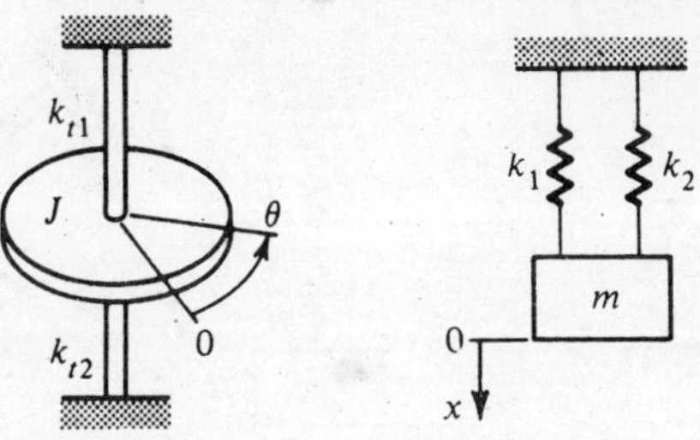

(a) Vibratory system (b) Equivalent system

FIG. 3-5. *Springs in parallel.*

Solution:

(a) Since both shafts tend to restore J to its equilibrium position, the equivalent system is as shown in Fig. 3-5(b). The restoring torque of a circular shaft is

$$T = \frac{\pi d^4 G}{32L}\theta = k_t\theta$$

where G is the shear modulus and d and L are the diameter and length of the shaft, respectively. If the disk J is rotated by an angle θ, the restoring torque T is the sum of the restoring torques of the individual shafts.

$$T = k_{eq}\theta = (k_{t1} + k_{t2})\theta$$

or

$$k_{eq} = k_{t1} + k_{t2} \tag{3-7}$$

$$k_{eq} = \frac{\pi}{32}\left(\frac{d_1^4 G_1}{L_1} + \frac{d_2^4 G_2}{L_2}\right)$$

(b) From the equivalent system, the natural frequency is

$$f_n = \frac{1}{2\pi}\sqrt{\frac{k_{eq}}{J}} = \frac{1}{2\pi}\sqrt{\frac{\pi}{32}\left(\frac{d_1^4 G_1}{L_1} + \frac{d_2^4 G_2}{L_2}\right)\frac{1}{J}}$$

Example 6. Effect of Orientation

The equivalent spring force could be due to a combination of the spring and gravitational forces. Determine the equation of motion of the systems shown in Fig. 3-6.

Solution:

Owing to the difference in the orientations of the systems, the restoring torque due to gravitation on the mass is different for the three systems. Assuming small oscillations and taking moments about 0, the equations of motion are

(a) $J_0\ddot{\theta} = \sum (\text{torque})_o$

$$mL^2\ddot{\theta} = -mgL\sin\theta - (ka\sin\theta)(a\cos\theta)$$

$$mL^2\ddot{\theta} + (mgL + ka^2)\theta = 0$$

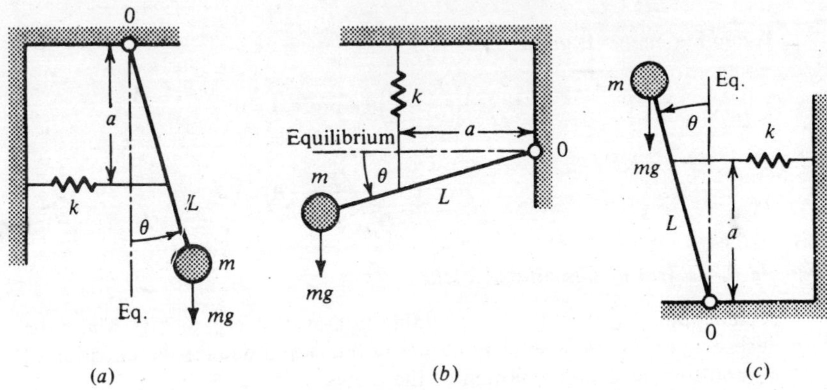

FIG. 3-6. *Effect of orientation.*

(b) $mL^2\ddot{\theta} = -(ka \sin \theta)(a \cos \theta)$

$$mL^2\ddot{\theta} + ka^2\theta = 0$$

(c) $mL^2\ddot{\theta} = mgL \sin \theta - (ka \sin \theta)(a \cos \theta)$

$$mL^2\ddot{\theta} + (ka^2 - mgL)\theta = 0$$

Note that the quantity mg does not appear in all the equations in the examples above. Furthermore, the pendulum could be at the slant; that is, its static equilibrium positions need not be vertical or horizontal. These considerations are left as home problems.

The next two examples are variations of the pendulum problem in which the equivalent spring force is due to gravitation and a centrifugal field.

Example 7

The equation of motion of the system in Fig. 2-4 was derived by the energy method in Example 2, Chap. 2. Derive the equation of motion by Newton's law of motion.

Solution:

Since θ_1 is the rotation of the cylinder relative to the curved surface, the absolute rotation of the cylinder is $(\theta_1 - \theta)$. Using the relation $R\theta = R_1\theta_1$ and taking moments about the instantaneous center of rotation b, the equation of motion is

$$J_b(\ddot{\theta}_1 - \ddot{\theta}) = -mgR_1 \sin \theta$$

$$J_b(R/R_1 - 1)\ddot{\theta} + mgR_1 \sin \theta = 0$$

where

$$J_b = (J_0 + mR_1^2) = \left(\frac{1}{2} mR_1^2 + mR_1^2\right) = \frac{3}{2} mR_1^2$$

Assuming $\sin\theta \simeq \theta$ gives

$$\left(\frac{3}{2}mR_1^2\right)\left(\frac{R}{R_1}-1\right)\ddot{\theta}+mgR_1\theta=0$$

$$\ddot{\theta}+\frac{2g}{3(R-R_1)}\theta=0$$

Example 8. Effect of Centrifugal Field

A helicopter blade and rotor assembly is shown in Fig. 3-7(a). Make the necessary assumptions to simplify the problem and deduce the equation of motion for the flapping motion of the blade.

Solution:

Assume (1) the blade of mass m is a uniform bar hinged at 0, (2) the rotor angular velocity Ω is constant, and (3) the gravitational field is negligible compared with the centrifugal field. Each element $d\xi$ of the blade is subjected to a centrifugal force $\Omega^2 R_1\rho\, d\xi$, where ρ is the mass/length of the blade. The corresponding moment about 0 is $(\Omega^2 R_1\rho\, d\xi)(\xi\sin\theta)$. Since $R_1 = R+\xi\cos\theta$, the total moment is

$$\int_0^L \rho\Omega^2(\sin\theta)(R+\xi\cos\theta)\xi\, d\xi = \rho\Omega^2(\sin\theta)\left(\frac{RL^2}{2}+\frac{L^3}{3}\cos\theta\right)$$

$$= m\Omega^2(RL/2+L^2/3)\theta$$

where $\sin\theta\simeq\theta$, $\cos\theta\simeq 1$, and $m=\rho L$. The mass moment of inertia of the blade about 0 is $J_0 = mL^2/3$. Taking moments about 0 gives

$$\frac{mL^2}{3}\ddot{\theta}+m\Omega^2\left(\frac{RL}{2}+\frac{L^2}{3}\right)\theta=0$$

$$\ddot{\theta}+\Omega^2(1+3R/2L)\theta=0$$

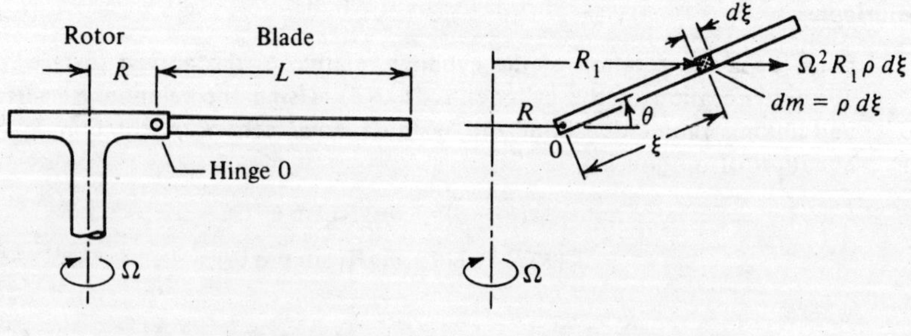

FIG. 3-7. *Helicopter rotor and blade.*

3-3 DAMPED-FREE VIBRATION

To a greater or lesser degree, all physical systems possess damping. For free vibration with damping, Eq. (3-1) becomes

$$m\ddot{x} + c\dot{x} + kx = 0 \tag{3-8}$$

If the system is underdamped, from Eq. (2-34) and for the initial conditions x_0 and $\dot{x}_0$, the solution of Eq. (3-8) is

$$x = e^{-\zeta\omega_n t}\left(x_0 \cos \omega_d t + \frac{\dot{x}_0 + \zeta\omega_n x_0}{\omega_d} \sin \omega_d t\right)$$
$$x = Ae^{-\zeta\omega_n t} \sin(\omega_d t + \psi) \tag{3-9}$$

where

$$A = \sqrt{(x_0\omega_d)^2 + (\dot{x}_0 + \zeta\omega_n x_0)^2}/\omega_d \tag{3-10}$$

$$\psi = \tan^{-1} \frac{x_0\omega_d}{\dot{x}_0 + \zeta\omega_n x_0} \tag{3-11}$$

and ζ, ω_n are defined in Eq. (2-27) and $\omega_d = \sqrt{1-\zeta^2}\,\omega_n$.

***Example* 9**

A component of a machine is represented schematically in Fig. 3-8. Derive its equation of motion.

Solution:

Assuming small oscillations and taking moments about 0, the equation of motion is

$$J_0\ddot{\theta} = \sum (\text{torque})_0$$
$$[m_1L_1^2 + m_2L_2^2 + m_3(L_3+L_4)^2]\ddot{\theta} = m_2gL_2\theta - cL_3^2\dot{\theta} - k(L_3+L_4)^2\theta$$
$$[m_1L_1^2 + m_2L_2^2 + m_3(L_3+L_4)^2]\ddot{\theta} + cL_3^2\dot{\theta} + [k(L_3+L_4)^2 - m_2gL_2]\theta = 0$$

which is of the same form as Eq. (3-8).

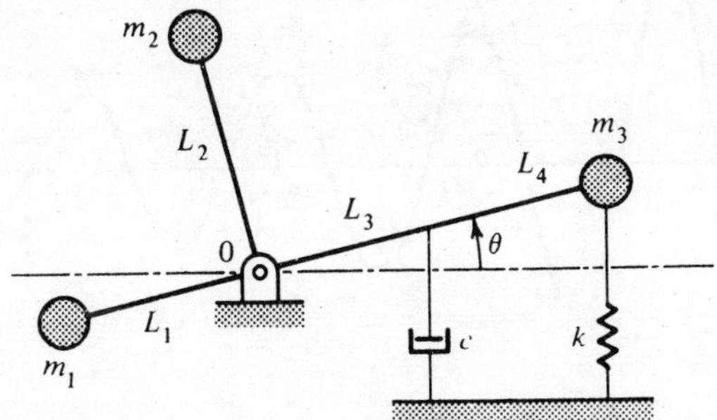

FIG. 3-8. *One-degree-of-freedom system with damping.*

Example 10. Logarithmic Decrement

A mass-spring system with viscous damping is shown in Fig. 2-7. The mass m is displaced by an amount x_0 from its static equilibrium position and then released with zero initial velocity. Determine the ratio of any two consecutive amplitudes.

Solution:

From Eq. (3-9), the maximum amplitude occurs when the product $Ae^{-\zeta\omega_n t}$ and $\sin(\omega_d t+\psi)$ is a maximum. Rewriting the equation with $(\omega_d t)$ as the independent variable and equating $dx/d(\omega_d t)=0$ for maximum, we have

$$x = Ae^{-\zeta\omega_d t/\sqrt{1-\zeta^2}}\sin(\omega_d t+\psi)$$

$$\frac{dx}{d\omega_d t} = Ae^{-\zeta\omega_d t/\sqrt{1-\zeta^2}}\left[-\frac{\zeta}{\sqrt{1-\zeta^2}}\sin(\omega_d t+\psi)+\cos(\omega_d t+\psi)\right]$$

$$=0$$

Hence the maximum amplitude occurs at

$$\tan(\omega_d t+\psi)=\sqrt{1-\zeta^2}/\zeta$$

Let the motion $x(t)$ be as illustrated in Fig. 3-9 and let $\omega_d t_1$ and $\omega_d t_2$ correspond to the maxima x_1 and x_2. The last equation indicates that $\tan(\omega_d t_1+\psi)=\tan(\omega_d t_2+\psi)$. Hence $(t_2-t_1)=2\pi/\omega_d$ is a period and $\sin(\omega_d t_1+\psi)=\sin(\omega_d t_2+\psi)$. The consecutive amplitude ratio is

$$\frac{x_1}{x_2}=\frac{Ae^{-\zeta\omega_n t_1}}{Ae^{-\zeta\omega_n t_2}}=e^{\zeta\omega_n(t_2-t_1)}$$

$$\frac{x_1}{x_2}=e^{\zeta\omega_n(2\pi/\omega_d)}=e^{2\pi\zeta/\sqrt{1-\zeta^2}}$$

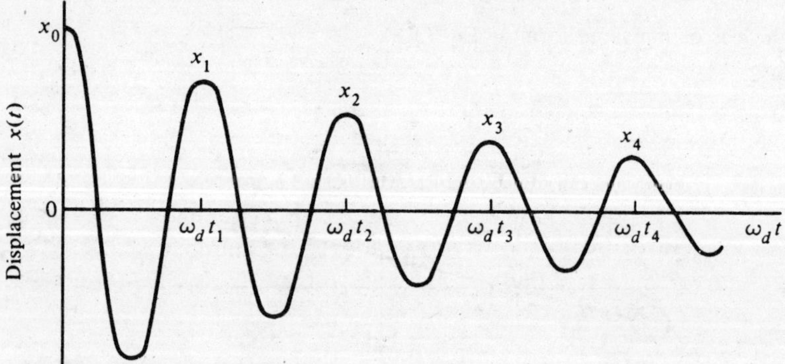

FIG. 3-9. *Free vibration with damping: initial conditions* $x(0)=x_0$ *and* $\dot{x}(0)=0$.

The natural logarithm of this ratio is called the *logarithmic decrement* δ, that is, $\ln(x_1/x_2) = \delta$. Hence

$$\delta = 2\pi\zeta/\sqrt{1-\zeta^2} \tag{3-12}$$

or

$$\delta \simeq 2\pi\zeta \qquad \text{for} \quad \zeta \ll 1 \tag{3-13}$$

The logarithmic decrement is a measure of the damping factor ζ and it gives a convenient method to measure the damping in a system.

It was pointed out in Chap. 2 that the rate of decay of the vibrations is a property of the system. Hence the logarithmic decrement must be independent of initial conditions. Furthermore, any two points on the curve in Fig. 3-9 one period apart may serve to evaluate the logarithmic decrement. The use of consecutive amplitudes, however, is a convenience.

Example 11

The following data are given for a system with viscous damping: mass $m = 4$ kg (9 lb_m), spring constant $k = 5$ kN/m (28 lb_f/in.), and the amplitude decreases to 0.25 of the initial value after five consecutive cycles. Find the damping coefficient of the damper.

Solution:

The amplitude ratio of any two consecutive amplitudes is

$$\frac{x_0}{x_1} = \frac{x_1}{x_2} = \frac{x_2}{x_3} = \frac{x_3}{x_4} = \frac{x_4}{x_5} = e^{\delta}$$

Hence

$$\frac{x_0}{x_5} = \frac{x_0}{x_1} \cdot \frac{x_1}{x_2} \cdot \frac{x_2}{x_3} \cdot \frac{x_3}{x_4} \cdot \frac{x_4}{x_5} = e^{5\delta} = \frac{1}{0.25}$$

$$\delta = \frac{1}{5} \ln 4 = 0.277 = \frac{2\pi\zeta}{\sqrt{1-\zeta^2}}$$

The damping factor ζ and the damping coefficient c are

$$\zeta = 0.044$$

$$c = 2\zeta\sqrt{km} = 2(0.044)\sqrt{5000(4)} = 12.5 \text{ s} \cdot \text{N/m}$$

Following the method in the example above, the number of cycles n required to reduce the amplitude by a factor of N is given by the expression

$$\frac{x_0}{x_n} = N = e^{n\delta} \quad \text{or} \quad \delta = \frac{1}{n} \ln N \tag{3-14}$$

Assuming $\delta \simeq 2\pi\zeta$, Eq. (3-14) is plotted in Fig. 3-10. Note that it takes

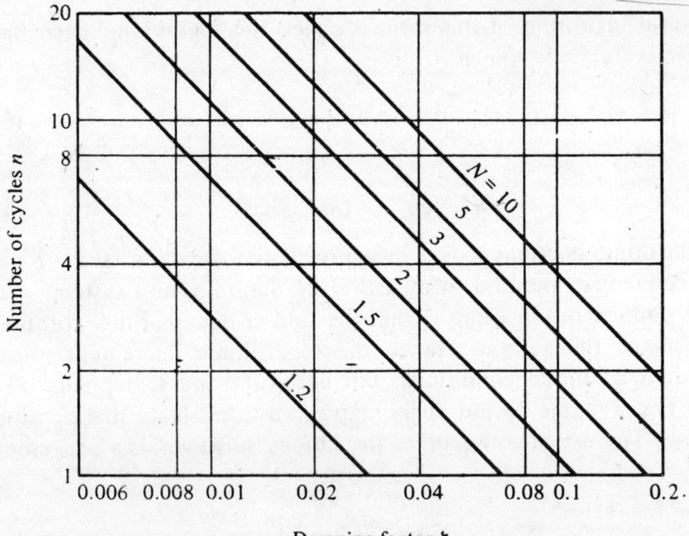

FIG. 3-10. *Number of cycles n to reduce the amplitude by a factor of N for small values of damping.*

less than four cycles to attenuate the amplitude by a factor of 10 when $\zeta = 0.1$. This implies that it does not take many cycles for the transient motion to die out even for lightly damped systems.

3-4 UNDAMPED FORCED VIBRATION—HARMONIC EXCITATION

The usual interest in the study of forced vibration with harmonic excitation is the steady-state response of the system. As discussed in the last section, the transient motion will soon die out, even for lightly damped systems. Except near resonance, the steady-state response can be approximated by that of an undamped system.

Neglecting the damping term in Eq. (3-1), the equation of motion for a system with harmonic excitation is

$$m\ddot{x} + kx = F \sin \omega t \tag{3-15}$$

If the transient motion is assumed to have died out, the steady-state response from Eq. (2-42) is

$$x = \frac{F/k}{1 - r^2} \sin \omega t \tag{3-16}$$

where $r = \omega/\omega_n$ and $\omega_n^2 = k/m$.

The equation above is plotted in Figs. 2-8 to 2-12. The corresponding curves are when $\zeta = 0$. Resonance occurs when the frequency ratio $r = 1$.

The amplification factor R is infinite at resonance and the amplitude of the displacement also becomes infinite.

Alternatively, the behavior at resonance can be deduced from the particular integral of Eq. (3-15). Dividing the equation by m and substituting ω_n for ω in the excitation term, we obtain

$$\ddot{x} + \omega_n^2 x = \frac{F}{m} \sin \omega_n t \tag{3-17}$$

From Example 2, App. D, the particular integral is of the form

$$x = A_1 t \sin \omega_n t + A_2 t \cos \omega_n t$$

where A_1 and A_2 are undetermined coefficients. Substituting this into Eq. (3-17) and solving for the coefficients, we get

$$x = -\frac{F}{2\sqrt{km}} t \cos \omega_n t \tag{3-18}$$

Thus, the amplitude increases proportionately with time and would theoretically become infinite.

Equation (3-18) indicates that it takes time for the amplitude to build up at resonance. Hence, if the resonance is passed through rapidly, it is possible to bring up the speed of a machine, such as a turbine, to beyond resonance or its *critical speed.* At frequencies considerably above resonance, Fig. 2-8 shows that the magnification factor is less than unity. It may be advantageous to operate the machine in this speed range. It should be cautioned that during the shut-down the machine would again pass through the critical speed and excessive vibration might be encountered.

Example 12. Determination of Natural Frequency

The control tab of an airplane elevator is shown schematically in Fig. 3-11. The mass moment of inertia J_0 of the control tab about the hinge point 0 is

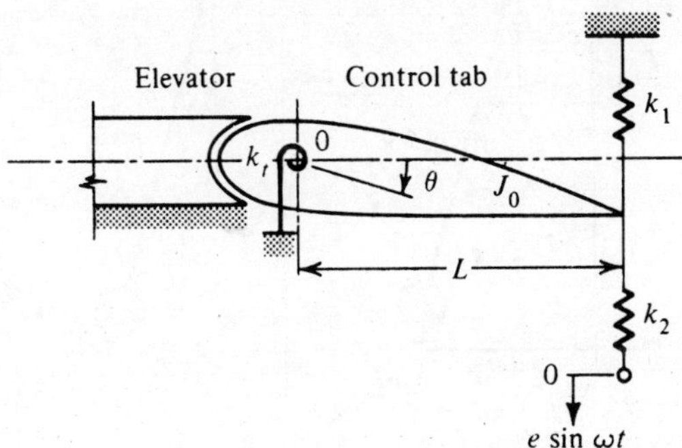

FIG 3-11. *Determination of natural frequency.*

known, but the torsional spring constant k_t due to control linkage is difficult to evaluate. To determine the natural frequency experimentally, the elevator is rigidly mounted and the tab is excited as illustrated. The excitation frequency is varied until resonance occurs. If the resonance frequency is ω_r, find the natural frequency $\omega_n = \sqrt{k_t/J_0}$ of the control tab.

Solution:

Taking moments about the hinge point 0, the equation of motion of the test system is

$$J_0\ddot{\theta} = -k_t\theta - k_1L^2\theta - k_2(L\theta - e\sin\omega t)L$$

where $(L\theta - e\sin\omega t)$ is the deformation of the spring k_2. Rearranging, this equation becomes

$$J_0\ddot{\theta} + [k_t + (k_1 + k_2)L^2]\theta = k_2eL\sin\omega t$$

At resonance,

$$\omega^2 = \omega_r^2 = \frac{k_t + (k_1 + k_2)L^2}{J_0} = \omega_n^2 + \frac{(k_1 + k_2)L^2}{J_0}$$

Hence

$$f_n = \frac{1}{2\pi}\omega_n = \frac{1}{2\pi}\sqrt{\omega_r^2 - \frac{(k_1 + k_2)L^2}{J_0}}\ \text{Hz}$$

The next three examples illustrate an application of the simple pendulum in a dynamic absorber for vibration control.

Example 13

The simple pendulum in Fig. 3-12 is hinged at the point 0. The hinge point 0 is given a horizontal motion $x(t) = e\sin\omega t$. Find **(a)** the angular displacement θ of the pendulum for frequency ratios $r \lesseqgtr 1$, and **(b)** the force required to move the hinge point.

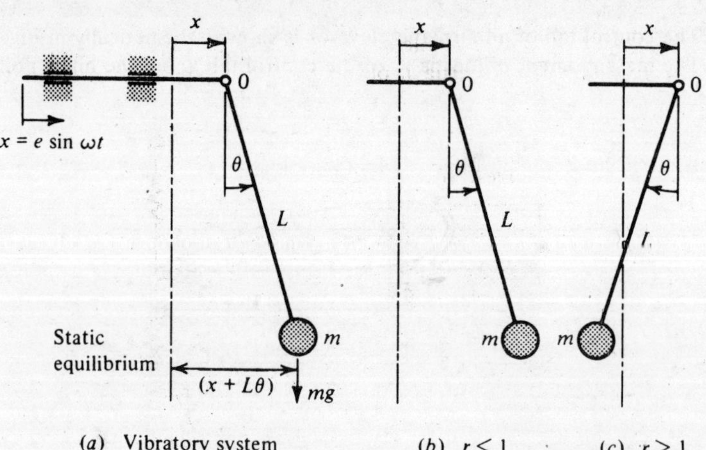

FIG. 3-12. *Pendulum excited at support.*

Solution:

(**a**) Assuming small oscillations, the horizontal acceleration of m in Fig. 3-12(a) is $(\ddot{x}+L\ddot{\theta})$ and the vertical acceleration is of second order. Taking moments about 0, the equation of motion is

$$m(\ddot{x}+L\ddot{\theta})L=-mgL\theta$$

$$mL^2\ddot{\theta}+mgL\theta=-mL\ddot{x}=me\omega^2L\sin\omega t \qquad \textbf{(3-19)}$$

$$mL^2\ddot{\theta}+mgL\theta=T_{eq}\sin\omega t$$

where $T_{eq}=me\omega^2L$ is the amplitude of the equivalent torque. This equation is of the same form as Eq. (3-15). From Eq. (3-16), the steady-state response is

$$\theta(t)=\frac{T_{eq}/mgL}{1-r^2}\sin\omega t=\frac{(e/L)r^2}{1-r^2}\sin\omega t$$

or

$$\theta(t)=\frac{(e/L)r^2}{|1-r^2|}\sin(\omega t-\phi)$$

where $r=\omega/\omega_n$ and $\omega_n=\sqrt{g/L}$. Note that $\phi=0°$ when $r<1$ and $\phi=180°$ when $r>1$. In other words, $\theta(t)$ and $x(t)$ are in phase with one another when $r<1$ and 180° out of phase when $r>1$. These phase relations are illustrated in Figs. 3-12(b) and (c).

(**b**) For dynamic equilibrium of the pendulum, the horizontal force at the hinge point 0 is equal to the horizontal component of the inertia force of the pendulum, that is,

$$F_x(t)=-m(\ddot{x}+L\ddot{\theta})=mg\theta=\frac{me\omega^2}{1-r^2}\sin\omega t$$

where $x(t)$ is positive if the motion is to the right of the static equilibrium position. The equation shows that near resonance when $r\simeq 1$ a large force $F_x(t)$ could associate with a small amplitude e at the hinge point 0.

Example 14. Centrifugal Pendulum Dynamic Absorber

A simple pendulum in a centrifugal field can be used to nullify the torsional disturbing moment on a rotating machine member, such as a crank shaft. A rotating disk with a pendulum hinged at B is shown in Fig. 3-13. The disk has an average speed ω and a superimposed oscillation $\gamma=\Gamma\sin n\omega t$, where n is the number of disturbing cycles per revolution of the disk. (**a**) Derive the equation of motion of the system. (**b**) Find the amplitude ratio $\Gamma:\Theta$. (**c**) Briefly discuss the application of the pendulum.

Solution:

The pendulum is under the influence of a centrifugal field when the system is in rotation. Assume the gravitation is negligible compared with the

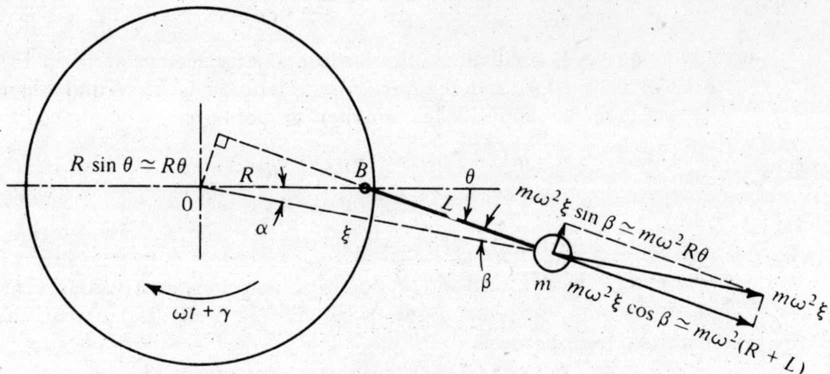

FIG. 3-13. *Centrifugal pendulum.*

centrifugal field at moderate and high speeds of rotation. As shown in the figure, the pendulum bob m is subjected to a centrifugal force $m\omega^2\xi$. The component of this force normal to L is $(m\omega^2\xi \sin\beta)$. From the triangle $0Bm$, we have

$$\frac{R}{\sin\beta} = \frac{\xi}{\sin(180-\theta)} = \frac{\xi}{\sin\theta} = \frac{L}{\sin\alpha}$$

or

$$\sin\beta = \frac{R}{\xi}\sin\theta \simeq \frac{R}{\xi}\theta$$

(**a**) For small oscillations, the tangential acceleration of m is $[L\ddot{\theta} + (R+L)\ddot{\gamma}]$. Taking moments about B, the equation of motion of the pendulum is

$$m[L\ddot{\theta} + (R+L)\ddot{\gamma}]L + m\omega^2 RL\theta = 0$$

Since $\ddot{\gamma} = -(n\omega)^2\Gamma \sin n\omega t$, the equation becomes

$$mL^2\ddot{\theta} + m\omega^2 RL\theta = m(R+L)L(n\omega)^2\Gamma \sin n\omega t$$

which is of the same form as Eq. (3-15).

(**b**) Comparing with Eq. (3-15), we have $m_{eq} = mL^2$, $k_{eq} = m\omega^2 RL$, and $F_{eq} = m(R+L)L(n\omega)^2\Gamma$ and $\theta(t)$ is analogous to $x(t)$. The steady-state solution of the equation is

$$\theta = \Theta \sin(n\omega t - \phi)$$

The amplitude ratio Γ/Θ can be obtained by substituting the corresponding eqivalent quantities into Eq. (3-16). Performing the substitution and simplifying, we obtain

$$\frac{\Gamma}{\Theta} = \frac{R/L - n^2}{n^2(R+L)/L} \tag{3-20}$$

(c) The equation above shows that if $n = \sqrt{R/L}$, the amplitude ratio $\Gamma/\Theta = 0$. Physically, this means that a finite value of Θ is possible for an arbitrarily small value of Γ. If the superposed oscillation $\gamma = \Gamma \sin n\omega t$ is due to a disturbing torque, the resultant oscillation $\gamma(t)$ of the machine member can be very small for a finite value of Θ. In other words, if the centrifugal pendulum is tuned such that $n = \sqrt{R/L}$, the disturbing torque can be balanced by the inertia torque of the pendulum.

The number of disturbing cycles n per revolution in a rotary machine, such as an internal combustion engine, is constant. When a centrifugal pendulum is tuned for $n = \sqrt{R/L}$, it is effective as a *dynamic absorber* for all speeds of operation of the machine. This is not a friction type damper, since it creates an equal and opposite torque to nullify the disturbing torque.

From Fig. 3-13, the reacting moment on the shaft is due to the tension $m\omega^2(R+L)$ of the pendulum. The moment arm $R \sin \theta \simeq R\theta$. Hence the magnitude of the reacting torque T_0 is

$$T_0 = m\omega^2(R+L)R\Theta \tag{3-21}$$

Example 15

An eight cylinder, four-stroke cycle engine operates at 1,800 rpm. The fluctuating torque is absorbed by the flywheel and a dynamic absorber. Assume the disturbing torque T_0 to be balanced by the absorber is 500 N · m (4,425 lb_f-in.). The most convenient length for R as shown in Fig. 3-13 is 96 mm (3.8 in.). If the maximum amplitude of oscillation of the pendulum is 10°, find the length L and the mass m of the pendulum.

Solution:

The eight cylinder engine has four power strokes per revolution. Hence $n = 4$. The length of a properly tuned pendulum is

$$L = R/n^2 = 96/16 = 6 \text{ mm}$$

From the expression for T_0 in Example 14, the required mass m of the

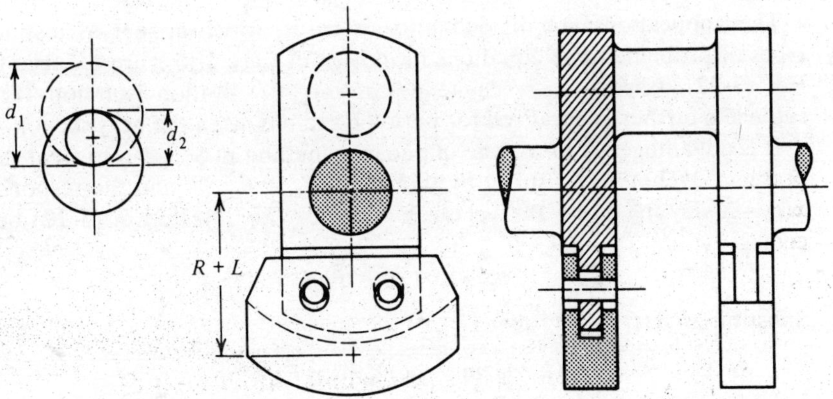

FIG. 3-14. *Bifilar-type centrifugal pendulum dynamic absorber.*

pendulum is

$$m = \frac{T_0}{\omega^2(R+L)R\Theta} = \frac{500}{(60\pi)^2(0.096+0.006)(0.096)(\pi/18)}$$
$$= 8.40 \text{ kg } (18.5 \text{ lb}_m)$$

The length of the pendulum is small; $L = 6$ mm. This poses a design problem in providing sufficient mass with a small L. The problem is solved by using a bifilar-type centrifugal pendulum as shown in Fig. 3-14. Let half of the pendulum mass be mounted on two loosely fitted pins on each side of the crank. The diameter of the pins is d_2 and that of the holes through the mass and the crank is d_1. Thus, each point on the mass moves in an arc of a circle of radius $(d_1 - d_2)$. The length of the pendulum is $L = (d_1 - d_2)$.

3-5 DAMPED FORCED VIBRATION—HARMONIC EXCITATION

Periodic excitation generally occurs in machines under steady-state operation. Considering one of the harmonic components of the periodic excitation, the equation of motion, Eq. (3-1), becomes

$$m\ddot{x} + c\dot{x} + kx = F_{eq} \sin \omega t \tag{3-22}$$

This is identical to Eq. (2-18) except for the substitution of F_{eq} for F.

The objective is to examine the sources of F_{eq} and the manner it affects the system response $x(t)$. In other words, the aim is in problem formulation and in reducing the equation of motion to the form shown above. The solution of the equation and the interpretation of its response were fully discussed in Chap. 2. Note that the phase angle of $x(t)$ relative to F_{eq} would be identical to the previous discussions as shown in Eq. (2-54). Unless it is justifiable, the phase relation will not be discussed.

The applications are divided into six cases, which appear as distinct types of problems. The equations of motion of Case 1 to 5 are reduced to Eq. (3-22). Case 6 shows a generalization of vibration isolation. The subject is further generalized for periodic excitations in the next section.

It is advantageous to use the impedance method in Sec. 2-6 to treat the problem. Denoting $F_{eq} \sin \omega t$ by the vector $\mathbf{F}_{eq} = \bar{F}_{eq} e^{j\omega t}$ and the response $x(t) = X \sin(\omega t - \phi)$ by the vector $\mathbf{X} = \bar{X}e^{j\omega t} = (Xe^{-j\phi})e^{j\omega t}$, Eq. (3-22) becomes

$$m\ddot{x} + c\dot{x} + kx = \bar{F}_{eq} e^{j\omega t}$$

Substituting $x(t) = \bar{X}e^{j\omega t}$ gives

$$(k - \omega^2 m + j\omega c)\bar{X}e^{j\omega t} = \bar{F}_{eq} e^{j\omega t} \tag{3-23}$$

where $\bar{F}_{eq} = F_{eq} e^{j\alpha}$ is the phasor of the excitation vector $\mathbf{F}_{eq}$. The magnitude of $\mathbf{F}_{eq}$ is $F_{eq} = |\bar{F}_{eq}|$ and its phase angle is α, relative to a reference vector. If the excitation is the reference, then α is zero. Similarly, $\bar{X} = Xe^{-j\phi}$ is the phasor of the vector **X**. The magnitude of **X** is $X = |\bar{X}|$ and its phase angle is $-\phi$ relative to the reference vector. It is understood that the force and the response must be along the same axis in a physical problem. Hence it is unnecessary to denote the real or the imaginary parts of the vectors in all subsequent discussions.

Eliminating the $e^{j\omega t}$ term in Eq. (3-23) and rearranging, we get

$$\bar{X} = \frac{1}{k - \omega^2 m + j\omega c} F_{eq} \tag{3-24}$$

which is identical to Eq. (2-52). Hence

$$X = |\bar{X}| = \frac{F_{eq}}{k} R = \frac{F_{eq}}{k} \frac{1}{\sqrt{(1-r^2)^2 + (2\zeta r)^2}} \tag{3-25}$$

$$-\phi = \angle\bar{X} = -\tan^{-1} \frac{2\zeta r}{1-r^2} \tag{3-26}$$

where $r = \omega/\omega_n$, $\omega_n^2 = k/m$, and $\zeta = \frac{1}{2}c/\sqrt{km}$.

Case 1. Rotating and Reciprocating Unbalance

A turbine, an electric motor, or any device with a rotor as a working part is a rotating machine. Unbalance exists if the mass center of the rotor does not coincide with the axis of rotation. The unbalance me is measured in terms of an equivalent mass m with an eccentricity e.

A rotating machine of total mass m_t with an unbalance me is shown in Fig. 3-15. The eccentric mass m rotates with the angular velocity ω and its vertical displacement is $(x + e \sin \omega t)$. The machine is constrained to move in the vertical direction and it has one degree of freedom. The

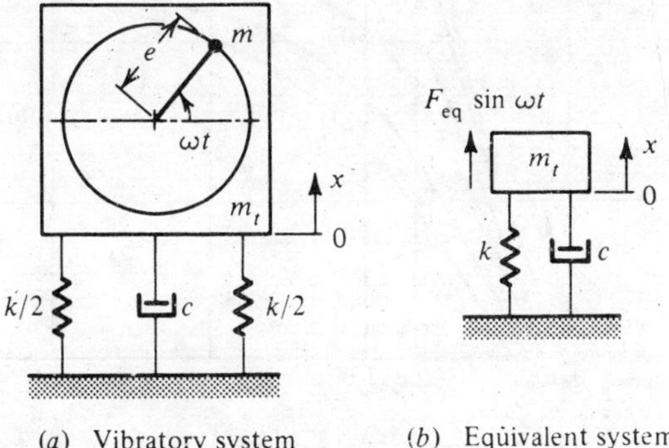

(*a*) Vibratory system (*b*) Equivalent system

FIG. 3-15. *Rotating unbalance.*

displacement of the mass $(m_t - m)$ is $x(t)$. Hence the equation of motion of the system is

$$(m_t - m)\ddot{x} + m\frac{d^2}{dt^2}(x + e \sin \omega t) + c\dot{x} + kx = 0$$

Rearranging the equation yields

$$m_t\ddot{x} + c\dot{x} + kx = me\omega^2 \sin \omega t = F_{eq} \sin \omega t \tag{3-27}$$

where $F_{eq} = me\omega^2$ is the amplitude of the excitation force. Hence the equivalent system is as shown in Fig. 3-15(*b*). The steady-state solution is given in Eqs. (3-25) and (3-26).

From Eq. (3-25), the amplitude of the harmonic response is

$$X = \frac{F_{eq}}{k}R = \frac{me\omega^2}{k}R$$

This can be expressed in a nondimensional form. Multiplying and dividing the equation by m_t, recalling $\omega_n^2 = k/m_t$, $r = \omega/\omega_n$, and simplifying, we obtain

$$\frac{m_t X}{me} = r^2 R = \frac{r^2}{\sqrt{(1-r^2)^2 + (2\zeta r)^2}} \tag{3-28}$$

This is plotted in Fig. 3-16.

At low speeds, when $r \ll 1$, the force $me\omega^2$ is small and the amplitude

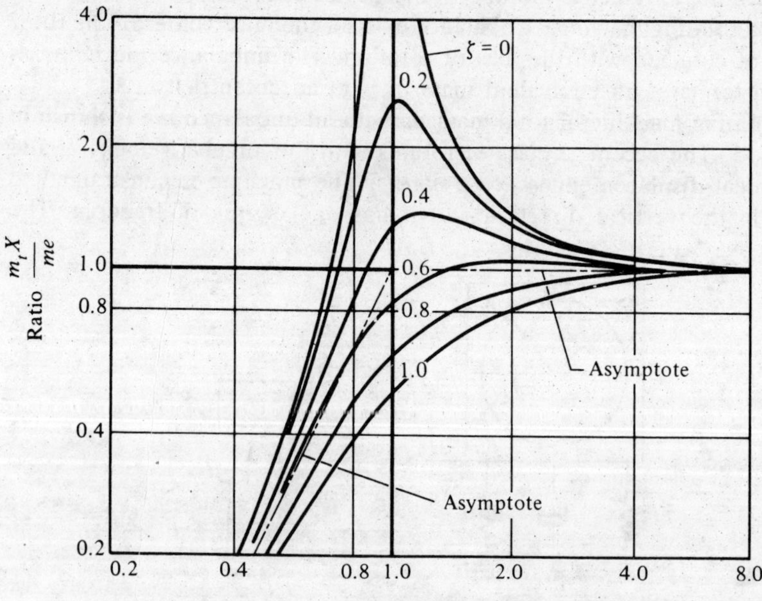

FIG. 3-16. *Harmonic response of systems with inertial excitation; system shown in Fig. 3-13.*

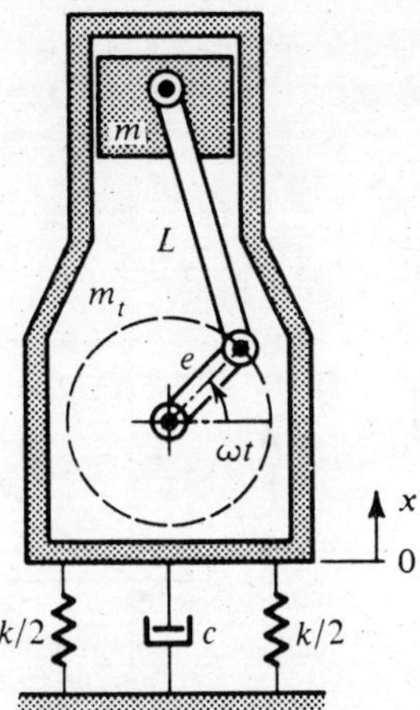

FIG. 3-17. *Reciprocating unbalance.*

of vibration X is nearly zero. At resonance, when $r = 1$ and the amplification factor $R = 1/2\zeta$, the mass $(m_t - m)$ has an amplitude equal to $X = me/(2\zeta m_t)$. Hence the amplitude of vibration is limited only by the presence of damping in the system. Furthermore, the mass $(m_t - m)$ is 90° out of phase with the unbalance mass m. For example, when $(m_t - m)$ is moving upward and passing its static equilibrium position, the mass m is directly above its center of rotation. At high speeds, when $r \gg 1$, the mass $(m_t - m)$ has an amplitude $X \simeq me/m_t$. In other words, the amplitude remains constant independent of the frequency of excitation or the damping in the system. The phase angle is 180°; that is, when $(m_t - m)$ is at its topmost position, m is directly below its center of rotation.

The discussion of rotating unbalance can be used to estimate a reciprocating unbalance. A reciprocating engine is illustrated in Fig. 3-17. The reciprocating mass m consists of the mass of the piston, the wrist pin, and part of the connecting rod. The exciting force is equal to the inertia force of the reciprocating mass, which is approximately equal to $me\omega^2[\sin \omega t + (e/L)\sin 2\omega t]$,* where e is the crank radius and L the length of the connecting rod. If the ratio e/L is small, the second-harmonic term, $(e/L)\sin 2\omega t$, can be neglected.† Thus, the problem reduces to that of the rotating unbalance.

Case 2. Critical Speed of Rotating Shafts

A rotating shaft carrying an unbalance disk at its midspan is shown in Fig. 3-18(a). *Critical speed* occurs when the speed of rotation of the shaft

* See, for example, R. T. Hinkle, *Kinematics of Machines*, 2d ed., Prentice-Hall, Inc., Englewood Cliffs, NJ, 1960, p. 107.

† It will be shown in Case 3 for vibration isolation that if an isolator is adequate for the fundamental frequency it would be adequate for the higher harmonics.

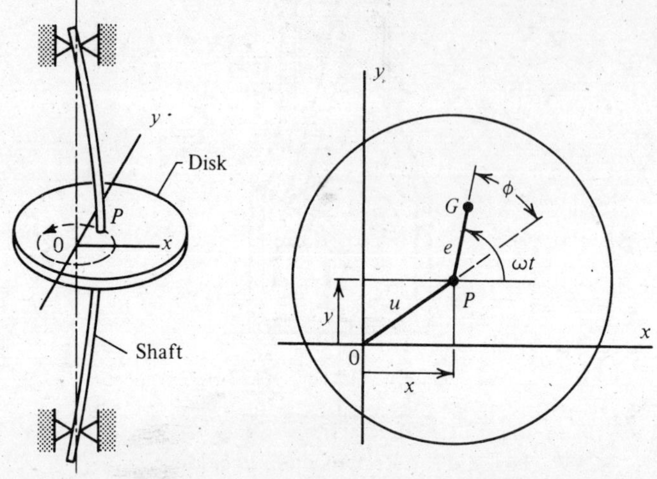

(*a*) Vibratory system (*b*) General position of disk.

FIG. 3-18. *Critical speed of rotating shaft.*

is equal to the natural frequency of lateral (beam) vibration of the shaft. Since the shaft has distributed mass and elasticity along its length, the system has more than one degree of freedom. We assume the mass of the shaft is negligible and its lateral stiffness is k.

The top view of a general position of the rotating disk of mass m is shown in Fig. 3-18(b). Let G be the *mass center* of the disk, P the geometric center, and O the center of rotation. Assume the damping force, such as air friction opposing the shaft whirl, is proportional to the linear speed of the geometric center P and that the flexibility of the bearings is negligible compared with that of the shaft.

Resolving the forces in the x and y directions gives

$$m\frac{d^2}{dt^2}(x + e\cos\omega t) = -kx - c\dot{x}$$

$$m\frac{d^2}{dt^2}(y + e\sin\omega t) = -ky - c\dot{y}$$

$$m\ddot{x} + c\dot{x} + kx = me\omega^2\cos\omega t = F_{eq}\cos\omega t$$

$$m\ddot{y} + c\dot{y} + ky = me\omega^2\sin\omega t = F_{eq}\sin\omega t$$

Applying the impedance method illustrated in Eq. (3-23), the equations above become

$$(k - \omega^2 m + j\omega c)\bar{X} = F_{eq}$$
$$(k - \omega^2 m + j\omega c)\bar{Y} = F_{eq}e^{j\pi/2} \tag{3-29}$$

The phase angle $\pi/2$ in the second equation above indicates that the

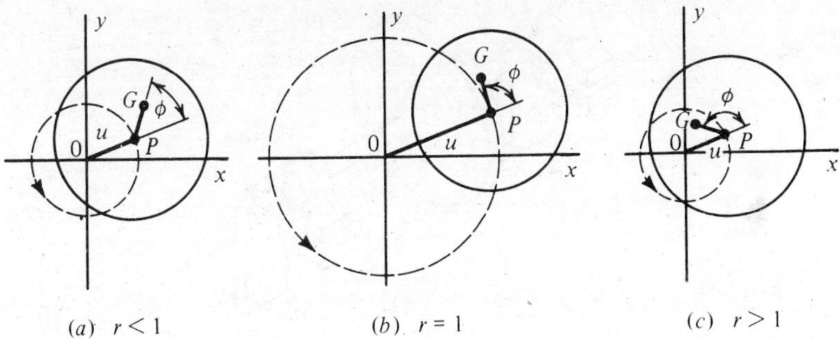

FIG. 3-19. *Phase relation of rotating shaft for* $r \lesseqgtr 1$; $r = \omega/\omega_n$.

displacements in the generic x and y directions are at 90° with one another. It is evident that the amplitudes X and Y are equal.

Since the two harmonic motions $x(t)$ and $y(t)$ are equal in magnitude, of the same frequency, and at 90° to each other, their sum is a circle. Thus, the radius u of the circle is equal to X or Y. The motion of the geometric center P of the disk in Fig. 3-18 describes a circle of radius u about the center of rotation O. From Eq. (3-25), we get

$$u = X = Y = \frac{F_{\text{eq}}}{k} R = \frac{me\omega^2}{k} R$$

Substituting $\omega_n^2 = k/m$ and $r = \omega/\omega_n$ and simplifying we obtain

$$\frac{u}{e} = r^2 R = \frac{r^2}{\sqrt{(1-r^2)^2 + (2\zeta r)^2}} \tag{3-30}$$

This is identical to Eq. (3-28) if $m_t = m$. This is true because the total mass is also the eccentric mass for a rotating disk. Hence the response curves in Fig. 3-16 for rotating unbalance also represents Eq. (3-30) for the whirling of rotating shafts.

The phase relation for various operating frequencies is shown in Fig. 3-19. It is interesting to note that, when the frequency ratio $r \gg 1$, the mass center G tends to coincide with the center of rotation O. This can be demonstrated readily. Assume that an unbalance rotor is rotating in a balancing machine and that a piece of chalk is moved towards the rotor until it barely touches. When the rotational speed is below critical, the chalk mark is found on the side closer to the mass center of the rotor. When the speed is above critical, the chalk mark is on the side away from the mass center.

Example 16. Elasticity of Bearings and Supports

Rigid bearings were assumed in the above discussion of critical speed. Figure 3-20 shows a pulley assembly, in which the supporting brackets can be deflected more easily in the vertical direction than in the lateral

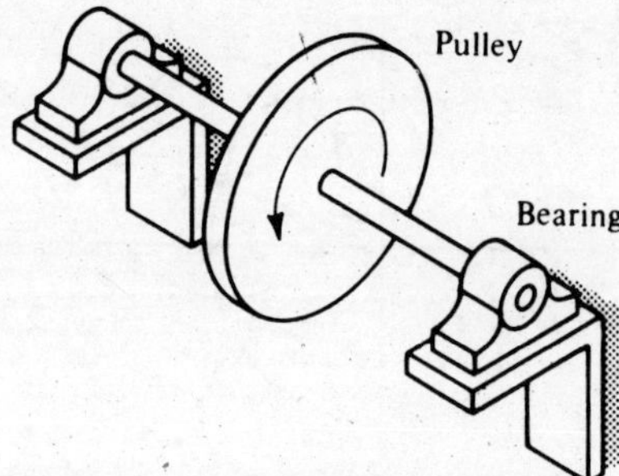

FIG. 3-20. *Pulley assembly with flexible bearing supports.*

direction. The effect of the elastic bearings is to render the system more flexible and therefore lowering the critical speed. The critical speed can be lowered by 25 percent in some installations. (**a**) Derive the equation of motion of the system and (**b**) briefly discuss the effect of the unequal elastic supports on the system performance.

Solution:

A schematic representation of the system is shown in Fig. 3-21(a). The elasticity of the bearings and supports is represented by springs mounted in rigid frames. The equivalent spring constants k_x and k_y are due to the stiffness of the shaft, the bearings, and the supports in the x and y directions. A general position of the disk is shown in Fig. 3-21(b), which may be compared with Fig. 3-18(b). P is the geometric center and G the mass center of the disk. O the center of rotation of the system corresponding to the static equilibrium position of the shaft.

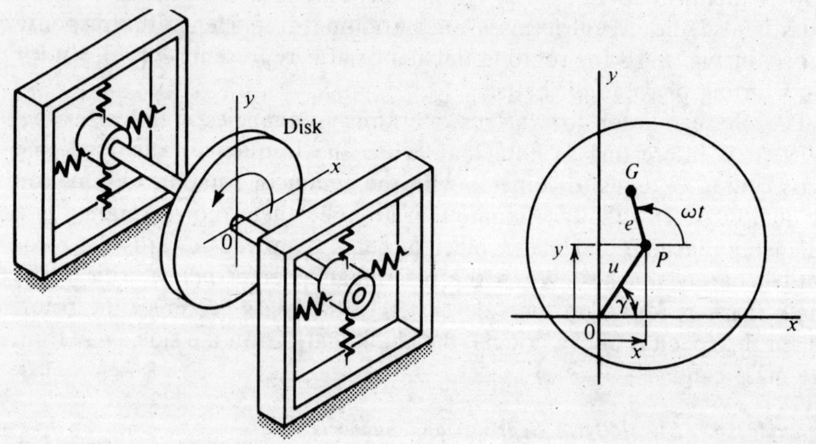

(a) Schematic (b) General position of rotating disk.

FIG. 3-21. *System with elastic bearing supports.*

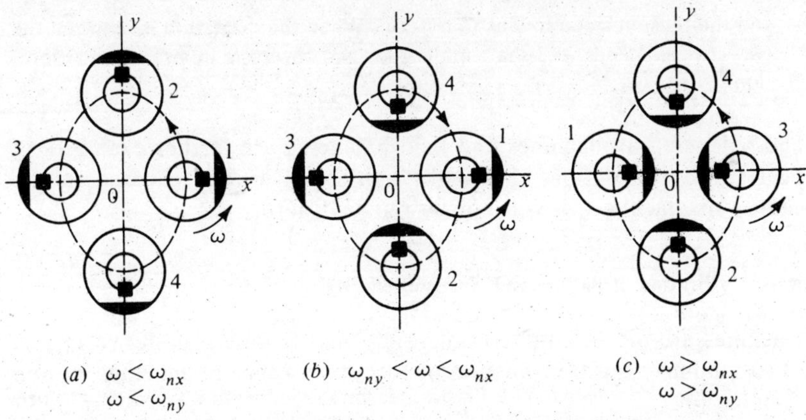

FIG. 3-22. *Rotation of disk about* 0 *for various frequencies.*

(a) For simplicity, the system is assumed undamped. The equations of motion are

$$m\ddot{x} + k_x x = me\omega^2 \cos \omega t$$

$$m\ddot{y} + k_y y = me\omega^2 \sin \omega t$$

Since $k_x \neq k_y$, the equations indicate that the system has two natural frequencies and therefore two critical speeds. We define $\omega_{nx} = \sqrt{k_x/m}$, $\omega_{ny} = \sqrt{k_y/m}$, $r_x = \omega/\omega_{nx}$, $r_y = \omega/\omega_{ny}$, and $F_{eq} = me\omega^2$. From Eq. (3-30), the amplitude ratios are

$$\frac{X}{e} = \frac{r_x^2}{|1 - r_x^2|} \quad \text{and} \quad \frac{Y}{e} = \frac{r_y^2}{|1 - r_y^2|}$$

(b) The two harmonic motions $x(t)$ and $y(t)$ are of the same frequency and at 90° to each other. Since their amplitudes are unequal, the sum of their motions is an ellipse. Thus, the geometric center P moves in an ellipse about O as shown in Fig. 3-22. By neglecting the damping in the system, the phase angle can be either zero for below the critical speed or 180° for above the critical speed. Since there are two natural frequencies, we may consider the operations at speeds above and below the critical.

When $\omega < \omega_{nx}$ and $\omega < \omega_{ny}$, both the disk and P rotate in the same direction with the same speed as shown in Fig. 3-22(*a*). The heavy side of the disk and the position of the shaft key are marked for purpose of identification. Assume $\omega_{nx} > \omega_{ny}$. When $\omega_{nx} > \omega > \omega_{ny}$, the disk and P rotate in opposite directions with the same speed as shown in Fig. 3-22(*b*). When ω is greater than both natural frequencies, again the disk and P rotate in the same direction with the same speed as shown in Fig. 3-22(*c*).

It is interesting to note that when the excitation is above or below the critical speeds, there is no reversal in stresses in the shaft; that is, while the shaft is revolving, the compression side of the shaft remains in compression and the tension side remains in tension. When the excitation is between the two critical speeds, the shaft undergoes two reversals in stress per revolution.

The balancing of machines and field balancing are further examples of critical speed calculations. Since the subject is usually covered in the dynamics of machines, it will not be pursued here.

Case 3. Vibration Isolation and Transmissibility

Machines are often mounted on springs and dampers as shown in Fig. 3-23 to minimize the transmission of forces between the machine m and its foundation.

We shall first consider the system illustrated in Fig. 3-23(a). If a harmonic force is applied to m and the deflection of the foundation is negligible, the equation of motion is identical to Eq. (3-22). The force transmitted to the foundation is the sum of the spring force kx and the damping force $c\dot{x}$.

$$\text{Force transmitted} = kx + c\dot{x}$$

If the excitation is harmonic, the magnitude and the phase angle of the excitation force F_{eq} and the other forces are as illustrated in Fig. 3-24. The phase angle γ is generally is of secondary interest. Using Eq. (3-24), the force transmitted $\bar{F}_T$ is

$$\bar{F}_T = k\bar{X} + j\omega c\bar{X} = \frac{k + j\omega c}{k - \omega^2 m + j\omega c}\,\bar{F}_{\text{eq}} \tag{3-31}$$

The ratio of the amplitude of the force transmitted F_T and the amplitude of the driving force F_{eq} is called the *transmissibility* TR. From the equation above, we have

$$\text{TR} = \frac{\sqrt{1 + (2\zeta r)^2}}{\sqrt{(1 - r^2)^2 + (2\zeta r)^2}} \tag{3-32}$$

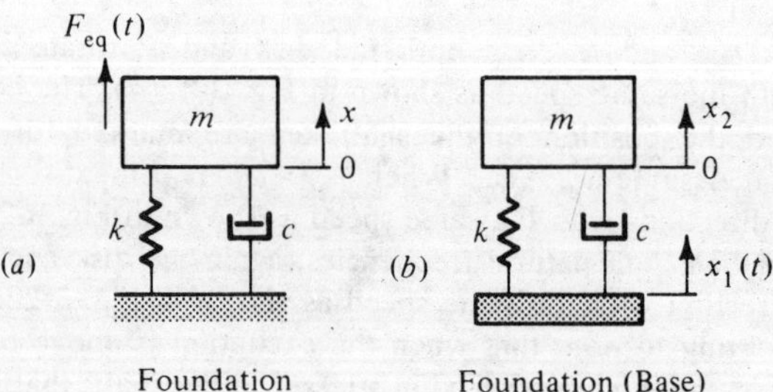

FIG. 3-23. *Vibration isolation.*

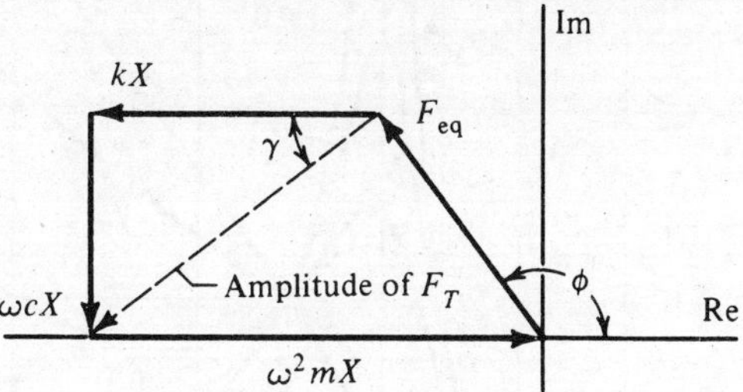

FIG. 3-24. *Relation of force transmitted and other force vectors.*

where $r = \omega/\omega_n$ and $c\omega/k = 2\zeta r$. The equation is plotted in Fig. 3-25. Note that all the curves in the figure cross at $r = \sqrt{2}$. Hence the transmitted force is greater than the driving force below this frequency ratio and less than the driving force when the machine is operated above this frequency ratio.

For a constant speed machine, the amplitude of the exciting force F_{eq} is constant. Hence the force transmitted is proportional to the value of the

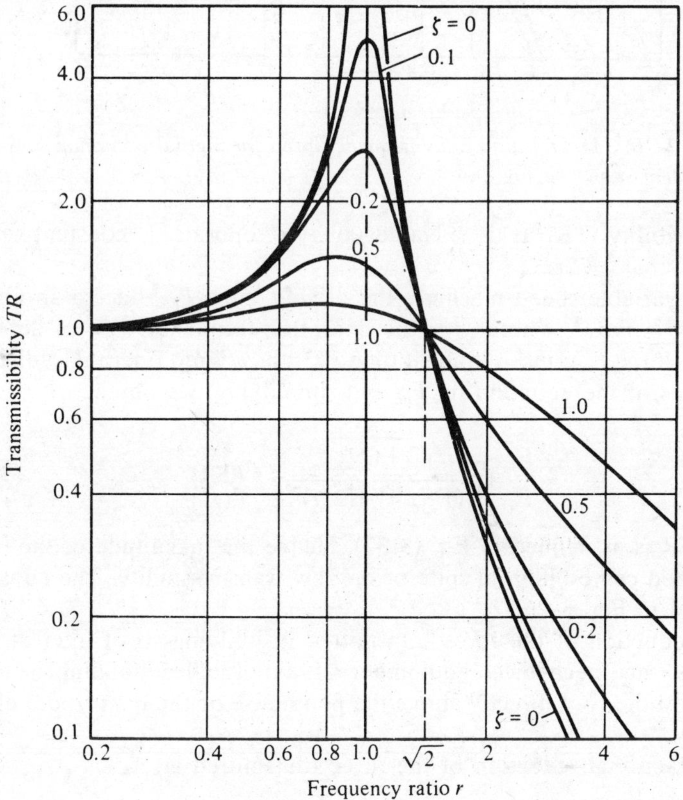

FIG. 3-25. *Transmissibility versus frequency ratio; system shown in Fig.* 3-23(*a*).

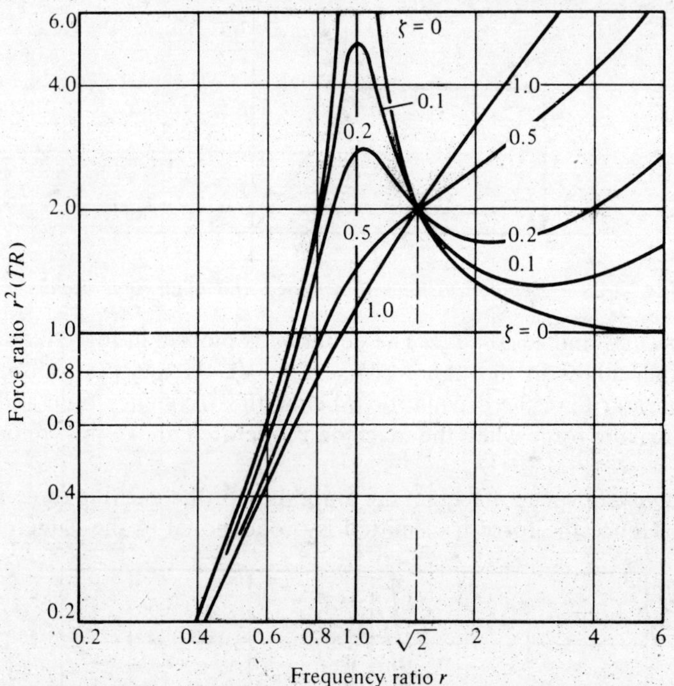

FIG. 3-26. *Force ratio versus frequency ratio for inertial excitation; system shown in Fig. 3-15.*

transmissibility TR. It is advantageous to operate a constant speed machine at $\omega > \sqrt{2}\omega_n$.

For a variable speed machine, the driving force F_{eq}, due to an unbalance me, is $me\omega^2$, where ω is the operating frequency. Let us define a constant force $F_n = me\omega_n^2$. Substituting $F_{eq} = me\omega^2$ into Eq. (3-31), dividing both sides of the equation by F_n, and simplifying, we obtain

$$\frac{F_T}{F_n} = \frac{r^2\sqrt{1+(2\zeta r)^2}}{\sqrt{(1-r^2)^2+(2\zeta r)^2}} = r^2(\text{TR}) \tag{3-33}$$

where TR is as defined in Eq. (3-32). Hence the magnitude of the force transmitted can be high in spite of the low transmissibility. The equation is plotted in Fig. 3-26.

The reduction of the force transmitted in buildings is of interest. For example, the mechanical equipment of a tall office building is often located on the roof directly above the penthouse or the boardroom of the company.

The fractional reduction of the force transmitted is

$$\text{Force reduction} = \frac{F_{eq} - F_T}{F_{eq}} = 1 - \text{TR}$$

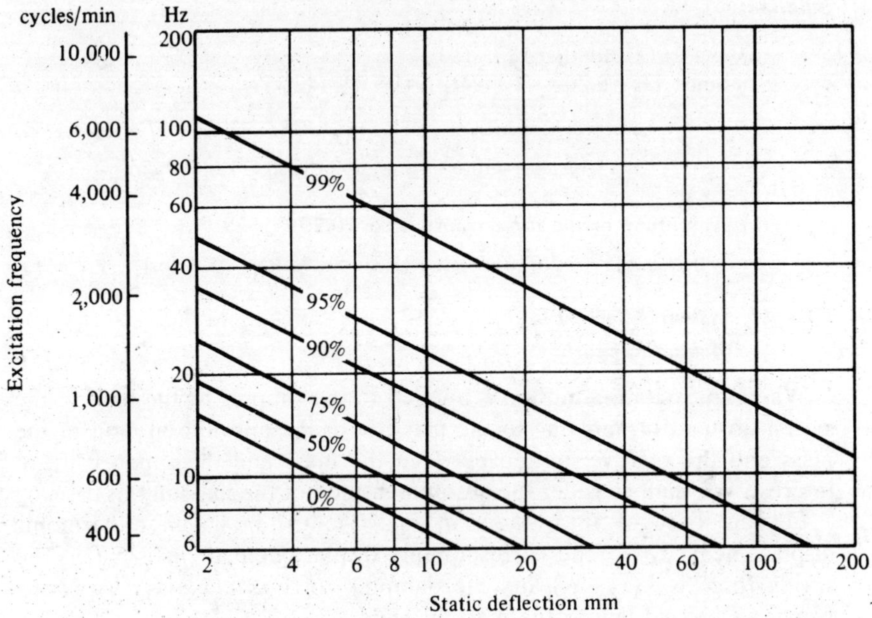

FIG. 3-27. *Percentage reduction in force transmitted to foundation in vibration isolation,* $\zeta = 0$.

where F_{eq} and F_T are the amplitudes of the excitation and the transmitted force, respectively. It is observed in Fig. 3-25 that low natural frequency and low damping are desirable for vibration isolation. Assume $\zeta \simeq 0$ and $r > 1$ in Eq. (3-32). Thus, $\text{TR} = 1/(r^2 - 1)$ and the force reduction becomes

$$\text{Force reduction} = \frac{r^2 - 2}{r^2 - 1}$$

Since $r^2 = (\omega/\omega_n)^2$, $\omega_n^2 = k/m$, and the static deflection of a spring $\delta_{st} = mg/k$, the equation above reduces to

$$\text{Force reduction} = \frac{\omega^2 \delta_{st} - 2g}{\omega^2 \delta_{st} - g} \tag{3-34}$$

The equation is plotted in Fig. 3-27.

Example 17

An air compressor of 450 kg mass (992 lb_m) operates at a constant speed of 1,750 rpm. The rotating parts are well balanced. The reciprocating parts are of 10 kg (22 lb_m). The crank radius is 100 mm (4 in.). If the damper for the mounting introduces a damping factor $\zeta = 0.15$, **(a)** specify the springs for the mounting such that only 20 percent of the unbalance force is transmitted to the foundation, and **(b)** determine the amplitude of the transmitted force.

Solution:

(**a**) $\omega = 2\pi(1750/60) = 183.3$ rad/s
Since $\text{TR} = 0.20 = \sqrt{1+(2\zeta r)^2}/\sqrt{(1-r^2)^2+(2\zeta r)^2}$

$$\therefore \quad r = 2.72 = \omega/\omega_n \qquad \text{or} \qquad \omega_n = 183.3/2.72 = 67.4 \text{ rad/s}$$

$$k = m_t\omega_n^2 = 450(67.4)^2 = 2042 \text{ kN/m } (11{,}660 \text{ lb}_f/\text{in.})$$

(**b**) Amplitude of the force transmitted $= 0.20\, F_{eq}$
$= 0.20 me\omega^2 = 0.20(10)(0.10)(183.3)^2 = 6.72$ kN (1,510 lb_f)

Case 4. System Attached to Moving Support

When an excitation motion is applied to the support or the base of the system instead of applying to the mass, both the absolute motion of the mass and the relative motion between the mass and the support are of interest. We shall consider the absolute motion of the mass in this case.

Let the base of the system in Fig. 3-23(*b*) be given a harmonic displacement $x_1(t)$. The corresponding displacement of m is $x_2(t)$. The spring force is $k(x_2 - x_1)$ and the damping force is $c(\dot{x}_2 - \dot{x}_1)$. Applying Newton's second law to the mass m yields

$$m\ddot{x}_2 = -k(x_2 - x_1) - c(\dot{x}_2 - \dot{x}_1) \tag{3-35}$$

which can be rearranged to

$$m\ddot{x}_2 + c\dot{x}_2 + kx_2 = kx_1 + c\dot{x}_1$$

Applying the impedance method gives $x_1 = \bar{X}_1 e^{j\omega t}$. The quantity $(kx_1 + c\dot{x}_1) = (k + j\omega c)\bar{X}_1 e^{j\omega t} = \bar{F}_{eq} e^{j\omega t}$ can be considered as an equivalent force as shown in Eq. (3-22). Hence the equation above reduces to

$$(k - \omega^2 m + j\omega c)\bar{X}_2 = (k + j\omega c)\bar{X}_1$$

or

$$\frac{\bar{X}_2}{\bar{X}_1} = \frac{k + j\omega c}{k - \omega^2 m + j\omega c} = \frac{X_2}{X_1} e^{j(\gamma - \phi)} \tag{3-36}$$

where

$$\frac{X_2}{X_1} = \frac{\sqrt{1+(2\zeta r)^2}}{\sqrt{(1-r^2)^2+(2\zeta r)^2}} \tag{3-37}$$

$$\gamma - \phi = \tan^{-1} 2\zeta r - \tan^{-1}\frac{2\zeta r}{1-r^2} \tag{3-38}$$

and $\omega_n^2 = k/m$, $r = \omega/\omega_n$, and $c/k = 2\zeta/\omega_n$.

Since Eq. (3-37) for X_2/X_1 is identical to Eq. (3-32) for F_T/F_{eq}, both can be called the transmissibility equation, although the former denotes the transmission of motion from the base to the mass and the latter the transmission of force from the mass to its foundation. Hence Fig. 2-25 is a plot of both equations.

Example 18. Mounting of Instruments

An instrument of mass m is mounted on a vibrating table as shown in Fig. 3-23(b). Find (**a**) the maximum acceleration of the instrument, and (**b**) the maximum force transmitted to the instrument. Assume the motion $x_1(t)$ of the table is harmonic at the frequency ω.

Solution:

(**a**) The equation of the mass m is as shown in Eq. (3-35). The maximum acceleration of m is $\ddot{x}_{2\max} = \omega^2 X_2$. Applying Eq. (3-37) yields

$$\frac{\ddot{x}_{2\max}}{X_1} = \frac{\omega^2 X_2}{X_1} = \frac{\omega^2\sqrt{1+(2\zeta r)^2}}{\sqrt{(1-r^2)^2+(2\zeta r)^2}}$$

where $r = \omega/\omega_n$ and $\omega_n^2 = k/m$. Alternatively, we can compare the acceleration of m with that of the table, that is,

$$\frac{\ddot{x}_{2\max}}{\ddot{x}_{1\max}} = \frac{\omega^2 X_2}{\omega^2 X_1} = \frac{\sqrt{1+(2\zeta r)^2}}{\sqrt{(1-r^2)^2+(2\zeta r)^2}}$$

Thus, the characteristics of the acceleration ratio is the same as the displacement ratio X_2/X_1 and plotted in Fig. 3-25.

(**b**) Force is transmitted to m through the spring and the damper. From Eq. (3-35), the sum of these forces is $m\ddot{x}_{2\max} = m\omega^2 X_2$. Applying Eq. (3-37), the maximum force transmitted $F_{T\max}$ is

$$\frac{F_{T\max}}{X_1} = m\omega^2 \frac{X_2}{X_1}$$

Comparing $F_{T\max}$ with the maximum acceleration of the support, we have $\ddot{x}_{1\max} = \omega^2 X_1$ and

$$\frac{F_{T\max}}{\ddot{x}_{1\max}} = m\frac{X_2}{X_1} = \frac{m\sqrt{1+(2\zeta r)^2}}{\sqrt{(1-r^2)^2+(2\zeta r)^2}}$$

Hence, with the exception of the constant m, this equation is represented in Fig. 3-25. Comparing $F_{T\max}$ with the maximum displacement of the support, we have

$$\frac{F_{T\max}}{X_1} = kr^2\frac{X_2}{X_1} = \frac{kr^2\sqrt{1+(2\zeta r)^2}}{\sqrt{(1-r^2)^2+(2\zeta r)^2}}$$

where $m\omega^2 = k(\omega/\omega_n)^2 = kr^2$. Hence, with the exception of the constant k, this equation is represented in Fig. 3-26.

Example 19. Vehicle Suspension

A vehicle is a complex system with many degrees of freedom. As a first approximation, Fig. 3-28 may be considered as a vehicle driven on a rough

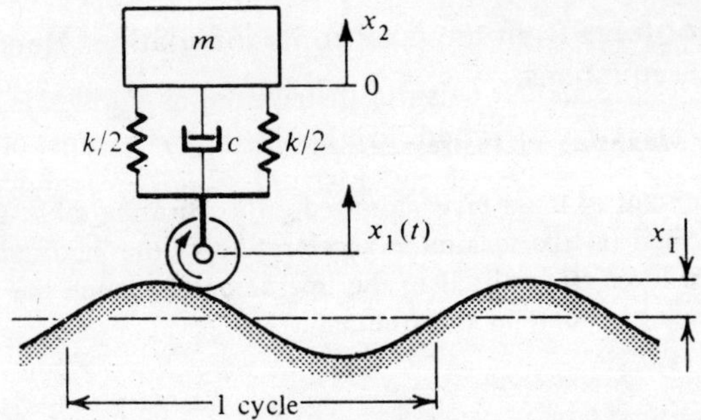

FIG. 3-28. *Schematic sketch of vehicle moving over rough road.*

road. It is assumed that (1) the vehicle is constrained to one degree of freedom in the vertical direction, (2) the spring constant of the tires is infinite, that is, the road roughness is transmitted directly to the suspension system of the vehicle, and (3) the tires do not leave the road surface. Assume a trailer has 1,000 kg mass (2,200 lb_m) fully loaded and 250 kg empty. The spring of the suspension is of 350 kN/m (2,000 lb_f/in.). The damping factor $\zeta = 0.50$ when the trailer is fully loaded. The speed is 100 km/hour (62 mph). The road varies sinusoidally with 5.0 m/cycle (16.4 ft/cycle). Determine the amplitude ratio of the trailer when fully loaded and empty.

Solution:

The excitation frequency is

$$f = \frac{100{,}000}{3{,}600} \times \frac{1}{5} = 5.56\text{ Hz}$$

From Eq. (2-27), the damping coefficient is $c = 2\zeta\sqrt{km}$. Since c and k are constant, ζ varies inversely with the square root of m. Thus, $\zeta = 1.0$ when the trailer is empty. Applying Eq. (3-37), the calculations are tabulated as follows:

ITEM	TRAILER, FULLY LOADED	TRAILER, EMPTY
Natural frequency		
$\omega_n = \sqrt{k/m}$	$\omega_n = \sqrt{(350{,}000)/1000}$	$\omega_n = 2(18.7)$
	$= 18.7$ rad/s	$= 37.4$ rad/s
	$= 2.98$ Hz	$= 5.96$ Hz
$r = \omega/\omega_n$	$r = 5.56/2.98$	$r = 5.56/5.96$
	$= 1.87$	$= 0.93$
Ratio of X_2/X_1		
$= \dfrac{\sqrt{1+(2\zeta r)^2}}{\sqrt{(1-r^2)^2+(2\zeta r)^2}}$	$X_2 = 2.12X_1/3.10$	$X_2 = 2.12X_1/1.87$
	$= 0.68X_1$	$= 1.13X_1$

The amplitude ratio when fully loaded and empty is $0.68/1.13 = 1/1.67$.

Case 5. Seismic Instruments

A schematic sketch of a seismic instrument is shown in Fig. 3-29. It consists of a mass m attached to the base by means of springs and dampers, that is, the base of the instrument is securely attached to a vibrating body, the motion of which is to be measured. Let $x_1(t)$ be the motion of the base and $x_2(t)$ the motion of m. The relative motion $(x_2 - x_1)$ is used to indicate $x_1(t)$. As illustrated, $(x_2 - x_1)$ is recorded by means of a pen and a rotating drum.

Since this system is essentially the same as that shown in Fig. 3-23(*b*), its equation of motion is identical to Eq. (3-35). Let $x_1 = X_1 \sin \omega t$. Defining $x(t) = x_2(t) - x_1(t)$ and substituting $\ddot{x}_2(t) = \ddot{x}(t) + \ddot{x}_1(t)$ into Eq. (3-35) yields

$$m\ddot{x} + c\dot{x} + kx = -m\ddot{x}_1 = m\omega^2 X_1 \sin \omega t \qquad \textbf{(3-39)}$$

which is of the same form as Eq. (3-22) with $F_{eq} = m\omega^2 X_1$.

Using the impedance method in Sec. 2-6, Eq. (3-39) gives

$$\bar{X} = \frac{\omega^2 m}{k - \omega^2 m + j\omega c} X_1 = X e^{-j\phi}$$

where $\phi = \tan^{-1} 2\zeta r/(1 - r^2)$, $\omega_n^2 = k/m$, and $r = \omega/\omega_n$. The amplitude ratio X/X_1 from the equation above is

$$\frac{X}{X_1} = r^2 R = \frac{r^2}{\sqrt{(1 - r^2)^2 + (2\zeta r)^2}} \qquad \textbf{(3-40)}$$

The right side of this equation is identical to Eqs. (3-28) and (3-30). Hence the characteristic of the equation is also portrayed in Fig. 3-16.

The relative motion between the mass m and its base in a seismic instrument, or vibration pickup, can be measured mechanically as illustrated in Fig. 3-29. For high speed operations and convenience, this motion is often converted into an electrical signal. The schemes illustrated in Figs. 3-30(*a*) and (*b*) for this conversion are self-evident. A typical piezoelectric accelerometer is illustrated in Fig. 3-30(*c*). The piezoelectric

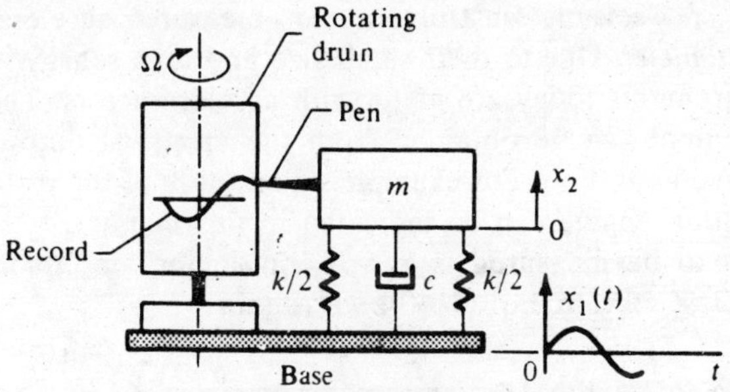

FIG. 3-29. *Schematic sketch of seismic instrument.*

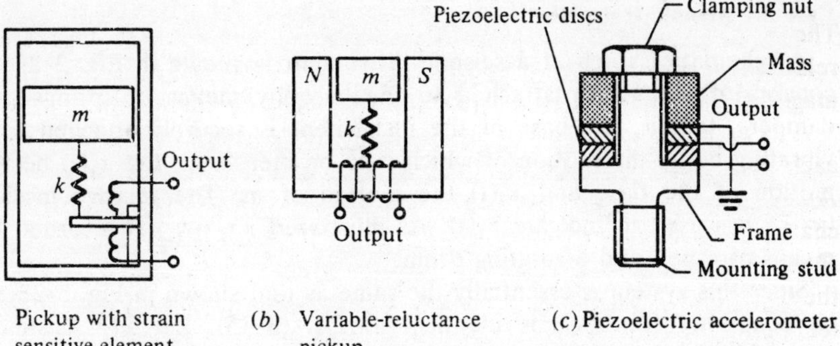

(a) Pickup with strain sensitive element (b) Variable-reluctance pickup (c) Piezoelectric accelerometer

FIG. 3-30. *Vibration pickups with electrical output.*

elements are sandwiched between the mass and the frame. The voltage output of the device is due to the cyclic deformation of the piezoelectric crystals. The effective spring, damping, and mass in this accelerometer, however, are not self-evident. Since vibration measurement is a separate study, we shall not pursue the subject further.

Vibrometer: Referring to Fig. 3-16, if $\omega \gg \omega_n$ or $r \gg 1$, the ratio X/X_1 in Eq. (3-40) approaches unity regardless of the value of ζ. In other words, the relative displacement $x(t)$ is equal to the displacement $x_1(t)$, which is the motion to be measured. The phase angle is approximately 180°. An instrument for displacement measurement is called a *vibrometer.* Since there is no advantage in introducing damping in the system, a vibrometer is designed with damping only to minimize the transient vibration.

The performance characteristics of a typical inductive type velocity pickup is illustrated in Fig. 3-31. The useful range is marked off in the figure with bold lines. Displacement and acceleration can be obtained from the electrical output of the velocity pickup by integration and differentiation. The natural frequency of this pickup is 8 Hz. The displacement amplitude ranges from 0.0025 to 10 mm.

Accelerometer: A seismic instrument to measure acceleration is called an *accelerometer.* Due to their small size and high sensitivity, most vibration measurements today are made with accelerometers. The velocity and displacement can be obtained from the electrical output of the accelerometer by integration. For example, the motion of the piston of an internal combustion engine can be indicated in this manner.

If the motion to be measured is $x_1 = X_1 \sin \omega t$, the amplitude of the acceleration is $\omega^2 X_1$. From Eq. (3-40), we obtain

$$X = \frac{R}{\omega_n^2}(\omega^2 X_1) = \frac{1}{\omega_n^2\sqrt{(1-r^2)^2+(2\zeta r)^2}}\,\omega^2 X_1 \qquad \textbf{(3-41)}$$

The quantity ω_n^2 is a constant, since ω_n is a property of the system. The relative motion $x(t)$ is proportional to the acceleration $\ddot{x}_1(t)$ if the magnification factor R is constant for all ranges of operation.

A periodic vibration generally has a number of harmonic components, each of which gives a corresponding value of r $(=\omega/\omega_n)$ in Eq. (3-41). *Amplitude distortion* occurs if the magnification factor $1/\sqrt{(1-r^2)^2+(2\zeta r)^2}$ changes with the harmonic components. In other words, the magnification R of each harmonic component must be identical in order to reproduce the input waveform. Since $R \simeq 1$ when r approaches zero, an accelerometer is constructed such that $r \ll 1$, or $\omega_n \gg \omega$. The percent amplitude distortion is defined as

$$\text{Amplitude distortion} = (R-1)\times 100\% \tag{3-42}$$

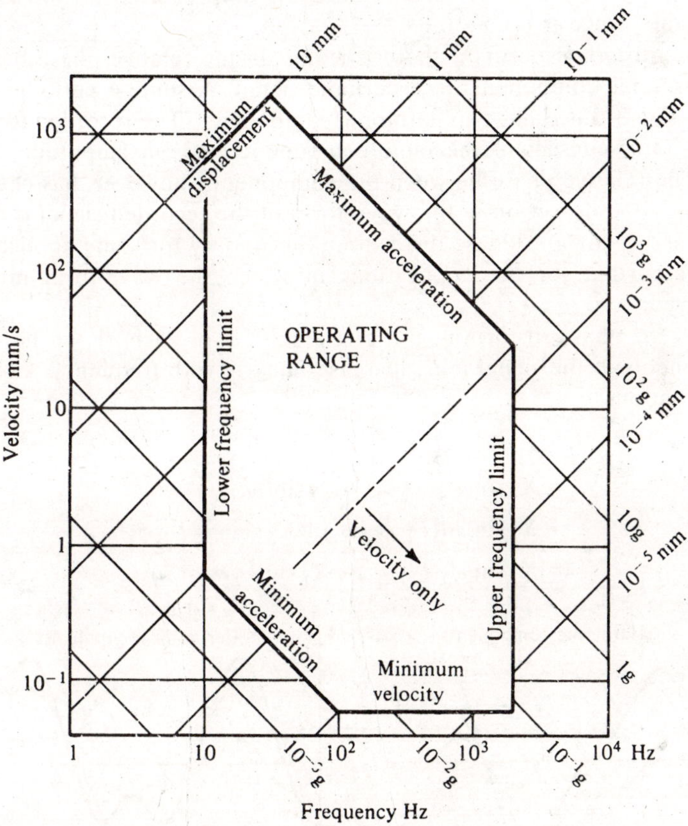

FIG. 3-31. *Performance characteristics of a typical inductive velocity pickup.*

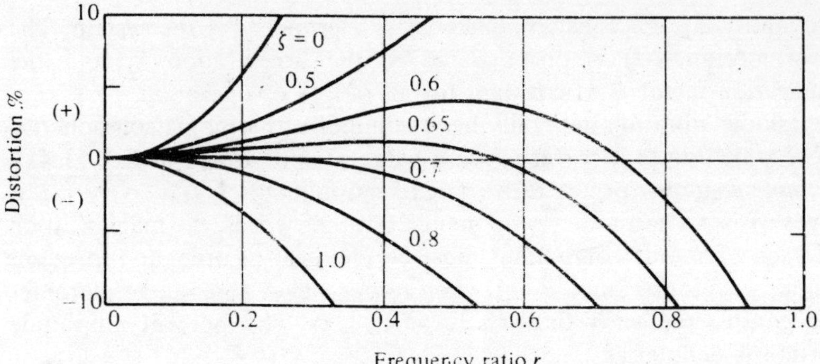

FIG. 3-32. *Amplitude distortion in accelerometer.*

This is plotted in Fig. 3-32. Note that (1) an accelerometer should be built with $0.6<\zeta<0.7$ in order to minimize the amplitude distortion and (2) the usable range is $0<r<0.6$.

Phase distortion occurs if there is a shift in the relative phase between the harmonic components in a periodic signal. Assume a periodic signal $x(t)$ in Fig. 3-33(a) has two harmonic components. The recorded signal in Fig. 3-33(b) consists of the same components without amplitude distortion. The relative phase between the components, however, has changed. Evidently the distortion in the wave-form of the recorded signal is due to the phase distortion. Phase distortion is secondary for some applications. It is important for the applications in which the wave-form must be preserved.

For zero phase distortion, the phase shift ϕ of each of the harmonic components in the signal must increase linearly with frequency. Consider the equations

$$\begin{aligned} x(t) &= x_1(t)+x_2(t) \\ &= X_1 \sin(\omega_1 t-\phi_1)+X_2 \sin(\omega t-\phi_2) \\ &= X_1 \sin \omega_1(t-\phi_1/\omega_1)+X_2 \sin \omega_2(t-\phi_2/\omega_2) \\ &= X_1 \sin \omega_1(t-t_{\phi 1})+X_2 \sin \omega_2(t-t_{\phi 2}) \end{aligned}$$

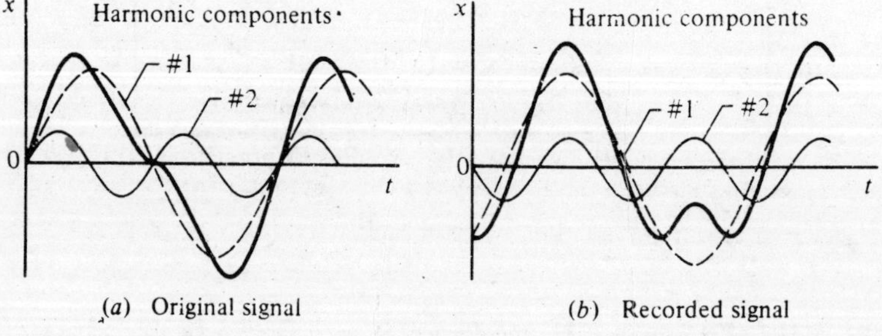

(a) Original signal (b) Recorded signal

FIG. 3-33. *Phase distortion.*

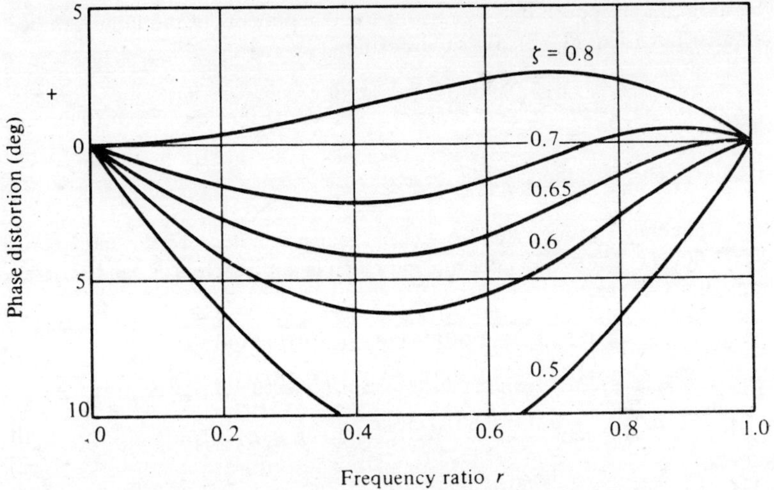

FIG. 3-34. *Phase distortion in accelerometer.*

The quantities $t_{\phi 1}$ and $t_{\phi 2}$ denote the delay or the shift of the signal along the positive time axis. If each harmonic component of a signal $x(t)$ is shifted by the same amount of time, the wave-form is preserved. This requires that $t_{\phi_1} = t_{\phi_2} = \text{constant}$, that is, $\phi_1/\omega_1 = \phi_2/\omega_2 = \text{constant}$. In other words, the phase angle varies linearly with the frequency ω.

It is observed in Fig. 2-9 that, for the phase angle ϕ to vary linearly with frequency over an acceptable range $0 < r < 1$, the phase shift is $(90° \times r)$. Hence the phase distortion of an accelerometer is defined as

$$\text{Phase distortion} = (\phi - 90r) \text{ deg} \qquad \textbf{(3-43)}$$

This is plotted in Fig. 3-34. Again, it is seen that an appropriate damping in an accelerometer is necessary in order to minimize the phase distortion.

Example 20

A machine component is vibrating with the motion

$$\begin{aligned} y &= y_1(t) + y_2(t) \\ &= Y_1 \sin \omega_1 t + Y_2 \sin \omega_2 t \\ &= 0.10 \sin 60\pi t + 0.05 \sin 120\pi t \end{aligned}$$

Determine the vibration record that would be obtained with an accelerometer. Assume $\zeta = 0.65$ and $f_n = \omega_n/2\pi = 1{,}500$ Hz.

Solution:

From Eq. (3-39), the equation of motion is

$$m\ddot{x} + c\dot{x} + kx = -m\ddot{y} = -m(\ddot{y}_1 + \ddot{y}_2)$$

The harmonic response due to each of the components in the input can be obtained from Eq. (3-40). By superposition, we have

$$x = r_1^2 R_1 Y_1 \sin(\omega_1 t - \phi_1) + r_2^2 R_2 Y_2 \sin(\omega_2 t - \phi_2)$$

where

$$\phi_1 = \tan^{-1}\frac{2\zeta r_1}{1 - r_1^2} \qquad \text{and} \qquad \phi_2 = \tan^{-1}\frac{2\zeta r_2}{1 - r_2^2}$$

The frequency ratios are $r_1 = \omega_1/\omega_n = 60\pi/3{,}000\pi = 1/50$ and $r_2 = 120\pi/3{,}000\pi = 2/50$. The values of the magnification factors R_1 and R_2 are almost unity. Thus,

$$r_1^2 R_1 Y_1 = (1/50)^2(1)(0.10) = 0.10/2500$$

$$r_2^2 R_2 Y_2 = (2/50)^2(1)(0.05) = 0.20/2500$$

$$\phi_1 = \tan^{-1}\frac{(2)(0.65)(1/50)}{1-(1/50)^2} = \tan^{-1} 0.026 = 1.49°$$

$$\phi_2 = \tan^{-1}\frac{(2)(0.65)(2/50)}{1-(2/50)^2} = \tan^{-1} 0.052 = 2.98°$$

Hence the acceleration record is

$$x = \frac{1}{2500}[0.10 \sin(60\pi t - 1.49°) + 0.20 \sin(120\pi t - 2.98°)]$$

The output of an accelerometer is usually converted to an electrical signal and amplified. Hence the value of the recorded acceleration depends on the amplification used in the data processing. If the equation above is integrated twice to give a displacement, the measured value of the given motion is

$$\underset{(\text{measured})}{y} = (\text{constant})[0.10 \sin(60\pi t - 1.49°) + 0.05 \sin(120\pi t - 2.98°)]$$

Note that the measured value of $y(t)$ has practically no amplitude distortion and only a slight shift in the phase angles. There is no phase distortion, however, because the phase shift is linear with the frequency of the harmonic components. There is a slight time delay between the input $y(t)$ and its measured value. The time delay is ϕ/ω. For the given values, we have time delay $= 1.49° (\pi/180°)/(60\pi) = 0.14$ ms. The results above are due to the high natural frequency of the accelerometer.

Case 6. Elastically Supported Damped Systems

The damping of a real system can be considerably more complex than a simple damper shown in Fig. 3-23. Equivalent viscous damping will be discussed in Sec. 3-8. We shall consider the elastically supported damper as illustrated in Fig. 3-35.

From the free body sketch, the equation of motion for the mass m is

$$m\ddot{x} + c(\dot{x} - \dot{x}_1) + kx = F \sin \omega t$$

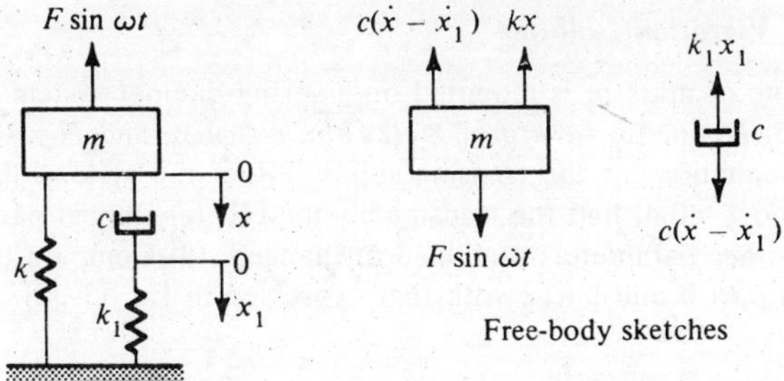

FIG. 3-35. *Elastically supported damper.*

Since the damper c and the spring k_1 are in series, the damping force is equal to the spring force.

$$c(\dot{x}-\dot{x}_1)=k_1x_1$$

Since the motions $x(t)$ and $x_1(t)$ are harmonic, the equations above can be solved readily by the mechanical impedance method discussed in Sec. 2-6. Substituting $\bar{F}e^{j\omega t}$ for $F\sin\omega t$, $j\omega$ for the time derivatives, and using phasor notation, the equations become

$$\begin{aligned}(k-\omega^2m+j\omega c)\bar{X}-j\omega c\bar{X}_1&=\bar{F}\\ -j\omega c\bar{X}+(k_1+j\omega c)\bar{X}_1&=0\end{aligned} \tag{3-44}$$

The phasors $\bar{X}$ and $\bar{X}_1$ can be solved for by Cramer's rule:

$$\begin{aligned}\bar{X}&=\frac{\bar{F}(k_1+j\omega c)}{k_1(k-\omega^2m)+j\omega c(k+k_1-\omega^2m)}=Xe^{-j\gamma}\\ \bar{X}_1&=\frac{j\omega c\bar{F}}{k_1(k-\omega^2m)+j\omega c(k+k_1-\omega^2m)}=X_1e^{-j\gamma_1}\end{aligned} \tag{3-45}$$

Defining the stiffness ratio of the springs as $N=k_1/k$, $\omega_n=\sqrt{k/m}$, $c/m=2\zeta\omega_n$, $r=\omega/\omega_n$, and

$$\begin{aligned}\Delta(\omega)&=|k_1(k-\omega^2m)+j\omega c(k+k_1-\omega^2m)|/(kk_1)\\ &=\sqrt{(1-r^2)^2+[2\zeta r(1+1/N-r^2/N)]^2}\end{aligned} \tag{3-46}$$

we obtain

$$\begin{aligned}X&=\frac{F}{k}\sqrt{1+(2\zeta r/N)^2}/\Delta(\omega)\\ X_1&=\frac{F}{k}(2\zeta r/N)/\Delta(\omega)\end{aligned} \tag{3-47}$$

$$\begin{aligned}\gamma&=\tan^{-1}\frac{2\zeta r(1+1/N-r^2/N)}{1-r^2}-\tan^{-1}\frac{2\zeta r}{N}\\ \gamma_1&=\tan^{-1}\frac{2\zeta r(1+1/N-r^2/N)}{1-r^2}-\frac{\pi}{2}\end{aligned} \tag{3-48}$$

Example 21. Vibration Isolation*

A machine of mass m is mounted on a spring-damper system as shown in Fig. 3-35. Define $\omega_n = \sqrt{k/m}$, $\zeta = c/2\sqrt{km}$, $r = \omega/\omega_n$, and $N = k_1/k$: (**a**) Derive the equation for the transmissibility TR of the system. (**b**) If $r = 0.6$, $N = 2$, and $\zeta = 0.4$, find the transmissibility TR. (**c**) Repeat part **b** if $r = 10$ and the other parameters remained unchanged. (**d**) Compare the values of TR from part **b** and part **c** with that expressed in Eq. (3-32).

Solution:

(**a**) The force transmitted F_T to the foundation is the sum of the forces transmitted through the springs k and k_1.

$$F_T = kx + k_1 x_1$$

The phasors $\bar{X}$ and $\bar{X}_1$ are given in Eq. (3-45). Using vectorial addition, the phasor $\bar{F}_T$ of the transmitted force is

$$\bar{F}_T = k\bar{X} + k_1\bar{X}_1$$

$$= \frac{F[k(k_1 + j\omega c) + j\omega k_1 c]}{k_1(k - \omega^2 m) + j\omega c(k + k_1 - \omega^2 m)}$$

or

$$\bar{F}_T = \frac{F\sqrt{1 + [2\zeta r(1 + 1/N)]^2}}{\Delta(\omega)} e^{-j\gamma_T} \tag{3-49}$$

where γ_T is the phase angle of the transmitted force relative to the excitation and $\Delta(\omega)$ is defined in Eq. (3-46). The transmissibility TR is the ratio of the magnitude of the transmitted force relative to that of the excitation; that is,

$$\text{TR} = F_T/F = \sqrt{1 + [2\zeta r(1 + 1/N)]^2}/\Delta(\omega)$$

(**b**) For $r = 0.6$, $N = 2$, and $\zeta = 0.4$, we obtain

$$\text{TR} = 1.23/0.90 = 1.37$$

(**c**) For $r = 10$, $N = 2$, and $\zeta = 0.4$, we have

$$\text{TR} = 12/400 = 0.03$$

(**d**) Substituting the corresponding values in Eq. (3-32), the values of TF for the system in Fig. 3-21(a) are

$$\text{TR} = \sqrt{1.23/0.64} = 1.39 \qquad \text{for} \qquad r = 0.6$$

$$TR = \sqrt{65/9865} = 0.081 \qquad \text{for} \qquad r = 10$$

* A detailed analysis is shown in J. C. Snowdon, *Vibration and Shock in Damped Mechan cal Systems*, John Wiley & Sons, Inc., New York, 1968, pp. 33–38.

Comparing parts (**b**), (**c**), and (**d**), the transmissibility for the twc isolation systems are approximately the same for $r = 0.6$. When operating at $r = 10$, the system in Fig. 3-35 seems to be superior to that in Fig. 3-23(a).

3-6 DAMPED FORCED VIBRATION—PERIODIC EXCITATION

Harmonic response of systems was presented in the last section. Forces arising from machinery are commonly periodic but seldom harmonic. In considering periodic excitations, we are in effect generalizing the previous applications. We shall briefly review the Fourier series and then illustrate the applications.

As illustrated in Sec. 1-2, a function is periodic if

$$F(t) = F(t \pm \tau) \tag{3-50}$$

where τ is the *period*, or the *minimum time* required for $F(t)$ to repeat itself. The Fourier series* expansion of $F(t)$ is

$$F(t) = \frac{a_0}{2} + \sum_{n=1}^{\infty} (a_n \cos n\omega t + b_n \sin n\omega t) \tag{3-51}$$

where n is a positive integer, a_n and b_n are the coefficients of the infinite series. Note that $a_0/2$ gives the average value of $F(t)$. The fundamental frequency of the periodic function is $\omega = 2\pi/\tau$, that is, when $n = 1$. The frequency of the nth harmonic is $n\omega = 2n\pi/\tau$, that is, when $n > 1$.

The following relations are used to evaluate a_n and b_n:

$$\begin{aligned}
&\int_0^\tau \cos m\omega t \cos n\omega t\, dt = \begin{cases} 0 & \text{if} \quad m \neq n \\ \tau/2 & \text{if} \quad m = n \end{cases} \\
&\int_0^\tau \sin m\omega t \sin n\omega t\, dt = \begin{cases} 0 & \text{if} \quad m \neq n \\ \tau/2 & \text{if} \quad m = n \end{cases} \\
&\int_0^\tau \cos m\omega t \sin n\omega t\, dt = 0 \qquad \text{whatever } m \text{ and } n
\end{aligned} \tag{3-52}$$

where m and n are integers and $\tau = 2\pi/\omega$ is the period of $F(t)$. Rewriting the series in an expanded form, we get

$$F(t) = \left(\frac{a_0}{2} + a_1 \cos \omega t + a_2 \cos 2\omega t + \cdots\right) + (b_1 \sin \omega t + b_2 \sin 2\omega t + \cdots)$$

A particular coefficient a_p can be obtained by multiplying both sides of

* See, for example, I. S. Sokolnikoff and R. M. Redheffer, *Mathematics of Physics and Modern Engineering*, McGraw-Hill Book Co., New York, 1958, p. 175.

this equation by (cos $p\omega t$) and integrating each term using the relations in Eq. (3-52). Except for the term containing a_p, all the integrals on the right side are identically zero. Thus,

$$\int_0^\tau F(t)\cos p\omega t\,dt = 0+\cdots+0+\int_0^\tau a_p \cos^2 p\omega t\,dt+0+\cdots$$

$$=\frac{a_p\tau}{2}$$

or

$$a_p=\frac{2}{\tau}\int_0^\tau F(t)\cos p\omega t\,dt$$

Similarly, a particular coefficient b_p can be obtained by multiplying the series by (sin $p\omega t$) and applying the relations in Eq. (3-52). Thus, the coefficients of the Fourier series in Eq. (3-51) are

$$a_0=\frac{2}{\tau}\int_0^\tau F(t)\,dt$$

$$a_n=\frac{2}{\tau}\int_0^\tau F(t)\cos n\omega t\,dt \qquad \textbf{(3-53)}$$

$$b_n=\frac{2}{\tau}\int_0^\tau F(t)\sin n\omega t\,dt$$

Let a periodic force $F(t)$ be applied to a one-degree-of-freedom system. $F(t)$ may represent the equivalent force in any of the five cases enumerated in the previous section. Expanding $F(t)$ in a Fourier series and applying Eq. (3-22), the equation of motion becomes

$$m\ddot{x}+c\dot{x}+kx=\frac{a_0}{2}+\sum_{n=1}^{\infty}(a_n\cos n\omega t+b_n\sin n\omega t) \qquad \textbf{(3-54)}$$

The steady-state response due to each of the components of the excitation can be calculated. By superposition, the steady-state response of the system is

$$x=\frac{a_0}{2k}+\sum_{n=1}^{\infty}\frac{a_n\cos(n\omega t-\phi_n)+b_n\sin(n\omega t-\phi_n)}{k\sqrt{(1-n^2r^2)^2+(2\zeta nr)^2}} \qquad \textbf{(3-55)}$$

where

$$\phi_n=\tan^{-1}\frac{2\zeta nr}{1-n^2r^2} \qquad \text{and} \qquad r=\frac{\omega}{\omega_n}$$

Example 22

Find the Fourier series of the square wave in Fig. 3-36(a).

Solution:

For any one cycle, the given periodic function is

$$F(t)=\begin{cases}1 & \text{for} \quad 0<t<\tau/2\\ -1 & \text{for} \quad \tau/2<t<\tau\end{cases}$$

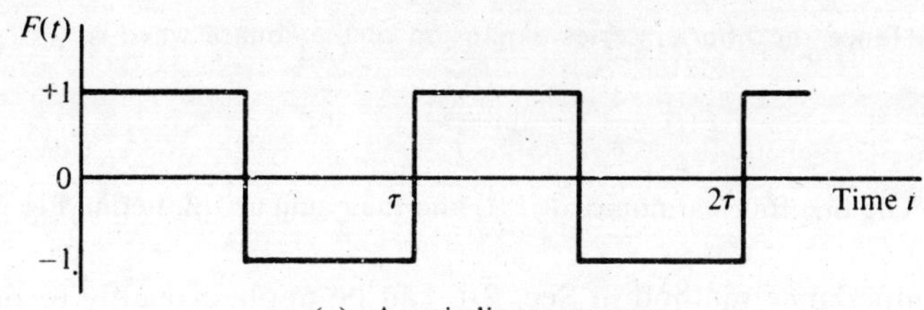

(a) A periodic square wave

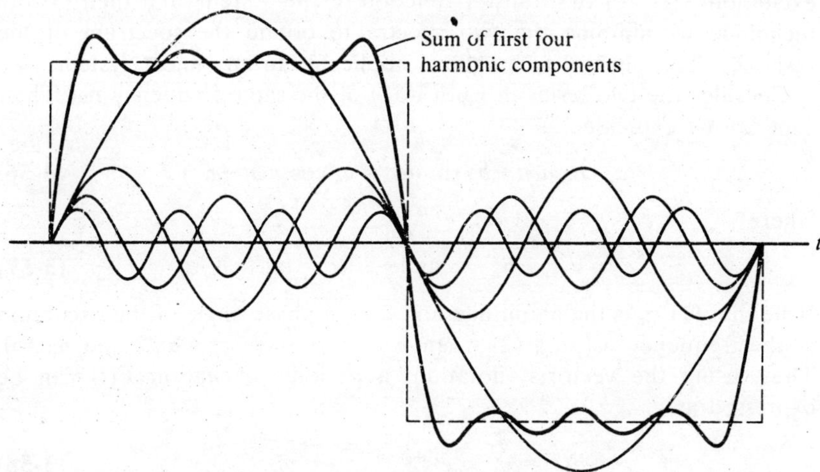

(b) Harmonic components of a square wave

FIG. 3-36. *Fourier series analysis of a square wave.*

Applying Eq. (3-53) and for $\omega = 2\pi/\tau$, the coefficients of the Fourier series of $F(t)$ are

$$a_0 = \frac{2}{\tau}\int_0^{\tau} F(t)\,dt = \frac{2}{\tau}\left[\int_0^{\tau/2} (1)\,dt - \int_{\tau/2}^{\tau} (1)\,dt\right] = 0$$

$$a_n = \frac{2}{\tau}\int_0^{\tau} F(t)\cos n\omega t\,dt$$

$$= \frac{2}{\tau}\left[\int_0^{\tau/2} \cos n\omega t\,dt - \int_{\tau/2}^{\tau} \cos n\omega t\,dt\right] = 0$$

$$b_n = \frac{2}{\tau}\int_0^{\tau} F(t)\sin n\omega t\,dt$$

$$= -\frac{2}{\tau}\frac{\tau}{2n\pi}\left(\cos\frac{2n\pi}{\tau}t\,\Big|_0^{\tau/2} - \cos\frac{2n\pi}{\tau}t\,\Big|_{\tau/2}^{\tau}\right)$$

$$= \begin{cases} \dfrac{4}{n\pi} & \text{for } n \text{ odd} \\ 0 & \text{for } n \text{ even} \end{cases}$$

Hence the Fourier series expansion of the square wave is

$$F(t) = \frac{4}{\pi} \sum_n \frac{1}{n} \sin \frac{2n\pi}{\tau} t \qquad \text{for} \qquad n = 1, 3, 5, \ldots$$

The first four harmonics of $F(t)$ and their sum are plotted in Fig. 3-36(b).

The impedance method in Sec. 2-6 can be applied readily to this type of problem. We shall consider (1) the Fourier spectrum of the periodic excitation $F(t)$, (2) the transfer function of the system, and then (3) the technique to combine the two spectra to obtain the spectrum of the response. This general technique is applicable to any linear system.

Consider the two terms in Eq. (3-51) of the same frequency $n\omega$. Their sum can be expressed as

$$a_n \cos n\omega t + b_n \sin n\omega t = c_n \cos(n\omega t - \alpha_n) \tag{3-56}$$

where

$$c_n = \sqrt{a_n^2 + b_n^2} \qquad \text{and} \qquad \alpha_n = \tan^{-1} b_n / a_n \tag{3-57}$$

Note that (1) c_n is the amplitude and α_n the phase angle of the excitation at the frequency $n\omega$ and (2) when $n = 0$, we have $c_n = a_0/2$ and $\alpha_n = 0$. Thus, using the vectorial notation, a periodic excitation $F(t)$ can be expressed as

$$F(t) = \sum_{n=0}^{\infty} c_n e^{j(n\omega t - \alpha_n)} = \sum_{n=0}^{\infty} \bar{c}_n e^{jn\omega t} \tag{3-58}$$

where $\bar{c}_n = c_n e^{-j\alpha_n}$ is the phasor of the harmonic component at the frequency $n\omega$. The plot of c_n versus frequency is called the *frequency spectrum* and α_n versus frequency the *phase spectrum* of $F(t)$. The two plots as shown in Fig. 3-37(a) are known as the *Fourier spectrum.*

From Eq. (2-57), the sinusoidal transfer function of a one-degree-of-freedom system is

$$\frac{X}{F}(j\omega) = \frac{1}{k - \omega^2 m + j\omega c} = \left|\frac{X}{F}\right| e^{-j\phi} \tag{3-59}$$

The plot of $|X/F|$ versus frequency is analogous to the frequency spectrum and ϕ versus frequency the phase spectrum. In other words, the transfer function plots in Fig. 3-37(b) are the continuous plots of the Fourier spectrum of the system for excitations of unit magnitude and zero phase angle for all frequencies.

The equation of motion of the system is obtained by substituting Eq. (3-58) into (3-22). Thus,

$$m\ddot{x} + c\dot{x} + kx = \sum_{n=0}^{\infty} \bar{c}_n e^{jn\omega t} \tag{3-60}$$

Using the impedance method and Eqs. (3-24) to (3-26), the response due

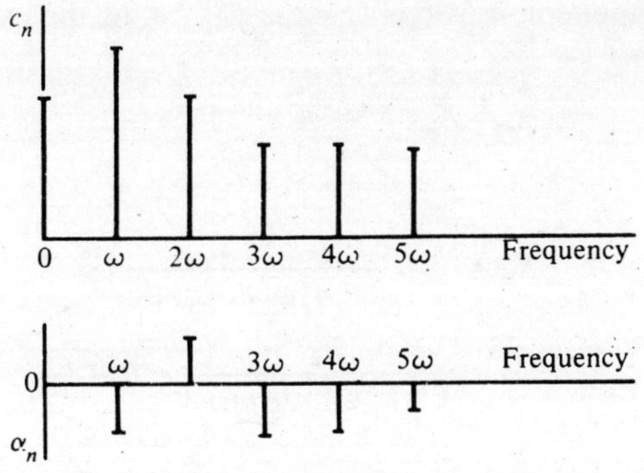

(*a*) Fourier spectrum of periodic input $F(t)$

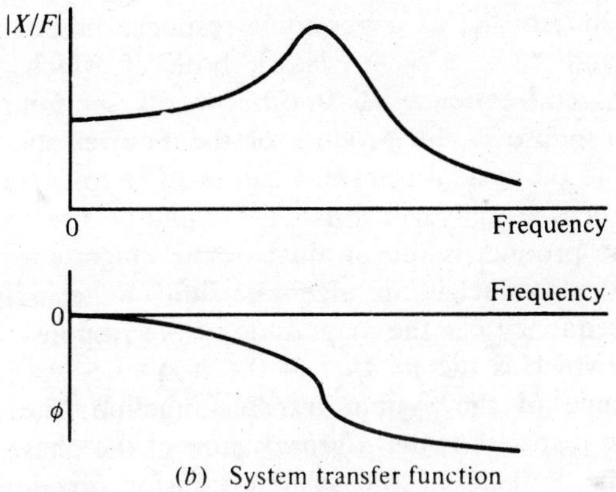

(*b*) System transfer function

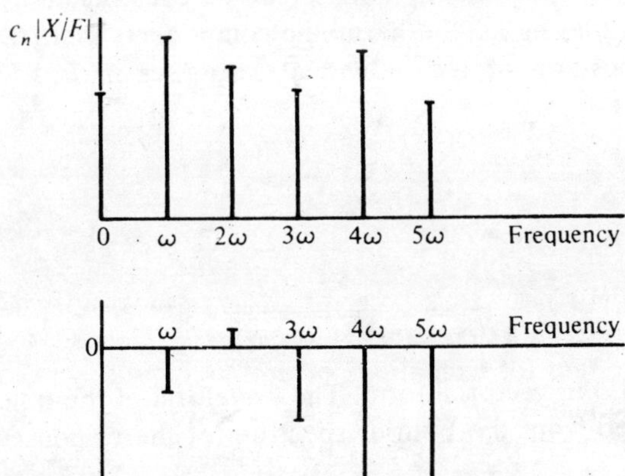

(*c*) Fourier spectrum of system response

FIG. 3-37. *Construction of response spectrum from input spectrum and transfer function.*

to a typical component, $\mathbf{c}_n = \bar{c}_n e^{jn\omega t} = c_n e^{j(n\omega t - \alpha_n)}$, of the excitation $F(t)$ at the frequency $n\omega$ is

$$x_n = \bar{X}_n e^{jn\omega t} = X_n e^{j(n\omega t - \alpha_n - \phi_n)} \tag{3-61}$$

where

$$\bar{X}_n = (\bar{c}_n)\left(\frac{1}{k - n^2\omega^2 m + jn\omega c}\right)$$

$$= \frac{c_n}{|k - n^2\omega^2 m + jn\omega c|} e^{-j(\alpha_n + \phi_n)} \tag{3-62}$$

and

$$\phi_n = \tan^{-1} \frac{jn\omega c}{k - n^2\omega^2 m} \tag{3-63}$$

Note that the phasor $\bar{X}_n$ of a harmonic response in Eq. (3-62) is the product of $\bar{c}_n$ and $1/(k - n^2\omega^2 m + jn\omega c)$, both of which are complex numbers at the given frequency $n\omega$. In other words, the Fourier spectrum of the system response is the product of the Fourier spectrum of the excitation $F(t)$ and the system transfer function. The rules for the product of complex numbers are given in Eqs. (1-14) and (1-15), that is, (1) the magnitude of the product is the product of the magnitudes and (2) the phase angle of the product is the algebraic sum of the individual phase angles. At the frequency $n\omega$, the magnitude of the response is $|\bar{c}_n|\,|1/(k - n^2\omega^2 m + jn\omega c)|$, which is the *product* of the frequency spectrum of $F(t)$ and the magnitude of the system transfer function; the phase angle $-(\alpha_n + \phi_n)$ of the response is the *algebraic sum* of the phase spectrum of $F(t)$ and the phase angle of the system transfer function. Thus, the Fourier spectrum of the system response can be constructed as shown in Fig. 3-37(c), considering all the harmonic components of $F(t)$.

By the superposition of the individual responses in Eq. (3-61), the system response is

$$x(t) = \sum_{n=0}^{\infty} \bar{X}_n e^{jn\omega t} \tag{3-64}$$

or

$$x(t) = \sum_{n=0}^{\infty} \frac{c_n}{k\sqrt{(1 - n^2r^2)^2 + (2\zeta nr)^2}} e^{j(n\omega t - \alpha_n - \phi_n)} \tag{3-65}$$

which is Eq. (3-55) in vectorial form. The waveform of the time response can be constructed from the Fourier spectrum of the response by superposition.

Example 23

A cam actuating a spring-mass system is shown in Fig. 3-38. The total cam lift of the sawtooth is 25 mm (1 in.). The cam speed is 60 rpm. Assume

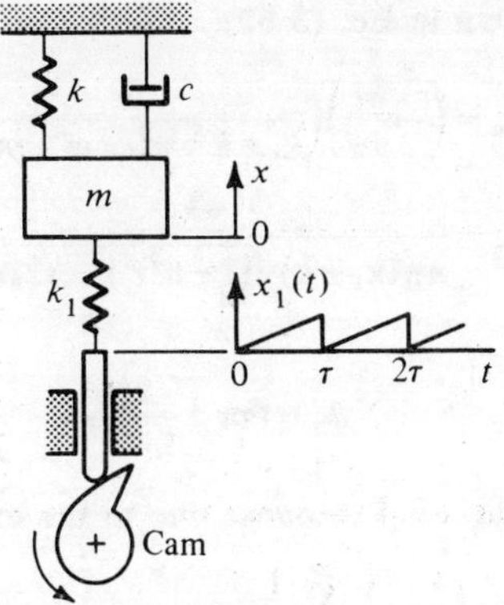

FIG. 3-38. *Periodic excitation.*

$m = 20$ kg (44 lb_m) and $k_1 = k = 3.5$ kN/m (20 lb_f/in.). The damping coefficient is $c = 0.2$ kN · s/m (1.14 lb_f-sec/in.). Find the response $x(t)$.

Solution:

A cycle of the sawtooth motion can be expressed as

$$x_1(t) = \frac{1}{\tau} t \quad \text{for} \quad 0 < t < \tau$$

Since the fundamental frequency is 1 Hz or $\omega = 2\pi$, the period τ is 1 sec/cycle. Applying Eq. (3-52), it can be verified readily that

$$a_0 = 1 \qquad a_n = 0 \qquad b_n = -\frac{1}{n\pi}$$

Hence the Fourier series expansion of $x_1(t)$ is

$$x_1(t) = \frac{1}{2} - \frac{1}{\pi} \sum_{n=1}^{\infty} \frac{1}{n} \sin 2n\pi t$$

The equation of motion of the system is

$$m\ddot{x} + c\dot{x} + (k + k_1)x = k_1 x_1(t) = k_1 \left(\frac{1}{2} - \frac{1}{\pi} \sum_{n=1}^{\infty} \frac{1}{n} \sin 2n\pi t \right)$$

We define $\omega_n^2 = (k + k_1)/m$, $2\zeta\omega_n = c/m$, and $r = \omega/\omega_n$. The response due to the constant excitation term $k_1/2$ is

$$x = \frac{k_1}{2(k + k_1)}$$

The response due to a typical harmonic excitation term at the frequency

$n\omega = 2n\pi$ is as shown in Eq. (3-62).

$$\bar{X}_n = \left(-\frac{k_1}{n\pi}\right)\left(\frac{1}{k_1 + k - n^2\omega^2 m + jn\omega c}\right)$$

$$\bar{X}_n = \frac{-k_1}{n\pi(k_1+k)\sqrt{(1-n^2r^2)^2 + (2\zeta nr)^2}}\, e^{-j\phi_n}$$

where

$$\phi_n = \tan^{-1}\frac{2\zeta nr}{1-n^2r^2}$$

By superposition, the total response due to the excitation $F(t)$ is

$$x(t) = \frac{k_1}{k_1+k}\left(\frac{1}{2} - \frac{1}{\pi}\sum_{n=1}^{\infty}\frac{1}{n}\frac{1}{\sqrt{(1-n^2r^2)^2+(2\zeta nr)^2}}\, e^{j(2n\pi t - \phi_n)}\right)$$

For the given data, we have

$$\omega_n^2 = (k_1+k)/m = (3500+3500)/20 = 350 = 18.7^2$$

$$r = \omega/\omega_n = 2\pi/18.7 = 0.107\pi$$

$$\zeta = \tfrac{1}{2}c/\sqrt{(k_1+k)m} = \tfrac{1}{2}(200)/\sqrt{(3500+3500)(20)} = 0.267$$

Thus, the response of the system is

$$x(t) = \frac{1}{2}\left(\frac{1}{2} - \frac{1}{\pi}\sum_{n=1}^{\infty}\frac{1}{n}\frac{1}{\sqrt{[1-(0.11n\pi)^2]^2+(0.05n\pi)^2}}\sin(2n\pi t - \phi_n)\right)$$

where

$$\phi_n = \tan^{-1}\frac{0.05n\pi}{1-(0.11n\pi)^2}$$

3-7 TRANSIENT VIBRATION—SHOCK SPECTRUM

The design of equipment to withstand shock is of concern to the engineer. Vibrations induced by the steady-state operation of a machine are generally periodic. This was discussed in the last two sections. Vibrations due to shock and transients usually originate from sources outside the machine or from a sudden change in the machine operation. The transient will die out, but the machine may be damaged or may malfunction momentarily, both of which should be well considered.

A *shock* is a transient excitation, the duration of which is short compared with the natural period (reciprocal of natural frequency) of oscillation of the system. The transient response due to a transient excitation was discussed in Sec. 2-7. The recording from a vibration test, in the form of a "wavering line" versus time, cannot be used directly by the designer. The shock spectrum is a common method to reduce the test data to a more usable form.

A *shock spectrum* (*response spectrum*) is a plot of the peak response versus frequency due to the applied shock. The peak response is that of a number of one-degree-of-freedom systems, each tuned to a different natural frequency. The frequency is that of the natural frequency of the individual systems. The response may be expressed in units of acceleration, velocity, or displacement.* For example, a vibrating reed shown in Fig. 3-3 is a simple mass-spring system. A *reed gage* consists of a number of reeds of different natural frequencies. Using a reed gage in a shock test, the maximum displacements of the tips of the reeds give the maximum response for the various natural frequencies. The reed gage can be replaced by a single accelerometer and computers employed to simulate the reeds.† The one-degree-of-freedom system is variously called a *resonator*, an *oscillator*, or a *simple structure.*

The objective of shock spectrum is to describe the effect of shock rather than the shock itself. Shocks are difficult to characterize and a specific pulse shape is difficult to obtain in a test machine. It is necessary to correlate test data from different laboratories. The shock spectrum is a "common denominator" on the assumption that shocks having the same spectrum would produce similar effect. The spectrum may be regarded as indicative of the potential for damage due to the shock. For example, the peak relative displacement between the mass m and its base for the system in Fig. 3-23(b) is related to the stress in the spring. In other words, the envelope of the spectrum establishes an upper bound of the stress induced, or the damage potential, by a specific shock on the equipment under test.

Types of shock excitation are usually categorized using the undamped resonator as the standard system. Types of shocks and methods of data reduction can be found in the literature.‡ It is paradoxical that a complex study like shock is treated in a seemingly simple manner. The reason is that the time and expense for a detail study must be justified by the past experience of the engineer. An undamped resonator is used, since the largest excursion occurs within the first cycle of the transient and the error introduced by neglecting damping tends to provide a margin of safety. Note that a system under test commonly has more than one degree of freedom. It will be shown in later chapters that a complex system has discrete modes of vibration and a natural frequency is associated with

* C. E. Crede, *Shock and Vibration Concepts in Engineering Design*, McGraw-Hill Book Co., New York, 1965, p. 138.

† C. T. Morrow, *Shock and Vibration Engineering*, John Wiley and Sons, Inc., New York, 1963, p. 111.

‡ R. S. Ayre, "Transient Response to Step and Pulse Functions," Chap. 8 of *Shock and Vibration Handbook*, vol. 1, C. M. Harris and C. E. Crede (eds.) McGraw-Hill Book, Co., New York, 1961.

S. Rubin, "Concepts in Shock Data Analysis," Chap. 23 of *Shock and Vibration Handbook*, vol. 2, C. M. Harris and C. E. Crede (eds.), McGraw-Hill Book Co., New York, 1961.

each of the modes. Hence a complex system can be described in terms of equivalent one-degree-of-freedom systems. Thus, a shock will excite all the modes of a complex system.

If a shock $F_{eq}(t)$ is due to a sudden change in the machine m in Fig. 3-23(a), the equation of motion of m is identical to Eq. (3-1). Assuming zero initial conditions, the response $x(t)$ from Eq. (2-71) is

$$x(t) = \int_0^t F_{eq}(\tau)h(t-\tau)\,d\tau \tag{3-66}$$

where

$$h(t) = \frac{1}{\omega_d m} e^{-\zeta\omega_n t} \sin \omega_d t \tag{3-67}$$

as defined in Eq. (2-69). On the other hand, if the excitation is applied to the base of the machine as shown in Fig. 3-23(b), the equation for the relative displacement $x(t)$ between m and its base is identical to Eq. (3-39).

$$m\ddot{x} + c\dot{x} + kx = -m\ddot{x}_1(t) \tag{3-68}$$

where $x(t) = x_2(t) - x_1(t)$ and $x_1(t)$ and $x_2(t)$ are the absolute motions indicated in the figure. Applying Eq. (2-71) yields

$$x(t) = -\int_0^t m\ddot{x}_1(\tau)h(t-\tau)\,d\tau \tag{3-69}$$

where $h(t)$ is as defined above. The maximum response and the corresponding shock spectrum can be obtained from Eq. (3-66) or (3-69) depending on the application. Computers can be used for the calculations.*

To illustrate a shock spectrum by hand calculation, let a one-half sine pulse $F(t)$ shown in Fig. 3-39 be applied to the mass m of the system in

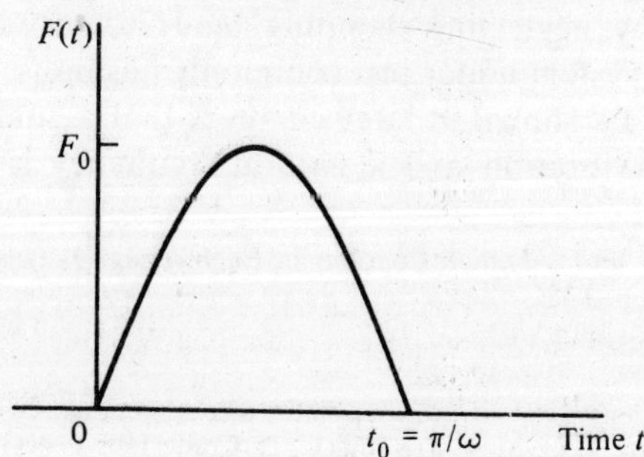

FIG. 3-39. *A half-sine pulse.*

* See, for example, J. B. Vernon, *Linear Vibration and Control System Theory*, John Wiley & Sons, Inc., New York, 1967, pp. 251.

Fig. 3-23. Assume the system is undamped. $F(t)$ is described by the equation

$$F(t)=\begin{cases}F_0 \sin \omega t & \text{for} \quad 0 \le t \le t_0 \\ 0 & \text{for} \quad t > t_0\end{cases}$$

For $0 \le t \le t_0$, the system response from Eq. (3-66) is

$$x=\frac{1}{\omega_n m}\int_0^t F_0 \sin \omega\tau \sin \omega_n(t-\tau)\, d\tau$$

which can be integrated to yield

$$x=\frac{F_0}{m(\omega_n^2-\omega^2)}\left(\sin \omega t-\frac{\omega}{\omega_n}\sin \omega_n t\right) \tag{3-70}$$

Equating $\dot{x}(t)=0$ to find x_{max}, we get

$$\cos \omega t_m-\cos \omega_n t_m=0$$

where t_m is the time when $x(t)$ is a maximum or minimum. The roots of this equation is deduced using the identity

$$\cos \omega t_m-\cos \omega_n t_m=-2 \sin \tfrac{1}{2}(\omega+\omega_n)t_m \sin \tfrac{1}{2}(\omega-\omega_n)t_m$$

Thus,

$$t_m=\frac{2n\pi}{\omega_n \pm \omega} \qquad \text{for} \qquad n=\text{integer}$$

Consider $t_m=2n\pi/(\omega_n+\omega)$. Defining $r=\omega/\omega_n$, the terms in Eq. (3-70) can be expressed as

$$\sin \omega_n t_m=\sin\left(\omega_n\frac{2n\pi}{\omega_n+\omega}\right)=\sin\frac{2n\pi}{1+r}$$

$$\begin{aligned}\sin \omega t_m&=\sin\left(2n\pi\frac{\omega}{\omega_n+\omega}\right)=\sin\left(2n\pi\frac{r}{1+r}\right)\\ &=\sin\left(2n\pi\frac{r+1-1}{1+r}\right)=\sin 2n\pi\left(1-\frac{1}{1+r}\right)\\ &=-\sin\frac{2n\pi}{1+r}\end{aligned}$$

Hence $\sin \omega_n t_m=-\sin \omega t_m$. Recalling $\omega_n^2=k/m$, from Eq. (3-70), we get

$$\begin{aligned}x_m&=\frac{F_0}{m(\omega_n^2-\omega^2)}\left(1+\frac{\omega}{\omega_n}\right)\sin\frac{2n\pi\omega}{\omega_n+\omega}\\ \frac{x_m}{F_0/k}&=\frac{1}{1-r}\sin\frac{2n\pi r}{1+r}\end{aligned} \tag{3-71}$$

Similarly, if $t_m = 2n\pi/(\omega_n - \omega)$, we obtain

$$\frac{x_m}{F_0/k} = \frac{1}{1+r} \sin \frac{2n\pi r}{1-r}$$

Comparing the last two equations, it is evident a value of n can be selected to have x_{max} occur at $t_m = 2n\pi/(\omega_n + \omega)$. Equation (3-71) is plotted in Fig. 3-40. The *initial shock spectrum*, defined by Eq. (3-71), gives the response within the duration of the shock pulse for $0 \le t \le t_0$. Note that there is no solution for $\omega_n/\omega < 1$, since the maximum response does not occur during the pulse if the natural frequency is smaller than the pulse frequency.

For $t > t_0$, the system response from Eq. (3-66) is

$$x = \int_0^{t_0} F(\tau) h(t-\tau)\, d\tau$$

The upper limit of the integration is t_0, because $F(t) = 0$ for $t > t_0$. Performing the integration, substituting $\omega t_0 = \pi$, and simplifying, we get

$$x = \frac{F_0}{2\omega_n m}\left(\frac{\sin \omega_n t - \sin \omega_n (t_0 - t)}{\omega_n + \omega} - \frac{\sin \omega_n t - \sin \omega_n (t_0 - t)}{\omega_n - \omega}\right)$$

It is convenient to define $t' = t - t_0$, a new origin for the time axis. Recalling $\omega_n^2 = k/m$, defining $r = \omega/\omega_n$ and $\omega_n t_0 = \pi/r$, the equation can be

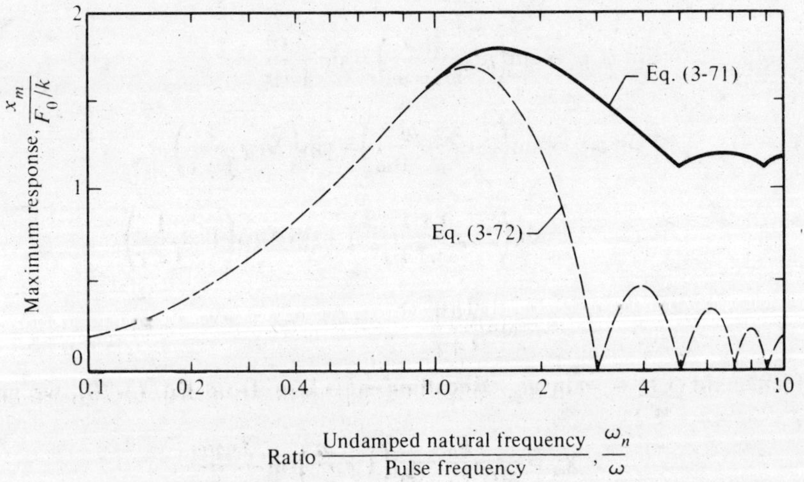

FIG. 3-40. *Shock spectra for half-sine pulse applied to m in Fig. 3-23(a): Eq. (3-71) shows initial spectrum; Eq. (3-72) residual spectrum (Crede).*

simplified to

$$x=\frac{F_0 r}{k(r^2-1)}\left(\sin\frac{\pi}{r}\cos\omega_n t'+\left(1+\cos\frac{\pi}{r}\right)\sin\omega_n t'\right)$$

The maximum value of $x(t)$ can be expressed as

$$\frac{x_{\max}}{F_0/k}=\frac{2r}{r^2-1}\cos\frac{\pi}{2r} \tag{3-72}$$

This gives the *residual shock spectrum*, as it occurs after the shock has terminated. The equation is plotted in Fig. 3-40 and shown as a dash line.*

Now consider the shock spectrum due to a one-half sine pulse $x_1(t)$ applied to the base of the system in Fig. 3-23(b). Assume the system is undamped. The equation of $x_1(t)$ is

$$x_1(t)=\begin{cases}X_1\sin\omega t & \text{for}\quad 0\le t\le t_0\\ 0 & \text{for}\quad t>t_0\end{cases}$$

For $0\le t\le t_0$, the equation for the absolute motion $x_2(t)$ is

$$m\ddot{x}_2+kx_2=kx_1(t)$$

Since the excitation is $kX_1\sin\omega t$, the response $x_2(t)$ can be obtained from Eq. (3-70) by substituting kX_1 for F_0, that is,

$$x_2=\frac{\omega_n^2X_1}{\omega_n^2-\omega^2}\left(\sin\omega t-\frac{\omega}{\omega_n}\sin\omega_n t\right) \tag{3-73}$$

Hence the initial shock spectrum is deduced directly from Eq. (3-71):

$$\frac{x_{2\max}}{X_1}=\frac{1}{1-r}\sin\frac{2n\pi r}{1+r}$$

The relative motion $x(t)$ between m and its base is given by the relation

$$x=x_2-x_1$$

where $x_2(t)$ is as defined in Eq. (3-73). The expression for $x(t)$ can be maximized to give the corresponding spectrum.

The system is unforced for $t>t_0$. The residual shock spectrum can be calculated as before. The shock spectra for the system shown in Fig. 3-23(b) excited by a half sine pulse is illustrated in Fig. 3-41.†

* C. E. Crede, *op. cit.*, p. 85.

† L. S. Jacobsen and R. S. Ayre, *Engineering Vibrations*, McGraw-Hill Book Co., New York, 1958, p. 163.

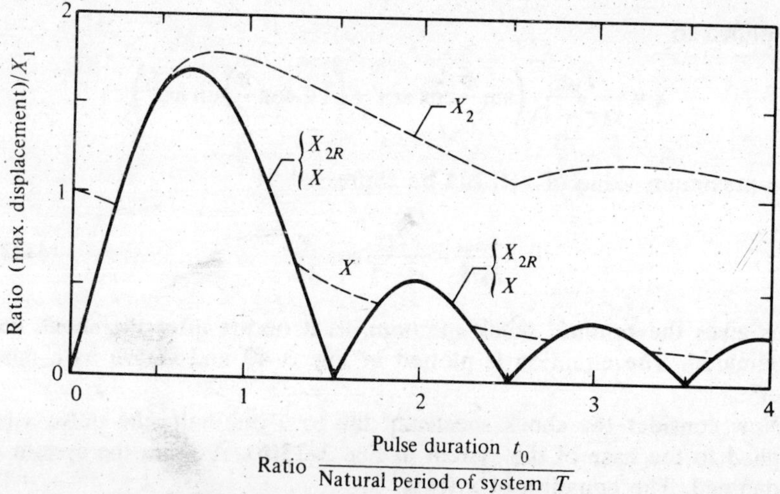

FIG. 3-41. *Shock spectra for half-sine pulse applied to the base in Fig. 3-23(b); X is relative displacement; X_2 the initial spectrum; X_{2R} the residual spectrum (Jacobsen & Ayre).*

3-8 EQUIVALENT VISCOUS DAMPING

Damping is a complex phenomenon.* It exists whenever there is energy dissipation. Viscous damping was assumed in the previous sections. This occurs only when the velocity between lubricated surfaces is sufficiently low to ensure laminar flow condition. More than one type of damping may exist in a problem. Since damping is generally nonlinear, the superposition of different types of damping in a calculation does not always give reliable results. Furthermore, damping may be dependent on the operating conditions and the past history of the damping mechanism or even the shape of the damper, such as in a visco-elastic material. In order to have a simple mathematical model, we shall examine the viscous equivalent of different types of damping.

Most mechanical systems are inherently lightly damped. The effect of damping may be insignificant for some problems, which may be treated as undamped except near resonance. Damping must be considered, however, in order to control (1) the near resonance conditions of a dynamic system, and (2) the performance of a machine, such as an accelerometer or the riding quality of an automobile. The control can be achieved by using (1) an energy transfer mechanism, such as a dynamic absorber

*The interest and knowledge in damping has increased exponentially in recent years. Between 1945 and 1965, two thousand papers were published in this area.

B. J. Lazan, *Damping of Materials and Members in Structural Mechanics*, Pergamon Press Ltd., 4 & 5 Fitzroy Square, London W1, 1968, p. 36.

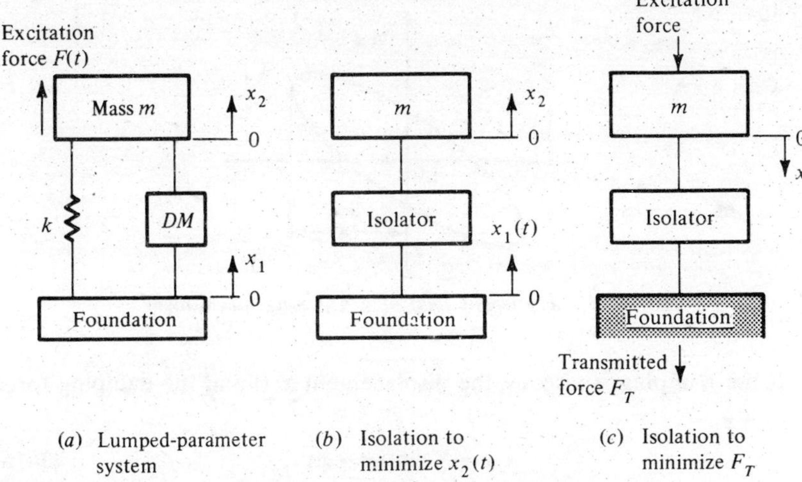

FIG. 3-42. *Model of a one-degree-of-freedom system.*

discussed in Examples 14 and 15, or (2) an energy dissipating mechanism, which is the present topic of discussion.

Consider the one-degree-of-freedom system in Fig. 3-42(*a*). Damping occurs in the *damping mechanism* DM, which can be viscous or otherwise. The elements are shown separate in order to form a model for the study. In reality, they may not be separable, such as in an isolator of visco-elastic material. An *isolator* is shown in Figs. 3-42(*b*) and (*c*) to minimize the transmission of forces between the machine m and its foundation.*

Let us find the equivalent damping from energy considerations. Under cyclic strain, the energy dissipation in a damper is measured by the area of the *hysteresis loop* shown in Fig. 3-43. Without energy dissipation, the cyclic stress strain curve is a line with zero enclosed area, that is, the loop degenerates into a single valued curve. From Fig. 3-42(*a*), if $x_1 = 0$, $x = x_2 - x_1$, and $F(t)$ is sinusoidal, the equation of motion is

$$m\ddot{x} + F_{DM} + kx = F \sin \omega t \tag{3-74}$$

where F_{DM} is the damping force. The energy dissipation per cycle ΔE is

$$\Delta E = \oint F_{DM}\, dx = \int_0^{\tau} F_{DM}\dot{x}\, dt \tag{3-75}$$

where τ is the period in time per cycle.

* See, for example, J. E. Ruzicka and T. F. Derkley, *Influence of Damping in Vibration Isolation*, SVM-7, The Shock and Vibration Information Center, U.S. Department of Defense, 1971.

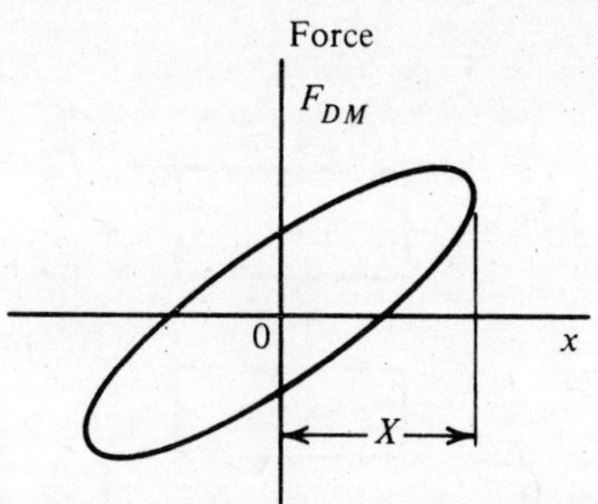

FIG. 3-43. *Hysteresis loop of a damping mechanism.*

If the damping is viscous, the displacement $x(t)$ and the damping force F_{DM} are

$$x = X\sin(\omega t - \phi) \tag{3-76}$$

$$F_{\text{DM}} = c\dot{x} = \omega c X\cos(\omega t - \phi) \tag{3-77}$$

where c is the viscous damping coefficient. Combining Eqs. (3-75) to (3-77) gives

$$\Delta E = \int_0^\tau (c\dot{x})\dot{x}\,dt = \pi\omega c X^2 \tag{3-78}$$

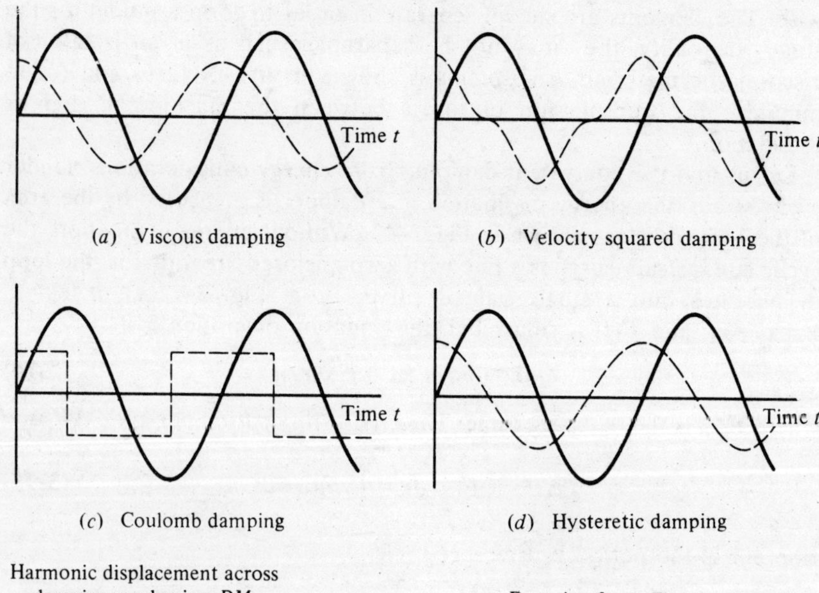

FIG. 3-44. *Wave form of damping force and relative displacement of damping mechanism.*

Combining Eqs. (3-76) and (3-77) yields

$$\left(\frac{x}{X}\right)^2 + \left(\frac{F_{\text{DM}}}{\omega cX}\right)^2 = 1 \tag{3-79}$$

Hence the hysteresis loop for viscous damping is an ellipse having the major and minor semiaxis of ωcX and X, respectively.

If the damping is nonviscous, the *equivalent viscous damping coefficient* c_{eq} is obtained from Eq. (3-78).

$$\Delta E = \pi\omega c_{\text{eq}} X^2 \tag{3-80}$$

Note that the criteria for equivalence are (1) equal energy dissipation per cycle of vibration, and (2) the same harmonic relative displacement. The assumption of harmonic motion is reasonable only for small nonviscous damping. The wave forms of damping forces for harmonic displacements across typical dampers are illustrated in Fig. 3-44.

Example 24. Coulomb Damping

A mass-spring system with Coulomb (dry friction) damping is shown in Fig. 3-45. Assume (1) the frictional force F_{DM} is proportional to the normal force F_{N}, and (2) the initial conditions are $x(0) = -x_0$ and $\dot{x}(0) = 0$. Find the motion $x(t)$ and the change in amplitude per cycle.

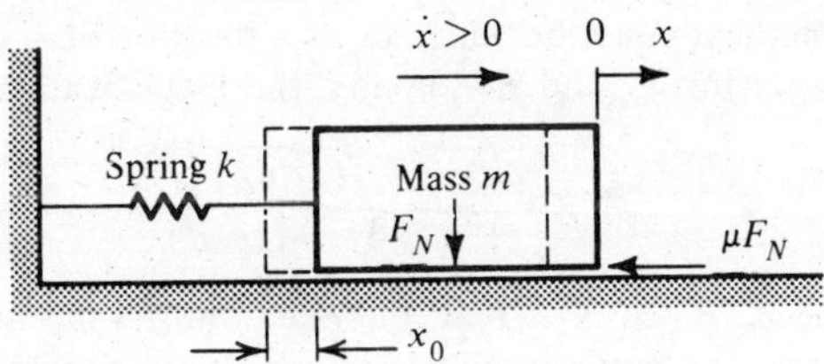

FIG. 3-45. *Coulomb damping.*

Solution:

Let $|F_{\text{DM}}| = \mu F_{\text{N}}$=constant, where μ is a frictional coefficient. If the motion is from left to right ($\dot{x} > 0$), the frictional force is $-\mu F_{\text{N}}$ and vice versa. From Eq. (3-74) the equation of motion is

$$m\ddot{x} + \mu F_N \operatorname{sgn}(\dot{x}) + kx = 0$$

where $\operatorname{sgn}(\dot{x}) = \dot{x}/|\dot{x}|$ accounts for the sign change. The corresponding solution is

$$x = A_1 \cos \omega_n t + A_2 \sin \omega_n t - \mu F_N \operatorname{sgn}(\dot{x})/k$$

where $\omega_n^2 = k/m$ and A_1 and A_2 are constants. Substituting the initial conditions, we have

$$x = [\mu F_N \operatorname{sgn}(\dot{x})/k - x_0] \cos \omega_n t - \mu F_N \operatorname{sgn}(\dot{x})/k$$

For the first half cycle ($\omega_n t = \pi$), the displacement is

$$x = x_0 - 2\mu F_N \operatorname{sgn}(\dot{x})/k$$

Hence the decrease in amplitude is $2\mu F_N/k$ per half cycle. The amplitude change per cycle for free vibration with Coulomb friction is $4\mu F_N/k$.

Example 25. Equivalent Damping Coefficient

If a force $F \sin \omega t$ is applied to the mass m in Fig. 3-45, find the equivalent viscous-damping coefficient c_{eq} and the magnification factor R.

Solution:

Since the damping force $F_{DM} = \mu F_N$ is constant and the total displacement per cycle is $4X$, the energy dissipation ΔE per cycle is

$$\Delta E = 4\mu F_N X$$

Comparing this with Eq. (3-80), we have

$$c_{eq} = 4\mu F_N/(\pi\omega X)$$

From Eq. (2-39), the amplitude X of the steady-state response is

$$X = \frac{F}{\sqrt{(k-\omega^2 m)^2 + (\omega c_{eq})^2}} \qquad \textbf{(3-81)}$$

This is an implicit equation, since c_{eq} is a function of ω and X. Substituting the expression for c_{eq} and simplifying, the magnification factor R is

$$R = \frac{X}{F/k} = \sqrt{\frac{1-(4\mu F_N/\pi F)^2}{(1-r^2)^2}}$$

where $r = \omega/\omega_n$. Since X is real, the equation is valid only if $4\mu F_N/\pi F < 1$. Note that the amplitude at resonance is always theoretically infinite. The resonance amplitude can also be viewed from energy considerations. The energy input per cycle* is πFX and the dissipation is $4\mu F_N X$. If $4\mu F_N X/\pi FX < 1$, the excess energy is used to build up the amplitude of oscillation.

Example 26. Quadratic Damping

Quadratic or velocity squared damping is encountered in the turbulent flow of a fluid. Determine the equivalent viscous-damping coefficient and the amplitude of the steady-state response. Assume $x = X \sin \omega t$.

* Let the force $= F \sin \omega t$ and the harmonic response be $x = X \sin(\omega t - \phi)$. At resonance $\phi = 90°$. The energy input per cycle is

$$\oint (\text{force})\, dx = \int_0^{\tau} (F \sin \omega t)\dot{x}\, dt$$

where τ = period. This is integrated to give πFX.

Solution:

Assume $|F_{DM}| = c_2\dot{x}^2$ where c_2 is constant and $\dot{x}$ is the relative velocity across the damper. Since damping always opposes the motion, the equation of motion from Eq. (3-74) is

$$m\ddot{x} + c_2\dot{x}^2 \operatorname{sgn}(\dot{x}) + kx = F \sin \omega t$$

where $\operatorname{sgn}(\dot{x}) = \dot{x}/|\dot{x}|$ accounts for the sign change.

The energy dissipation per cycle from Eq. (3-75) is

$$\Delta E = 2\int_{-X}^{X} c_2\dot{x}^2\, dx = 2X^3 \int_{-\pi/2}^{\pi/2} c_2\omega^2 \cos^3 \omega t\, d(\omega t)$$
$$= \tfrac{8}{3}\omega^2 c_2 X^3$$

Substituting the ΔE in Eq. (3-80) gives

$$c_{eq} = \frac{8}{3\pi}\, \omega c_2 X$$

Again, c_{eq} is not a constant as assumed for viscous damping.

The amplitude X for the steady-state response is obtained by substituting c_{eq} into Eq. (3-81). Simplifying the resultant equation yields

$$X = \frac{3\pi m}{8c_2r^2}\sqrt{-\frac{(1-r^2)^2}{2} + \sqrt{\frac{(1-r^2)^4}{4} + \left(\frac{8c_2r^2F}{3\pi km}\right)^2}}$$

Example 27. Velocity–nth Power Damping

Determine the c_{eq} for the velocity–nth power or exponential damping.

Solution:

The energy dissipation ΔE per cycle is

$$\Delta E = 2\int_{-X}^{X} (c_n\dot{x}^n)\, dx = 2\omega^n c_n X^{n+1} \int_{-\pi/2}^{\pi/2} \cos^{n+1} \omega t\, d(\omega t)$$

Substituting the ΔE into Eq. (3-80), we obtain

$$c_{eq} = \omega^{n-1} c_n X^{n-1} \Phi_n$$

where

$$\Phi_n = \frac{2}{\pi}\int_{-\pi/2}^{\pi/2} \cos^{n+1} \omega t\, d(\omega t)$$

Example 28. Hysteretic Damping

Assume the damping force in a one-degree-of-freedom system is in phase with the relative velocity but is proportional to the relative displacement across the damper. Find the harmonic response of this system and its equivalent viscous-damping coefficient.

Solution:

Writing the equation of motion in the vectorial form as shown in Sec. 3-5, we have

$$m\ddot{x} + jhx + kx = Fe^{j\omega t}$$
$$m\ddot{x} + k(1 + j\eta)x = Fe^{j\omega t} \tag{3-82}$$

where (jhx) is the damping force, h a constant, $\eta = h/k$ is called the *loss coefficient*, and $k(1 + j\eta)$ the *complex stiffness*. Solving Eq. (3-82) by the impedance method, the amplitude X of the harmonic response is

$$X = \frac{F}{|k - \omega^2 m + j\eta k|} = \frac{F/k}{\sqrt{(1 - r^2)^2 + \eta^2}} \tag{3-83}$$

where $r = \omega/\omega_n$. Comparing Eqs. (3-81) and (3-83), the equivalent viscous-damping coefficient c_{eq} is

$$c_{eq} = h/\omega \tag{3-84}$$

Solid damping, hysteretic damping, and *structural damping* are terms commonly used to describe the internal damping of material. It is assumed that the energy dissipation per cycle is independent of frequency and is proportional to the square of the strain amplitude. Substituting Eq. (3-84) into (3-80), we obtain $\Delta E = h\pi X^2$, which confirms the assumption. The nature of structural damping is rather complex. For mild steel, the energy dissipation is proportional to $X^{2.3}$. For other cases, the value of the amplitude exponent may range from 2 to 3. The damping of a material may decrease slightly with increasing frequency instead of being constant. In contrast, the common viscous-damping theory assumes that the loss coefficient increases linearly with frequency.

Owing to their high damping characteristics, visco-elastic materials have gained importance in vibration control in recent years. The physical properties of such materials are more complex than those of metals. The properties are influenced by the operating conditions, the past history, and the geometry of the damping mechanism. The variation of properties with temperature and frequency of a typical visco-elastic material is shown in Fig. 3-46.* The complex modulus $E(1 + j\eta)$ is defined by the relation

$$(\text{Stress}) = E(1 + j\eta)(\text{strain})$$

where η is the loss coefficient. Note that high damping can be achieved in the transitional region. Furthermore, if the spring constant of a damping mechanism increases with preloading, it is feasible to tune a dynamic absorber by adjusting its preload.

* D. J. Jones, *Material Damping*, ASA Damping Conference, Cleveland, Ohio, Nov. 21, 1968.

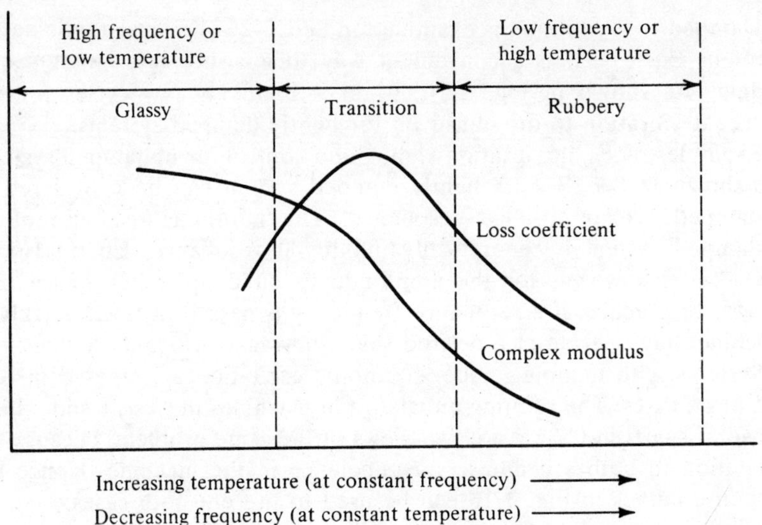

FIG. 3-46. *Temperature and frequency effect of a "typical" polymer.*

The frequency dependence of the complex modulus is often considered as a disadvantage in using visco-elastic material for isolation mounting. However, for a tuned dynamic absorber with damping (see Sec. 4-8), the frequency dependency also moves the "resonant frequency" of the absorber progressively higher or lower with the exciting frequency. Thus, the absorber is effective for a greater frequency range of operation.*

3-9 SUMMARY

The theory of one-degree-of-freedom systems is applied to a variety of problems, the model of which is as shown in Fig. 2-6. The generalized equation of motion is

$$m_{eq}\ddot{x} + c_{eq}\dot{x} + k_{eq}x = F_{eq}(t) \tag{3-1}$$

The examples in the chapter are grouped to illustrate the equivalent quantities m_{eq}, c_{eq}, k_{eq}, and $F_{eq}(t)$. The emphasis is on problem formulation and interpretation, since the general theory was developed in the last chapter.

The equivalent mass m_{eq} and equivalent spring k_{eq} are illustrated in Examples 1 to 8.

* J. C. Snowdon, *Vibration and Shock in Damped Mechanical Systems*, John Wiley & Sons, Inc., New York, 1968, p. 96.

Damped free vibration is examined in Sec. 3-3. The logarithmic decrement in Eq. (3-12) is a convenient way to measure the damping in a system. As shown in Fig. 3-10, it takes relatively few cycles for the transient vibration to die out, even for lightly damped systems.

Examples of F_{eq} in vibration testing and control for undamped systems are shown in Sec. 3-4. A lightly damped system can be considered as undamped, except at near resonance. The harmonic response of an undamped system is theoretically infinite at resonance, Eq. (3-16). It takes time, however, for the amplitude to build up at resonance, Eq. (3-18). In practice, if a resonance frequency is passed through quickly, a machine may operate at a desired speed between resonant frequencies.

Systems with damping under harmonic excitation are treated in Sec. 3-5 in six cases. The rotating unbalance in machines in Case 1 and critical speed of shafts in Case 2 are two views of the same problem, because the excitation in both are due to an unbalance in the machine. Hence the response curves in Fig. 3-16 can be used to present both cases.

Vibration isolation and transmissibility in Case 3 and the response to moving support in Case 4 are again two aspects of the same problem as evident by comparing Eqs. (3-32) and (3-37). The response of both cases are shown in Fig. 3-25.

The seismic instrument in Case 5 uses the relative displacement $x(t)$ between a mass and its base to measure the motion $x_1(t)$ of the base. The equation of motion in Eq. (3-39) is of the same form as those for Cases 1 and 2. Hence the response can also be represented in Fig. 3-16, although they are distinct types of problems. Two types of instruments can be constructed from this general theory, depending on the excitation frequency ω and the natural frequency ω_n of the system. If $\omega \gg \omega_n$, we have a vibrometer for displacement measurement and $x(t) = -x_1(t)$, regardless of the damping as shown in Fig. 3-16. If $\omega \ll \omega_n$, we have an accelerometer and $x(t)$ is proportional to $\ddot{x}_1(t)$. Appropriate damping is necessary in an accelerometer in order to minimize the amplitude and phase distortions.

A periodic excitation can be expressed as a Fourier series. The system response due to each harmonic component of the periodic excitation can be calculated as illustrated in Fig. 3-37. The total response is obtained by superposition as shown in Eq. (3-65). Hence this is a generalization of the harmonic excitation in Sec. 3-5.

Shock is a transient phenomenon. The shock spectrum in Sec. 3-7 describes the effect of shock on the assumption that shock excitations having the same spectrum would produce similar effect on the system.

Damping is seldom purely viscous in a real system. For nonviscous damping, an equivalent viscous damping coefficient c_{eq} is defined in Eq. (3-80). The criteria for equivalence are (1) equal energy dissipation per cycle of vibration, and (2) harmonic oscillations of the same amplitude.

PROBLEMS

Assume all the systems in the figures to follow are shown in their static equilibrium positions.

3-1 Find the natural frequency of the system in Fig. P3-1(a). Assume that (1) the cantilevers are of negligible mass and (2) their equivalent spring constants are k_1 and k_3.

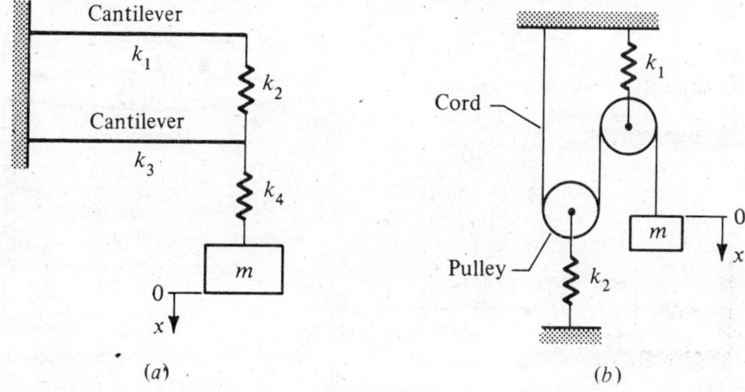

FIG. P3-1.

3-2 Neglecting the mass of the pulleys, find the natural frequency of the system in Fig. P3-1(b).

3-3 A mass m is attached to a rigid bar of negligible mass as shown in Fig. P3-2(a). Find the natural frequency of the system, if (**a**) the bar is constrained to remain horizontal while m oscillates vertically; (**b**) the bar is free to pivot at the hinges A and B. (**c**) Show that the natural frequency determined in part **a** is higher than that of part **b**.

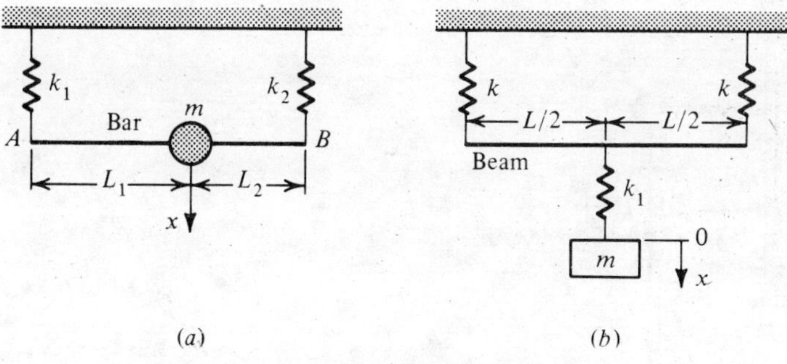

FIG. P3-2.

3-4 A mass m is suspended as shown in Fig. P3-2(b). If the beam is of negligible mass and its deflection δ is given by the equation $\delta = PL^3/48EI$, find the natural frequency of the system.

3-5 Referring to Fig. 2-1(a), let $k = 7$ kN/m and $m = 18$ kg. (**a**) Find the natural frequency of the system. (**b**) If the ends of the spring are fixed and the mass

m is attached to the midpoint of the spring, find the natural frequency. (**c**) If the ends of the spring are fixed and m is attached to some intermediate point of the spring, show that the natural frequency of this configuration is higher than that of part **b**.

3-6 A rocker arm assembly is shown in Fig. P3-3(a). Let m_A = mass of rocker arm and J_A = the mass moment of inertia about the pivot A. Find the equivalent mass m_{eq} and the equivalent spring k_{eq} of the system referring to the x coordinate.

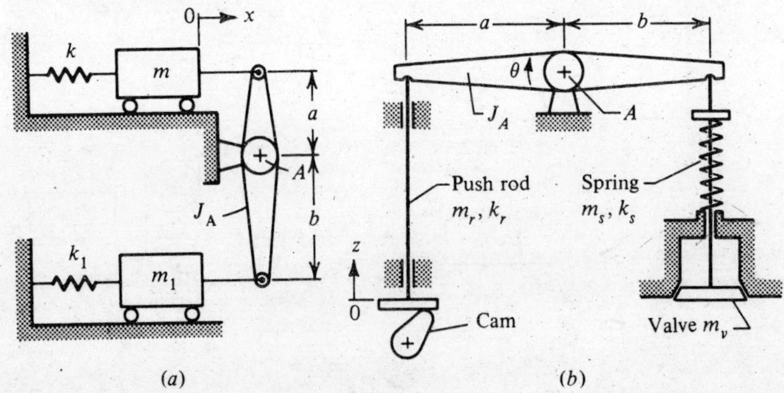

FIG. P3-3.

3-7 An engine valve arrangement is shown in Fig. P3-3(b), where J_A is the mass moment of inertia of the rocker arm about the pivot A. Assume the effective mass m_r and the effective stiffness k_r of the pushrod are known. Reduce the valve arrangement to an equivalent mass-spring system.

3-8 A mechanism is shown schematically in Fig. P3-4(a). Assuming that the tension of the spring k_3 is constant, derive the equation of motion of the system.

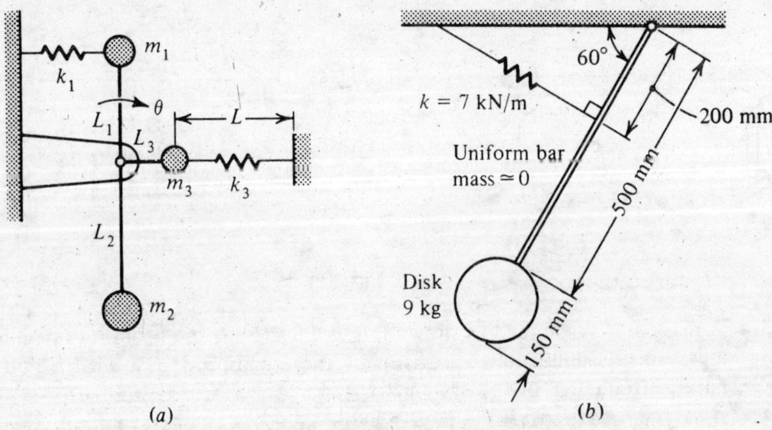

FIG. P3-4.

3-9 A machine component is depicted as a pendulum shown in Fig. P3-4(*b*). Determine its natural frequency by: (**a**) Newton's second law; (**b**) the energy method.

3-10 The mass moment of inertia J_{cg} of a connecting rod of mass m is determined by placing the rod on a horizontal platform of mass m_1 and timing the periods of oscillation. The platform, shown in Fig. P3-5(*a*), is suspended by equally spaced wires. With the platform empty and an amplitude of 6°, the period is τ_1. With the mass center of the rod coinciding with that of the platform and an amplitude of 8°, the period is τ_2. Find J_{cg} of the connecting rod.

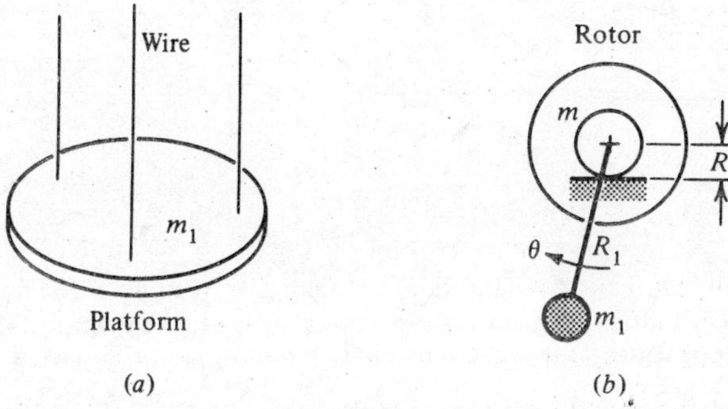

Fig. P3-5.

3-11 The mass moment of inertia J_0 of the rotor of an electrical generator of mass m is found by attaching a small mass m_1 at a distance R_1 from its longitudinal axis and timing the periods of oscillation. The test setup is shown in Fig. P3-5(*b*). (**a**) Find J_0 of the rotor. (**b**) Show that small variation of R will have the least effect when $R_1 = (m/m_1 + 1)R$.

3-12 A machine component is shown in Fig. P3-6(*a*). The mass m is constrained by rails to move only in the x direction. Neglecting the mass of the arm, find the equation of motion.

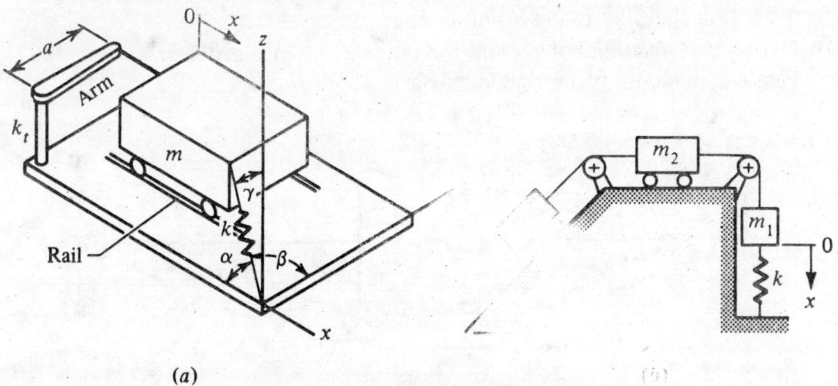

Fig. P3-6.

3-13 A machine part is schematically shown in Fig. P3-6(b). Find the equation of motion.

3-14 Find the equation of motion of the mass m for the system shown in Fig. P3-7(a). Assume that the horizontal bar is rigid and is of negligible mass.

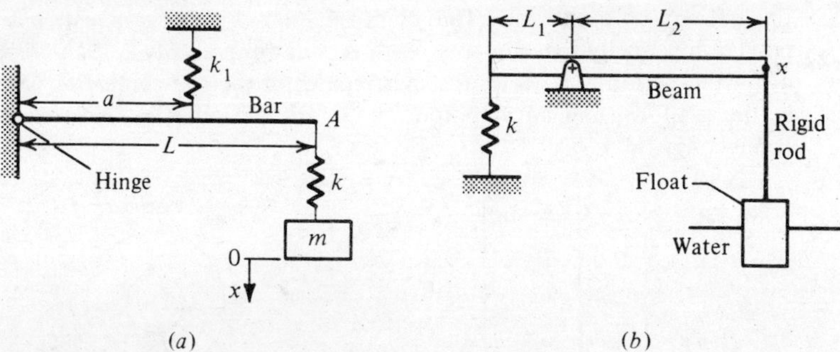

FIG. P3-7.

3-15 A walking-beam configuration, consisting of a uniform beam of mass m_1 and a cylindrical float of cross-sectional area A, is shown in Fig. P3-7(b). If the mass of the float and the rod is m_2, determine the equation of motion.

3-16 For a one-degree-of-freedom system, if $m = 7$ kg, $k = 6$ kN/m, and $c =$ 35 N · s/m, find: (**a**) the damping factor ζ, (**b**) the logarithmic decrement and (**c**) the ratio of any two consecutive amplitudes.

3-17 From the data in Prob. 2-14, find the logarithmic decrement for each of the given sets of initial conditions.

3-18 A device bought from a surplus store is depicted as a one-degree-of-freedom system. It is desired to find: (**a**) its natural frequency, (**b**) the mass moment of inertia of the rotor, and (**c**) the damping required for it to be critically damped. It is not possible, however, to disassemble the device. It is found that (1) when the rotor is turned 30°, a torque of 0.35 N · m is needed to maintain this position, (2) when the rotor is held in this position and released, it swings to −5.5° and then to 1°, and (3) the time of the swing is 1.0 s. Calculate the required information.

3-19 Derive the equations of motion for the systems in Fig. P3-8. Assume the bars are rigid and of negligible mass.

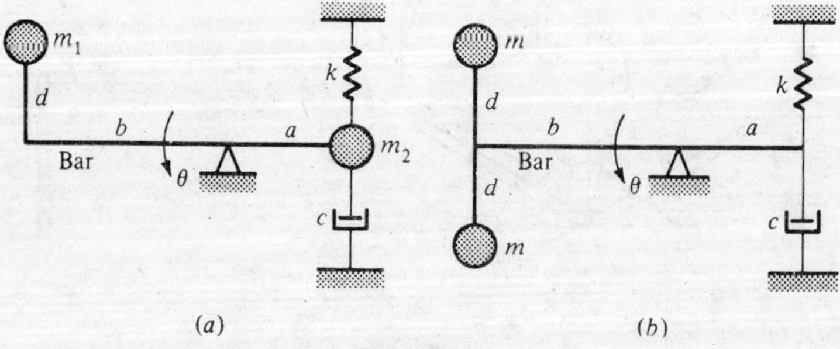

FIG. P3-8.

3-20 Derive the equations of motion for the system shown in Fig. P3-7(a) if an excitation force $F \sin \omega t$ is applied to: (**a**) the mass m (**b**) the free end A of the bar.

3-21 A force $F \sin \omega t$ is applied to the mass m of the system shown in Fig. 2-1(a). If $\omega = (1+\varepsilon)\omega_n$, determine the motion of m. Assume zero initial conditions and $\varepsilon \ll 1$.

3-22 A harmonic motion is applied to each of the systems shown in Fig. P3-9. Derive the equations of motion.

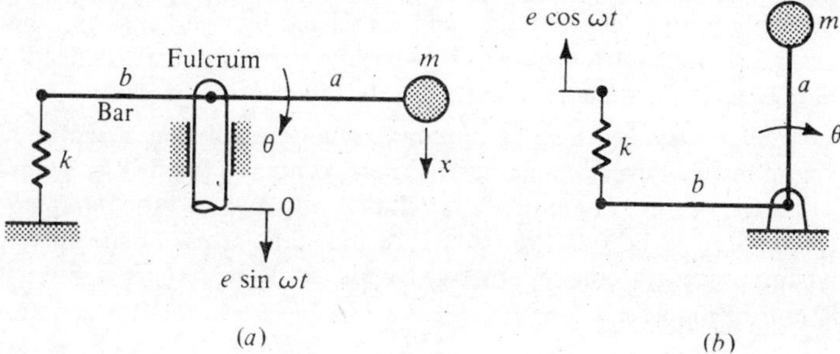

FIG. P3-9.

3-23 Referring to Fig. P3-9(a), the general position of the fulcrum can be above or below the static equilibrium position of the system and the bar can be rotated clockwise or counterclockwise. Show that the equation of motion derived in Prob. 3-22 is true for all positions of the system.

3-24 A wide-flange I beam is cantilevered from the foundation of a building. The beam is 2 m in length with a total mass of 60 kg. The I of the beam section is 30×10^6 mm^4. A construction worker places a small electric motor of 4 kg at the end of the beam. The mass of the armature of the motor is 1.5 kg with an eccentricity of 0.05 mm. If the motor speed is 3,600 rpm, estimate the amplitude of vibration at the end of the beam.

3-25 Show analytically that the maximum value of the curves in Fig. 2-8 occur at $r<1$. Sketch a locus through the maxima of the curves.

3-26 Show analytically that the maximum values of the curves in Fig. 3-16 occur at $r>1$. Sketch a locus through the maxima of the curves.

3-27 A table for sorting seeds requires a reciprocating motion with a stroke of 1.0 mm and frequency from 2 to 20 Hz. The excitation is provided by an eccentric weight shaker. The total mass of the table and shaker is 200 kg. (**a**) Propose a scheme for mounting the table. (**b**) Specify the spring constant, the damping coefficient, and the unbalance of the exciter.

3-28 A machine of 100 kg mass has a 20 kg rotor with 0.5 mm eccentricity. The mounting springs have $k = 85$ kN/m and the damping is negligible. The operating speed is 600 rpm and the unit is constrained to move vertically. (**a**) Determine the dynamic amplitude of the machine. (**b**) Redesign the mounting so that the dynamic amplitude is reduced to one half of the original value, but maintaining the same natural frequency.

3-29 A variable-speed counter-rotating eccentric-weight exciter is attached to a machine to determine its natural frequency. With the exciter at 1,000 rpm, a stroboscope shows that the eccentric weights of the exciter are at the top the instant the machine is moving upward through its static equilibrium position. The amplitude of the displacement is 12 mm. The mass of the machine is 500 kg and that of the exciter is 20 kg with an unbalance of 0.1 kg · m. Find (**a**) the natural frequency of the machine and its mounting and (**b**) the damping factor of the system.

3-30 A rotating machine for research has an annular clearance of 0.8 mm between the rotor and the stator. The mass of the rotor is 36 kg with an unbalance of 3×10^{-3} kg · m. The rotor is mounted symmetrically on a round shaft, 300 mm in length and supported by two bearings. The operating speed ranges from 600 to 6,000 rpm. If the dynamic deflection of the shaft is to be less than 0.1 mm, specify the size of the shaft.

3-31 A circular disk of 18 kg is mounted symmetrically on a shaft, 0.75 m in length and 20 mm in diameter. The mass center of the disk is 3 mm from its geometric center. The unit is rotated at 1,000 rpm and the damping factor ζ is estimated to be 0.05. (**a**) Compare the static stress of the shaft with the dynamic stress at the operating speed. (**b**) Repeat part **a** for a shaft of 30 mm in diameter.

3-32 It is proposed to use a three-cylinder two-stroke-cycle diesel engine to drive an electric generator at 600 rpm. The generator consists of a 2×10^3 kg rotor mounted on a hollow shaft, 2 m in length with a 200-mm OD and a 100-mm bore. A preliminary test shows that, when the rotor is suspended horizontally with its axis 0.95 m from the point of suspension, the period of oscillation is $2\pi/3$ s. If you are the consulting engineer, would you approve this proposal?

3-33 A turbine at 6,000 rpm is mounted as shown in Fig. P3-10(a). The 14 kg rotor has an unbalance of 2.8×10^{-3} kg · m. (**a**) Neglecting the mass of the shaft, find the amplitude of vibration and the force at each bearing for a 20-mm diameter shaft. (**b**) Repeat for a 25-mm shaft. (**c**) Repeat for a 30-mm shaft. (**d**) Estimate the error due to neglecting the mass of the shaft.

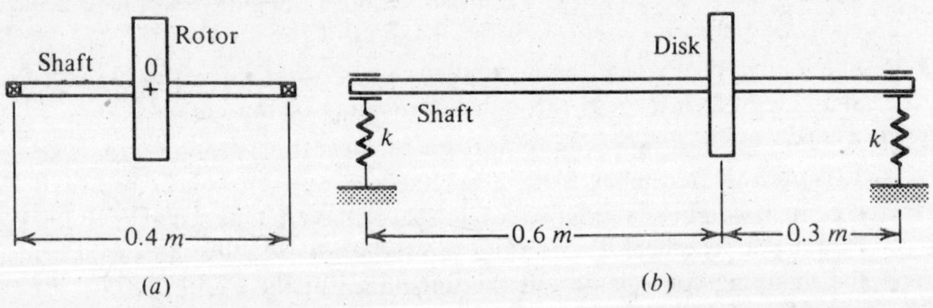

FIG. P3-10.

3-34 A 180-kg steel disk is mounted on a 100-mm OD and 75-mm ID shaft as shown in Fig. P3-10(b). (**a**) Neglecting the flexibility of the bearing supports, find the critical speed of the assembly. (**b**) If the bearings are flexible with a spring constant $k = 70\times10^6$ N/m in any direction normal to the shaft axis, find the change in critical speed. Assume that the mass of the shaft and the gyroscopic effect of the disk are negligible.

3-35 A 140-kg disk is mounted on a 75-mm diameter shaft as shown in Fig. P3-10(*b*). The bearing supports are essentially rigid in the vertical direction but are flexible with a spring constant $k = 50 \times 10^6$ N/m each in the horizontal direction. Find the critical speeds of the assembly.

3-36 Show analytically that the crossover points of the transmissibility curves in Figs. 3-25 and 3-26 occur at $r = \sqrt{2}$.

3-37 A refrigeration unit of 30-kg mass operates at 700 rpm. The unit is supported by three equal springs. (**a**) Specify the springs if 10 percent or less of the unbalance is transmitted to the foundation. (**b**) Verify the calculation, using Fig. 3-27.

3-38 A vertical single-cylinder diesel engine of 500-kg mass is mounted on springs with $k = 200$ kN/m and dampers with $\zeta = 0.2$. The rotating parts are well-balanced. The mass of the equivalent reciprocating parts is 10 kg and the stroke is 200 mm. Find the dynamic amplitude of the vertical motion, the transmissibility, and the force transmitted to the foundation, if the engine is operated at (**a**) 200 rpm; (**b**) at 600 rpm.

3-39 A 50-kg rotor is mounted as shown in Fig. P3-11(*a*). It has an unbalance of 0.06 kg · m and operates at 800 rpm. If the dynamic amplitude of the rotor is to be less than 6 mm and it is desired to have low transmissibility, specify the springs and the dampers for the mounting.

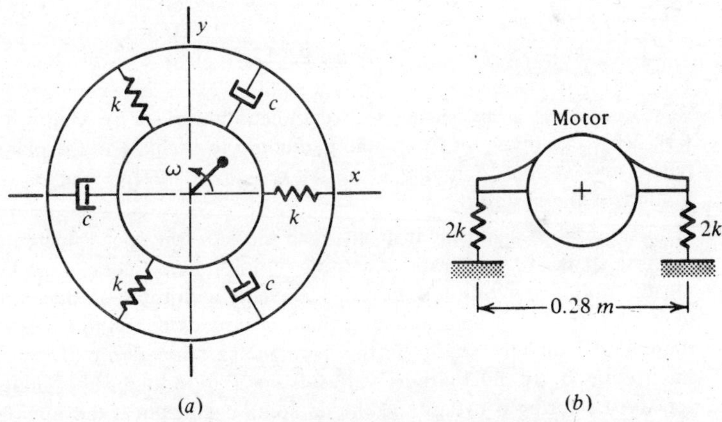

Fig. P3-11.

3-40 A 15-kg electric motor is supported by four equal springs as shown in Fig. P3-11(*b*). The stiffness of each spring is 2.5 kN/m. The radius of gyration of the motor assembly about its shaft axis is 100 mm. The operating speed is 1,800 rpm. Find the transmissibility for the vertical and the torsional vibrations.

3-41 An instrument in an aircraft is to be isolated from the engine vibrations, ranging from 1,800 to 3,600 cycles per minute. If the damping is negligible and the instrument is of 20-kg mass, specify the springs for the mounting for 80 percent isolation.

3-42 A 250-kg table for repairing instruments is isolated from the floor by springs with $k = 20$ kN/m and dampers with $c = 4$ kN · s/m. If the floor vibrates vertically ± 2.5 mm at a frequency of 10 Hz, find the motion of the table.

3-43 Referring to the vehicle suspension problem in Example 19, if $X_1 = 0.1$ m, find the amplitude X_2 when the speed of the trailer is: (**a**) 70 km/hr; (**b**) 120 km/hr.

3-44 A body m, mounted as shown in Fig. P3-12(a), is dropped on a floor. Assume that, when the base first contacts the floor, the spring is unstressed and the body has dropped through a height of 1.5 m. Find the acceleration $\ddot{x}(t)$ of m. If $m = 18$ kg, $c = 72$ N · s/m, and $k = 1.8$ kN/m, determine the maximum acceleration of m.

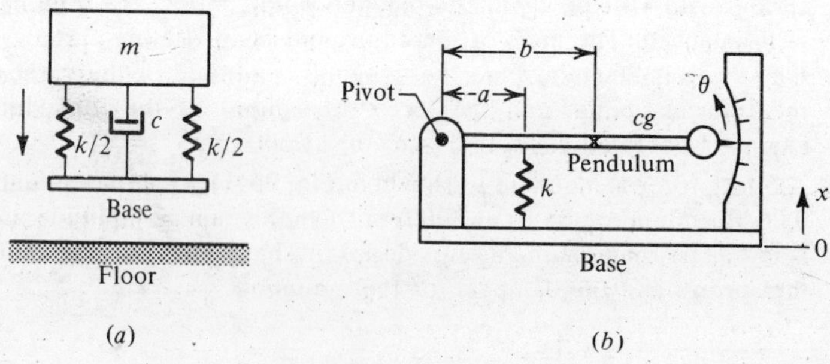

FIG. P3-12.

3-45 A vibrometer for measuring the rectilinear motion $x(t)$ is shown in Fig. P3-12(b). The pivot constrains the pendulum to oscillate in the plane of the paper and viscous damping exists at the pivot. Derive the equation of motion of the system.

3-46 A torsiograph is a seismic instrument to measure the speed fluctuation of a rotating shaft. A torsiograph consisting of a hollow cylinder of 0.5 kg with a 40-mm radius of gyration is mounted coaxially with the shaft and connecting to it by a spiral spring. Assuming that (1) viscous damping exists between the cylinder and the shaft, (2) the average shaft speed is 600 rpm, and (3) the frequency of fluctuations varies from 4 to 8 times the shaft speed, specify the spring constant and the damping coefficient if the torsiograph is to measure relative displacement.

3-47 A vibrometer to measure the vibrations of a variable speed engine is schematically shown in Fig. 3-29. The vibrations consist of a fundamental and a second harmonic. The operating speed ranges from 500 to 1500 rpm. It is desired to have the amplitude distortion less than 4 percent. Determine the natural frequency of the vibrometer if: (**a**) the damping is negligible; (**b**) the damping factor $\zeta = 0.6$.

3-48 An accelerometer with $\zeta = 0.6$ is used to measure the vibrations described in Prob. 4-47. The amplitude distortion is to be less than 4 percent. (**a**) From Fig. 3-32, find the natural frequency of the accelerometer. (**b**) If the engine operates at 1,000 rpm, find the amplitude distortion of the second harmonic. (**c**) Find the phase distortion from Fig. 3-34 and calculate the phase shift in unit of time.

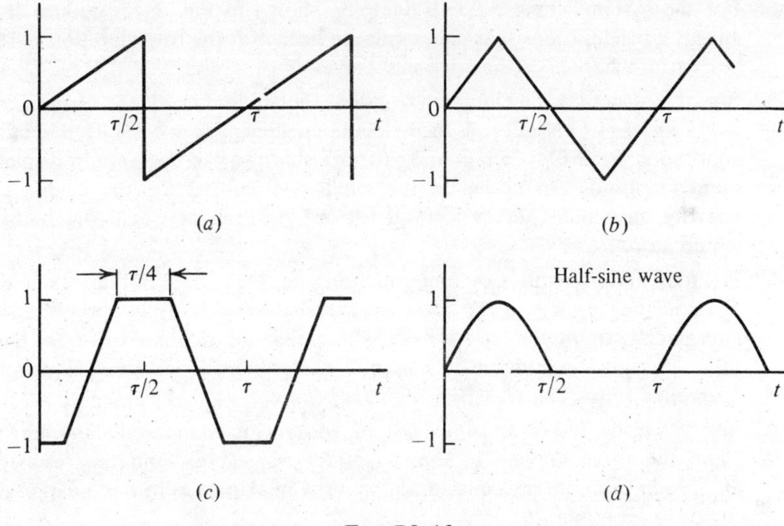

FIG. P3-13.

3-49 Find the Fourier series expansions of each of the periodic functions shown in Fig. P3-13.

3-50 From the answers in Prob. 3-49, find the Fourier spectrum of the periodic functions in Figs. P3-13(a) and (b).

3-51 Find the Fourier series expansions of the periodic functions in Fig. P3-13 if each of the functions is delayed by an amount $\tau/4$.

3-52 From the answers in Prob. 3-51, find the Fourier spectrum of the periodic functions of parts **a** and **b**.

3-53 The excitation of a system has two harmonic components.

$$3\ddot{x}+18\dot{x}+675x=6\sin 10t+3\sin(20t+3\pi)$$

(**a**) Sketch the wave form of the excitation. (**b**) Use the impedance method in Eq. (3-62) and (3-63) to find the steady-state response due to each of the harmonic components. (**c**) Sketch the wave form of the composite steady-state response.

3-54 If the system in Fig. 3-38 is actuated by a cam with the profile as shown in Fig. P3-13(c), find the steady-state response of the system. Assume that $m=170$ kg, $k_1=k=7$ kN/m, $c=1.7$ kN · s/m, total cam lift $=50$ mm, and the cam speed $=60$ rpm.

3-55 Derive the transmissibility equation for each of the systems shown in Fig. P2-5. (See Prob. 2-24, Chap. 2).

3-56 A periodic force, with the waveform as shown in Fig. P3-13(b), is applied to a mass-spring system. Will there be a resonance if the fundamental frequency of the excitation is one-half of the natural frequency of the system?

3-57 Find the relative motion $x=(x_2-x_1)$ of the mass m in Fig. 3-23, if the base is given an excitation: (**a**) $\ddot{x}_1=g$; (**b**) $\dot{x}_1=Ce^{-\alpha t}$, where C and α are constants. Assume that the damping is negligible and the system is initially at rest.

3-58 For the system with Coulomb damping shown in Fig. 3-45, deduce from energy considerations that the amplitude decay for the free vibration is $4F/k$ per cycle, where F is the frictional force.

3-59 For the system with Coulomb friction shown in Fig. 3-45, assume that $m = 9$ kg, $k = 7$ kN/m, and the friction coefficient $\mu = 0.15$. If the initial conditions are $x(0) = 25$ mm and $\dot{x}(0) = 0$, find: (**a**) the decrease in displacement amplitude per cycle, (**b**) the maximum velocity, (**c**) the decrease in velocity amplitude per cycle, and (**d**) the position at which the body m would stop.

3-60 For the system with Coulomb damping in Fig. 3-45, let an excitation $F_0 \sin \omega t$ be applied to the mass m. (**a**) Use Eq. (3-75) to show that the energy dissipation per cycle is $4FX$, where F is the frictional force. (**b**) Show that the transmissibility TR is infinite at resonance. (**c**) Find TR for the frequency ratios $r \gtrsim \sqrt{2}$, where $r = \omega/\omega_n$.

3-61 For a one-degree-of-freedom system with velocity-squared damping, assume the force applied to the mass is $F_0 \sin \omega t$. (**a**) Find the resonance amplitude from energy considerations. (**b**) Check part **a** from the expression for X in Example 26.

3-62 A machine of 350-kg mass and 1.8-kg · m eccentricity is mounted on springs and a damper with velocity squared damping. The damper consists of a 70-mm diameter cylinder-piston arrangement. The piston has a nozzle for the passage of the damping fluid, the density of which is $\rho = 960\ \text{kg/m}^3$. The natural frequency of the system is 5 Hz. Assuming that the equivalent viscous-damping factor $\zeta_{eq} = 0.2$ at resonance, determine: (**a**) the resonance amplitude; (**b**) the diameter of the nozzle if the pressure drop across the nozzle is $p = (\rho/2)(\text{velocity})^2$, where (velocity) is that at the throat of the nozzle.

Computer problems:

3-63 Use the program TRESP1 listed in Fig. 9-1(a) to find the free vibration of the system $\ddot{x} + 2\zeta\omega_n\dot{x} + \omega_n^2 x = 0$. Choose appropriate values for ζ, ω_n, $x(0)$, and $\dot{x}(0)$. Select about two cycles for the duration of the run.

3-64 (**a**) Repeat Prob. 3-63 by modifying the program for plotting, as illustrated in Fig. 9-4(a). (**b**) Plot the results using the program PLOTFILE listed in Fig. 9-5(a).

3-65 Consider the equation of motion and the harmonic response

$$m\ddot{x} + c\dot{x} + kx = F_{eq} \sin \omega t$$

$$x = X \sin(\omega t - \phi)$$

The response can be expressed as shown in Eqs. (3-25) and (3-26). (**a**) Modify the program FRESP1, listed in Fig. 9-6(a), to write the values of the amplitude ratio $X/(F_{eq}/k)$ and the phase angle in separate files for plotting. Let the damping factor $\zeta = 0.2$ and $0.3 \le r \le 4.5$. (**b**) Plot the results using the program PLOTFILE listed in Fig. 9-5(a).

3-66 Repeat Prob. 3-65 but plot for a set of values of ζ, as illustrated in Figs. 2-10 and 2-11. Let $\zeta = 0.1$, 0.2, 0.3, 0.4, 0.55, 0.7, and 1.0.

3-67 (**a**) Classify and tabulate the harmonic response equations for Cases 1 to 5 in Sec. 3-5. (**b**) Write a program to plot the amplitude-ratio versus frequency-ratio r for each type of response for a range of damping factor ζ. In other words, the object is to plot the response curves illustrated in Figs. 3-16, 3-25, and 3-26. *Hint:* Modify the program FRESP1, as listed in Fig. 9-6(a), and use PLOTFILE, listed in Fig. 9-5(a).

3-68 Repeat Prob. 2-32, 2-33, 2-34, or 2-35, but (1) modify the program for plotting, as illustrated in Fig. 9-4(a) and (2) write the program such that only every nth data point is plotted. Use $n = 2$ for this program.

4

Systems with More Than One Degree of Freedom

4-1 INTRODUCTION

The degree of freedom of discrete systems was defined in Chap. 2. Since there is no basic difference in concept between systems with two or more degrees of freedom, we shall introduce multi-degree-of-freedom systems from the generalization of systems with two degrees of freedom. Computers, however, are mandatory for the numerical solution of problems with more than two degrees of freedom.

An n-degree-of-freedom system is described by a set of n simultaneous ordinary differential equations of the second-order. The system has as many natural frequencies as the degrees of freedom. A mode of vibration is associated with each natural frequency. Since the equations of motion are coupled, the motion of the masses are the combination of the motions of the individual modes. If the equations are uncoupled by the proper choice of coordinates, each mode can be examined as an independent one-degree-of-freedom system.

To implement the topics enumerated above, computer sub-routines are listed in App. C, such as for (1) finding the characteristic equation from the equations of motion, (2) solving the characteristic equation for the natural frequencies, (3) obtaining the modal matrix in order to uncouple the equations of motion, and (4) solving the corresponding one-degree-of-freedom system. Computer solutions are given as home problems. For purpose of organization, typical computer programs are grouped and illustrated in Chap. 9.

We shall begin by formulating the equations of motion from Newton's second law, and then discuss natural frequencies, coordinate coupling and transformation, modal analysis, and applications. The method of influence coefficients is presented in the latter part of the chapter. Orthogonality of the modes and additional concepts for a better understanding of the material will be discussed in Chap. 6.

4-2 EQUATIONS OF MOTION: NEWTON'S SECOND LAW

The equations of motion for the two-degree-of-freedom system in Fig. 4-1(a) can be derived by applying Newton's second law to *each* of the masses. Assume the damping is viscous and the displacements $x_1(t)$ and $x_2(t)$ measured from the static equilibrium positions of the masses. Summing the *dynamic forces* in the vertical direction on each mass shown in the free-body sketches, we get

$$m_1\ddot{x}_1 = -k_1x_1 - k(x_1 - x_2) - c_1\dot{x}_1 - c(\dot{x}_1 - \dot{x}_2) + F_1(t)$$
$$m_2\ddot{x}_2 = -k_2x_2 - k(x_2 - x_1) - c_2\dot{x}_2 - c(\dot{x}_2 - \dot{x}_1) + F_2(t)$$

which can be rearranged to

$$\begin{aligned} m_1\ddot{x}_1 + (c + c_1)\dot{x}_1 + (k + k_1)x_1 - c\dot{x}_2 - kx_2 &= F_1(t) \\ -c\dot{x}_1 - kx_1 + m_2\ddot{x}_2 + (c + c_2)\dot{x}_2 + (k + k_2)x_2 &= F_2(t) \end{aligned} \tag{4-1}$$

where $F_1(t)$ and $F_2(t)$ are the excitation forces applied to the respective masses. Note that the equations are not independent, because the equation for m_1 contains terms in x_2 and $\dot{x}_2$. Hence the coupling terms in the first equation in Eq. (4-1) are $-(c\dot{x}_2 + kx_2)$. Similarly, the coupling terms in the second equation are $-(c\dot{x}_1 + kx_1)$. In other words, the motion $x_1(t)$ of m_1 is influenced by the motion $x_2(t)$ of m_2 and vice versa. Coordinate coupling will be discussed in detail in Sec. 4-4.

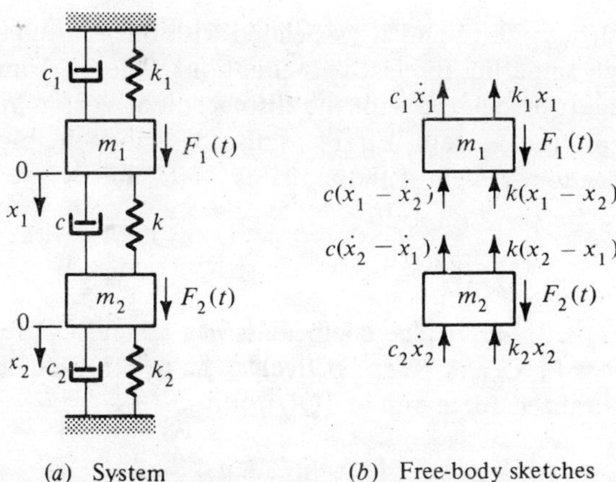

FIG. 4-1. *A two-degree-of-freedom system.*

For conciseness, Eq. (4-1) can be expressed in matrix notations as

$$\begin{bmatrix} m_1 & 0 \\ 0 & m_2 \end{bmatrix}\begin{bmatrix} \ddot{x}_1 \\ \ddot{x}_2 \end{bmatrix}+\begin{bmatrix} c+c_1 & -c \\ -c & c+c_2 \end{bmatrix}\begin{bmatrix} \dot{x}_1 \\ \dot{x}_2 \end{bmatrix}$$
$$+\begin{bmatrix} k+k_1 & -k \\ -k & k+k_2 \end{bmatrix}\begin{bmatrix} x_1 \\ x_2 \end{bmatrix}=\begin{bmatrix} F_1(t) \\ F_2(t) \end{bmatrix} \quad \textbf{(4-2)}$$

or

$$M\{\ddot{x}\}+C\{\dot{x}\}+K\{x\}=\{F(t)\} \quad \textbf{(4-3)}$$

By simple matrix operations, it can be shown that Eqs. (4-1) and (4-2) are equivalent. The quantities in Eq. (4-3) can be identified by comparing with Eq. (4-2). The 2×2 matrices M, C, and K are called the *mass matrix*, *damping matrix*, and *stiffness matrix*, respectively. The 2×1 matrix $\{x\}$ is called the *displacement vector*. The corresponding *velocity vector* is $\{\dot{x}\}$ and the *acceleration vector* is $\{\ddot{x}\}$. The 2×1 matrix $\{F(t)\}$ is the *force vector*.

It will be shown in Sec. 4-4 that if another set of coordinates $\{q_1q_2\}$ is used to describe the motion of the same system, the values of the elements in the matrices M, C, and K will differ from those shown in Eq. (4-2). The inherent properties of the system, such as natural frequencies, must be independent of the coordinates used to describe the system. Hence the general form of the equations of motion of a two-degree-of-freedom system is

$$\begin{bmatrix} m_{11} & m_{12} \\ m_{21} & m_{22} \end{bmatrix}\begin{bmatrix} \ddot{q}_1 \\ \ddot{q}_2 \end{bmatrix}+\begin{bmatrix} c_{11} & c_{12} \\ c_{21} & c_{22} \end{bmatrix}\begin{bmatrix} \dot{q}_1 \\ \dot{q}_2 \end{bmatrix}+\begin{bmatrix} k_{11} & k_{12} \\ k_{21} & k_{22} \end{bmatrix}\begin{bmatrix} q_1 \\ q_2 \end{bmatrix}=\begin{bmatrix} Q_1(t) \\ Q_2(t) \end{bmatrix} \quad \textbf{(4-4)}$$

or

$$M\{\ddot{q}\}+C\{\dot{q}\}+K\{q\}=\{Q(t)\} \quad \textbf{(4-5)}$$

The 2×2 matrices M, C, and K associated with the coordinates $\{q\}$ can be identified by comparing the last two equations. The 2×1 matrix $\{Q(t)\}$ is the force vector associated with the displacement vector $\{q\}$.

Generalizing the concept, Eq. (4-5) also describes the motion of an n-degree-of-freedom system if the matrices M, C, and K are of nth-order, that is,

$$M=[m_{ij}], \qquad C=[c_{ij}], \qquad K=[k_{ij}] \quad \textbf{(4-6)}$$

where $i, j=1, 2, 3, \ldots, n$. The coefficients m_{ij}, c_{ij}, and k_{ij} are the elements of the matrices M, C, and K, respectively. The generalized coordinates $\{q\}$ and the generalized force vector $\{Q(t)\}$ are

$$\{q\}=\{q_1 \ldots q_n\} \quad \textbf{(4-7)}$$

$$\{Q(t)\}=\{Q_1(t) \cdots Q_n(t)\} \quad \textbf{(4-8)}$$

Hence Eq. (4-5) is also the general form of the equations of motion of an n-degree-of-freedom system.

4-3 UNDAMPED FREE VIBRATION: PRINCIPAL MODES

A dynamic system has as many natural frequencies and modes of vibration as the degrees of freedom. The general motion is the superposition of the modes. We shall discuss (1) a method to find the natural frequencies, and (2) the modes of vibration of an undamped system at its natural frequencies.

In the absence of damping and excitation, the system in Fig. 4-1 reduces to that shown in Fig. 4-2(a). Hence the equations of motion from Eq. (4-2) are

$$\begin{bmatrix} m_1 & 0 \\ 0 & m_2 \end{bmatrix}\begin{bmatrix} \ddot{x}_1 \\ \ddot{x}_2 \end{bmatrix}+\begin{bmatrix} k+k_1 & -k \\ -k & k+k_2 \end{bmatrix}\begin{bmatrix} x_1 \\ x_2 \end{bmatrix}=\begin{bmatrix} 0 \\ 0 \end{bmatrix} \tag{4-9}$$

The equations are linear and homogeneous and are in the form of Eq. (D-47), App. D. Hence the solutions can be expressed as

$$\begin{aligned} x_1 &= B_1 e^{st} \\ x_2 &= B_2 e^{st} \end{aligned} \tag{4-10}$$

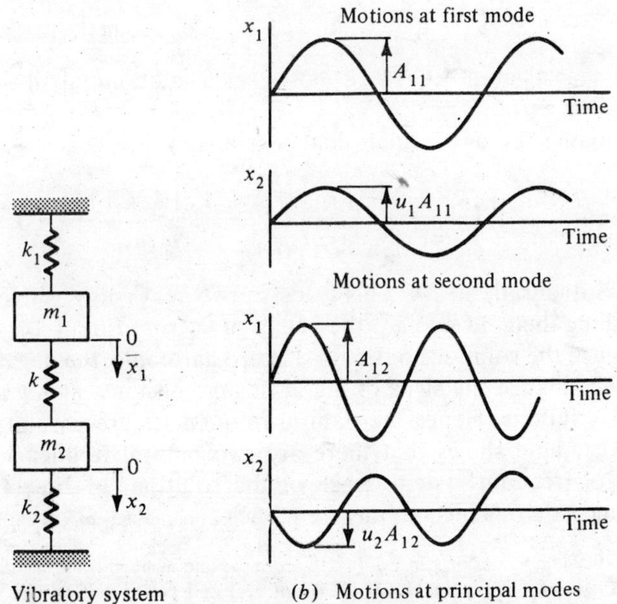

(a) Vibratory system (b) Motions at principal modes

FIG. 4-2. *Modes of vibration.*

where B_1, B_2, and s are constants. Since the system is undamped, it can be shown that the values of s are imaginary, $s=\pm j\omega$. By Euler's formula, $e^{\pm j\omega t}=\cos\omega t \pm j\sin\omega t$, and recalling that the x's are real, the solutions above must be harmonic and the general solution must consist of a number of harmonic components.

Assume one of the harmonic components is

$$\begin{aligned} x_1 &= A_1 \sin(\omega t+\psi) \\ x_2 &= A_2 \sin(\omega t+\psi) \end{aligned} \tag{4-11}$$

where A_1, A_2, and ψ are constants and ω is a natural frequency of the system. If the motions are harmonic, the choice of sine or cosine functions is arbitrary.

Substituting Eq. (4-11) in (4-9), dividing out the factor $\sin(\omega t+\psi)$, and rearranging, we have

$$\begin{aligned} (k+k_1-\omega^2 m_1)A_1 - kA_2 &= 0 \\ -kA_1+(k+k_2-\omega^2 m_2)A_2 &= 0 \end{aligned} \tag{4-12}$$

which are homogeneous linear algebraic equations in A_1 and A_2. The determinant $\Delta(\omega)$ of the coefficients of A_1 and A_2 is called the *characteristic determinant.* If $\Delta(\omega)$ is equated to zero, we obtain the *characteristic* or the *frequency equation* of the system from which the values of ω are found, that is,

$$\Delta(\omega)=\begin{vmatrix} k+k_1-\omega^2 m_1 & -k \\ -k & k+k_2-\omega^2 m_2 \end{vmatrix}=0 \tag{4-13}$$

From linear algebra, Eq. (4-12) possesses a solution only if the determinant $\Delta(\omega)$ is zero.

Expanding the determinant and rearranging, we get

$$\omega^4-\left(\frac{k+k_1}{m_1}+\frac{k+k_2}{m_2}\right)\omega^2+\frac{k_1k_2+k_1k+k_2k}{m_1m_2}=0 \tag{4-14}$$

which is quadratic in ω^2. This leads to two real and positive values* for ω^2. Calling them ω_1^2 and ω_2^2, the values of ω from Eq. (4-14) are $\pm\omega_1$ and $\pm\omega_2$. Since the solutions in Eq. (4-11) are harmonic, the negative signs for ω merely change the signs of the arbitrary constants and would not lead to new solutions. Hence the natural frequencies are ω_1 and ω_2.

The example shows that there are two natural frequencies in a two-degree-of-freedom system. Each of the solutions of Eq. (4-9) has two harmonic components at the frequencies ω_1 and ω_2, respectively. By

* Note that the values of s in Eq. (4-10) are $\pm j\omega_1$ and $\pm j\omega_2$ in order to have the periodic solutions assumed in Eq. (4-11). If ω^2 is not real and positive, it can be shown that the solutions by Eq. (4-10) would either diminish to zero or increase to infinity with increasing time.

superposition, the solutions from Eq. (4-11) are

$$\begin{bmatrix} x_1 \\ x_2 \end{bmatrix} = \begin{bmatrix} A_{11} \\ A_{21} \end{bmatrix} \sin(\omega_1 t + \psi_1) + \begin{bmatrix} A_{12} \\ A_{22} \end{bmatrix} \sin(\omega_2 t + \psi_2) \tag{4-15}$$

where the A's and ψ's are arbitrary constants. The lower frequency term is called the *fundamental* and the others are the *harmonics.* Double subscripts are assigned to the amplitudes; the first subscript refers to the coordinate and the second to the frequency. For example, A_{12} is the amplitude of $x_1(t)$ at the frequency $\omega = \omega_2$.

The relative amplitudes of the harmonic components in Eq. (4-15) are defined in Eq. (4-12). Substituting ω_1 and ω_2 in Eq. (4-12) and rearranging, we obtain

$$\begin{aligned} \frac{A_{11}}{A_{21}} &= \frac{k}{k + k_1 - \omega_1^2 m_1} = \frac{k + k_2 - \omega_1^2 m_2}{k} = \frac{u_{11}}{u_{21}} \triangleq \frac{1}{u_1} \\ \frac{A_{12}}{A_{22}} &= \frac{k}{k + k_1 - \omega_2^2 m_1} = \frac{k + k_2 - \omega_2^2 m_2}{k} = \frac{u_{12}}{u_{22}} \triangleq \frac{1}{u_2} \end{aligned} \tag{4-16}$$

where the u's are constants, defining the relative amplitudes of x_1 and x_2 at each of the natural frequencies ω_1 and ω_2. Thus, Eq. (4-15) becomes

$$\begin{bmatrix} x_1 \\ x_2 \end{bmatrix} = \begin{bmatrix} 1 \\ u_1 \end{bmatrix} A_{11} \sin(\omega_1 t + \psi_1) + \begin{bmatrix} 1 \\ u_2 \end{bmatrix} A_{12} \sin(\omega_2 t + \psi_2) \tag{4-17}$$

where A_{11}, A_{12}, ψ_1, and ψ_2 are the constants of integration, to be determined by the initial conditions. There are four constants because the system is described by two second-order differential equations. Note that (1) from the homogeneous equation in Eq. (4-12), only the ratios $1:u_1$ and $1:u_2$ can be found, and (2) the relative amplitudes at a given natural frequency are invariant, regardless of the initial conditions.

A *principal* or natural *mode* of vibration occurs when the entire system executes synchronous harmonic motion at one of the natural frequencies as illustrated in Fig. 4-2(*b*). For example, the first mode occurs if $A_{12} = 0$ in Eq. (4-17), that is,

$$\begin{bmatrix} x_1 \\ x_2 \end{bmatrix} = \begin{bmatrix} 1 \\ u_1 \end{bmatrix} A_{11} \sin(\omega_1 t + \psi_1) \quad \text{or} \quad \{x\} = \begin{bmatrix} u_{11} \\ u_{21} \end{bmatrix} p_1(t) \triangleq \{u\}_1 p_1(t) \tag{4-18}$$

where $\{u\}_1$ is called a *modal vector* or eigenvector. Note that $\{u\}_1 \triangleq \{u_{11}\ u_{21}\} = \{1\ u_1\}$ as shown above. It represents the *relative amplitude,* or the *mode shape,* of the motions $x_1(t)$ and $x_2(t)$ at $\omega = \omega_1$. Hence a principal mode is specified by the modal vector at the given natural frequency. The quantity $p_1(t) = A_{11} \sin(\omega_1 t + \psi_1)$ is harmonic. It shows that the entire system executes synchronous harmonic motion at a principal mode. Similarly, the second mode occurs if A_{11} in Eq. (4-17) is zero, that is,

$$\begin{bmatrix} x_1 \\ x_2 \end{bmatrix} = \begin{bmatrix} 1 \\ u_2 \end{bmatrix} A_{12}\sin(\omega_2 t + \psi_2) \quad \text{or} \quad \{x\} = \begin{bmatrix} u_{12} \\ u_{22} \end{bmatrix} p_2(t) \triangleq \{u\}_2 p_2(t) \quad \textbf{(4-19)}$$

The modal vector for the second mode is $\{u\}_2$.

The harmonic functions of the motions $x_1(t)$ and $x_2(t)$ in Eq. (4-17) can be expressed as

$$\begin{bmatrix} x_1 \\ x_2 \end{bmatrix} = \begin{bmatrix} 1 & 1 \\ u_1 & u_2 \end{bmatrix} \begin{bmatrix} A_{11}\sin(\omega_1 t + \psi_1) \\ A_{12}\sin(\omega_2 t + \psi_2) \end{bmatrix} \triangleq \begin{bmatrix} u_{11} & u_{12} \\ u_{21} & u_{22} \end{bmatrix} \begin{bmatrix} p_1(t) \\ p_2(t) \end{bmatrix} \quad \textbf{(4-20)}$$

or

$$\{x\} = [u]\{p\} \quad \textbf{(4-21)}$$

where the *modal matrix* $[u]$ is

$$[u] = \begin{bmatrix} u_{11} & u_{12} \\ u_{21} & u_{22} \end{bmatrix} = \begin{bmatrix} 1 & 1 \\ u_1 & u_2 \end{bmatrix}$$

or

$$[u] = [u_{ij}] = [\{u\}_1 \quad \{u\}_2] \quad \textbf{(4-22)}$$

and $p_1(t) = A_{11}\sin(\omega_1 t + \psi_1)$ and $p_2(t) = A_{12}\sin(\omega_2 t + \psi_2)$. Note that in Eqs. (4-18) through (4-20) only the relative values in a modal vector can be defined, as shown in Eq. (4-16). A modal matrix $[u]$ in Eq. (4-22) is simply a combination of the modal vectors. The actual motions $\{x\}$ in Eq. (4-20) are specified by the constants A's and ψ's, which are determined by the initial conditions.

The vector $\{p\}$ consists of a set of harmonic functions at the frequencies ω_1 and ω_2. The vector $\{p\}$ is called the *principal coordinates.* Each principal coordinate $p_i(t)$ and its associated modal vector $\{u\}_i$ describe a mode of vibration as shown in Eqs. (4-18) and (4-19). Principal coordinates will be further discussed in Sec. 4-5 and Chap. 6. The coordinate transformation between the $\{x\}$ and $\{p\}$ coordinates is shown in Eq. (4-21).

The extension of the concept to n-degree-of-freedom systems is immediate. For example, a system may be described by the $\{x\} = \{x_1\, x_2 \cdots x_n\}$ coordinates. Analogous to Eq. (4-17), each of the motions $x_i(t)$ has n harmonic components. A principal mode occurs if the entire system executes synchronous harmonic motion at one of the natural frequencies. The corresponding principal coordinates is $\{p\} = \{p_1\, p_2 \cdots p_n\}$. The modal matrix $[u]$ consists of n modal vectors $\{u\}_i$.

$$[u] = [\{u\}_1 \quad \{u\}_2 \cdots \{u\}_n] = [u_{ij}] \quad \textbf{(4-23)}$$

where $i, j = 1, 2, \ldots, n$. The transformation between the $\{x\}$ and the $\{p\}$ coordinates is analogous to that in Eq. (4-21).

Example 1

Referring to Fig. 4-2(*a*), let $m_1 = m_2 = m$ and $k_1 = k_2 = k$. If the initial conditions are $\{x(0)\} = \{1 \quad 0\}$ and $\{\dot{x}(0)\} = \{0 \quad 0\}$, find the natural frequencies of the system and the displacement vector $\{x\}$.

Solution:

From Eq. (4-14), the natural frequencies are

$$\omega_1 = \sqrt{k/m} \qquad \text{and} \qquad \omega_2 = \sqrt{3k/m}$$

Substituting ω_1 and ω_2 in Eq. (4-16), we obtain $u_1 = 1$ and $u_2 = -1$. Hence the displacement vector $\{x\}$ from Eq. (4-20) is

$$\{x\} = \begin{bmatrix} 1 & 1 \\ 1 & -1 \end{bmatrix} \begin{bmatrix} A_{11} \sin(\omega_1 t + \psi_1) \\ A_{12} \sin(\omega_2 t + \psi_2) \end{bmatrix}$$

For the initial conditions $\{x(0)\} = \{1 \quad 0\}$, we get

$$\begin{bmatrix} 1 \\ 0 \end{bmatrix} = \begin{bmatrix} 1 & 1 \\ 1 & -1 \end{bmatrix} \begin{bmatrix} A_{11} \sin \psi_1 \\ A_{12} \sin \psi_2 \end{bmatrix}$$

Premultiplying the equation by the inverse $[u]^{-1}$ of $[u]$ gives

$$\begin{bmatrix} A_{11} \sin \psi_1 \\ A_{12} \sin \psi_2 \end{bmatrix} = \frac{1}{2} \begin{bmatrix} 1 & 1 \\ 1 & -1 \end{bmatrix} \begin{bmatrix} 1 \\ 0 \end{bmatrix} = \frac{1}{2} \begin{bmatrix} 1 \\ 1 \end{bmatrix}$$

or

$$A_{11} = \frac{1}{2 \sin \psi_1} \qquad \text{and} \qquad A_{12} = \frac{1}{2 \sin \psi_2}$$

For the initial conditions $\{\dot{x}(0)\} = \{0 \quad 0\}$, we have

$$\begin{bmatrix} 0 \\ 0 \end{bmatrix} = \begin{bmatrix} 1 & 1 \\ 1 & -1 \end{bmatrix} \begin{bmatrix} \omega_1 A_{11} \cos \psi_1 \\ \omega_2 A_{12} \cos \psi_2 \end{bmatrix}$$

Premultiplying the equation by the inverse $[u]^{-1}$ of $[u]$ gives

$$\begin{bmatrix} \omega_1 A_{11} \cos \psi_1 \\ \omega_2 A_{12} \cos \psi_2 \end{bmatrix} = \frac{1}{2} \begin{bmatrix} 1 & 1 \\ 1 & -1 \end{bmatrix} \begin{bmatrix} 0 \\ 0 \end{bmatrix} = \begin{bmatrix} 0 \\ 0 \end{bmatrix}$$

Since the A's and ω's are nonzero, we have $\cos \psi_1 = \cos \psi_2 = 0$. Let $\psi_1 = m\pi/2$ and $\psi_2 = n\pi/2$, where m and n are odd integers. It can be shown that the choice of m and n other than 1 will not lead to new solutions. Thus, $A_{11} = A_{12} = 1/2$.

From Eq. (4-17), we obtain

$$\begin{bmatrix} x_1 \\ x_2 \end{bmatrix} = \frac{1}{2} \begin{bmatrix} 1 \\ 1 \end{bmatrix} \cos \sqrt{\frac{k}{m}}\, t + \frac{1}{2} \begin{bmatrix} 1 \\ -1 \end{bmatrix} \cos \sqrt{\frac{3k}{m}}\, t$$

The motions are plotted in Fig. 4-3 for $\sqrt{k/m} \doteq 2\pi$. The example can be repeated for different initial conditions to show that the relative amplitudes of the principal modes remain unchanged. This is left as an exercise.

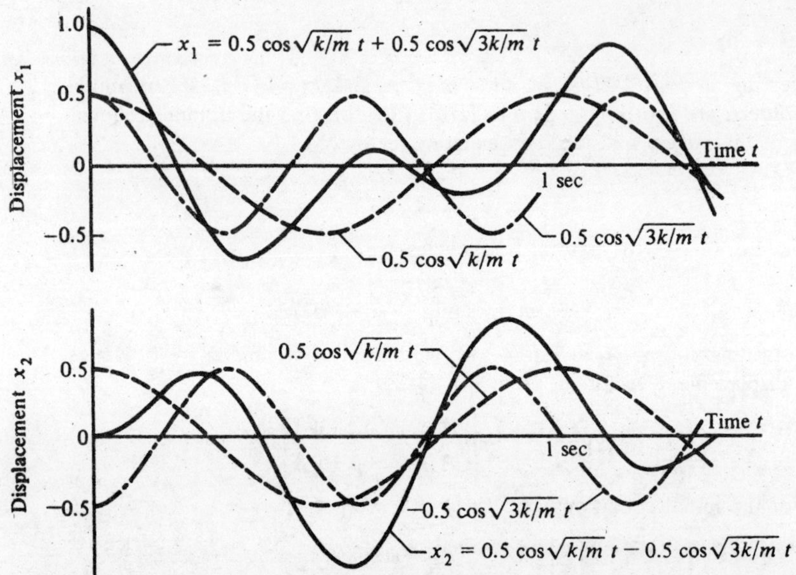

FIG. 4-3. *Superposition of modes of vibration: Example 1.*

Example 2. Natural modes

Find the initial conditions that would set a two-degree-of-freedom into its natural modes of vibration, that is, A_{11} or A_{12} in Eq. (4-17) becomes zero.

Solution:

From Eq. (4-20), we have

$$\begin{bmatrix} x_1 \\ x_2 \end{bmatrix} = \begin{bmatrix} 1 & 1 \\ u_1 & u_2 \end{bmatrix} \begin{bmatrix} A_{11}\sin(\omega_1 t + \psi_1) \\ A_{12}\sin(\omega_2 t + \psi_2) \end{bmatrix}$$

Applying the initial conditions $\{x(0)\} = \{x_{10} \quad x_{20}\}$, we get

$$\begin{bmatrix} x_{10} \\ x_{20} \end{bmatrix} = \begin{bmatrix} 1 & 1 \\ u_1 & u_2 \end{bmatrix} \begin{bmatrix} A_{11}\sin\psi_1 \\ A_{12}\sin\psi_2 \end{bmatrix}$$

Premultiplying the equation by the inverse $[u]^{-1}$ of $[u]$ gives

$$\begin{bmatrix} A_{11}\sin\psi_1 \\ A_{12}\sin\psi_2 \end{bmatrix} = \frac{1}{u_2 - u_1} \begin{bmatrix} u_2 & -1 \\ -u_1 & 1 \end{bmatrix} \begin{bmatrix} x_{10} \\ x_{20} \end{bmatrix}$$

or

$$A_{11} = \frac{u_2 x_{10} - x_{20}}{(u_2 - u_1)\sin\psi_1} \qquad \text{and} \qquad A_{12} = \frac{x_{20} - u_1 x_{10}}{(u_2 - u_1)\sin\psi_2}$$

Similarly, using the initial conditions $\{\dot{x}(0)\} = \{\dot{x}_{10} \quad \dot{x}_{20}\}$, we get

$$A_{11} = \frac{u_2 \dot{x}_{10} - \dot{x}_{20}}{\omega_1(u_2 - u_1)\cos\psi_1} \qquad \text{and} \qquad A_{12} = \frac{\dot{x}_{20} - u_1 \dot{x}_{10}}{\omega_2(u_2 - u_1)\cos\psi_2}$$

The first mode occurs at ω_1 if $A_{12}=0$, that is,

$$x_{20}=u_1x_{10} \quad \text{and} \quad \dot{x}_{20}=u_1\dot{x}_{10}$$

In other words, the system will vibrate at its first mode if the initial conditions are $\{x_{10} \quad x_{20}\}=\{1 \quad u_1\}$ with zero initial velocities. Alternatively, the initial conditions can be $\{\dot{x}_{10} \quad \dot{x}_{20}\}=\{1 \quad u_1\}$ with zero initial displacements. Any combination of the above conditions would also set the system in its first mode. It is only necessary to set the initial values of $\{x\}$ and/or $\{\dot{x}\}$ to conform to their relative values for the first mode, as indicated by the corresponding modal vector $\{u\}_1=\{1 \quad u_1\}$.

Similarly, the second mode occurs when $A_{11}=0$, that is, $x_{20}=u_2x_{10}$ and $\dot{x}_{20}=u_2\dot{x}_{10}$. Any combination of these conditions will give the second mode.

Example 3. Vehicle suspension

An automobile is shown schematically in Fig. 4-4. Find the natural frequencies of the car body.

Solution:

An automobile has many degrees of freedom. Simplifying, we assume that the car moves in the plane of the paper and the motion consists of (1) the vertical motion of the car body, (2) the rotational pitching motion of the body about its mass center, and (3) the vertical motion of the wheels. Even then, the system has more than two degrees of freedom.

When the excitation frequency due to the road roughness is high, the wheels move up and down with great rapidity but little of this motion is transmitted to the car body. In other words, the natural frequency of the car body is low and only the low frequency portion of the road roughness is being transmitted. (See Case 4 in Sec. 3-5.) Because of this large separation of natural frequencies between the wheels and the car body, the problem can be further simplified by neglecting the wheels as shown in Fig. 4-5.

Assuming small oscillations, the equations of motion in the $x(t)$ and $\theta(t)$ coordinates are

$$m\ddot{x}=\sum \text{(forces)}_x$$
$$m\ddot{x}=-k_1(x-L_1\theta)-k_2(x+L_2\theta)$$

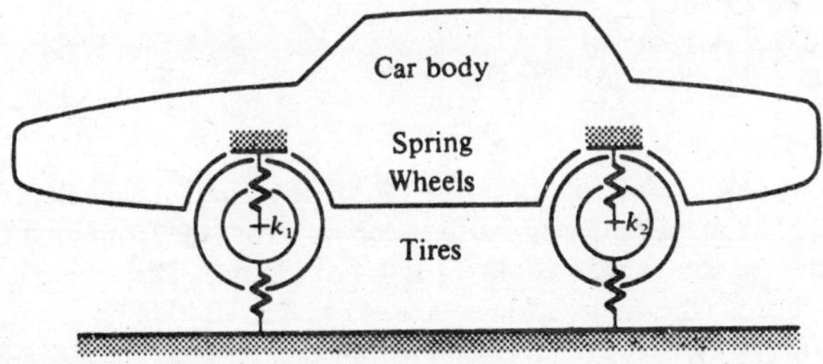

FIG. 4-4. *Schematic of an automobile.*

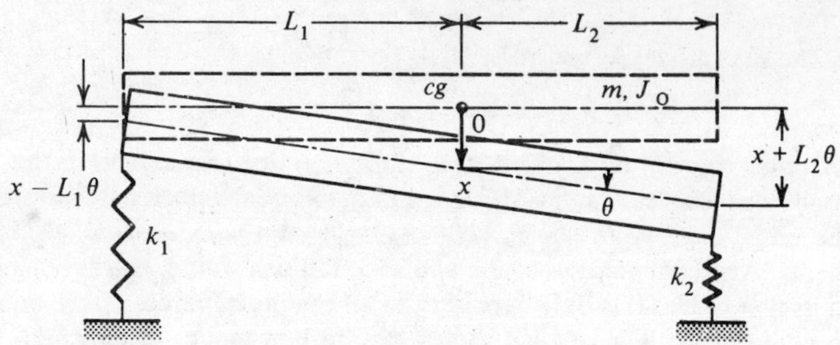

FIG. 4-5. *Simplified representation of an automobile body.*

and

$$J_0\ddot{\theta} = \sum (\text{moments})_0$$

$$J_0\ddot{\theta} = k_1(x - L_1\theta)L_1 - k_2(x + L_2\theta)L_2$$

Rearranging, we obtain

$$\begin{bmatrix} m & 0 \\ 0 & J_0 \end{bmatrix}\begin{bmatrix} \ddot{x} \\ \ddot{\theta} \end{bmatrix} + \begin{bmatrix} k_1 + k_2 & -(k_1L_1 - k_2L_2) \\ -(k_1L_1 - k_2L_2) & k_1L_1^2 + k_2L_2^2 \end{bmatrix}\begin{bmatrix} x \\ \theta \end{bmatrix} = \begin{bmatrix} 0 \\ 0 \end{bmatrix}$$

which is of the same form as Eq. (4-9). The frequency equation from Eq. (4-13) is

$$\Delta(\omega) = \begin{vmatrix} k_1 + k_2 - \omega^2 m & k_2L_2 - k_1L_1 \\ k_2L_2 - k_1L_1 & k_1L_1^2 + k_2L_2^2 - \omega^2 J_0 \end{vmatrix} = 0$$

Expanding the determinant and solving the equation, we get

$$\omega_{1,2}^2 = \frac{1}{2}\left[\frac{k_1 + k_2}{m} + \frac{k_1L_1^2 + k_2L_2^2}{J_0} \mp \sqrt{\left(\frac{k_1 + k_2}{m} + \frac{k_1L_1^2 + k_2L_2^2}{J_0}\right)^2 - \frac{4k_1k_2(L_1 + L_2)^2}{mJ_0}}\right]$$

The natural frequencies are $\omega_1/2\pi$ and $\omega_2/2\pi$ Hz.

Example 4

A vehicle has a mass of 1,800 kg (4,000 lb_m) and a wheelbase of 3.6 m (140 in.). The mass center *cg* is 1.6 m (63 in.) from the front axle. The radius of gyration of the vehicle about *cg* is 1.4 m (55 in.). The spring constants of the front and the rear springs are 42 kN/m (240 lb_f/in.) and 48 kN/m (275 lb_f/in.), respectively. Determine **(a)** the natural frequencies, **(b)** the principal modes of vibration, and **(c)** the motion $x(t)$ and $\theta(t)$ of the vehicle.

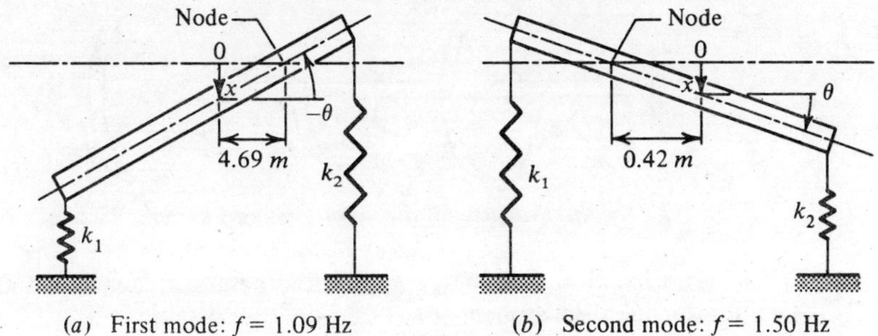

(a) First mode: $f = 1.09$ Hz (b) Second mode: $f = 1.50$ Hz

FIG. 4-6. *Principal modes of vibration of a car body (not to scale).*

Solution:

(a) From the given data and the equations in Example 3, we have

$$\frac{k_1 + k_2}{m} = 50 \qquad \frac{k_1 L_1 - k_2 L_2}{m} = -16.0$$

$$\frac{k_1 L_1^2 + k_2 L_2^2}{J_0} = 84.9 \qquad \frac{4 k_1 k_2 (L_1 + L_2)^2}{m J_0} = 16\,457$$

$$\omega_{1,2}^2 = \tfrac{1}{2}\left[50 + 84.9 \mp \sqrt{(50 + 84.9)^2 - 16\,457}\right] = \begin{cases} 46.6 \\ 88.3 \end{cases}$$

$$\omega_{1,2} = \begin{cases} 6.83 \text{ rad/s} = 1.09 \text{ Hz} \\ 9.40 \text{ rad/s} = 1.50 \text{ Hz} \end{cases}$$

(b) The amplitude ratios for the two modes of vibration are

$$\frac{X}{\Theta} = \frac{(k_1 L_1 - k_2 L_2)/m}{(k_1 + k_2)/m - \omega_{1,2}^2} = \frac{-16}{50 - \omega_{1,2}^2} = \begin{cases} -4.69 \text{ m/rad} \\ 0.42 \text{ m/rad} \end{cases}$$

The two principal modes of vibration are shown schematically in Fig. 4-6. The mode shape at 1.09 Hz is $\{X \quad \Theta\} = \{1 \quad -1/4.69\}$. Thus, when $x(t)$ is positive, $\theta(t)$ is negative from the assumed direction of rotation. When $x(t) = 1$ m, $\theta(t) = -1/4.69$ rad, that is, the *node* is 4.69 m from the cg of the car body. Similarly, at 1.50 Hz, the mode shape is $\{X \quad \Theta\} = \{1 \quad 1/0.42\}$.

(c) From Eq. (4-17), the $x(t)$ and $\theta(t)$ motions are

$$\begin{bmatrix} x \\ \theta \end{bmatrix} = \begin{bmatrix} 1 \\ -1/4.69 \end{bmatrix} A_{11} \sin(6.83t + \psi_1) + \begin{bmatrix} 1 \\ 1/0.42 \end{bmatrix} A_{12} \sin(9.40t + \psi_2)$$

where A_{11}, A_{12}, ψ_1, and ψ_2 are the constants of integration.

Example 5. A three-degree-of-freedom system

A torsional system with three degrees of freedom is shown in Fig. 4-7. **(a)** Determine the equations of motion and the frequency equation. **(b)** If

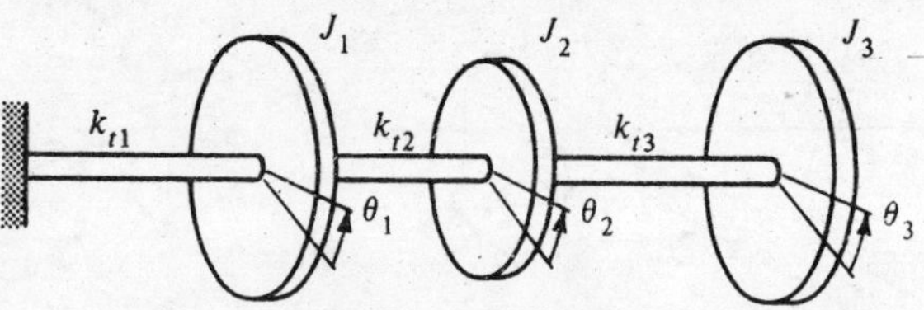

FIG. 4-7. *A three-degree-of-freedom torsional system; Example 5.*

$J_1 = J_2 = J_3 = J$ and $k_{t1} = k_{t2} = k_{t3} = k_t$, find the natural frequencies and the equation for the displacement $\{\theta\}$.

Solution:

(a) From Newton's second law, the equations of motion are

$$\begin{aligned} J_1\ddot{\theta}_1 &= -k_{t1}\theta_1 - k_{t2}(\theta_1 - \theta_2) \\ J_2\ddot{\theta}_2 &= -k_{t2}(\theta_2 - \theta_1) - k_{t3}(\theta_2 - \theta_3) \\ J_3\ddot{\theta}_3 &= -k_{t3}(\theta_3 - \theta_2) \end{aligned} \tag{4-24}$$

Substituting $\theta_i = \Theta_i \sin(\omega t + \psi)$, for $i = 1, 2,$ and 3, in Eq. (4-24), factoring out the $\sin(\omega t + \psi)$ term, and rearranging, we have

$$\begin{aligned} (k_{t1} + k_{t2} - \omega^2 J_1)\Theta_1 - k_{t2}\Theta_2 &= 0 \\ -k_{t2}\Theta_1 + (k_{t2} + k_{t3} - \omega^2 J_2)\Theta_2 - k_{t3}\Theta_3 &= 0 \\ -k_{t3}\Theta_2 + (k_{t3} - \omega^2 J_3)\Theta_3 &= 0 \end{aligned} \tag{4-25}$$

The frequency equation is obtained by equating the determinant $\Delta(\omega)$ of the coefficients of Θ_1, Θ_2, and Θ_3 to zero.

$$\Delta(\omega) = \begin{vmatrix} k_{t1} + k_{t2} - \omega^2 J_1 & -k_{t2} & 0 \\ -k_{t2} & k_{t2} + k_{t3} - \omega^2 J_2 & -k_{t3} \\ 0 & -k_{t3} & k_{t3} - \omega^2 J_3 \end{vmatrix} = 0$$

(b) If $J_1 = J_2 = J_3 = J$ and $k_{t1} = k_{t2} = k_{t3} = k_t$, the frequency equation is

$$\omega^6 - 5\left(\frac{k_t}{J}\right)\omega^4 + 6\left(\frac{k_t}{J}\right)^2\omega^2 - \left(\frac{k_t}{J}\right)^3 = 0$$

The roots of the equation are $\omega^2 = 0.198(k_t/J)$, $1.55(k_t/J)$, and $3.25(k_t/J)$. The corresponding frequency vector $\{f\}$ is

$$\{f\} = \frac{1}{2\pi}\cdot\sqrt{\frac{k_t}{J}}\begin{bmatrix} \sqrt{0.198} \\ \sqrt{1.55} \\ \sqrt{3.25} \end{bmatrix} = \sqrt{\frac{k_t}{J}}\begin{bmatrix} 0.071 \\ 0.198 \\ 0.288 \end{bmatrix} \text{Hz}$$

From Eq. (4-25), the amplitude ratios are

$$\frac{\Theta_1}{\Theta_2} = \frac{k_{t2}}{k_{t1} + k_{t2} - \omega^2 J_1} \quad \text{and} \quad \frac{\Theta_2}{\Theta_3} = \frac{k_{t3} - \omega^2 J_3}{k_{t3}}$$

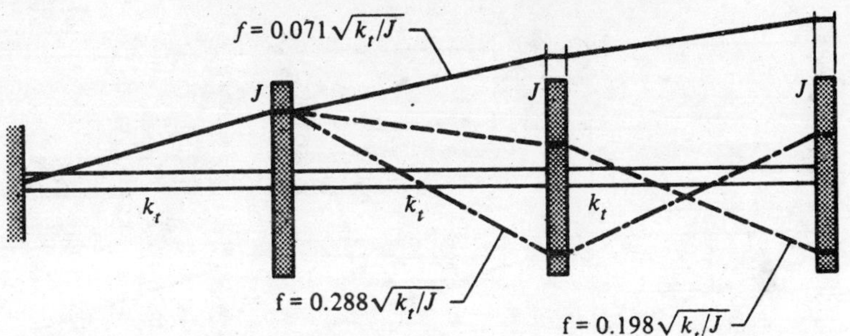

FIG. 4-8. *Principal modes of vibration; amplitudes plotted normal to axis of rotation; Example 5.*

Thus, a modal vector $\{\Theta_1 \quad \Theta_2 \quad \Theta_3\}_i$ can be calculated for each of the natural frequencies ω_i. The modal matrix $[\theta_{ij}]$, for $i, j = 1, 2$, and 3, is formed from a combination of the modal vectors $\{\theta\}_i$ as in Eq. (4-21).

$$[\theta_{ij}] = [\{\theta\}_1 \quad \{\theta\}_2 \quad \{\theta\}_3] = \begin{bmatrix} 1 & 1 & 1 \\ 1.802 & 0.445 & -1.247 \\ 2.25 & -0.802 & 0.555 \end{bmatrix}$$

where $\{\theta\}_1$, $\{\theta\}_2$, and $\{\theta\}_3$ are the modal vectors for the frequencies $\omega_1 = \sqrt{0.198k_t/J}$, $\omega_2 = \sqrt{1.55k_t/J}$, and $\omega_3 = \sqrt{3.25k_t/J}$, respectively. The principal modes of vibration are illustrated in Fig. 4-8.

By superposition of the principal modes, the motions $\{\theta\}$ of the rotating disks are

$$\{\theta\} = \begin{bmatrix} 1 \\ 1.802 \\ 2.25 \end{bmatrix} \Theta_{11} \sin(\omega_1 t + \psi_1) + \begin{bmatrix} 1 \\ 0.445 \\ -0.802 \end{bmatrix} \Theta_{12} \sin(\omega_2 t + \psi_2) + \begin{bmatrix} 1 \\ -1.247 \\ 0.555 \end{bmatrix} \Theta_{13} \sin(\omega_3 t + \psi_3)$$

The Θ's and ψ's are to be determined by the initial conditions.

4-4 GENERALIZED COORDINATES AND COORDINATE COUPLING

The general form of the equations of motion of a two-degree-of-freedom system is shown in Eq. (4-4). For undamped free vibration, we have

$$\begin{bmatrix} m_{11} & m_{12} \\ m_{21} & m_{22} \end{bmatrix} \begin{bmatrix} \ddot{x}_1 \\ \ddot{x}_2 \end{bmatrix} + \begin{bmatrix} k_{11} & k_{12} \\ k_{21} & k_{22} \end{bmatrix} \begin{bmatrix} x_1 \\ x_2 \end{bmatrix} = \begin{bmatrix} 0 \\ 0 \end{bmatrix} \tag{4-26}$$

The system is described by the coordinates x_1 and x_2, which are the elements of the displacement vector $\{x\}$. The coupling terms in the equations are m_{12}, m_{21}, k_{12}, and k_{21}. We shall show that the values of the

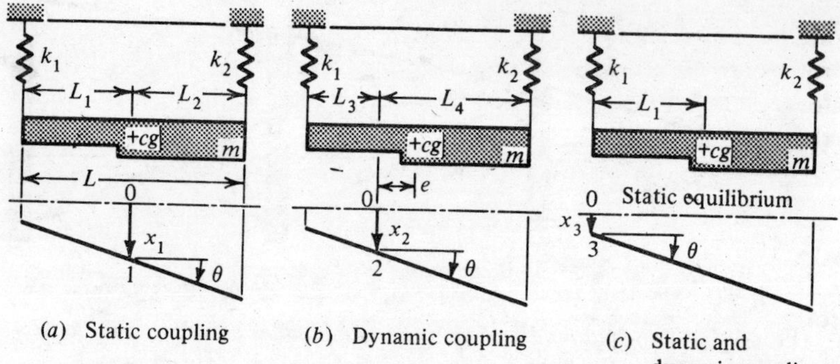

(*a*) Static coupling (*b*) Dynamic coupling (*c*) Static and dynamic coupling

FIG. 4-9. *Generalized coordinates and coordinate coupling: a two-degree-of-freedom system described by the* (x_1,θ), (x_2,θ), *and* (x_3,θ) *coordinates.*

elements in the matrices M and K are dependent on the coordinates selected for the system description.

A vibratory system can be described by more than one set of independent spatial coordinates, each of which can be called a set of *generalized coordinates.* We often use the displacements from the static equilibrium positions of the masses and the rotations about the mass centers for the coordinates. This choice is convenient, but, nonetheless, arbitrary. We shall describe the system in Fig. 4-9 by the displacement vectors $\{x_1 \quad \theta\}$, $\{x_2 \quad \theta\}$, and $\{x_3 \quad \theta\}$.

Referring to Fig. 4-9(a) and assuming small oscillations, the equations of motion in the (x_1,θ) coordinates are

$$m\ddot{x}_1 = -k_1(x_1 - L_1\theta) - k_2(x_1 + L_2\theta)$$
$$J_1\ddot{\theta} = k_1(x_1 - L_1\theta)L_1 - k_2(x_1 + L_2\theta)L_2$$

Rearranging, we obtain

$$\begin{bmatrix} m & 0 \\ 0 & J_1 \end{bmatrix}\begin{bmatrix} \ddot{x}_1 \\ \ddot{\theta} \end{bmatrix} + \begin{bmatrix} (k_1+k_2) & -(k_1L_1-k_2L_2) \\ -(k_1L_1-k_2L_2) & (k_1L_1^2+k_2L_2^2) \end{bmatrix}\begin{bmatrix} x_1 \\ \theta \end{bmatrix} = \begin{bmatrix} 0 \\ 0 \end{bmatrix} \quad \textbf{(4-27)}$$

The coupling term, $-(k_1L_1 - k_2L_2)$, occurs only in the stiffness matrix, and the system is said to be *statically* or *elastically coupled* (see Examples 3 and 4). If a static force is applied through the cg at point 1, the body will rotate as well as translate in the x_1 direction. Conversely, if a torque is applied at point 1, the body will translate as well as rotate in the θ direction.

The same system is described by the (x_2,θ) coordinates in Fig. 4-9(b). The distance e is selected to give $k_1L_3 = k_2L_4$. If a static force is applied at point 2 to cause a displacement x_2, the body will not rotate. Hence no static coupling is anticipated in the equations of motion. During vibration, however, the inertia force $m\ddot{x}_2$ through cg will create a moment $m\ddot{x}_2e$

about point 2, tending to rotate the body in the θ direction. Conversely, a rotation θ about point 2 will give a displacement $e\theta$ at cg and therefore a force $me\ddot{\theta}$ in the x_2 direction. Hence dynamic coupling is anticipated in the equations.

The equations of motion in the (x_2,θ) coordinates are

$$m\ddot{x}_2 = -k_1(x_2 - L_3\theta) - k_2(x_2 + L_4\theta) - me\ddot{\theta}$$
$$J_2\ddot{\theta} = k_1(x_2 - L_3\theta)L_3 - k_2(x_2 + L_4\theta)L_4 - me\ddot{x}_2$$

or

$$\begin{bmatrix} m & me \\ me & J_2 \end{bmatrix}\begin{bmatrix} \ddot{x}_2 \\ \ddot{\theta} \end{bmatrix} + \begin{bmatrix} k_1 + k_2 & 0 \\ 0 & k_1L_3^2 + k_2L_4^2 \end{bmatrix}\begin{bmatrix} x_2 \\ \theta \end{bmatrix} = \begin{bmatrix} 0 \\ 0 \end{bmatrix} \qquad \textbf{(4-28)}$$

The coupling terms are associated with the inertia forces and the system is said to be *dynamically*, or *inertia, coupled.*

Lastly, let the same system be described by the (x_3,θ) coordinates as shown in Fig. 4-9(c). It can be shown that the equations of motion are

$$\begin{bmatrix} m & mL_1 \\ mL_1 & J_3 \end{bmatrix}\begin{bmatrix} \ddot{x}_3 \\ \ddot{\theta} \end{bmatrix} + \begin{bmatrix} k_1 + k_2 & k_2L \\ k_2L & k_2L^2 \end{bmatrix}\begin{bmatrix} x_3 \\ \theta \end{bmatrix} = \begin{bmatrix} 0 \\ 0 \end{bmatrix} \qquad \textbf{(4-29)}$$

For the (x_3,θ) description, the equations are statically and dynamically coupled.

Note that (1) the choice of coordinates for the system description is a mere convenience, (2) the system will vibrate in its own natural way regardless of the coordinate description, (3) the equations for one coordinate description can be obtained from those for other descriptions, and (4) coupling in the equations is *not* an inherent property of the system, such as natural frequencies.

The examples above show that the matrices M and K are symmetric, that is, $m_{12} = m_{21}$ and $k_{12} = k_{21}$. Symmetry is assured if the deflections are measured from a fixed position in space. This can be deduced from Maxwell's reciprocity theorem in Sec. 4-9. Let us select a set of generalized coordinates based on relative deflections to illustrate the *nonsymmetric matrices* in the equations of motion.

Example 6

Consider the system shown in Fig. 4-2 and assume the generalized coordinates $q_1 = x_1$ and $q_2 = x_2 - x_1$, that is, q_2 is proportional to the spring force due to k. Find the equations of motion of the system.

Solution:

The coordinates $\{q\}$ and $\{x\}$ are related by

$$\begin{bmatrix} q_1 \\ q_2 \end{bmatrix} = \begin{bmatrix} 1 & 0 \\ -1 & 1 \end{bmatrix}\begin{bmatrix} x_1 \\ x_2 \end{bmatrix}$$

or

$$\begin{bmatrix} x_1 \\ x_2 \end{bmatrix} = \begin{bmatrix} 1 & 0 \\ 1 & 1 \end{bmatrix} \begin{bmatrix} q_1 \\ q_2 \end{bmatrix}$$

Replacing the displacement and acceleration vectors $\{x\}$ and $\{\ddot{x}\}$ by $\{q\}$ and $\{\ddot{q}\}$ in Eq. (4-9) for the same system, we get

$$\begin{bmatrix} m_1 & 0 \\ m_2 & m_2 \end{bmatrix} \begin{bmatrix} \ddot{q}_1 \\ \ddot{q}_2 \end{bmatrix} + \begin{bmatrix} k_1 & -k \\ k_2 & k+k_2 \end{bmatrix} \begin{bmatrix} q_1 \\ q_2 \end{bmatrix} = \begin{bmatrix} 0 \\ 0 \end{bmatrix}$$

4-5 PRINCIPAL COORDINATES

It was shown in the last section that the elements of the matrices M and K depend on the coordinates selected for the system description. It is possible to select a particular set of coordinates, called the *principal coordinates,* such that there is no coupling terms in the equations of motion, that is, the matrices M and K become diagonal matrices. Hence each of the uncoupled equations can be solved independently. In other words, when the system is described in terms of the principal coordinates, the equations of motion are uncoupled, and the modes of vibration are mathematically separated. Thus, each of the uncoupled equations can be solved independently, as if for systems with one degree of freedom.

Assume an undamped two-degree-of-freedom system is uncoupled by the principal coordinates $\{p\}$. The corresponding equations of motion from Eq. (4-4) are

$$\begin{bmatrix} m_{11} & 0 \\ 0 & m_{22} \end{bmatrix} \begin{bmatrix} \ddot{p}_1 \\ \ddot{p}_2 \end{bmatrix} + \begin{bmatrix} k_{11} & 0 \\ 0 & k_{22} \end{bmatrix} \begin{bmatrix} p_1 \\ p_2 \end{bmatrix} = \begin{bmatrix} 0 \\ 0 \end{bmatrix} \tag{4-30}$$

Expanding the equations gives

$$m_{11}\ddot{p}_1 + k_{11}p_1 = 0$$
$$m_{22}\ddot{p}_2 + k_{22}p_2 = 0$$

The solutions of the equations are

$$\begin{aligned} p_1 &= A_{11}\sin(\omega_1 t + \psi_1) \\ p_2 &= A_{12}\sin(\omega_2 t + \psi_2) \end{aligned} \tag{4-31}$$

where $\omega_1^2 = k_{11}/m_{11}$, $\omega_2^2 = k_{22}/m_{22}$, and the A's and ψ's are constants. Evidently, each of the solutions above represents a mode of vibration as discussed in Sec. 4-3. At a given mode, the system resembles an independent one-degree-of-freedom system.

Now, assume the same system is described by the generalized coordinates $\{q\}$ and the equations of motion are coupled. From Eq. (4-17), the motions in the $\{q\}$ coordinates are

$$\begin{bmatrix} q_1 \\ q_2 \end{bmatrix} = \begin{bmatrix} u_{11} \\ u_{21} \end{bmatrix} A_{11} \sin(\omega_1 t + \psi_1) + \begin{bmatrix} u_{12} \\ u_{22} \end{bmatrix} A_{21} \sin(\omega_2 t + \psi_2) \qquad \textbf{(4-32)}$$

where $\{u_{11} \quad u_{21}\}$ and $\{u_{12} \quad u_{22}\}$ are the modal vectors for the frequencies ω_1 and ω_2, respectively.

Substituting Eq. (4-31) in (4-32) and simplifying, we get

$$\begin{bmatrix} q_1 \\ q_2 \end{bmatrix} = \begin{bmatrix} u_{11} & u_{12} \\ u_{21} & u_{22} \end{bmatrix} \begin{bmatrix} p_1 \\ p_2 \end{bmatrix} \qquad \textbf{(4-33)}$$

or

$$\{q\} = [u]\{p\} \qquad \text{and} \qquad \{p\} = [u]^{-1}\{q\} \qquad \textbf{(4-34)}$$

where $[u]^{-1}$ is the inverse of the modal matrix $[u]$, as defined in Eq. (4-22). The transformation between the $\{p\}$ and the $\{q\}$ coordinates in Eq. (4-34) is identical to that shown in Eq. (4-21).*

The discussion implies that the equations of motion can be uncoupled by means of a coordinate transformation. In other words, given the coupled equations in the $\{q\}$ coordinates, the equations can be uncoupled by substituting $\{p\}$ for $\{q\}$ as shown in Eq. (4-34). This can be done, but the general theory in Sec. 6-4 will be needed. In the mean time, we shall illustrate with another example and then show the general technique in the next section.

Example 7

Determine the principal coordinates for the system shown in Fig. 4-2(a) if $m_1 = m_2 = m$ and $k_1 = k_2 = k$.

Solution:

From Example 1, we have $u_1 = 1$ and $u_2 = -1$. Hence Eq. (4-33) becomes

$$\begin{bmatrix} x_1 \\ x_2 \end{bmatrix} = \begin{bmatrix} 1 & 1 \\ 1 & -1 \end{bmatrix} \begin{bmatrix} p_1 \\ p_2 \end{bmatrix} \qquad \text{and} \qquad \begin{bmatrix} p_1 \\ p_2 \end{bmatrix} = \frac{1}{2} \begin{bmatrix} 1 & 1 \\ 1 & -1 \end{bmatrix} \begin{bmatrix} x_1 \\ x_2 \end{bmatrix}$$

The transformation above indicates that, if $p_1 = \frac{1}{2}(x_1 + x_2)$ and $p_2 = \frac{1}{2}(x_1 - x_2)$, the equations of motion in the $\{p\}$ coordinates are uncoupled. Let us further examine this statement.

The equations of motion for the same system from Eq. (4-9) are

$$m\ddot{x}_1 + 2kx_1 - kx_2 = 0$$
$$m\ddot{x}_2 + 2kx_2 - kx_1 = 0$$

* We assume that the matrices M and K in the generalized coordinates are symmetric. An inherent symmetry in the system can be assumed. We shall not discuss nonsymmetric matrices as illustrated in Example 6.

Adding and subtracting the equations, we obtain

$$\begin{aligned} m(\ddot{x}_1+\ddot{x}_2)+k(x_1+x_2)&=0 \\ m(\ddot{x}_1-\ddot{x}_2)+3k(x_1-x_2)&=0 \end{aligned} \quad \text{or} \quad \begin{aligned} m\ddot{p}_1+kp_1&=0 \\ m\ddot{p}_2+3kp_2&=0 \end{aligned}$$

Again, the equations are uncoupled if we define $p_1=(x_1+x_2)$ and $p_2=(x_1-x_2)$. Since the amplitudes of oscillation are arbitrary, the factor $(\frac{1}{2})$ between the two definitions of p_1 and p_2 in the problem is secondary.

4-6 MODAL ANALYSIS: TRANSIENT VIBRATION OF UNDAMPED SYSTEMS

Consider the steps to solve the equations of motion of an undamped system. From Eq. (4-5), we get

$$M\{\ddot{q}\}+K\{q\}=\{Q(t)\} \tag{4-35}$$

(1) The equations can be uncoupled by means of the modal matrix $[u]$ and expressed in the principal coordinates $\{p\}$ as shown in Example 7. (2) Each of the uncoupled equations can be solved as an independent one-degree-of-freedom system. (3) Applying the coordinate transformation in Eq. (4-34), the solution can be expressed in the $\{p\}$ or $\{q\}$ coordinates as desired. The steps enumerated are conceptually simple. Except for the formula to uncouple the equations of motion, we have the necessary information for the modal analysis of transient vibration of undamped systems. For systems with more than two degrees of freedom, however, computer solutions, as illustrated in Chap. 9, are manditory to alleviate the numerical computations.

The modal matrix of undamped systems can be found by the method described in the previous sections. From Eq. (4-35), the equations of motion of an unforced system are

$$M\{\ddot{q}\}+K\{q\}=\{0\} \tag{4-36}$$

A principal mode occurs if the entire system executes synchronous harmonic motion at a natural frequency ω. Thus, the acceleration of q_i is $\ddot{q}_i=-\omega^2 q_i$, or $\{\ddot{q}\}=\{-\omega^2 q\}=-\omega^2\{q\}$. Substituting $-\omega^2\{q\}$ for $\{\ddot{q}\}$ in Eq. (4-36) and simplifying, we get

$$[-\omega^2 M+K]\{q\}=\{0\} \tag{4-37}$$

Since the system is at a principal mode, the displacement vector $\{q\}$ is also a modal vector at the natural frequency ω. In other words, $\{q\}$ in Eq. (4-37) gives the relative amplitude of vibration of the masses of the system at the given natural frequency.

Since Eq. (4-37) is a set of homogeneous algebraic equations, it possesses a solution only if the characteristic determinant $\Delta(\omega)$ is zero;

that is,

$$\Delta(\omega) = |K - \omega^2 M| = 0 \tag{4-38}$$

This is the characteristic of the frequency equation, which may be compared with Eq. (4-13). Previously, the frequency equation was solved by hand calculations as shown in Eq. (4-14). Computer solutions, however, are necessary for systems with more than two degrees of freedom. Likewise, instead of solving for the modal vector by hand calculations as shown in Eq. (4-16), computers can be used to solve for $\{q\}$ in Eq. (4-37). A modal vector is found for each natural frequency. The modal matrix $[u]$ is formed from a combination of the modal vectors as shown in Eq. (4-23).

Example 8

Find the coefficients of the frequency equation for the system shown in Fig. 4-10(a).

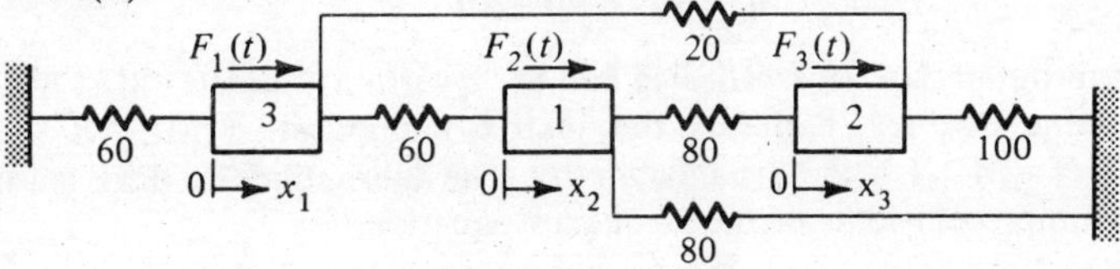

(a) Vibratory system

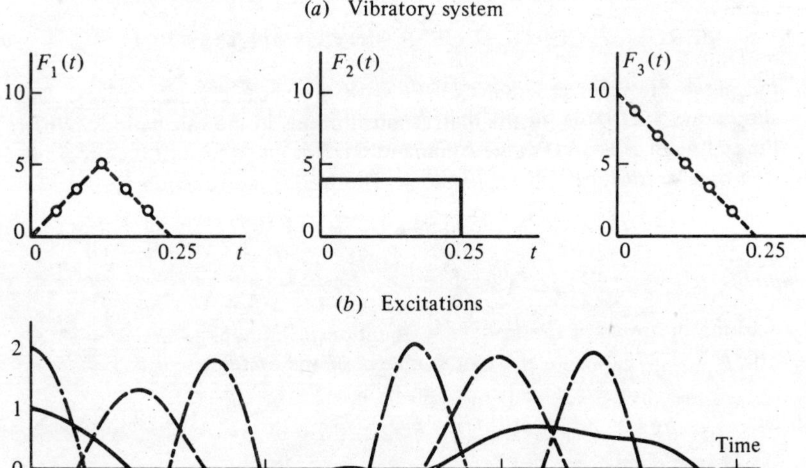

(b) Excitations

(c) Transient response

FIG. 4-10. *Transient vibration of undamped system: Examples 8 to 10.*

Solution:

Applying Newton's second law to each of the masses of the system, we have

$$3\ddot{x}_1 = -60x_1 - 60(x_1 - x_2) - 20(x_1 - x_3) + F_1(t)$$
$$1\ddot{x}_2 = -80x_2 - 60(x_2 - x_1) - 80(x_2 - x_3) + F_2(t)$$
$$2\ddot{x}_3 = -100x_3 - 20(x_3 - x_1) - 80(x_3 - x_2) + F_3(t)$$

$$\begin{bmatrix} 3 & 0 & 0 \\ 0 & 1 & 0 \\ 0 & 0 & 2 \end{bmatrix} \begin{bmatrix} \ddot{x}_1 \\ \ddot{x}_2 \\ \ddot{x}_3 \end{bmatrix} + \begin{bmatrix} 140 & -60 & -20 \\ -60 & 220 & -80 \\ -20 & -80 & 200 \end{bmatrix} \begin{bmatrix} x_1 \\ x_2 \\ x_3 \end{bmatrix} = \begin{bmatrix} F_1(t) \\ F_2(t) \\ F_3(t) \end{bmatrix} \qquad \textbf{(4-39)}$$

or

$$M\{\ddot{x}\} + K\{x\} = \{F(t)\}$$

The program COEFF to find the coefficients of the frequency equation is listed in Fig. 4-11(a). The values of the matrices M and K are entered (READ) and verified (WRITE) in Par. #I. The computations in Par. #II first change Eq. (4-36) into the form

$$\{\ddot{q}\} + M^{-1}K\{q\} = \{0\}$$

The subroutine $INVS is called to find the inverse M^{-1} (MINVS) of M. The subroutine $MPLY performs the matrix multiplication MINVS*K = H, where $H = M^{-1}K$ is the dynamic matrix. The subroutine $COEFF is called to give the coefficients of the frequency equation.

The print-out is listed in Fig. 4-11(b). We first give the command

MERGE COEFF, $COEFF, $INVS, $MPLY, $SUBN

to merge the main program COEFF with the necessary subroutines. The subroutine $SUBN is for the matrix substitutions in the calculations. When the computer is READY, the command RUN is given to start the program. The data entries are self-explanatory. The frequency equation is

$$1 - 45.685 \times 10^{-3}\omega^2 + 515.94 \times 10^{-6}\omega^4 - 1.4071 \times 10^{-6}\omega^6 = 0$$

Example 9

Assume the roots of the frequency equation in Example 8 are $\omega^2 = 33.23$, 86.67, and 246.8. Find the modal matrix of the system.

Solution:

For the given values of M and K and $\omega^2 = 33.23$, the direct application of Eq. (4-37) gives

$$\begin{bmatrix} 140 - 33.23 \times 3 & -60 & -20 \\ -60 & 220 - 33.23 \times 1 & -80 \\ -20 & -80 & 200 - 33.23 \times 2 \end{bmatrix} \begin{bmatrix} x_1 \\ x_2 \\ x_3 \end{bmatrix} = \{0\}$$

It can be shown that the solution of the homogeneous algebraic equation is $\{x\} = \{2.172 \quad 1.126 \quad 1.0\} = \{u\}_1$, where $\{u\}_1$ is the modal vector for $\omega^2 =$

```
COEFF

"   *** COEFFICIENTS OF CHARACTERISTIC POLYNOMIAL ***
"       SUBROUTINES REQUIRED: (1) $COEFF (2) $INVS (3) $MPLY (4) $SUBN
"
"   *** #I. FORMAT AND INPUT. ***
    REAL*8  C(12), H(10,10), K(10,10), M(10,10), MINVS(10,10)
80  FORMAT (' COEFFICIENTS OF CHARACTERISTIC POLYNOMIAL.',/)
81  FORMAT (' ENTER: (1) ORDER OF MATRICES M & K    ** N <= 10',/,8X,%
    '(2) MATRIX-M BY ROW',/,8X,'(3) MATRIX-K BY ROW')
82  FORMAT (' N = ', I3)
83  FORMAT (5D14.5)
84  FORMAT (' *** IS THIS CORRECT?   1 = YES;  2 = NO.')
85  FORMAT (' CHARACTERISTIC POLYNOMIAL:',/, %
    '    F(X) = C0 + C1*X + C2*X↑2 + C3*X↑3 + . . + CN*X↑N',/, %
    '    THE VALUES OF C0 TO CN ARE:-',/)
    WRITE (6,80)
10  WRITE (6,81)
    READ (5,*) N, ((M(I,J), J=1,N), I=1,N), ((K(I,J), J=1,N), I=1,N)
    WRITE (6,82) N
    DO 40 I=1,N
40    WRITE (6,83) (M(I,J), J=1,N)
    DO 41 I=1,N
41    WRITE (6,83) (K(I,J), J=1,N)
    WRITE (6,84)
    READ (5,*) IANS
    IF (IANS = 2) GOTO 10
    NP2 = N + 2
"
"   *** #II. CALCULATIONS AND OUTPUT. ***
    CALL $INVS (M, MINVS, N)
    CALL $MPLY (MINVS, K, H, N)
    CALL $COEFF (H, C, N)
    WRITE (6,85)
    WRITE (6,83) (C(I), I=2,NP2)
    STOP
    END
```

(a) Main program

```
MERGE COEFF, $COEFF, $INVS, $MPLY, $SUBN
READY

RUN

COEFFICIENTS OF CHARACTERISTIC POLYNOMIAL.

ENTER: (1) ORDER OF MATRICES M & K    ** N <= 10
       (2) MATRIX-M BY ROW
       (3) MATRIX-K BY ROW
?3   3 0 0  0 1 0  0 0 2   140 -60 -20  -60 220 -80  -20 -80 200

N =   3
  0.30000D+01   0.0           0.0
  0.0           0.10000D+01   0.0
  0.0           0.0           0.20000D+01
  0.14000D+03  -0.60000D+02  -0.20000D+02
 -0.60000D+02   0.22000D+03  -0.80000D+02
 -0.20000D+02  -0.80000D+02   0.20000D+03
*** IS THIS CORRECT?   1 = YES;  2 = NO.
?1

CHARACTERISTIC POLYNOMIAL:
  F(X) = C0 + C1*X + C2*X↑2 + C3*X↑3 + . . + CN*X↑N
  THE VALUES OF C0 TO CN ARE:-

  0.10000D+01  -0.45685D-01   0.51594D-03  -0.14071D-05
```

(b) Print-out

FIG. 4-11. *Program to find coefficients of frequency equation: Example 8.*

33.23. Similarly, for $\omega^2 = 86.67$ and 246.8, the modal vectors are $\{u\}_2 = \{-0.381 \quad 0.429 \quad 1.0\}$ and $\{u\}_3 = \{0.342 \quad -3.755 \quad 1.0\}$, respectively. Hence the modal matrix is

$$[u] = [\{u\}_1 \quad \{u\}_2 \quad \{u\}_3] = \begin{bmatrix} 2.172 & -0.381 & 0.342 \\ 1.126 & 0.429 & -3.755 \\ 1.0 & 1.0 & 1.0 \end{bmatrix}$$

The equations of motion of undamped systems can be uncoupled by the orthogonal relations, which will be derived in Sec. 6-4. These are

$$[u]^T M[u] = \lceil M \rfloor \qquad [u]^T K[u] = \lceil K \rfloor \tag{4-40}$$

where $[u]^T$ is the transpose of the modal matrix $[u]$ and $\lceil M \rfloor$ and $\lceil K \rfloor$ are diagonal matrices. Substituting $\{q\} = [u]\{p\}$ from Eq. (4-34) in (4-35), we have

$$M[u]\{\ddot{p}\} + K[u]\{p\} = \{Q(t)\}$$

Premultiplying this by the transpose $[u]^T$ of $[u]$ gives

$$[u]^T M[u]\{\ddot{p}\} + [u]^T K[u]\{p\} = [u]^T\{Q(t)\}$$

or

$$\lceil M \rfloor\{\ddot{p}\} + \lceil K \rfloor\{p\} \triangleq \{N(t)\} \tag{4-41}$$

where $\{N(t)\} = [u]^T\{Q(t)\}$ is the excitation associated with the $\{p\}$ coordinates. Since $\lceil M \rfloor$ and $\lceil K \rfloor$ are diagonal, Eq. (4-41) gives the uncoupled equations.

Example 10

Describe a procedure to find the transient response of the system in Fig. 4-10(a). Assume the excitation for each of the masses is as shown in Fig. 4-10(b). The initial conditions are $\{x(0)\} = \{1 \quad 2 \quad -1\}$ and $\{\dot{x}(0)\} = \{-3 \quad 4 \quad 1\}$.

Solution:

We shall first show the orthogonal relations in Eq. (4-40) and then describe the procedure. Substituting the values of M, K, and $[u]$ from Example 9 in Eq. (4-40) yields

$$[u]^T M[u] = \begin{bmatrix} 2.172 & 1.126 & 1.0 \\ -0.381 & 0.429 & 1.0 \\ 0.342 & -3.755 & 1.0 \end{bmatrix} \begin{bmatrix} 3 & 0 & 0 \\ 0 & 1 & 0 \\ 0 & 0 & 2 \end{bmatrix}$$

$$\begin{bmatrix} 2.172 & -0.381 & 0.342 \\ 1.126 & 0.429 & -3.755 \\ 1.0 & 1.0 & 1.0 \end{bmatrix}$$

$$= \begin{bmatrix} 174.3 & 0 & 0 \\ 0 & 2.619 & 0 \\ 0 & 0 & 16.45 \end{bmatrix} = \lceil M \rfloor.$$

Similarly, it can be shown that the K matrix gives

$$[u]^T K[u] = \begin{bmatrix} 579.0 & 0 & 0 \\ 0 & 227.0 & 0 \\ 0 & 0 & 4059 \end{bmatrix} = \lceil K \rfloor$$

Hence, from Eq. (4-41), the uncoupled equations of motion are

$$\begin{aligned} 1.743\ddot{p}_1 + 579p_1 &= N_1(t) \\ 2.619\ddot{p}_2 + 227p_2 &= N_2(t) \\ 16.45\ddot{p}_3 + 4059p_3 &= N_3(t) \end{aligned} \tag{4-42}$$

From Eq. (4-34), the initial conditions expressed in terms of the principal coordinates $\{p\}$ are

$$\{p(0)\} = [u]^{-1}\{x(0)\} \quad \text{and} \quad \{\dot{p}(0)\} = [u]^{-1}\{\dot{x}(0)\} \tag{4-43}$$

The uncoupled equations in Eq. (4-42) and the initial conditions from Eq. (4-43) are used to solve the problem in terms of the principal coordinates $\{p\}$. The response of the masses in the $\{x\}$ coordinates are obtained by means of the transformation $\{x\} = [u]\{p\}$ in Eq. (4-34).

Since the excitations $\{F(t)\}$ shown in Fig. 4-10(*b*) are arbitrary, they are quantized and assumed to have constant values for each time interval $\Delta t = 0.05$. The solutions are plotted in Fig. 4-10(*c*). Although the procedure is conceptually simple, computer solutions are mandatory. The problem is solved by the program TRESPUND shown in Sec. 9-8.

4-7 SEMIDEFINITE SYSTEMS

A special case of practical importance occurs when a root of the frequency equation vanishes. When a natural frequency is zero, there is no relative motion in the system. The system can move as a rigid body and is called *semidefinite.*

Two semidefinite systems are shown in Fig. 4-12. The rectilinear system consists of a number of masses coupled by springs. It may be used to represent the vibration of a train. The rotational system may represent a rotating machine, such as a diesel engine for marine propulsion. One of the disks may represent the propeller, another disk the flywheel, and the remaining disks the rotating and the equivalent reciprocating parts of the engine.

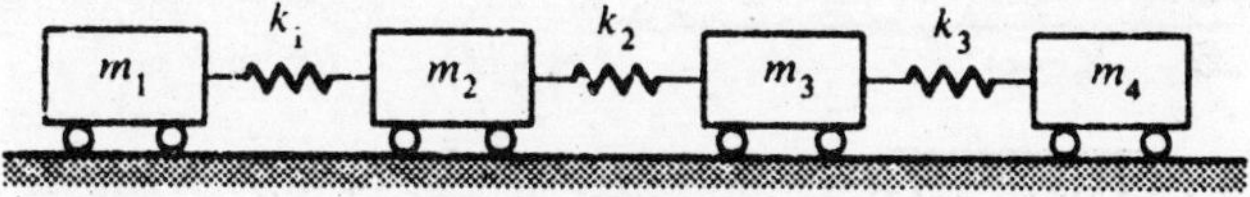

(*a*) Rectilinear system

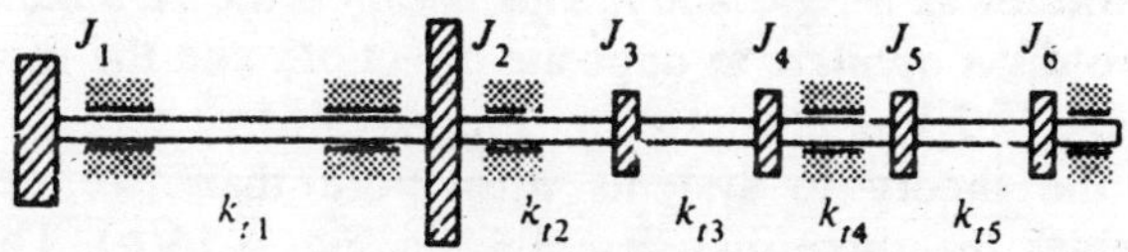

(*b*) Rotational system

FIG. 4-12. *Semidefinite systems.*

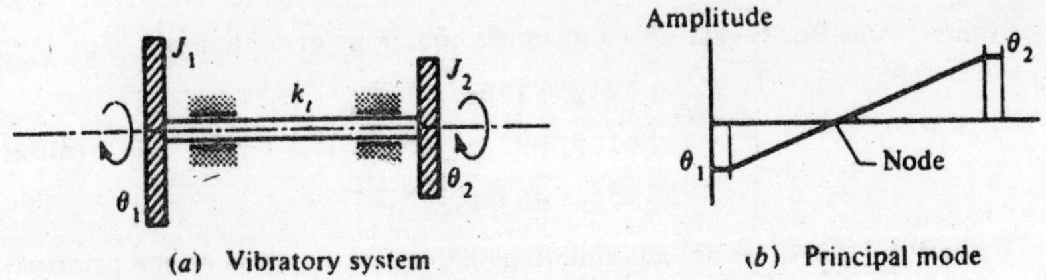

(a) Vibratory system (b) Principal mode

FIG. 4-13. *Schematic of a motor-generator set.*

Let an electrical motor-generator set be represented by a two-disk system shown in Fig. 4-13(a). The rotors J_1 and J_2 are connected by a shaft of spring constant k_t. Summing the torques for each rotor about the shaft axis, the equations of motion are

$$J_1\ddot{\theta}_1 = -k_t(\theta_1 - \theta_2)$$
$$J_2\ddot{\theta}_2 = -k_t(\theta_2 - \theta_1)$$

$$\begin{bmatrix} J_1 & 0 \\ 0 & J_2 \end{bmatrix}\begin{bmatrix} \ddot{\theta}_1 \\ \ddot{\theta}_2 \end{bmatrix} + \begin{bmatrix} k_t & -k_t \\ -k_t & k_t \end{bmatrix}\begin{bmatrix} \theta_1 \\ \theta_2 \end{bmatrix} = \begin{bmatrix} 0 \\ 0 \end{bmatrix} \tag{4-44}$$

This is of the same form as Eq. (4-9). Hence the frequency equation from Eq. (4-13) is

$$\Delta(\omega) = \begin{vmatrix} k_t - \omega^2 J_1 & -k_t \\ -k_t & k_t - \omega^2 J_2 \end{vmatrix} = 0 \tag{4-45}$$

Expanding the determinant and dividing by J_1J_2, we get

$$\omega^2\left[\omega^2 - \left(\frac{k_t}{J_1} + \frac{k_t}{J_2}\right)\right] = 0 \tag{4-46}$$

The roots of the equation are $\omega^2 = 0$ and $\omega^2 = k_t(1/J_1 + 1/J_2)$.

Following the procedure in Eq. (4-16), the relative amplitude of the disks at the principal modes are

$$\frac{\Theta_1}{\Theta_2} = \frac{k_t}{k_t - \omega^2 J_1} = \frac{k_t - \omega^2 J_2}{k_t} = \begin{cases} 1 & \text{for } \omega = 0 \\ -J_2/J_1 & \text{for } \omega = \omega_1 \end{cases} \tag{4-47}$$

where $\omega_1 = \sqrt{k_t(1/J_1 + 1/J_2)}$ as indicated in Eq. (4-46). When $\omega = 0$ and $\Theta_1/\Theta_2 = 1$, the two disks have identical angular displacements. Since there is no relative displacement between the disks, the shaft is not stressed and the assembly rotates as a rigid body. This is called the *zero mode*. When $\omega = \omega_1$, the two disks oscillate in opposite directions and the mode shape is $\{\Theta_1 \quad \Theta_2\} = \{J_2 \quad -J_1\}$ as illustrated in Fig. 4-13(b).

To extend the theory to systems with more than two degrees of freedom, consider the three-disk assembly in Fig. 4-14(a). The system could be used to represent, to the first approximation, a multidegree-of-freedom system, such as a diesel engine for marine propulsion. From

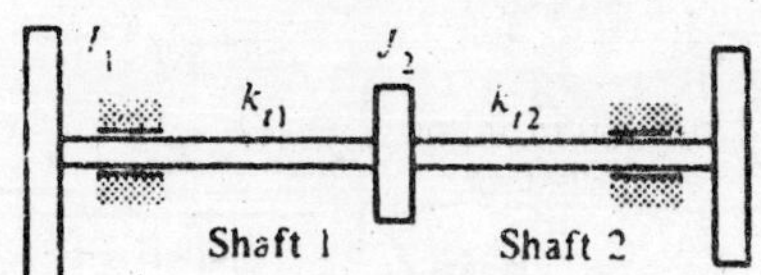

(a) Vibratory system

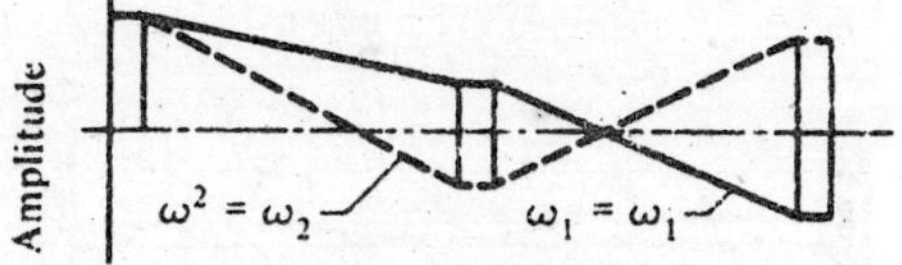

(b) Principal modes: $\omega_1 < \omega_2$

FIG. 4-14. *A semidefinite system.*

Newton's second law, the equations of motion of the disks are

$$\begin{aligned} J_1\ddot{\theta}_1 &= -k_{t1}(\theta_1 - \theta_2) \\ J_2\ddot{\theta}_2 &= -k_{t1}(\theta_2 - \theta_1) - k_{t2}(\theta_2 - \theta_3) \\ J_3\ddot{\theta}_3 &= -k_{t2}(\theta_3 - \theta_2) \end{aligned} \tag{4-48}$$

Similar to Eq. (4-46), the frequency equation of the system is

$$\omega^2\left\{\omega^4 - \left[k_{t1}\left(\frac{1}{J_1}+\frac{1}{J_2}\right)+k_{t2}\left(\frac{1}{J_2}+\frac{1}{J_3}\right)\right]\omega^2 + k_{t1}k_{t2}\frac{J_1+J_2+J_3}{J_1J_2J_3}\right\}=0 \tag{4-49}$$

The relative amplitudes at the principal modes can be obtained from Eq. (4-48) and expressed as

$$\frac{\Theta_1}{\Theta_2} = \frac{k_{t1}}{k_{t1} - \omega^2 J_1} \quad \text{and} \quad \frac{\Theta_2}{\Theta_3} = \frac{k_{t2} - \omega^2 J_3}{k_{t2}} \tag{4-50}$$

When $\omega^2 = 0$, the relative amplitudes of the disks are $\Theta_1/\Theta_2 = \Theta_2/\Theta_3 = 1$. This indicates that the whole assembly may rotate as a rigid body. The relative amplitudes of the principal modes are shown in Fig. 4-14(*b*). Note that there is one sign change in the amplitudes for the first mode and two sign changes for the second mode.

Example 11. Geared systems

Let a two-shaft system in Fig. 4-15(*a*) be connected by a pair of gears. (a) Neglecting the inertial effect of the gears, determine the frequencies of the system. (b) Repeat part a but include the inertial effect of the gears.

Solution:

Since the two shafts are at different rotational speeds, it is expedient to find an equivalent system referring to a common shaft. Let N_1 be the number of

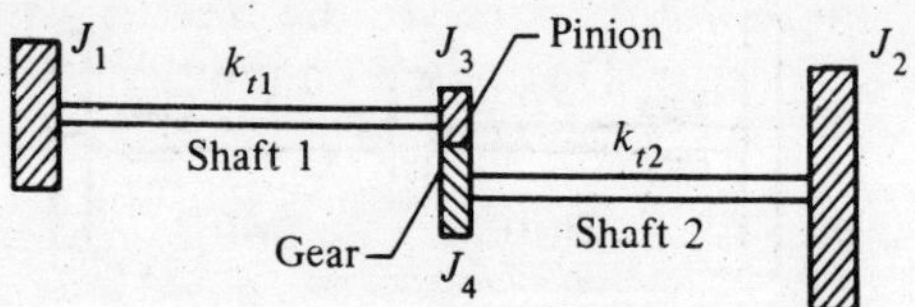

(a) Vibratory system

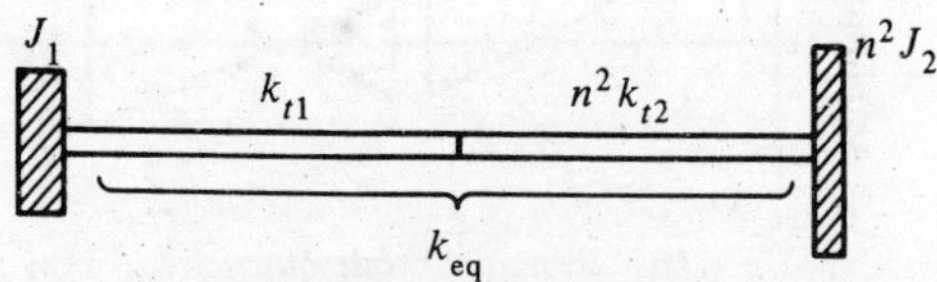

(b) Equivalent system, refer to shaft 1, neglect inertial effect of gears

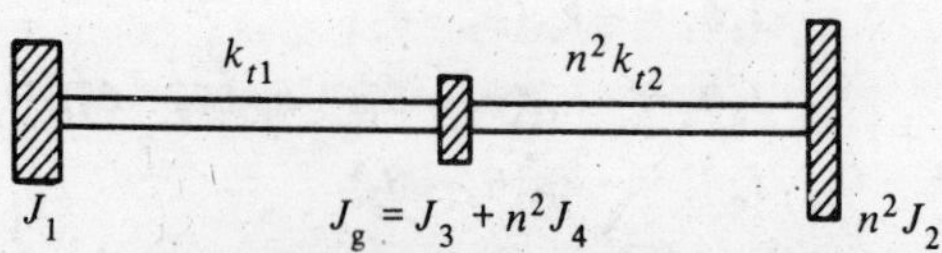

(c) Equivalent system, refer to shaft 1, including inertial effect of gears

FIG. 4-15. *Semidefinite geared system: Example 11.*

teeth on the pinion and N_2 that of the gear. Choosing shaft 1 as reference, the equivalent inertia of J_2 is $J_2^* = n^2 J_2$ as shown in Example 1, Chap. 3. Similarly, the equivalent spring constant of shaft 2 referring to shaft 1 is $k_{t2}^* = n^2 k_{t2}$.

(a) Referring to shaft 1 and neglecting the inertial effect of the gears, the equivalent system is shown in Fig. 4-15(*b*). The shafts are in series and the equivalent spring constant is

$$\frac{1}{k_{eq}} = \frac{1}{k_{t1}} + \frac{1}{k_{t2}^*} \qquad \text{or} \qquad k_{eq} = \frac{n^2 k_{t1} k_{t2}}{k_{t1} + n^2 k_{t2}}$$

The natural frequencies from Eq. (4-46) are

$$f_n = \begin{cases} 0 \\ \dfrac{1}{2\pi}\sqrt{k_{eq}\left(\dfrac{1}{J_1} + \dfrac{1}{n^2 J_2}\right)} \end{cases}$$

(b) Including the gears, the system has four disks and therefore four equations of motion. J_3 of the pinion and J_4 of the gear can be combined to give $J_g = (J_3 + n^2 J_4)$, referring to shaft 1. Thus, the equivalent system consists of three disks and two shafts as shown in Fig. 4-15(*c*). The frequency equation would be identical to Eq. (4-49).

Note that the natural frequencies are the frequencies of oscillation of one disk relative to another, superposed on the rigid body rotation of the assembly. The natural frequencies can be calculated referring to one shaft or the other. The proof of this statement is left as an exercise.

4-8 FORCED VIBRATION—HARMONIC EXCITATION

The general form of the equations of motion of a two-degree-of-freedom system is shown in Eq. (4-4). If the excitation is harmonic, the equations can be solved readily by the impedance method developed in Sec. 2-6. The numerical solutions for systems with damping, however, are tedious. Computers can be used to alleviate the calculations.

Applying the impedance method to Eq. (4-4), we substitute the harmonic force vector $\{\mathbf{F}\}$ for the generalized force $\{Q(t)\}$, where $\{\mathbf{F}\}=\{\bar{F}e^{j\omega t}\}=\{\bar{F}\}e^{j\omega t}$ and the 2×1 matrix $\{\bar{F}\}$ is the phasor of $\{\mathbf{F}\}$. All the harmonic components in $\{\mathbf{F}\}$ are assumed of the same frequency ω. If not, one frequency can be treated at a time and the resultant response obtained by superposition. Let $\{\mathbf{X}\}$ be the harmonic response, where $\{\mathbf{X}\}=\{\bar{X}e^{j\omega t}\}=\{\bar{X}\}e^{j\omega t}$ and $\{\bar{X}\}$ is the phasor of $\{\mathbf{X}\}$. Applying the impedance method and factoring out $e^{j\omega t}$, Eq. (4-4) becomes

$$-\omega^2\begin{bmatrix} m_{11} & m_{12} \\ m_{21} & m_{22}\end{bmatrix}\begin{bmatrix}\bar{X}_1\\ \bar{X}_2\end{bmatrix}+j\omega\begin{bmatrix} c_{11} & c_{12} \\ c_{21} & c_{22}\end{bmatrix}\begin{bmatrix}\bar{X}_1\\ \bar{X}_2\end{bmatrix}$$
$$+\begin{bmatrix} k_{11} & k_{12} \\ k_{21} & k_{22}\end{bmatrix}\begin{bmatrix}\bar{X}_1\\ \bar{X}_2\end{bmatrix}=\begin{bmatrix}\bar{F}_1\\ \bar{F}_2\end{bmatrix} \qquad \textbf{(4-51)}$$

or

$$[K-\omega^2 M+j\omega C]\{\bar{X}\}=\{\bar{F}\} \qquad \textbf{(4-52)}$$

where the matrices M, C, and K can be identified readily.

The equations above can be alternatively expressed as

$$\begin{bmatrix} k_{11}-\omega^2 m_{11}+j\omega c_{11} & k_{12}-\omega^2 m_{12}+j\omega c_{12} \\ k_{21}-\omega^2 m_{21}+j\omega c_{21} & k_{22}-\omega^2 m_{22}+j\omega c_{22}\end{bmatrix}\begin{bmatrix}\bar{X}_1\\ \bar{X}_2\end{bmatrix}=\begin{bmatrix}\bar{F}_1\\ \bar{F}_2\end{bmatrix} \qquad \textbf{(4-53)}$$

or

$$\begin{bmatrix} z_{11} & z_{12} \\ z_{21} & z_{22}\end{bmatrix}\begin{bmatrix}\bar{X}_1\\ \bar{X}_2\end{bmatrix}=\begin{bmatrix}\bar{F}_1\\ \bar{F}_2\end{bmatrix} \qquad \textbf{(4-54)}$$

or

$$Z(\omega)\{\bar{X}\}=\{\bar{F}\} \qquad \textbf{(4-55)}$$

where

$$z_{ij}=(k_{ij}-\omega^2 m_{ij}+j\omega c_{ij}) \qquad \text{for} \qquad i,j=1,2 \qquad \textbf{(4-56)}$$

and $Z(\omega) = [z_{ij}]$ is the *impedance matrix*. In other words, Eqs. (4-51) to (4-55) are different forms of the same equation, all of which can be summarized by Eq. (4-55).

The solution $\{\bar{X}\}$ gives the amplitude and phase angle of the response relative to the excitation $\{\bar{F}\}$. Premultiplying both sides of Eq. (4-55) by the inverse $Z(\omega)^{-1}$ of $Z(\omega)$ gives

$$\{\bar{X}\} = Z(\omega)^{-1}\{\bar{F}\} \tag{4-57}$$

For a two-degree-of-freedom system this can be written explicitly as

$$\bar{X}_1 = \frac{z_{22}\bar{F}_1 - z_{12}\bar{F}_2}{z_{11}z_{22} - z_{12}z_{21}} \quad \text{and} \quad \bar{X}_2 = \frac{-z_{21}\bar{F}_1 + z_{11}\bar{F}_2}{z_{11}z_{22} - z_{12}z_{21}} \tag{4-58}$$

Equations (4-55) to (4-57) are equally applicable to n-degree-of-freedom systems. The elements of the impedance matrix $Z(\omega)$ in Eq. (4-55) become

$$z_{ij} = (k_{ij} - \omega^2 m_{ij} + j\omega c_{ij}) \quad \text{for} \quad i, j = 1, 2, \ldots, n \tag{4-59}$$

and $Z(\omega)$ is of order n. Note that each z_{ij} is identical in form to the mechanical impedance in Eq. (2-51) and $Z(\omega)$ is symmetric by the proper choice of coordinates as discussed in Sec. 4-4.

Example 12. Undamped dynamic absorber

Excessive vibration, due to near resonance conditions, is encountered in a constant speed machine shown in Fig. 4-16(a). The original system consists of m_1 and k_1. It is not feasible to change m_1 and k_1. **(a)** Show that a dynamic absorber, consisting of m_2 and k_2, will remedy the problem. **(b)** Plot the response curves of the system, assuming $m_2/m_1 = 0.3$. **(c)** Investigate the effect of the mass ratio m_2/m_1.

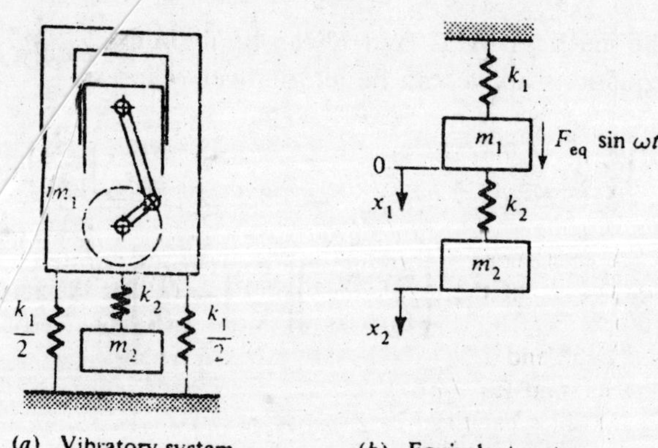

(a) Vibratory system (b) Equivalent system

FIG. 4-16. *Undamped dynamic absorber: Example 12.*

Solution:

The equations of motion for the equivalent system in Fig. 4-16(*b*) are

$$m_1\ddot{x}_1 = -k_1x_1 - k_2(x_1 - x_2) + F_{eq}\sin\omega t$$

$$m_2\ddot{x}_2 = -k_2(x_2 - x_1)$$

The impedance method can be applied directly, since the excitation is harmonic. From Eq. (4-53), we have

$$\begin{bmatrix} k_1 + k_2 - \omega^2 m_1 & -k_2 \\ -k_2 & k_2 - \omega^2 m_2 \end{bmatrix}\begin{bmatrix} \bar{X}_1 \\ \bar{X}_2 \end{bmatrix} = \begin{bmatrix} F_{eq} \\ 0 \end{bmatrix}$$

(a) Following Eq. (4-13), the frequency equation is obtained by equating the characteristic determinant $\Delta(\omega)$ of the coefficient matrix of $\{\bar{X}\}$ to zero, that is,

$$\Delta(\omega) = (k_1 + k_2 - \omega^2 m_1)(k_2 - \omega^2 m_2) - k_2^2 = 0 \qquad \textbf{(4-60)}$$

From Eq. (4-58), the phasors of the responses are

$$\bar{X}_1 = \frac{1}{\Delta(\omega)}(k_2 - \omega^2 m_2)F_{eq} \quad \text{and} \quad \bar{X}_2 = \frac{1}{\Delta(\omega)}k_2F_{eq} \qquad \textbf{(4-61)}$$

where $\Delta(\omega)$ is the characteristic determinant. Note that the amplitude $\bar{X}_1$ becomes zero at the excitation frequency $\omega = \sqrt{k_2/m_2}$. An undamped dynamic absorber is "tuned" for $k_1/m_1 = k_2/m_2$, such that $\bar{X}_1$ approaches zero at the resonance frequency of the original system.

(b) The frequency equation in Eq. (4-60) can be expressed as

$$\frac{m_1m_2}{k_1k_2}\omega^4 - \left[\left(1 + \frac{k_2}{k_1}\right)\frac{m_2}{k_2} + \frac{m_1}{k_1}\right]\omega^2 + 1 = 0$$

Since $k_2/k_1 = m_2/m_1$, a frequency ratio r is defined as $r = \omega/\sqrt{k_1/m_1} = \omega/\sqrt{k_2/m_2}$. The frequency equation reduces to

$$r^4 - (2 + m_2/m_1)r^2 + 1 = 0 \qquad \textbf{(4-62)}$$

From Eq. (4-61), the responses can be expressed as

$$\frac{\bar{X}_1}{F_{eq}/k_1} = \frac{1 - r^2}{r^4 - (2 + m_2/m_1)r^2 + 1}$$

$$\frac{\bar{X}_2}{F_{eq}/k_1} = \frac{1}{r^4 - (2 + m_2/m_1)r^2 + 1} \qquad \textbf{(4-63)}$$

The equations are plotted in Fig. 4-17 for $m_2/m_1 = 0.3$. The plus or minus sign of the amplitude ratio denotes that the response is either in-phase or 180° out-of-phase with the excitation. Resonances occur at $r = 0.762$ and 1.311. Note that $x_1(t) = 0$ when $k_2 - \omega^2 m_2 = 0$. It can be shown from Eq. (4-61) that this condition occurs when the excitation $F_{eq}\sin\omega t$ is balanced by the spring force $-k_2x_2$.

(c) The frequency equation, Eq. (4-62), is plotted in Fig. 4-18 to show the effect of the mass ratio m_2/m_1. When m_2/m_1 is small, the resonant

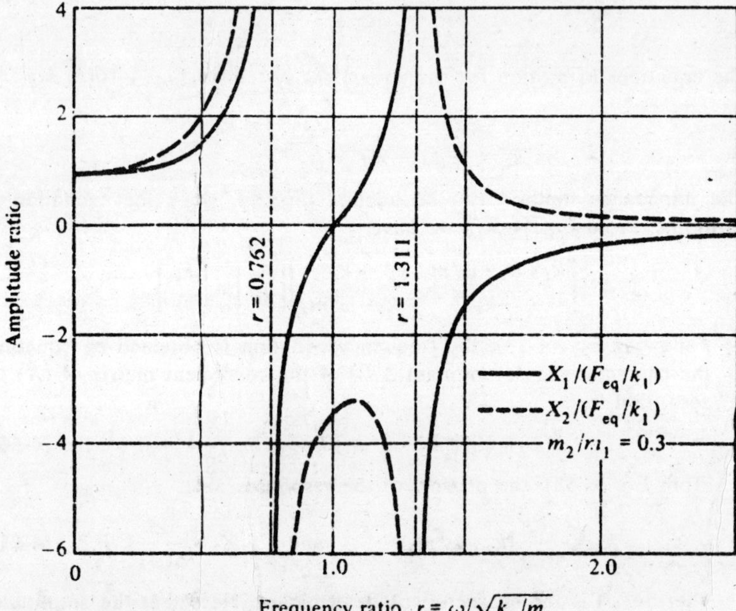

FIG. 4-17. *Typical harmonic response of a two-degree-of-freedom system: Example 12.*

frequencies are close together about the resonance frequency of the original system. This means that there is little tolerance for variations in the excitation frequency, although $x_1(t) = 0$ when $r = 1$. Furthermore, it is observed in Eq. (4-63) that the amplitude X_2 of the absorber at $r = 1$ can be large for small values of m_2/m_1. When m_2/m_1 is appreciable, the resonant frequencies are separated. For example, when $m_2/m_1 = 0.4$,

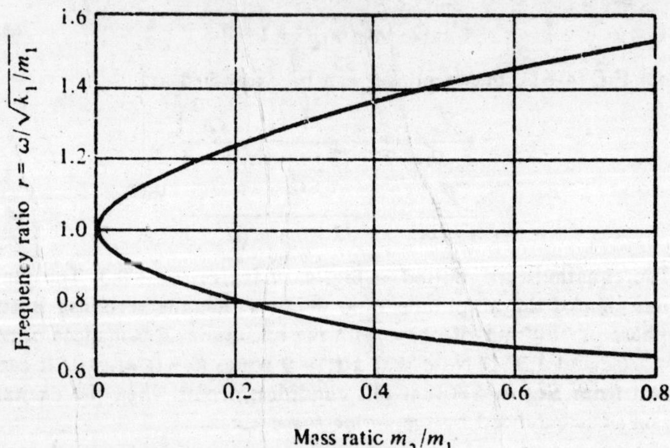

FIG. 4-18. *Undamped dynamic absorber: effect of mass ratio m_2/m_1; Example 12.*

resonances occur at r equal to 0.73 and 1.37 times that of the original system. The amplitude X_2 of the absorber mass is correspondingly reduced at $r = 1$ for larger mass ratios.

Example 13. Dynamic absorber with damping

Consider the dynamic absorber in Example 12 in which a viscous damper c is installed in parallel with the spring k_2 as shown in Fig. 4-19(b). Briefly discuss the problem.

Solution:

From Eq. (4-53), the equations of motion in phasor notations are

$$\begin{bmatrix} k_1 + k_2 - \omega^2 m_1 + j\omega c & -k_2 - j\omega c \\ -k_2 - j\omega c & k_2 - \omega^2 m_2 + j\omega c \end{bmatrix} \begin{bmatrix} \bar{X}_1 \\ \bar{X}_2 \end{bmatrix} = \begin{bmatrix} F_{eq} \\ 0 \end{bmatrix}$$

From Eq. (4-13), the corresponding frequency equation is

$$\Delta(\omega) = \begin{vmatrix} k_1 + k_2 - \omega^2 m_1 + j\omega c & -k_2 - j\omega c \\ -k_2 - j\omega c & k_2 - \omega^2 m_2 + j\omega c \end{vmatrix} = 0$$

From Eq. (4-58), the phasors of the responses are

$$\bar{X}_1 = \frac{1}{\Delta(\omega)}(k_2 - \omega^2 m_2 + j\omega c)F_{eq} \quad \text{and} \quad \bar{X}_2 = \frac{1}{\Delta(\omega)}(k_2 + j\omega c)F_{eq}$$

where $\Delta(\omega)$ is the characteristic determinant. The values of $\bar{X}_1$ and $\bar{X}_2$ can be calculated readily using the programs in Chap. 9.

The response curve of a properly tuned* dynamic absorber with appropriate damping is shown in Fig. 4-19(a). Curve 1 is that of an undamped system and curve 2 corresponding to $c = \infty$. Curve 3 of intermediate damping must pass through the intersections of these curves.

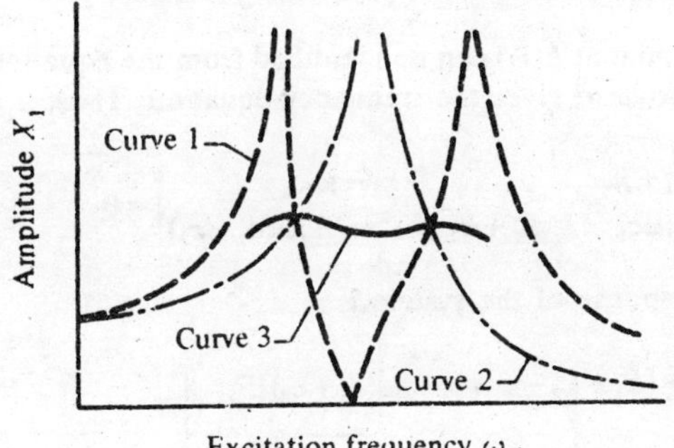

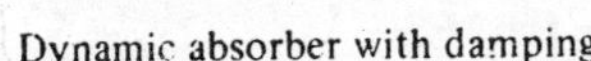

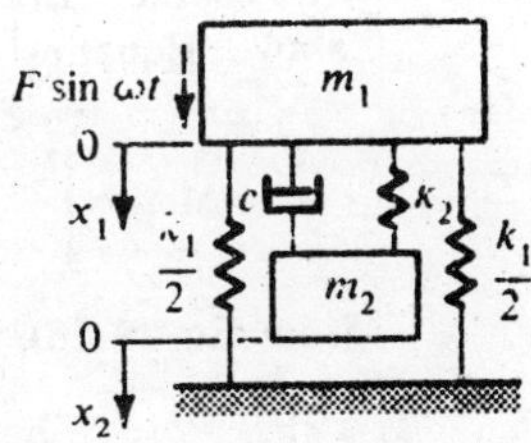

(a) Dynamic absorber with damping (b) Vibratory system

FIG. 4-19. *Dynamic absorber with damping: Example 13.*

* J. D. Den Hartog, *Mechanical Vibrations*, 4th ed., McGraw-Hill Book Company, New York, 1956, pp. 93–102. Note that the "tuned" condition of $k_1/m_1 = k_2/m_2$ in Example 12 is only for undamped absorbers. See Prob. 4-27 for discussion of dynamic absorbers with viscous damping.

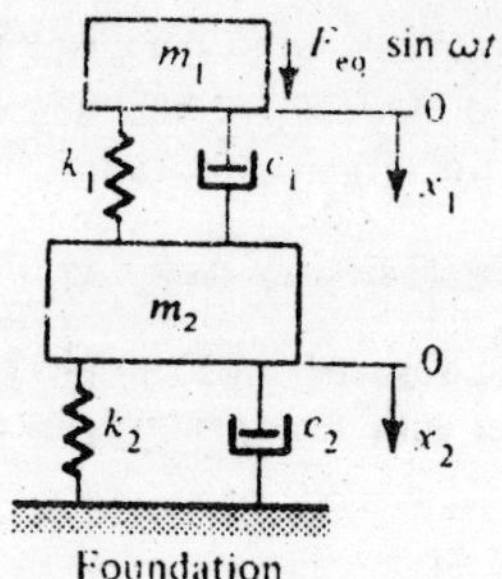

FIG. 4-20. *Vibration isolation: Example 14.*

Example 14. Vibration isolation

A constant speed machine is isolated as shown in Fig. 3-23(*a*) and the girls in the office complain of the annoying vibration transmitted from the machine. It is proposed (1) to mount the machine m_1 on a cement block m_2 as shown in Fig. 4-20, or (2) to bolt m_1 rigidly to m_2. Assume $m_2/m_1 = 4$, $k_1 = k_2$, and the excitation frequency $\omega = 2\omega_{n1} = 2\sqrt{k_1/m_1}$. (a) Find the magnitudes of $x_1(t)$ and $x_2(t)$ and the force F_T transmitted to the rigid foundation. (b) Neglecting the damping in the system for the estimation, would you approve proposal 1 or 2?

Solution:

(a) From Eq. (4-53), the equations of motion in phasor notations are

$$\begin{bmatrix} k_1-\omega^2 m_1+j\omega c_1 & -k_1-j\omega c_1 \\ -k_1-j\omega c_1 & k_1+k_2-\omega^2 m_2+j\omega(c_1+c_2) \end{bmatrix}\begin{bmatrix} \bar{X}_1 \\ \bar{X}_2 \end{bmatrix}=\begin{bmatrix} F_{eq} \\ 0 \end{bmatrix}$$

The characteristic determinant $\Delta(\omega)$ can be identified from the equation above. Equating $\Delta(\omega)$ to zero gives the frequency equation. Thus,

$$\Delta(\omega)=\begin{vmatrix} k_1-\omega^2 m_1+j\omega c_1 & -k_1-j\omega c_1 \\ -k_1-j\omega c_1 & k_1+k_2-\omega^2 m_2-j\omega(c_1+c_2) \end{vmatrix}=0$$

From Eq. (4-58), the response of the system is

$$\bar{X}_1=\frac{1}{\Delta(\omega)}[k_1+k_2-\omega^2 m_2+j\omega(c_1+c_2)]F_{eq}$$

$$\bar{X}_2=\frac{1}{\Delta(\omega)}(k_1+j\omega c_1)F_{eq}$$

The force $\bar{F}_T$ is transmitted to the foundation through the spring k_2 and the damper c_2. Thus,

$$\bar{F}_T=\bar{X}_2(k_2+j\omega c_2)=\frac{1}{\Delta(\omega)}(k_1+j\omega c_1)(k_2+j\omega c_2)F_{eq}$$

The solution $\{\bar{X}\}$ and the force transmitted $\bar{F}_T$ can be calculated from the equations above.

(b) *Proposal* 1: The frequency equation of the undamped system is

$$\Delta(\omega)=\begin{vmatrix} k_1-\omega^2 m_1 & -k_1 \\ -k_1 & k_1+k_2-\omega^2 m_2 \end{vmatrix}=0$$

Substituting the given conditions $m_2/m_1=4$, etc., the frequency equation can be expressed as

$$\Delta(\omega)=k_1k_2(1-6r^2+4r^4)=0$$

where $r=\omega/\omega_{n1}=\omega/\sqrt{k_1/m_1}$. Correspondingly, we get

$$\bar{X}_1=\frac{1}{\Delta(\omega)}(k_1+k_2-\omega^2 m_2)F_{eq} \qquad \text{or} \qquad \frac{X_1}{F_{eq}/k_1}=\frac{2-4r^2}{1-6r^2+4r^4}$$

$$\bar{X}_2=\frac{1}{\Delta(\omega)}k_1F_{eq} \qquad \text{or} \qquad \frac{\bar{X}_2}{F_{eq}/k_1}=\frac{1}{1-6r^2+4r^4}$$

$$\bar{F}_T=\frac{1}{\Delta(\omega)}k_1k_2F_{eq} \qquad \text{or} \qquad \frac{\bar{F}_T}{F_{eq}}=\frac{1}{1-6r^2+4r^4}$$

For the given values, resonance occurs at $r=0.437$ and 1.14. At $\omega=2\omega_{n1}$ or $r=2$, we have $X_1=|\bar{X}_1|=0.341F_{eq}/k_1$, $X_2=|\bar{X}_2|=0.024F_{eq}/k_1$, and $F_T=|\bar{F}_T|=0.024F_{eq}$.

Proposal 2: If m_1 is attached to m_2, the system has one degree of freedom. The equation of motion in phasor notations is

$$[k_2-\omega^2(m_1+m_2)]\bar{X}_2=F_{eq}$$

Using the given data and normalizing the results by k_1, we get

$$\frac{\bar{X}_2}{F_{eq}/k_1}=\frac{1}{1-5r^2}$$

$$\bar{F}_T=k_2\bar{X}_2 \qquad \text{and} \qquad \frac{\bar{F}_T}{F_{eq}}=\frac{1}{1-5r^2}$$

Resonance occurs at $r=0.45$. At the excitation frequency $\omega=2\omega_{n1}$ or $r=2$, $X_2=|\bar{X}_2|=0.053F_{eq}/k_1$ and $F_T=|\bar{F}_T|=0.053F_{eq}$. Hence the force transmitted is higher than that in proposal 1.

4-9 INFLUENCE COEFFICIENTS

The *method of influence coefficients* gives an alternative procedure to formulate the equations of motion of a dynamic system. It is widely used in the analysis of structures, such as an aircraft. A spring can be described by its stiffness or its *compliance*, which is synonymous to the flexibility influence coefficient. We shall first (1) show Maxwell's reciprocity theorem, (2) relate the stiffness and flexibility matrices, and then (3) illustrate the method of influence coefficients.

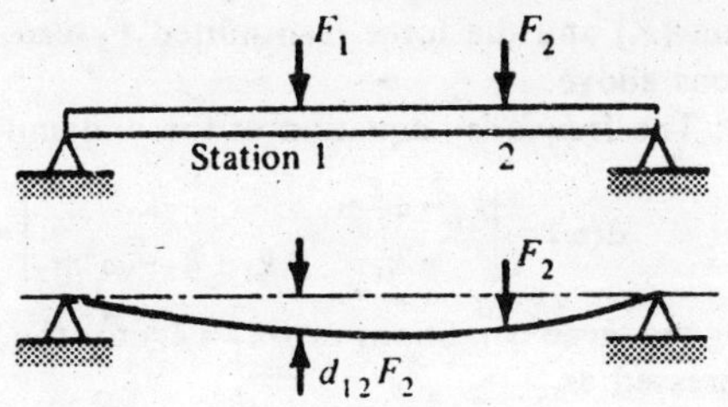

FIG. 4-21. *Method of influence coefficients.*

An *influence coefficient* d_{ij} defines the static elastic property of a system. The quantity d_{ij} is the deflection at station i owing to a unit force applied at station j, when this force is the *only* force applied. Consider the beam shown in Fig. 4-21. The vertical force F_1 is applied at station 1 and F_2 at station 2.

First, let F_1 be applied to station 1 and then F_2 to station 2. When F_1 is applied alone, the deflection at station 1 is $F_1 d_{11}$. The potential energy in the beam, by virtue of its deflection, is $\frac{1}{2}F_1^2 d_{11}$. Now, when F_2 is applied, the additional deflection at station 1 due to F_2 is $F_2 d_{12}$. The work by F_1 corresponding to this deflection is $F_1(F_2 d_{12})$. Thus, the total potential energy U of the system due to F_1 and F_2 is

$$U = \tfrac{1}{2}F_1^2 d_{11} + F_1(F_2 d_{12}) + \tfrac{1}{2}F_2^2 d_{22}$$

Secondly, let F_2 be applied to station 2 and then F_1 to station 1. It can be shown that the potential energy U due to F_2 and F_1 is

$$U = \tfrac{1}{2}F_2^2 d_{22} + F_2(F_1 d_{21}) + \tfrac{1}{2}F_1^2 d_{11}$$

The potential energies for the two methods of loading must be the same, since the final states of the system are identical. Comparing the expressions for U, we deduce that $d_{12} = d_{21}$ for the system with two loads. This is called ***Maxwell's reciprocity theorem.***

For the general case, we have

$$d_{ij} = d_{ji} \qquad \text{for} \qquad i, j = 1, 2, \ldots, n \tag{4-64}$$

which holds for all linear systems. When the force F is generalized to represent a force or a moment, the influence coefficient d_{ij} correspondingly represents a rectilinear or an angular displacement. Furthermore, when the deflections due to the inertia forces are considered, we obtain the equations of motion of the system.

During vibration, the inertia force associated with each mass is transmitted throughout the system to cause a motion at each of the other masses. For the undamped free vibration of a two-degree-of-freedom system, we have

$$\begin{bmatrix} q_1 \\ q_2 \end{bmatrix} = \begin{bmatrix} d_{11} & d_{12} \\ d_{21} & d_{22} \end{bmatrix} \begin{bmatrix} -m_1\ddot{q}_1 \\ -m_2\ddot{q}_2 \end{bmatrix} \tag{4-65}$$

or

$$\{q\} = [d_{ij}]\{-m\ddot{q}\} \tag{4-66}$$

where $\{q\}$ is a generalized displacement vector, $\{-m\ddot{q}\}$ the generalized force vector of the inertia forces, and $[d_{ij}]$ is the *flexibility* (influence coefficient) *matrix.* For example, the deflection q_1 at station 1 is due to the combined effect of the inertia forces $-m_1\ddot{q}_1$ and $-m_2\ddot{q}_2$. The total deflection q_1 is $[d_{11}(-m_1\ddot{q}_1) + d_{12}(-m_2\ddot{q}_2)]$.

From Eq. (4-26), in the absence of dynamic coupling, the equations of motion for the free vibration of an undamped system can be expressed as

$$\{m\ddot{q}\} = -[k_{ij}]\{q\} \tag{4-67}$$

where $[k_{ij}]$ is the stiffness matrix. Premultiplying Eq. (4-66) by the inverse $[d_{ij}]^{-1}$ of $[d_{ij}]$ and rearranging, we get

$$\{m\ddot{q}\} = -[d_{ij}]^{-1}\{q\}$$

Comparing the last two equations, it is evident that

$$[k_{ij}] = [d_{ij}]^{-1} \quad \text{or} \quad [k_{ij}][d_{ij}] = I \tag{4-68}$$

where I is a unit matrix. This is to say that $[k_{ij}]$ is the inverse of $[d_{ij}]$ and vice versa.

Example 15

Write the equations of motion for the system shown in Fig. 4-2(a) by the method of influence coefficients and find the frequency equation.

Solution:

To find the influence coefficients, let a unit static force be applied to m_1. The springs k and k_2 are in series and their combination is in parallel with k_1. Thus,

$$k_{eql} = k_1 + \frac{kk_2}{k + k_2}$$

The corresponding deflection of m_1 is d_{11}.

$$d_{11} = \frac{1}{k_{eql}} = \frac{k + k_2}{k_1k_2 + k(k_1 + k_2)}$$

Since the deflection of a spring is inversely proportional to its stiffness and the deflection of m_1 is d_{11}, it can be shown that the corresponding deflection

d_{21} of m_2 is

$$d_{21}=\frac{k}{k+k_2}\,d_{11}=\frac{k}{k_1k_2+k(k_1+k_2)}=d_{12}$$

Similarly, considering a unit static force at m_2, we get

$$d_{22}=\frac{k+k_1}{k_1k_2+k(k_1+k_2)}$$

Combining the influence coefficients yields

$$[d_{ij}]=\begin{bmatrix} d_{11} & d_{12} \\ d_{21} & d_{22}\end{bmatrix}=\frac{1}{k_1k_2+k(k_1+k_2)}\begin{bmatrix} k+k_2 & k \\ k & k+k_1\end{bmatrix}$$

The equations of motion from Eq. (4-66) are

$$\{x\}=[d_{ij}]\{-m\ddot{x}\} \tag{4-69}$$

$$\begin{bmatrix} x_1 \\ x_2\end{bmatrix}=\frac{1}{k_1k_2+k(k_1+k_2)}\begin{bmatrix} k+k_2 & k \\ k & k+k_1\end{bmatrix}\begin{bmatrix} -m_1\ddot{x}_1 \\ -m_2\ddot{x}_2\end{bmatrix} \tag{4-70}$$

Note that the stiffness matrix from Eq. (4-9) is

$$[k_{ij}]=\begin{bmatrix} k+k_1 & -k \\ -k & k+k_2\end{bmatrix}$$

It can be shown readily that $[k_{ij}]=[d_{ij}]^{-1}$. Moreover, the premultiplication of Eq. (4-70) by $[d_{ij}]^{-1}$ will give Eq. (4-9), which are the equations of motion of the same system by Newton's second law.

To find the frequency equation, we substitute $(j\omega)^2$ for the second time derivative and express Eq. (4-69) in phasor notation as

$$\begin{bmatrix} \bar{X}_1 \\ \bar{X}_2\end{bmatrix}=\begin{bmatrix} d_{11} & d_{12} \\ d_{21} & d_{22}\end{bmatrix}\begin{bmatrix} \omega^2 m_1\bar{X}_1 \\ \omega^2 m_2\bar{X}_2\end{bmatrix}$$

or

$$\begin{bmatrix} 1-d_{11}\omega^2 m_1 & -d_{12}\omega^2 m_2 \\ -d_{21}\omega^2 m_1 & 1-d_{22}\omega^2 m_2\end{bmatrix}\begin{bmatrix} \bar{X}_1 \\ \bar{X}_2\end{bmatrix}=\begin{bmatrix} 0 \\ 0\end{bmatrix} \tag{4-71}$$

The frequency equation is obtained by equating the determinant $\Delta(\omega)$ of the coefficient matrix of $\{\bar{X}\}$ to zero.

$$\Delta(\omega)=\begin{vmatrix} 1-d_{11}\omega^2 m_1 & -d_{12}\omega^2 m_2 \\ -d_{21}\omega^2 m_1 & 1-d_{22}\omega^2 m_2\end{vmatrix}=0 \tag{4-72}$$

Substituting the values for d_{ij}, expanding the determinant $\Delta(\omega)$, and simplifying, the frequency equation becomes

$$m_1m_2\omega^4-[(k+k_1)m_2+(k+k_2)m_1]\omega^2+[k_1k_2+k(k_1+k_2)]=0$$

This is identical to the frequency equation in Eq. (4-14) by Newton's second law for the same system.

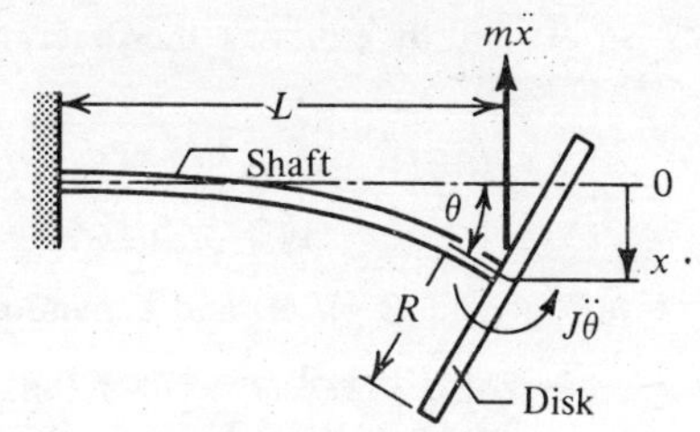

(a) Vibratory system

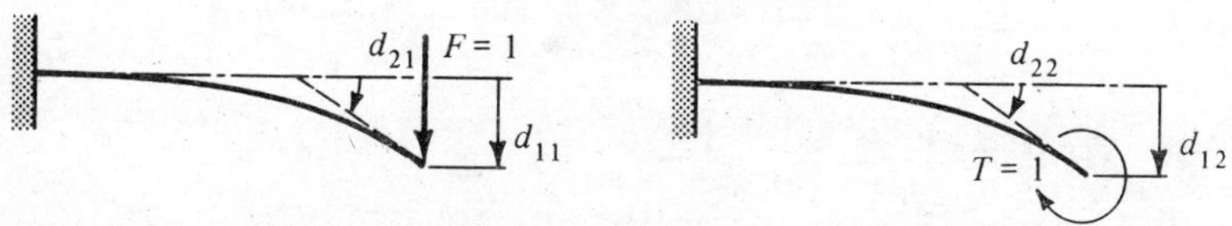

(b) Determination of influence coefficients

FIG. 4-22. *Influence coefficients due to force and moment; Example 16.*

Example 16

Determine the natural frequency of the system shown in Fig. 4-22(*a*). Assume that (1) the flexural stiffness of the shaft is *EI*, (2) the inertia effect of the shaft is negligible, (3) the shaft is horizontal in its static equilibrium position, and (4) the mass moment of inertia of the disk is $J = mR^2/4$ where $R = L/4$.

Solution:

The inertia forces are as shown in Fig. 4-22(*a*) and the influence coefficients are defined in Fig. 4-22(*b*). From elementary beam theory, it can be shown that the influence coefficients are

$$d_{11} = L^3/3EI, \qquad d_{22} = L/EI, \qquad d_{12} = d_{21} = L^2/2EI$$

The equations of motion from Eq. (4-65) are

$$\begin{bmatrix} x \\ \theta \end{bmatrix} = \begin{bmatrix} d_{11} & d_{12} \\ d_{21} & d_{22} \end{bmatrix} \begin{bmatrix} -m\ddot{x} \\ -J\ddot{\theta} \end{bmatrix}$$

or

$$\begin{bmatrix} x \\ \theta \end{bmatrix} = -\frac{L}{6EI} \begin{bmatrix} 2L^2 & 3L \\ 3L & 6 \end{bmatrix} \begin{bmatrix} m\ddot{x} \\ J\ddot{\theta} \end{bmatrix}$$

Following the last example, this can be expressed in phasor notations as shown in Eq. (4-71).

$$\begin{bmatrix} 6EI - 2\omega^2 mL^3 & -3\omega^2 JL^2 \\ -3\omega^2 mL^2 & 6EI - 6\omega^2 JL \end{bmatrix} \begin{bmatrix} \bar{X} \\ \bar{\Theta} \end{bmatrix} = \begin{bmatrix} 0 \\ 0 \end{bmatrix}$$

The frequency is obtained by equating the determinant $\Delta(\omega)$ of the coefficients of $\{\bar{X} \quad \bar{\Theta}\}$ to zero.

$$\Delta(\omega) = \begin{vmatrix} 6EI - 2\omega^2 mL^3 & -3\omega^2 JL^2 \\ -3\omega^2 mL^2 & 6EI - 6\omega^2 JL \end{vmatrix} = 0$$

Substituting $J = mR^2/4$ and $R = L/4$, and expanding $\Delta(\omega)$, we get

$$\omega^4 - 268(EI/mL^3)\omega^2 + 768(EI/mL^3)^2 = 0$$

Hence

$$\omega = 1.70\sqrt{EI/mL^3} \quad \text{and} \quad 16.3\sqrt{EI/mL^3}$$

4-10 SUMMARY

The chapter introduces the theory of discrete systems from the generalization of a two-degree-of-freedom system shown in Fig. 4-1. The equations of motion in Eq. (4-1) through (4-4) are coupled, because the equation for one mass is influenced by the motion of the other mass of the system.

The modes of vibration are examined in Sec. 4-3 for undamped free vibrations. The natural frequencies are obtained from the characteristic equation in Eq. (4-13). A mode of vibration, called the principal mode, is associated with each natural frequency. At a principal mode, (1) the entire system executes synchronous harmonic motion at a natural frequency and (2) the relative amplitudes of the masses are constant, as shown in Eq. (4-16) and illustrated in Fig. 4-2(*b*). The relative amplitudes define the modal vector for the given mode. The general motion is the superposition of the modes, as shown in Eq. (4-17).

A system can be described by more than one set of generalized coordinates $\{q\}$. In Eqs. (4-27) through (4-29), it is shown that the elements of the mass matrix and the stiffness matrix as well as the type of coupling in the equations of motion are dependent on the coordinates selected for the system description. Hence coordinate coupling is not an inherent property of the system. The coordinates that uncouple the equations are called the principal coordinates $\{p\}$. The coordinates $\{p\}$ and $\{q\}$ are related by the modal matrix $[u]$ as shown in Eq. (4-34).

A method for finding the modal matrix is shown in Sec. 4-6. The equations of motion can be uncoupled by means of the modal matrix. Thus, each uncoupled equation can be treated as an independent one-degree-of-freedom system. The results can be expressed in the $\{p\}$ or $\{q\}$ coordinates as desired. The technique is conceptually simple, but computers are necessary for the numerical solutions.

Many practical problems can be represented as semidefinite systems as discussed in Sec. 4-7. A system is semidefinite if it can move as a rigid body. Correspondingly, at least one of its natural frequencies is zero.

The harmonic response of discrete systems can be found readily by the mechanical impedance method. Using phasor notations, the equations of motion can be expressed as $Z(\omega)\{\bar{X}\}=\{\bar{F}\}$ in Eq. (4-55) and the response as $\{\bar{X}\}=Z(\omega)^{-1}\{\bar{F}\}$ in Eq. (4-57).

The method of influence coefficients in Sec. 4-9 gives an alternative procedure to formulate the equations of motion. From Maxwell's reciprocity theorem, the flexibility matrix $[d_{ij}]$ is symmetric. The inverse $[d_{ij}]^{-1}$ of the flexibility matrix is the stiffness matrix $[k_{ij}]$. Thus, except for the technique in obtaining the equations of motion and certain advantages in its application, the concepts of vibration in the previous sections can be applied readily in this method.

PROBLEMS

Assume all the systems in the figures to follow are shown in their static equilibrium positions.

4-1 Consider the system in Fig. 4-2(a). Let $m_1=m_2=10$ kg, $k_1=k_2=40$ N/m, and $k=60$ N/m. (**a**) Write the equations of motion and the frequency equation. (**b**) Find the natural frequencies, the principal modes, and the modal matrix. (**c**) Assume $\{x(0)\}=\{1 \quad 0\}$ and $\{\dot{x}(0)\}=\{0 \quad 1\}$. Plot $x_1(t)$ and $x_2(t)$ and their harmonic components. (**d**) Assume $\{x(0)\}=\{0 \quad 0\}$ and $\{\dot{x}(0)\}=\{1 \quad -1\}$. Find $x_1(t)$ and $x_2(t)$.

4-2 Repeat Prob. 4-1 if $m_1=m_2=10$ kg, $k_1=40$ N/m, $k_2=140$ N/m, and $k=60$ N/m. Are the motions $\{x(t)\}$ periodic?

4-3 A 200-kg uniform bar is supported by springs at the ends as illustrated in Fig. 4-5. The total length is $L=1.5$ m, $k_1=18$ kN/m, and $k_2=22$ kN/m. (**a**) Write the equations of motion and the frequency equation. (**b**) Find the natural frequencies, the principal modes, and the modal matrix. (**c**) If $x(0)=1$, $\dot{x}(0)=\theta(0)=\dot{\theta}(0)=0$, find the motions $x(t)$ and $\theta(t)$. (**d**) Illustrate the principal modes, such as shown in Fig. 4-6.

4-4 For the three-degree-of-freedom system in Fig. 4-7, if $J_1=2J_2$, $J_2=2J_3$, $k_{t1}=2k_{t2}$, and $k_{t2}=2k_{t3}$, find the motions $\theta_1(t)$, $\theta_2(t)$, and $\theta_3(t)$.

4-5 For each of the systems shown in Fig. P4-1, specify the coordinates to describe the system, write the equations of motion, and find the frequency equation.

(**a**) A double pendulum.

(**b**) The arm is horizontal in its static equilibrium position.

(**c**) Three identical pendulums.

(**d**) A double compound pendulum.

(**e**) A schematic representation of an overhead crane.

(**f**) The system is constrained to move in the plane of the paper.

(**g**) The bar and the shaft are initially horizontal. The shaft deflects vertically. The bar moves vertically as well as rotates in a vertical plane.

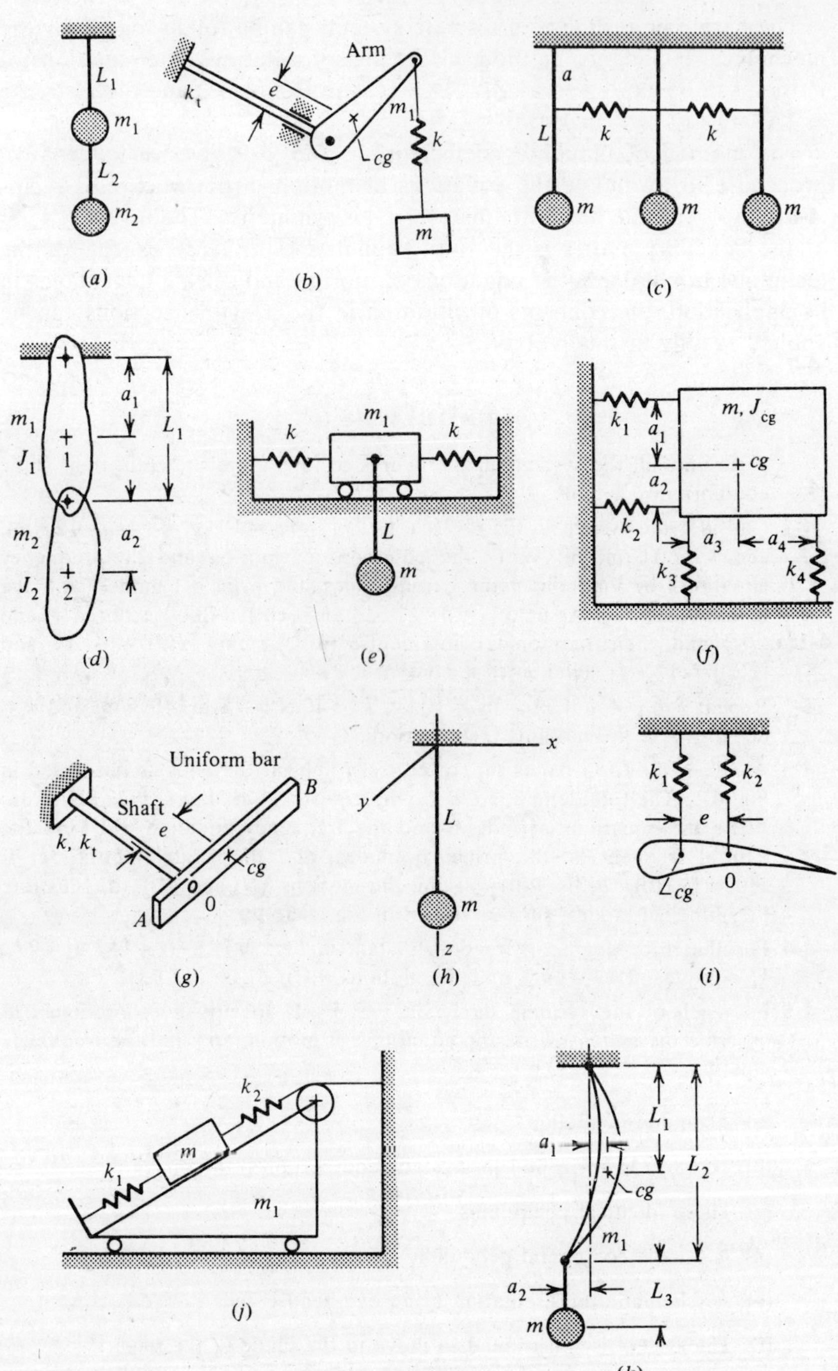

FIG. P4-1. *Vibratory systems.*

(**h**) A spherical pendulum.

(**i**) The airfoil moves vertically and pivots about *cg*.

(**j**) Assume that there is no friction between m and m_1.

(**k**) The pendulums are constrained to move in the plane of the paper.

4-6 For the double pendulum in Fig. P4-1(a), let $m_1 = m_2$ and $L_1 = L_2$. (**a**) If x_1 is the horizontal displacement of m_1 and x_2 that of m_2, write the equations of motion in terms of (x_1, x_2) and find the natural frequencies. (**b**) If θ_1 and θ_2 are the angular displacements of the pendulums, write the equations of motion in terms of (θ_1, θ_2) and find the natural frequencies.

4-7 Referring to Fig. 4-9 on coordinate coupling, (**a**) convert Eq. (4-28) to (4-27), using the relations $x_2 = x_1 - e\theta$, $L_3 = L_1 - e$, $L_4 = L_2 + e$, and $J_2 = J_1 + me^2$; and (**b**) convert Eq. (4-29) to (4-27), using the relations $x_3 = x_1 - L_1\theta$, $L = L_1 + L_2$, and $J_3 = J_1 + mL_1^2$.

4-8 Referring to Fig. P4-1(g), write the equations of motion of the system if the vertical displacement of the bar is measured from: (**a**) the mass center *cg*; (**b**) the point 0; (**c**) the point A; (**d**) the point B.

4-9 Show that the frequency equation for the case of non-symmetrical matrices in Example 6 is identical to Eq. (4-14).

4-10 A company crates its products for shipping as shown in Fig. P4-2(a). The skid is securely mounted on a truck. Experience indicates that this method of crating is satisfactory. To cut the shipping cost, it is proposed to put two items in a crate as shown in Fig. P4-2(b). Would you approve this proposal?

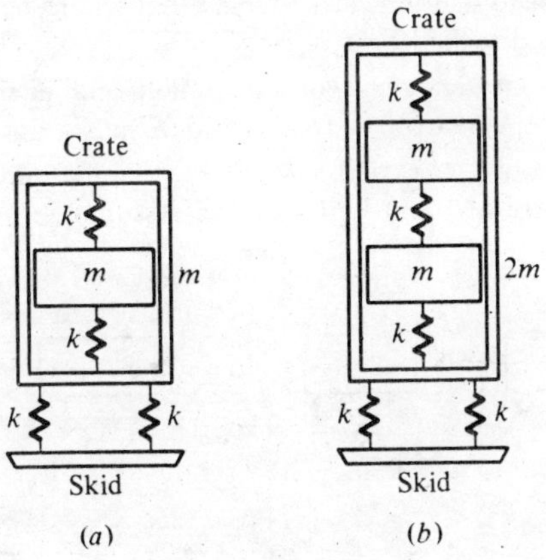

FIG. P4-2.

4-11 Consider an undamped three-degree-of-freedom system

$$\begin{bmatrix} 2 & 0 & 0 \\ 0 & 1 & 0 \\ 0 & 0 & 2 \end{bmatrix} \begin{bmatrix} \ddot{x}_1 \\ \ddot{x}_2 \\ \ddot{x}_3 \end{bmatrix} + \begin{bmatrix} 4 & -1 & 0 \\ -1 & 2 & -1 \\ 0 & -1 & 4 \end{bmatrix} \begin{bmatrix} x_1 \\ x_2 \\ x_3 \end{bmatrix} = \begin{bmatrix} F_1(t) \\ F_2(t) \\ F_3(t) \end{bmatrix}$$

where $\{F(t)\}$ is a vector of transient excitations. (**a**) Find the frequency equation and the natural frequencies. (**b**) Determine the modal vectors and the modal matrix. (**c**) Verify that the modal vectors are orthogonal relative to the matrices M and K as shown in Eq. (4-40). (**d**) Write the uncoupled equations as indicated in Eq. (4-41).

4-12 Repeat Prob. 4-11 for the equations

$$\begin{bmatrix} 1 & 0 & 0 \\ 0 & 1 & 0 \\ 0 & 0 & 2 \end{bmatrix} \begin{bmatrix} \ddot{x}_1 \\ \ddot{x}_2 \\ \ddot{x}_3 \end{bmatrix} + \begin{bmatrix} 2 & -1 & 0 \\ -1 & 2 & -1 \\ 0 & -1 & 3 \end{bmatrix} \begin{bmatrix} x_1 \\ x_2 \\ x_3 \end{bmatrix} = \begin{bmatrix} F_1(t) \\ F_2(t) \\ F_3(t) \end{bmatrix}$$

4-13 Determine the motions $\theta_1(t)$ and $\theta_2(t)$ of the semidefinite system in Fig. 4-13 for the initial conditions: (**a**) $\{\theta(0)\} = \{\theta_{10} \quad \theta_{20}\}$ and $\{\dot{\theta}(0)\} = \{0 \quad 0\}$; (**b**) $\{\theta(0)\} = \{0 \quad 0\}$ and $\{\dot{\theta}(0)\} = \{0 \quad \theta_{20}\}$.

4-14 For the semidefinite system shown in Fig. 4-14, if $J_1 = 1.2\ \text{m}^2\cdot\text{kg}$, $J_2 = J_3 = 2J_1$, $k_{t1} = 25 \times 10^3\ \text{N}\cdot\text{m/rad}$, and $k_{t2} = 2k_{t1}$, find the natural frequencies and the relative amplitudes at the principal modes.

4-15 For the system in Prob. 4-14, find the motions $\{\theta(t)\}$ if the initial conditions are: (**a**) $\{\theta(0)\} = \{0.1 \quad 0 \quad 0\}$ and $\{\dot{\theta}(0)\} = \{0\}$; (**b**) $\{\theta(0)\} = \{0\}$ and $\{\dot{\theta}(0)\} = \{10 \quad 0 \quad 0\}$.

4-16 Neglecting the inertial effect of the pinion and the gear in Fig. 4-15, let $J_1 = 0.2\ \text{m}^2\cdot\text{kg}$, $J_2 = 4J_1$, $k_{t1} = 60 \times 10^3\ \text{N}\cdot\text{m/rad}$, $k_{t2} = 7k_{t1}$, and the gear ratio $= 3:1$. Find the natural frequencies of the system: (**a**) referring to shaft 1; (**b**) referring to shaft 2.

4-17 Assume a variable speed engine with four impulses per revolution is attached to J_1 of the gear system described in Prob. 4-16. Find the resonance speed of the gear system. What would be the resonance speed if the engine is attached to J_2?

4-18 Assume the inertial effect of the pinion and gear in Prob. 4-16 is not negligible. Repeat Prob. 4-16 if $J_3 = 0.02\ \text{m}^2\cdot\text{kg}$ and $J_4 = 20J_3$.

4-19 Find the motions $x_1(t)$ and $x_2(t)$ of the semidefinite system shown in Fig. P4-3(a), where $\delta(t)$ is a unit impulse. Assume zero initial conditions.

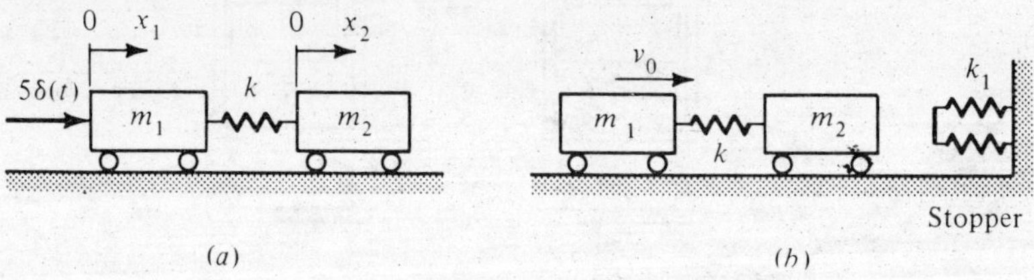

FIG. P4-3.

4-20 A semidefinite system strikes a stopper as shown in Fig. P4-3(b). Find the maximum force transmitted to the base of the stopper. Assume the velocity v_0 is constant and the springs are initially unstressed. Assume $m_1 = m_2$ and $k_1 = 2k$.

4-21 A branched-geared system is shown in Fig. P4-4. Assume the inertial effect of the shafts and the coupling is negligible. The gear ratio of the gears $J_b : J_c = 1:2$ and $J_b : J_d = 1:3$. The data as shown are in the SI units. (**a**)

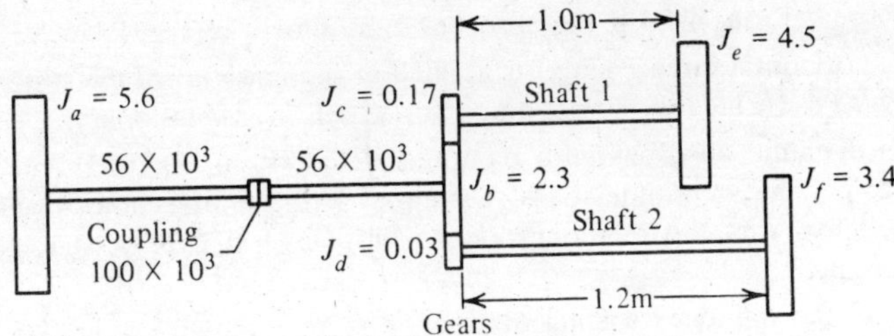

FIG. P4-4. *Branched-geared system.*

Specify the diameters of the shafts 1 and 2 such that the system has only two numerically distinct nonzero natural frequencies. (**b**) Find the natural frequencies.

4-22 Assuming harmonic excitations, find the steady-state response of each of the systems in Fig. P4-5.

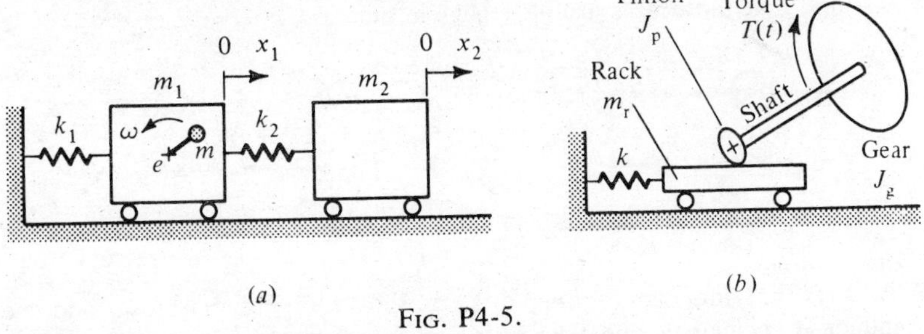

FIG. P4-5.

4-23 An air compressor of 270 kg mass is mounted as shown in Fig. 4-16. The normal operating speed is 1,750 rpm. (**a**) If the resonant frequencies should be at least ±20 percent from the operating speed, specify k_1, k_2, and m_2. (**b**) What is the amplitude of m_2 at the operating speed?

4-24 A torque $T \sin \omega t$ is applied to J_1 of the torsional system in Fig. P4-6(a). If $J_1 = 0.5 \text{ m}^2\cdot\text{kg}$, $k_{t1} = 560 \times 10^3 \text{ m}\cdot\text{N/rad}$, $T = 226 \text{ m}\cdot\text{N}$, and $\omega = 10^3$ rad/s, specify J_2 and k_{t2} of the absorber such that the resonant frequencies are 20 percent from the excitation frequency.

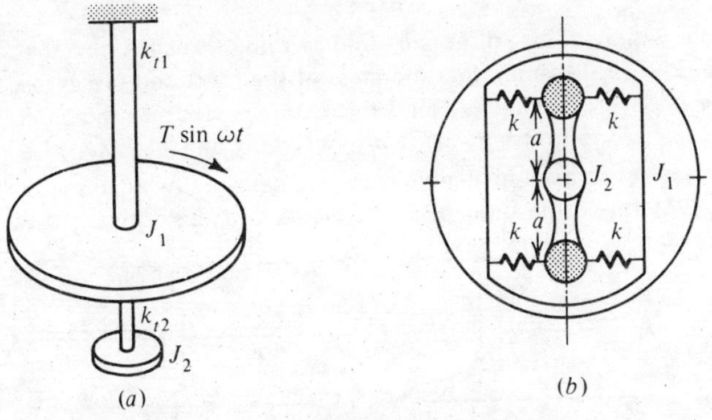

FIG. P4-6.

4-25 Repeat Prob. 4-24 if the absorber is as shown in Fig. P4-6(*b*).

4-26 A horizontal force $F \sin \omega t$ is applied to the mass m_1 of the system shown in Fig. P4-1(*e*). Find the condition for which m_1 is stationary.

4-27 A dynamic absorber with damping is shown in Fig. 4-19. For optimum design, the amplitudes of X_1 are equal at the intersections of curves 1 and 2. Show that the relation $k_2/k_1 = m_1m_2/(m_1+m_2)^2$ is satisfied for this optimum.

4-28 Find the influence coefficients for the system shown in Fig. 4-5. Write the dynamic equations. Show that the frequency equation can be reduced to that obtained from Newton's second law.

4-29 Repeat Prob. 4-28 for the system shown in Fig. 4-7. Assume $J_1 = J_2 = J_3$ and $k_{t1} = k_{t2} = k_{t3}$.

4-30 Find the influence coefficients and the frequency equations for each of the systems shown in Fig. P4-1(*a*) to (*c*).

4-31 Find the influence coefficients for each of the systems shown in Fig. P4-7. Assume that the beams are of negligible mass.

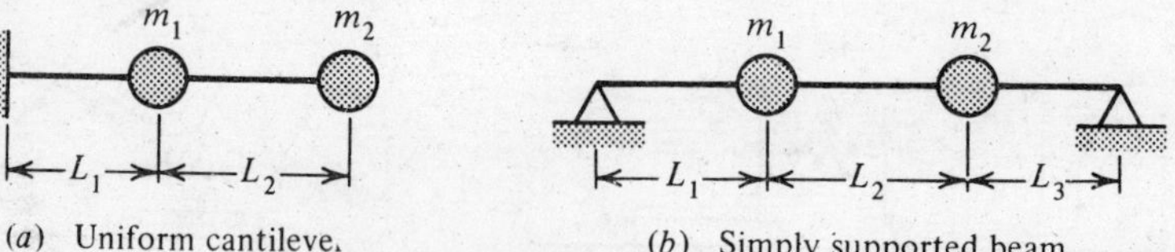

(*a*) Uniform cantileve. (*b*) Simply supported beam

Fig. P4-7.

4-32 A shaft carrying two rotating disks is shown in Fig. P4-8(*a*). Find tne influence coefficients and the critical speeds of the assembly. Assume that (1) the deflections of the bearings and the gyroscopic effect of the disks are negligible, and (2) $L_1 = 150$ mm and $L = 600$ mm.

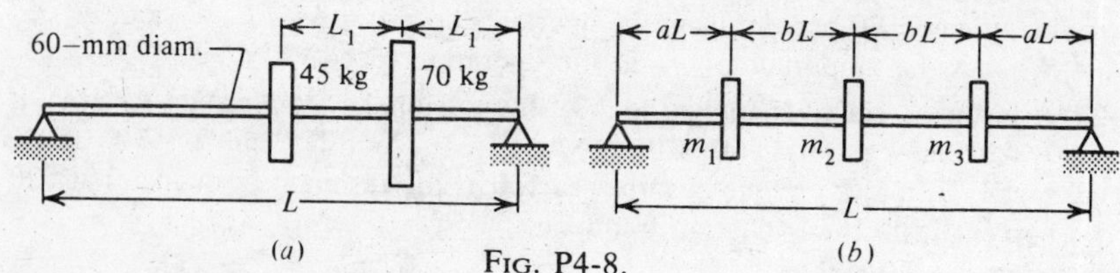

Fig. P4-8.

4-33 A shaft, with bending stiffness EI and carrying three rotating disks, is shown in Fig. P4-8(*b*). Assume that the mass of the shaft and the gyroscopic effect of the disks are negligible. Find the critical speeds of the assembly: (**a**) if $m_1 = m_3 = 2m_2$ and $a = b$; (**b**) if $m_1 = m_2 = m_3$ and $b = 2a$.

4-34 A continuous shaft of negligible mass and carrying two disks is shown in Fig. P4-9. Determine the influence coefficients and the critical speeds.

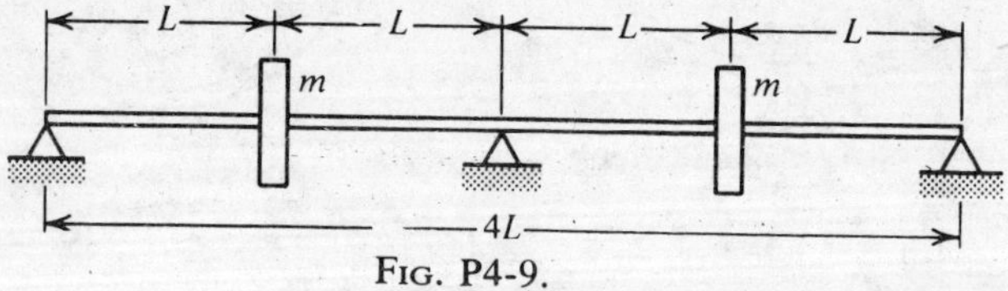

Fig. P4-9.

Computer problems:

Remarks: Transient response and frequency response are examined in the problems. The modal analysis for the transient response of positive-definite undamped systems with distinct frequencies can be examined by parts or by a combined program. Problems 4-35 to 4-39 show the parts of the modal analysis, following the theory developed in the chapter. Using the program TRESPUND, Probs. 4-40 to 4-42 show the combined program, which is a collection of subroutines. The remaining problems deal with the frequency response of discrete systems.

4-35 *Characteristic equation.* Use the program COEFF listed in Fig. 4-11 to find the coefficients of the characteristic equation of each of the following systems:

(**a**) $$\begin{bmatrix} 2 & 0 & 0 \\ 0 & 1 & 0 \\ 0 & 0 & 2 \end{bmatrix} \begin{bmatrix} \ddot{x}_1 \\ \ddot{x}_2 \\ \ddot{x}_3 \end{bmatrix} + \begin{bmatrix} 4 & -1 & 0 \\ -1 & 2 & -1 \\ 0 & -1 & 4 \end{bmatrix} \begin{bmatrix} x_1 \\ x_2 \\ x_3 \end{bmatrix} = \begin{bmatrix} 0 \\ 0 \\ 0 \end{bmatrix}$$

(**b**) $$\begin{bmatrix} 1 & 0 & 0 \\ 0 & 1 & 0 \\ 0 & 0 & 2 \end{bmatrix} \begin{bmatrix} \ddot{x}_1 \\ \ddot{x}_2 \\ \ddot{x}_3 \end{bmatrix} + \begin{bmatrix} 2 & -1 & 0 \\ -1 & 3 & -2 \\ 0 & -2 & 2 \end{bmatrix} \begin{bmatrix} x_1 \\ x_2 \\ x_3 \end{bmatrix} = \begin{bmatrix} 0 \\ 0 \\ 0 \end{bmatrix}$$

4-36 *Natural frequencies.* The roots ω^2 of the characteristic equation in Eq. (4-38) yield the natural frequencies ω of the system. Write a program to find the roots of the characteristic equation in Prob. 4-35. *Hint:* Use the subroutine $ROOT in Fig. C-7.

4-37 *Modal matrix.* The steps to obtain a modal matrix are: (1) substituting an eigenvalue $\lambda(=\omega^2)$ in Eq. (4-37), $[-\omega^2 M + K]\{q\} = \{0\}$; (2) finding the solutions of the homogeneous equations, as shown in Example 9, to obtain a modal vector; (3) combining the modal vectors to form a modal matrix. For the data in Prob. 4-35, write a program to

(**a**) convert the equations of motion $M\{\ddot{x}\} + K\{x\} = \{0\}$ to the form $\{\ddot{x}\} + H\{x\} = \{0\}$, where $H = M^{-1}K$ is a dynamic matrix,

(**b**) use the subroutine $ROOT in Fig. C-7 to find the eigenvalues,

(**c**) use the subroutine $HOMO in Fig. C-9 to solve the homogeneous equations in order to find the modal vectors, and

(**d**) combine the modal vectors to obtain the modal matrix.

4-38 *Modal matrix.* Repeat Prob. 4-37 by using the subroutine $MODL listed in Fig. C-10. The program gives the roots of the characteristic equation and the modal matrix.

4-39 *Principal modes.* The equations of motion $M\{\ddot{x}\} + K\{\dot{x}\} = \{0\}$ can be uncoupled and expressed in terms of the principal coordinates as shown in Eq. (4-41). Assuming appropriate initial conditions for the systems in Prob. 4-35, (**a**) write a program to uncouple the equations, (**b**) compute the vibrations in the original coordinates and the principal coordinates, and **(c)** plot the results for $t \simeq 1.5\tau$, where τ is the period of the first mode.

4-40 *Modal analysis.* Use the program TRESPUND in Fig. 9-8(a) to find the transient response of the system in Fig. 4-10 for $0 \leq t \leq 1$. Choose the appropriate initial conditions and consider the problem in three parts as follows:

(a) $\{F(t)\} = \{0\}, \{x(0)\} \neq \{0\}$, and $\{\dot{x}(0)\} \neq \{0\}$.

(b) $\{F(t)\} \neq \{0\}, \{x(0)\} = \{0\}$, and $\{\dot{x}(0)\} = \{0\}$.

(c) $\{F(t)\} \neq \{0\}, \{x(0)\} \neq \{0\}$, and $\{\dot{x}(0)\} \neq \{0\}$.

Verify from the computer print-out that the value of $\{x(t)\}$ and $\{\dot{x}(t)\}$ from part **c** is the sum of that of parts **a** and **b**.

4-41 *Transient response plot.*

(a) Modify the program TRESPUND in Fig. 9-8(a) such that the values of the displacement $\{x(t)\}$ are stored in one file and the velocity $\{\dot{x}(t)\}$ in another.

(b) Execute the program for the system in Fig. 4-10 and use the program PLOTFILE in Fig. 9-5(a) to plot the results.

4-42 *Dynamic absorber.* The undamped dynamic absorber shown in Fig. 4-16 was analyzed in Example 12. **(a)** Write a program to implement Eq. (4-64) for the harmonic response and store the results in a data file. **(b)** Use the program PLOTFILE in Fig. 9-5(a) to plot the results as illustrated in Fig. 4-17.

4-43 Dynamic absorber with damping was described in Example 13. Assume $k_1/m_1 = k_2/m_2$ as for undamped absorbers. Let $m_2/m_1 = 0.3$. Select values for k_1, k_2, m_1, m_2, and five values for the damping coefficient c. Write a program to store the data in a file for **(a)** amplitude X_1 versus frequency, with c as a parameter, and **(b)** amplitude X_2 versus frequency with c as a parameter. Use the program PLOTFILE in Fig. 9-5(a) to plot the results.

4-44 Repeat Prob. 4-43, but for the optimum condition $k_1/k_2 = m_1 m_2/(m_1 + m_2)^2$ as shown in Prob. 4-27.

4-45 *Vibration isolation.* Vibration isolation for the system shown in Fig. 4-20 was discussed in Example 14. Let $m_1 = 180$ kg, $k_1 = 162$ kN/m, $m_2 = 4m_1$, $k_2 = k_1$, $c_2 = c_1$, and the mean excitation frequency $f = 10$ Hz. Assume $1.0 \leq c_1 \leq 4.0$ kN·s/m. Calculate the amplitudes X_1, X_2, and the force transmitted F_T to the foundation with c_1 as a parameter for $5 \leq f \leq 15$ Hz. **(a)** Write a program to store the calculated values of X_1 versus f in a data file, X_2 versus f in a second file, and F_T versus f in a third file. **(b)** Use the program PLOTFILE in Fig. 9-5(a) to plot the results.

4-46 *Frequency response.* Consider an n-degree-of-freedom system with viscous damping

$$M\{\ddot{x}\} + C\{\dot{x}\} + K\{x\} = \{Q(t)\}$$

where M is the mass matrix, C the damping matrix, K the stiffness matrix, and $\{Q(t)\}$ the excitation vector. Assume each element of $\{Q(t)\}$ is harmonic and of the same frequency ω as discussed in Sec. 4-8. (**a**) Write a program for $n \leq 10$ to store the information of the amplitude versus ω and phase angles versus ω of the response in separate data files. (**b**) Use the program PLOTFILE in Fig. 9-5(a) to plot the results. For purpose of illustration, assume $0.1 \leq \omega \leq 2.0$ rad/s and the equations in the *SI* units are

$$\begin{bmatrix} 10 & 0 \\ 0 & 3 \end{bmatrix} \begin{bmatrix} \ddot{x}_1 \\ \ddot{x}_2 \end{bmatrix} + \begin{bmatrix} 2 & -1 \\ -1 & 1 \end{bmatrix} \begin{bmatrix} \dot{x}_1 \\ \dot{x}_2 \end{bmatrix} + \begin{bmatrix} 13 & -3 \\ -3 & 3 \end{bmatrix} \begin{bmatrix} x_1 \\ x_2 \end{bmatrix} = \begin{bmatrix} 10 \sin \omega t \\ 0 \end{bmatrix}$$

5

Methods for Finding Natural Frequencies

5-1 INTRODUCTION

The natural frequency of a vibratory system is perhaps the first item of interest in dynamic analysis. Many methods can be used to find the natural frequencies. The simplified Rayleigh method and the frequency equation were examined previously. Additional techniques will be discussed in Chaps. 6 and 7. We shall present (1) simple methods that are suitable for estimating the fundamental frequency by hand calculations, and (2) the transfer matrix technique, which is applicable to large or complex problems. The transfer matrix technique would inevitably necessitate the use of computers.

5-2 DUNKERLEY'S EQUATION*

Dunkerley advanced a method to approximate the fundamental frequency of multirotor systems. It gives good results if damping is negligible and the frequencies of the harmonics are much higher than that of the fundamental.

Consider the free vibration of an undamped system. By the method of influence coefficients in Eq. (4-66), we have

$$\{q\} = [d_{ij}]\{-m\ddot{q}\}$$

where $\{q\}$ is the displacement vector, $[d_{ij}]$ the flexibility matrix, and

* Dunkerley, S., "Whirling and Vibration of Shafts," *Trans. Royal Soc.* (London), vol. 185A, pt. 1, 1894, pp. 279–360.

$\{-m\ddot{q}\}$ the vector of inertia forces. At a principal mode of vibration, the deflections $\{q\}$ are harmonic with $\{\ddot{q}\}=\{-\omega^2 q\}$. Substituting this in the equation above gives

$$\{q\}=[d_{ij}]\{\omega^2 mq\} \tag{5-1}$$

A frequency equation can be found from Eq. (5-1) as outlined in Chap. 4. Dunkerley's equation is deduced from the frequency equation by retaining only the fundamental frequency.

Let us illustrate the method for a two-degree-of-freedom system. From Eq. (5-1) we have

$$\begin{bmatrix} q_1 \\ q_2 \end{bmatrix}=\begin{bmatrix} d_{11} & d_{12} \\ d_{21} & d_{22} \end{bmatrix}\begin{bmatrix} \omega^2 m_1 q_1 \\ \omega^2 m_2 q_2 \end{bmatrix}$$

Dividing by ω^2 and rearranging, we obtain

$$\begin{bmatrix} \frac{1}{\omega^2}-d_{11}m_1 & -d_{12}m_2 \\ -d_{21}m_1 & \frac{1}{\omega^2}-d_{22}m_2 \end{bmatrix}\begin{bmatrix} q_1 \\ q_2 \end{bmatrix}=\begin{bmatrix} 0 \\ 0 \end{bmatrix}$$

Hence the frequency equation is

$$\Delta(\omega)=\begin{vmatrix} \frac{1}{\omega^2}-d_{11}m_1 & -d_{12}m_2 \\ -d_{21}m_1 & \frac{1}{\omega^2}-d_{22}m_2 \end{vmatrix}=0$$

or

$$\left(\frac{1}{\omega^2}\right)^2-(d_{11}m_1+d_{22}m_2)\left(\frac{1}{\omega^2}\right)+m_1m_2(d_{11}d_{22}-d_{12}d_{21})=0 \tag{5-2}$$

Assume ω_1 and ω_2 are the natural frequencies. The factored form of Eq. (5-2) is

$$\left(\frac{1}{\omega^2}-\frac{1}{\omega_1^2}\right)\left(\frac{1}{\omega^2}-\frac{1}{\omega_2^2}\right)=0$$

or

$$\left(\frac{1}{\omega^2}\right)^2-\left(\frac{1}{\omega_1^2}+\frac{1}{\omega_2^2}\right)\left(\frac{1}{\omega^2}\right)+\frac{1}{\omega_1^2\omega_2^2}=0 \tag{5-3}$$

Equating the coefficients of the $1/\omega^2$ terms in Eqs. (5-2) and (5-3) yields

$$\frac{1}{\omega_1^2}+\frac{1}{\omega_2^2}=d_{11}m_1+d_{22}m_2$$

If the fundamental frequency ω_1 is much lower than that of the harmonic

ω_2, we have $1/\omega_2^2 \ll 1/\omega_1^2$ and

$$\frac{1}{\omega_1^2} \simeq d_{11}m_1 + d_{22}m_2$$

which is Dunkerley's equation. The corresponding equation for a multi-degree-of-freedon system is

$$\frac{1}{\omega_1^2} \simeq d_{11}m_1 + d_{22}m_2 + \cdots + d_{nn}m_n = \sum_{i=1}^{n} d_{ii}m_i \tag{5-4}$$

The influence coefficient d_{ii} is the deflection at station i of the system due to a unit force applied at the same location. The quantity $d_{ii}m_i$ is due to m_i acting alone, that is, the inertial effect of all other masses is not considered. Since d_{ii} is the reciprocal of the stiffness k_{ii}, we have

$$d_{ii}m_i = \frac{m_i}{k_{ii}} = \frac{1}{\omega_{ii}^2}$$

where ω_{ii} is the natural frequency of an equivalent mass-spring system with m_i acting alone at station i. Hence an alternative form of Dunkerley's equation is

$$\frac{1}{\omega_1^2} \simeq \frac{1}{\omega_{11}^2} + \frac{1}{\omega_{22}^2} + \cdots + \frac{1}{\omega_{nn}^2} = \sum_{i=1}^{n} \frac{1}{\omega_{ii}^2} \tag{5-5}$$

Note that the estimated fundamental frequency is always lower than the exact value, since the harmonics are neglected in the equation.

Example 1

Find the fundamental frequency of the torsional system shown in Fig. 4-8. Assume $k_{t1} = k_{t2} = k_{t3} = k_t$ and $J_1 = J_2 = J_3 = J$.

Solution:

It is evident that the influence coefficients are $d_{11} = 1/k_t$, $d_{22} = 2/k_t$, and $d_{33} = 3/k_t$. Applying Eq. (5-4) gives

$$\frac{1}{\omega_1^2} \simeq \frac{J}{k_t} + \frac{2J}{k_t} + \frac{3J}{k_t} = \frac{6J}{kt}$$

or

$$\omega_1 \simeq 0.408\sqrt{k_t/J}$$

From Example 5, Chap. 4, the exact value is $\omega_1 = 0.445\sqrt{k_t/J}$. The error is about 8 percent and ω_1 is less than the exact value.

Example 2

A mass m is attached to a uniform cantilever beam of mass m_1 as shown in Fig. 5-1. Assume $m = m_1$. Find the fundamental frequency of the system.

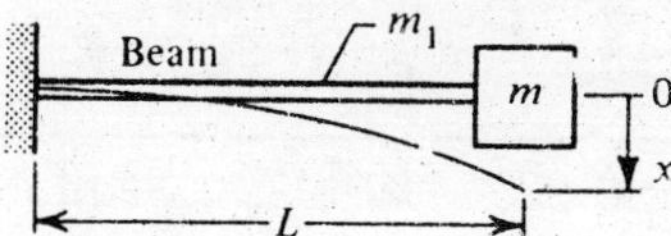

FIG 5-1. *Cantilever beam with attached mass.*

Solution:

From Example 3, Chap. 3, the natural frequency of a cantilever beam of negligible mass with a concentrated mass m attached is

$$\omega_{11}^2 = \frac{3EI}{mL^3}$$

It is estimated (see Prob. 2-4) that the natural frequency of a cantilever beam of mass m_1 is

$$\omega_{22}^2 = 12.7\frac{EI}{m_1L^3}$$

Substituting ω_{11}^2 and ω_{22}^2 in Eq. (5-5) yields

$$\frac{1}{\omega_1^2} \simeq \frac{mL^3}{3EI} + \frac{mL^3}{12.7EI} = 0.412\frac{mL^3}{EI}$$

or

$$\omega_1 \simeq 1.56\sqrt{EI/(mL^3)}$$

5-3 RAYLEIGH METHOD

Rayleigh method was applied to one-degree-of-freedom systems in Chaps. 2 and 3. The same technique is applicable to discrete systems. The method assumes that (1) the system is conservative, and (2) at a principal mode, the maximum potential energy of the system is equal to its maximum kinetic energy.

A discrete system has as many modes of vibration as its degrees of freedom. It is necessary to assume the dynamic mode shape, or the modal vector, in order to estimate the natural frequencies. Generally, the Rayleigh method is used to find the fundamental frequency, since the modal vectors for the higher frequencies are more difficult to estimate. If an exact mode shape is assumed, the frequency calculated will be exact. If the assumed mode shape is not the exact dynamic mode shape, it is equivalent to the application of additional constraints to the vibratory system. Hence the calculated frequency is higher than the true value.* Thus, the Rayleigh method tends to give a higher value for the estimated frequency. It will be shown in Sect. 6-7 that a reasonable assumed mode shape will give satisfactory results.

* J. P. Den Hartog, *Mechanical Vibrations*, 4th ed., McGraw-Hill Book Co., New York, 1956, p. 161.

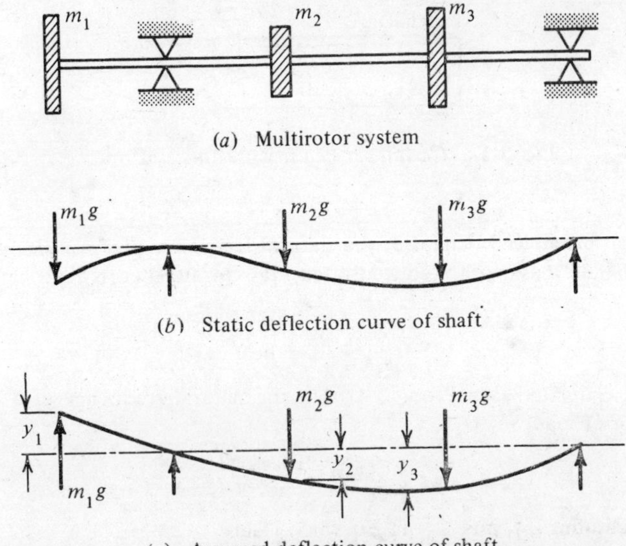

FIG. 5-2. *Fundamental frequency by Rayleigh method.*

Usually the dynamic deflections are estimated from the static deflections. This assumption, however, should not be automatic. Consider a multirotor system shown in Fig. 5-2(a) and the corresponding static deflection of the shaft in Fig. 5-2(b). Evidently, the dynamic deflection of the shaft for the fundamental frequency should be assumed as shown in Fig. 5-2(c).

Assume the shaft in Fig. 5-2(a) is of negligible mass. The potential energy of the system is the strain energy in the bent shaft, which is the work done by the static loads. The maximum potential energy U_{max} is

$$U_{max} = \tfrac{1}{2}(m_1y_1 + m_2y_2 + m_3y_3)g \tag{5-6}$$

where m_ig is the static load due to a rotor and y_i the deflection at the rotor. For harmonic oscillation, the maximum kinetic energy T_{max} due to rotors is

$$T_{max} = \frac{\omega^2}{2}(m_1y_1^2 + m_2y_2^2 + m_3y_3^2) \tag{5-7}$$

where ω is the frequency of oscillation. Equating U_{max} and T_{max} and simplifying, we obtain

$$\omega^2 = \frac{g(m_1y_1 + m_2y_2 + m_3y_3)}{(m_1y_1^2 + m_2y_2^2 + m_3y_3^2)} \tag{5-8}$$

or

$$\omega^2 = \frac{g\sum_{i=1}^{n} m_i y_i}{\sum_{i=1}^{n} m_i y_i^2} \tag{5-9}$$

Note that the equation is derived from the lateral (beam) deflection of a shaft. It was shown in Chap. 3 that the frequency for the transverse vibration of the system is also the critical speed at which the shaft whirl takes place.

Example 3

A uniform shaft with two disks and supported by bearings is shown in Fig. 5-3(*a*). Assume the static shaft deflections are as indicated in Fig. 5-3(*b*). Find the fundamental frequency of the system.

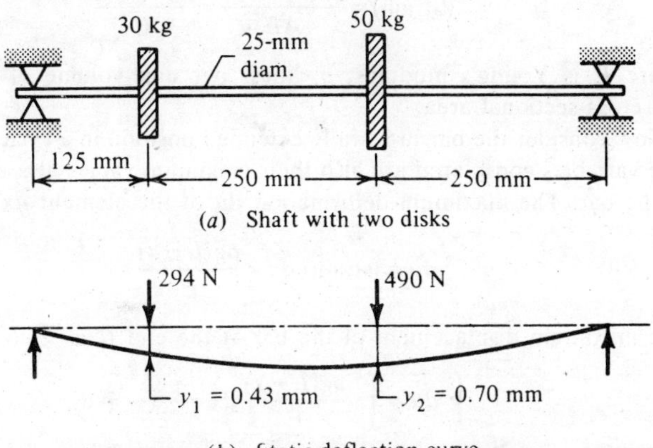

(*a*) Shaft with two disks

(*b*) Static deflection curve

FIG. 5-3. *Critical speed of shaft.*

Solution:

Assume the shaft assembly is a simply supported beam with two concentrated masses. Neglecting the mass of the shaft and applying Eq. (5-9), we get

$$\omega^2 = 9.81\frac{[30(0.43)+50(0.70)]10^{-3}}{[30(0.43)^2+50(0.70)^2]10^{-6}} = 15\ 640$$

$$\omega = 125 \text{ rad/s} \qquad \text{or} \qquad 1{,}194 \text{ rpm}$$

Example 4

Consider a vertical uniform bar shown in Fig. 5-4. Use Rayleigh method to find the fundamental frequency for the longitudinal vibration.

Solution:

We first find the potential and kinetic energies of an element dx of the bar and then apply Rayleigh's method by equating the maximum potential and kinetic energies of the bar.

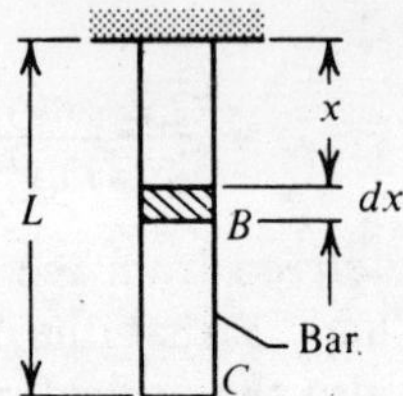

FIG. 5-4. *Fundamental frequency by Rayleigh method.*

The load on an element dx is due to the weight of the bar of length $(L-x)$ below x. For longitudinal loading, the normal stress and normal strain relation is

$$(\text{Stress}) = E\ (\text{strain})$$

$$(\text{Strain}) = \frac{\rho g A(L-x)}{AE} = \frac{\rho g(L-x)}{E}$$

where E is Young's modulus, ρ = mass per unit volume of the bar, and A = cross-sectional area.

Now consider the bar in its fully extended position in a cycle of vibration. The variables considered are also their maximum values at a given location of the bar. The maximum deformation du of the element dx is

$$du = (\text{strain})\ dx = \frac{\rho g(L-x)}{E}\ dx$$

The maximum displacement of the bar at the end C is

$$\int_0^C du = \int_0^L \frac{\rho g(L-x)}{E}\ dx = \frac{\rho g L^2}{2E} \triangleq u_c$$

The maximum displacement at an intermediate point B is

$$\int_0^B du = \int_0^x du = 2u_c\left(1-\frac{x}{2L}\right)\frac{x}{L} \triangleq u$$

We call this u instead of u_b for simplicity. Hence the oscillation at an intermediate point B of the bar is

$$u \sin \omega t = 2u_c\left(1-\frac{x}{2L}\right)\frac{x}{L} \sin \omega t$$

The maximum kinetic energy of the system is

$$T_{\max} = \int_0^L \frac{1}{2}\omega^2 u^2(\rho A\ dx)$$

$$= \int_0^L 2\omega^2 \rho A\left(1-\frac{x}{2L}\right)^2\left(\frac{x}{L}\right)^2 u_c^2\ dx = \frac{4}{15}\omega^2 \rho A L u_c^2$$

From elementary beam theory, the maximum potential energy is

$$U_{\max} = \int_0^L \frac{1}{2} EA\left(\frac{\partial u}{\partial x}\right)^2 dx = \frac{2}{3}\frac{EA}{L}u_c^2$$

Equating T_{max} and U_{max}, we obtain

$$\omega^2 = \frac{5}{2}\frac{E}{\rho L^2}$$

The exact value of ω^2 from Example 2, Chap. 7, is

$$\omega^2 = \frac{\pi^2}{4}\frac{E}{\rho L^2} \simeq 2.47\frac{E}{\rho L^2}$$

5-4 HOLZER METHOD

Holzer method is essentially a systematic tabulation of the frequency equation of the system. The method has general applications, including systems with rectilinear and angular motions, damped* or undamped, semidefinite, branched, or systems with fixed ends. The procedure can be programmed for computer applications as shown in the next section.

The method assumes a trial frequency. A solution is found when the assumed frequency satisfies the constraints of the problem. Usually, this requires several trials. Depending on the trial frequency employed, the fundamental as well as the harmonic frequencies can be determined. The tabulation also gives the mode shape of the system. We shall illustrate the method for an undamped semidefinite system.

Consider a three-disk semidefinite system shown in Fig. 5-5. The equations of motion from Newton's second law are

$$J_1\ddot{\theta}_1 = -k_{t1}(\theta_1 - \theta_2)$$
$$J_2\ddot{\theta}_2 = -k_{t1}(\theta_2 - \theta_1) - k_{t2}(\theta_2 - \theta_3)$$
$$J_3\ddot{\theta}_3 = -k_{t2}(\theta_3 - \theta_2)$$

The motions are harmonic at a principal mode of vibration. Substituting $\theta_i = \Theta_i \sin \omega t$ and simplifying, we get

$$\begin{aligned} -\omega^2 J_1\Theta_1 &= -k_{t1}(\Theta_1 - \Theta_2) \\ -\omega^2 J_2\Theta_2 &= -k_{t1}(\Theta_2 - \Theta_1) - k_{t2}(\Theta_2 - \Theta_3) \\ -\omega^2 J_3\Theta_3 &= -k_{t2}(\Theta_3 - \Theta_2) \end{aligned} \qquad \textbf{(5-10)}$$

FIG. 5-5. *A three-disk torsional system.*

*J. P. Den Hartog and J. P. Li, "Forced Torsional Vibrations with Damping: An Extension of Holzer's Method," *J. Appl. Mechanics*, Dec., 1946, pp. A–276–280.

Summing the equations gives

$$\sum_{i=1}^{3} J_i\Theta_i\omega^2 = 0 \tag{5-11}$$

Correspondingly, for an n disk system, we have

$$\sum_{i=1}^{n} J_i\Theta_i\omega^2 = 0 \tag{5-12}$$

The equation states that the sum of the inertia torque of a semidefinite system must be zero. The trial frequency ω must satisfy this constraint. Hence Eq. (5-12) is another form of the frequency equation.

To begin the tabulation, (1) assume a trial frequency ω and let $\Theta_1 = 1$, arbitrarily, (2) calculate Θ_2 from the first equation in Eq. (5-10), and (3) Θ_3 from the second equation, that is,

$$\begin{aligned}
\Theta_1 &= 1 \\
\Theta_2 &= \Theta_1 - \omega^2 J_1\Theta_1/k_{t1} \\
\Theta_3 &= \Theta_2 - \omega^2(J_1\Theta_1 + J_2\Theta_2)/k_{t2}
\end{aligned}$$

The values of Θ_1, Θ_2, and Θ_3 are substituted in Eq. (5-11) to check whether the constraint is satisfied. If not, a new value of ω is assumed and the process repeated. Note that the equations for Θ_2 and Θ_3 can be generalized for an n disk system as

$$\Theta_j = \Theta_{j-1} - \frac{\omega^2}{k_{t(j-1)}} \sum_{i=1}^{j-1} J_i\Theta_i \qquad j = 2, 3, \ldots, n \tag{5-13}$$

In summary, the method consists of the repeated application of Eqs. (5-12) and (5-13) for different trial frequencies. If the trial frequency is not a natural frequency of the system, Eq. (5-12) will not be satisfied. The residual torque in Eq. (5-12) represents a torque applied at the last disk. This is equivalent to a condition of steady-state forced vibration. A typical residual torque versus ω^2 plot is shown in Fig. 5-6.

The amplitudes Θ_i, $i = 1, 2, \ldots, n$, also give the mode shape for the given natural frequency. Recalling from Sec. 4-7 that there is one node

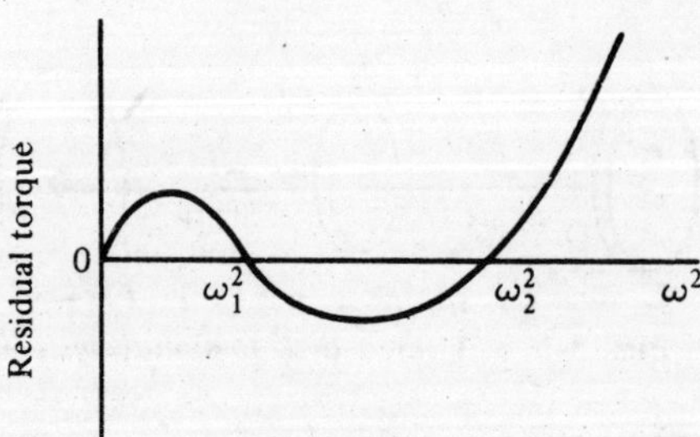

FIG. 5-6. *Residual torque versus ω^2.*

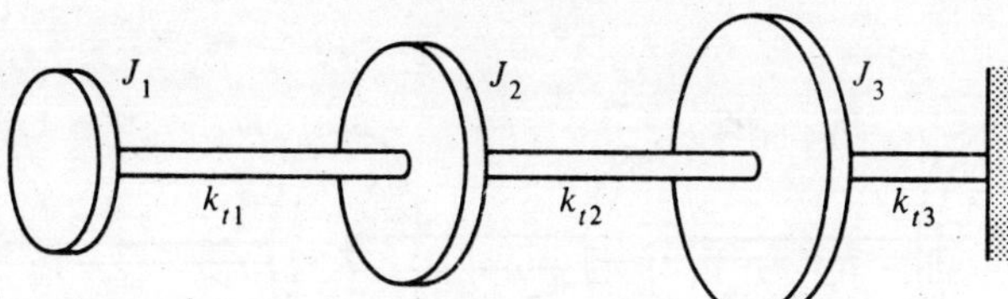

FIG. 5-7. *Torsional system with one fixed end.*

for the first mode, two for the second mode, etc., the mode of vibration can be determined from the sign changes in Θ_i. For example, if there are two sign changes in Θ_i, the frequency must be the second harmonic.

To show another example, consider a three-disk system with one fixed end shown in Fig. 5-7. The constraint is that the displacement at the fixed end must be zero. With slight modifications, the procedure above can be applied to this case. The applications are left as exercises.

Example 5

Find the natural frequencies and locate the nodes for the torsional system shown in Fig. 5-8(a).

Solution:

There is no standard procedure for estimating a trial frequency in applying Eqs. (5-12) and (5-13). As an initial trial, assume an equivalent system, consisting of J_1 and (J_3+J_4) at the two ends connected by a shaft with k_{t1} and k_{t2} in series. The spring constant k_t of this equivalent system is

$$k_t = \frac{1}{1/k_{t1}+1/k_{t2}} = \frac{10^3}{1/10+1/20} = \frac{20}{3}\ kN \cdot \text{m/rad}$$

From Eq. (4-46), the trial frequency is

$$\begin{aligned}\omega &= \sqrt{k_t/J_1 + k_t/(J_3+J_4)} \\ &= \sqrt{(20/3)(10^3)[1/0.4+1/(0.4+0.1)]} = 173 \text{ rad/s}\end{aligned}$$

Now we procede with the calculations as shown in Table 5-1. First, the values of J and k_t are entered in columns 1 and 5, respectively. The remaining values are determined as follows:

Row 1: Col. 2: Assume $\Theta_1 = 1$ rad
Col. 3: Compute $J_1\Theta_1\omega^2 = 0.4(1)(173)^2 = 11.97$ kN · m
Col. 4: Sum values in Col. 3 = 11.97 kN · m
Col. 6: Divide Col. 4 by Col. 5
$= 11.97/10 = 1.197$ rad

Row 2: Col. 2: Compute $\Theta_2 = \Theta_1 - (\text{twist in } k_{t1})$
$= 1.0 - 1.197 = -0.197$ rad
Col. 3: Compute $J_2\Theta_2\omega^2 = 0.1(-0.197)(173)^2$
$= -0.59$ kN · m

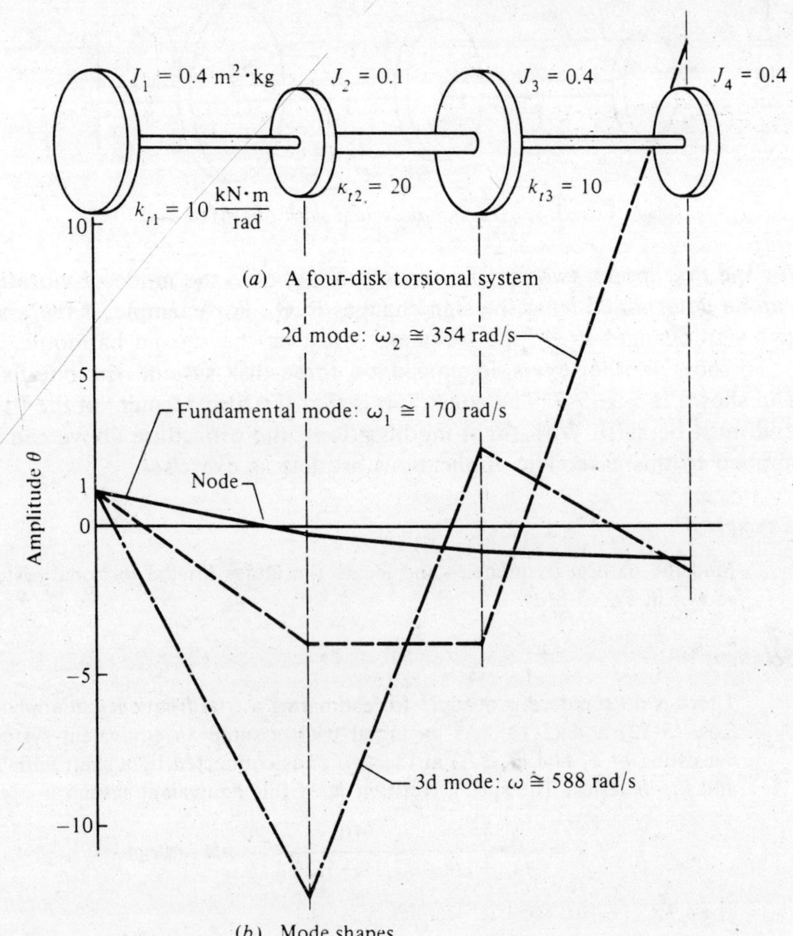

FIG. 5-8. *Natural frequencies of a semidefinite system; Example 6.*

Col. 4: Sum values in Col. $3 = 11.97 - 0.59$
$= 11.38$ kN · m

Col. 6: Divide Col. 4 by Col. $5 = 11.38/20$
$= 0.569$ rad (twist in k_{t2})

The values in rows 3 and 4 are determined in like manner. The residual torque is -0.92 kN · m, which is not zero. The negative sign indicates that the trial frequency is too high for the first mode as shown in Fig. 5-6.

Similarly, a frequency $\omega = 165$ rad/s is used for the second trial and the residual torque is 1.3 kN · m. A linear interpolation between the first and the second trial gives $\omega = 170$ rad/s. The calculated values for this frequency are shown in Table 5-1(*c*). The residual torque is -0.01 kN · m, which may

TABLE 5-1. *Holzer Method*

(*a*) First trial: *First mode:* $\omega = 173$ rad/s

COLUMN	1	2	3	4	5	6
ROW	J ($m^2 \cdot kg$)	Θ (rad)	$J\Theta\omega^2$ ($kN \cdot m$)	$\sum J\Theta\omega^2$ ($kN \cdot m$)	k_t ($kN \cdot m/rad$)	$\frac{1}{k_t}\sum J\Theta\omega^2$ (rad)
1	0.4	1.000	11.97	11.97	10	1.197
2	0.1	−0.197	−0.59	11.38	20	0.569
3	0.4	−0.766	−9.17	2.21	10	0.218
4	0.1	−1.043	−3.12	−0.92	residual torque	

(*b*) Second trial: *First mode:* $\omega = 165$ rad/s

ROW	1	2	3	4	5	6
1	0.4	1.000	10.89	10.89	10	1.089
2	0.1	−0.089	−0.24	10.65	20	0.532
3	0.4	−0.621	−6.77	3.88	10	0.388
4	0.1	−1.009	−2.74	1.13	residual torque	

(*c*) Third trial: *First mode:* $\omega = 170$ rad/s

ROW	1	2	3	4	5	6
1	0.4	1.000	11.56	11.56	10	1.156
2	0.1	−0.156	−0.45	11.11	20	0.555
3	0.4	−0.711	−8.22	2.88	10	0.288
4	0.1	−1.000	−2.89	−0.01	residual torque	

(*d*) *Second mode:* $\omega = 354$ rad/s

ROW	1	2	3	4	5	6
1	0.4	1.00	50.1	50.1	10	5.01
2	0.1	−4.01	−50.3	−0.2	20	−0.01
3	0.4	−4.00	−200.7	−200.9	10	−20.09
4	0.1	16.09	201.6	−0.6	residual torque	

(*e*) *Third mode:* $\omega = 588$ rad/s

ROW	1	2	3	4	5	6
1	0.4	1.00	138.3	138.3	10	13.83
2	0.1	−12.83	−443.6	−305.3	20	−15.26
3	0.4	2.43	336.7	31.4	10	3.14
4	0.1	−0.70	−24.4	7.0	residual torque	

be sufficiently accurate for the purpose. The mode shape for the fundamental mode is shown in Fig. 5-8(*b*).

The natural frequencies for the second and third harmonics are found in like manner. The calculations are tabulated in Table 5-1(*d*) and (*e*) and the mode shapes illustrated in Fig. 5-8(*b*).

5-5 TRANSFER MATRIX

We shall introduce the concepts of state vectors and transfer matrices and then apply the technique to Holzer-type problems in this section.

A *state vector* is a column of numbers, each of which is the value of a variable at a given station in the system. The numbers describe the state of the problem. Hence each variable, or an element of a state vector, is called a *state variable.*

The typical variables in a problem and their corresponding state vectors are shown in Fig. 5-9. For example, the variables in the last example are Θ and the torque T. The corresponding state vector $\{Z\}_n$ at station n is

$$\{Z\}_n = \begin{bmatrix} \Theta \\ T \end{bmatrix}_n \tag{5-14}$$

If the combination of shear and bending occurs at a station-n, the corresponding state vector is

$$\{Z\}_n = \begin{bmatrix} Y \\ \Phi \\ M \\ V \end{bmatrix}_n \tag{5-15}$$

where Y, Φ, M, and V are as defined in Fig. 5-9. The order of the elements in a state vector is arbitrary, provided it is consistent in a problem.

The sign convention of the variables in Fig. 5-9 will be further illustrated in the applications to follow. For the torsional problem in Fig. 5-9(b), the positive face of station n has its outward normal towards the positive coordinate direction. The positive angular displacement and

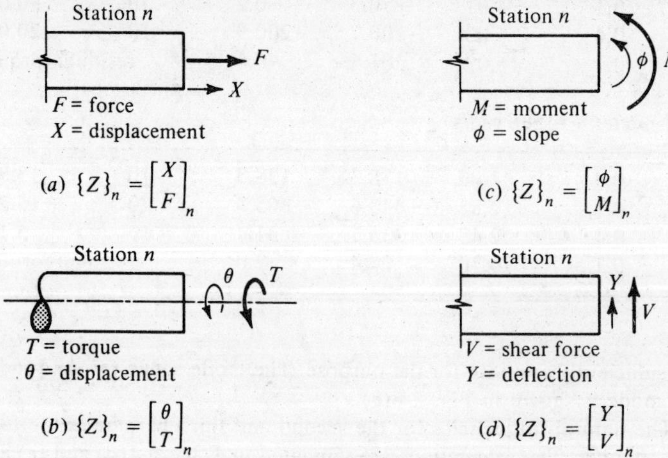

FIG. 5-9. *State vectors and generalized force and displacement.*

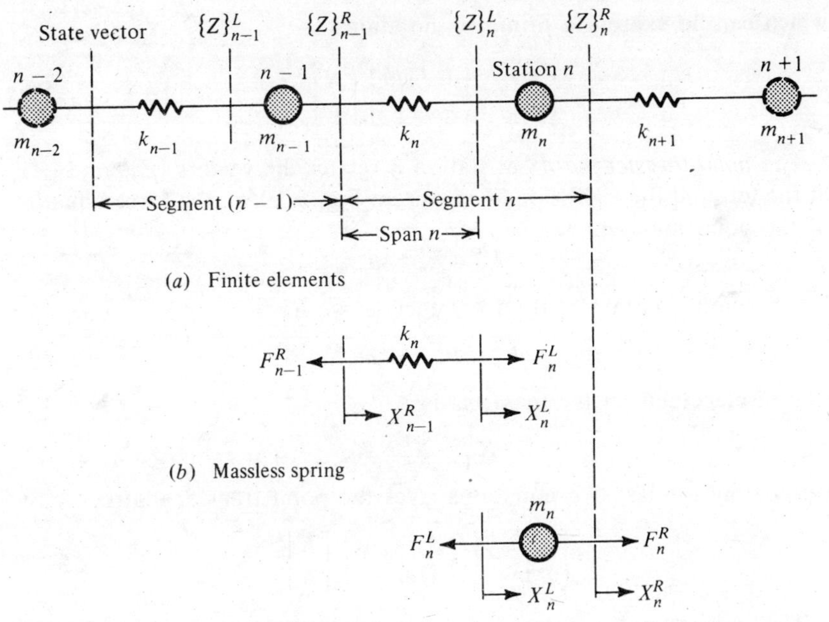

FIG. 5-10. *Finite element representation of a structure.*

positive torque are defined according to the right hand screw rule. As shown, the quantities are positive.

A *transfer matrix* transfers the state variables from one station to another. A complex system, such as a structure, can be divided into *finite elements** or *segments.* Thus, a complex or a continuous system can be approximated by an equivalent discrete system. Consider the spring-mass system in Fig. 5-10(a) for purpose of illustration. The positive direction of the x coordinate is from left to right. A typical *segment* of the system consists of a massless *span* with a spring k and a point mass m. The spring k_n is to the left of m_n in the segment n. The *superscripts* L and R denote the left and the right side of the mass m_n. Hence the transfer matrix of a segment consists of two parts, (1) the field transfer matrix of the elastic member, due to k_n, and (2) the point transfer matrix of the inertial mass, due to to m_n.

The *field transfer matrix* due to k_n relates the state vectors $\{Z\}_{n-1}^R$ and $\{Z\}_n^L$ at the beginning and the end of the span-n. From Fig. 5-10(b), the force equations of the span are

$$F_n^L = F_{n-1}^R \qquad \text{and} \qquad F_{n-1}^R = k_n(X_n^L - X_{n-1}^R)$$

* See, for example, R. E. D. Bishop, G. M. L. Gladwell, and S. Michaelson, *The Matrix Analysis of Vibration,* Cambridge University Press, London, 1965, p. 212.

which can be expressed in matrix notations as

$$\begin{bmatrix} X \\ F \end{bmatrix}_n^L = \begin{bmatrix} 1 & 1/k \\ 0 & 1 \end{bmatrix}_n \begin{bmatrix} X \\ F \end{bmatrix}_{n-1}^R \tag{5-16}$$

The *point transfer matrix* at station n relates the vectors $\{Z\}_n^L$ and $\{Z\}_n^R$ at the left and the right side of m_n. From Fig. 5-10(c), the force equation of the point mass m_n is

$$F_n^R - F_n^L = m_n \ddot{x}_n$$

For harmonic motion at the frequency ω, we have

$$F_n^R = F_n^L - \omega^2 m_n X_n$$

The displacement at the mass m_n is

$$X_n^R = X_n^L = X_n$$

Combining the last two equations gives the point transfer matrix.

$$\begin{bmatrix} X \\ F \end{bmatrix}_n^R = \begin{bmatrix} 1 & 0 \\ -\omega^2 m & 1 \end{bmatrix}_n \begin{bmatrix} X \\ F \end{bmatrix}_n^L \tag{5-17}$$

The *transfer matrix*, relating the state vectors $\{X \quad F\}_{n-1}^R$ and $\{X \quad F\}_n^R$ at the beginning and the end of the segment-n, is obtained by substituting $\{X \quad F\}_n^L$ from Eq. (5-16) in (5-17).

$$\begin{bmatrix} X \\ F \end{bmatrix}_n^R = \begin{bmatrix} 1 & 0 \\ -\omega^2 m & 1 \end{bmatrix}_n \begin{bmatrix} 1 & 1/k \\ 0 & 1 \end{bmatrix}_n \begin{bmatrix} X \\ F \end{bmatrix}_{n-1}^R$$

$$\begin{bmatrix} X \\ F \end{bmatrix}_n^R = \begin{bmatrix} 1 & 1/k \\ -\omega^2 m & 1-\omega^2 m/k \end{bmatrix}_n \begin{bmatrix} X \\ F \end{bmatrix}_{n-1}^R \tag{5-18}$$

or

$$\{Z\}_n^R = H_n \{Z\}_{n-1}^R \tag{5-19}$$

where H_n is a transfer matrix. For the spring-mass system in Fig. 5-10, H_n is defined by comparing Eqs. (5-18) and (5-19). Using the *recurrence formula* in Eq. (5-18), the state vector $\{Z\}_n^R$ at a typical station n, can be related to the state vector $\{Z\}_o^R$ at the boundary of the problem, that is,

$$\{Z\}_n^R = [H_n H_{n-1} \cdots H_2 H_1]\{Z\}_o^R \tag{5-20}$$

Example 6

Repeat Example 5 using the transfer matrix technique.

Solution:

Referring to Fig. 5-8 and noting the subscript usage in this section, k_1 is to the left of J_1. Since the system is semidefinite, $k_1 = 0$. We define $k_{t1} = k_2$.

$k_{t2} = k_3$, and $k_{t3} = k_4$ to conform to the subscript convention. Thus,

$$J_1 = J_3 = 0.4 \text{ m}^2 \cdot \text{kg} \qquad k_2 = k_4 = 10 \text{ kN} \cdot \text{m/rad}$$

$$J_2 = J_4 = 0.1 \text{ m}^2 \cdot \text{kg} \qquad k_3 = 20 \text{ kN} \cdot \text{m/rad}$$

Starting at disk J_1, assuming $\Theta_1 = 1$ rad and $T_1^L = 0$, and using Eq. (5-17), we get

$$\begin{bmatrix} \Theta \\ T \end{bmatrix}_1^R = \begin{bmatrix} 1 & 0 \\ -\omega^2 J & 1 \end{bmatrix}_1 \begin{bmatrix} \Theta \\ T \end{bmatrix}_1^L = \begin{bmatrix} 1 & 0 \\ -0.4\omega^2 & 1 \end{bmatrix} \begin{bmatrix} 1 \\ 0 \end{bmatrix}$$

$$= \begin{bmatrix} 1 \\ -0.4\omega^2 \end{bmatrix}$$

From Eq. (5-18), the state vector at station 2 is

$$\begin{bmatrix} \Theta \\ T \end{bmatrix}_2^R = \begin{bmatrix} 1 & 1/k \\ -\omega^2 J & 1-\omega^2 J/k \end{bmatrix}_2 \begin{bmatrix} \Theta \\ T \end{bmatrix}_1^R$$

$$= \begin{bmatrix} 1 & 10^{-4} \\ -0.1\omega^2 & 1-\omega^2(10^{-5}) \end{bmatrix} \begin{bmatrix} 1 \\ -0.4\omega^2 \end{bmatrix}$$

Similarly, using the appropriate values of J and k, the state vectors at stations 3 and 4 can be calculated.

For a semidefinite system, the boundary conditions require that the external torque T_4^R at the right of the disk J_4 be zero. It can be shown that this requires

$$T_4^R = 0.8(10^{-15})\omega^8 - 4(10^{-10})\omega^6 + 4.55(10^{-5})\omega^4 - \omega^2 = 0$$

which is the frequency equation of the system. The algebra is straightforward and computers can be used to alleviate the tedious calculations.

The values of the state vectors for the computed frequencies $\omega = 170$, 354, and 588 rad/s are tabulated in Table 5-2. These correspond to the fundamental, second, and third mode of vibration, respectively.

TABLE 5-2. *Natural Frequencies and State Vectors** (*System shown in Fig.* 5-8)

ω (rad/s)	$\begin{bmatrix} \Theta \\ T \end{bmatrix}_1^R$	$\begin{bmatrix} \Theta \\ T \end{bmatrix}_2^R$	$\begin{bmatrix} \Theta \\ T \end{bmatrix}_3^R$	$\begin{bmatrix} \Theta \\ T \end{bmatrix}_4^R$
170	1.00	−0.156	−0.711	−1.00
	−11.56	−11.11	−2.88	0.01
354	1.00	−4.01	−4.00	16.09
	−50.1	0.2	200.9	0.6
588	1.00	−12.83	2.43	−0.70
	−138.3	305.3	−31.4	−7.0

* Θ in rad and T in kN · m

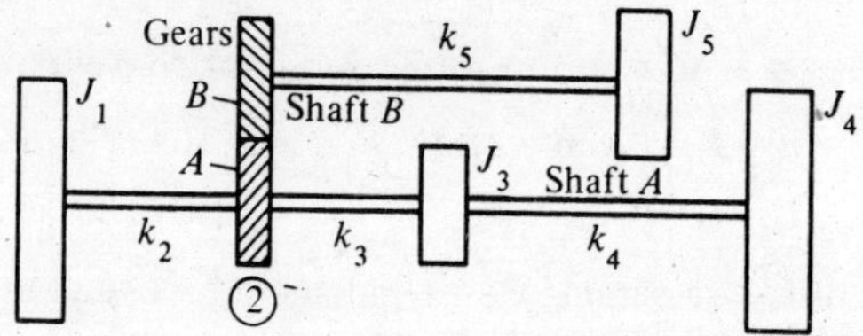

FIG. 5-11. *Branched gear system; Example 7.*

Example 7

A branched gear system consisting of shafts A and B and a number of inertia disks is shown in Fig. 5-11. Assume (1) the values of k_5 and J_5 are adjusted to refer to shaft A, and (2) the inertia effect of the gears are negligible. Find the transfer matrix for the segment consisting of k_2 and the gears.

Solution:

Let the additional subscripts A and B refer to shafts A and B, respectively. Consider the assembly of shaft A. The only difference between this problem and Example 6 is the additional torque at gear A due to gear B. In other words, if the torque T_{B2}^R is part of the transfer matrix H_2 in the recurrence formula in Eq. (5-19), then the two examples can be analyzed in like manner. Let

$$\begin{bmatrix} \Theta_A \\ T_A \end{bmatrix}_2^R = H_2 \begin{bmatrix} \Theta_A \\ T_A \end{bmatrix}_1^R$$

From Eq. (5-18), for the segment consisting of k_5 and J_5, we get

$$\begin{bmatrix} \Theta_B \\ T_B \end{bmatrix}_5^R = \begin{bmatrix} 1 & 1/k_5 \\ -\omega^2 J_5 & 1-\omega^2 J_5/k_5 \end{bmatrix} \begin{bmatrix} \Theta_B \\ T_B \end{bmatrix}_2^R$$

Premultiplying this by the inverse of the transfer matrix and noting that $T_{B5}^R = 0$ at the free end, we obtain

$$\begin{bmatrix} \Theta_B \\ T_B \end{bmatrix}_2^R = \begin{bmatrix} 1-\omega^2 J_5/k_5 & -1/k_5 \\ \omega^2 J_5 & 1 \end{bmatrix} \begin{bmatrix} \Theta_B \\ T_B \end{bmatrix}_5^R$$

$$= \begin{bmatrix} 1-\omega^2 J_5/k_5 \\ \omega^2 J_5 \end{bmatrix} \Theta_{B5}^R$$

or

$$T_{B2}^R = \omega^2 J_5 \Theta_{B5}^R = \frac{\omega^2 J_5}{1-\omega^2 J_5/k_5} \Theta_B^R \tag{5-21}$$

Thus, the torque balance equation is

$$T_{A2}^R = T_{A2}^L + T_{B2}^R \tag{5-22}$$

Since the gear ratio 1:1 is assumed, we have

$$\Theta_{B2}^R = -\Theta_{A2}^L = -\Theta_{A2}^R \tag{5-23}$$

Substituting T_{B2}^R from Eq. (5-21) in (5-22) and then write the resultant equation and Eq. (5-23) in matrix notations, we get

$$\begin{bmatrix} \Theta_A \\ T_A \end{bmatrix}_2^R = \begin{bmatrix} 1 & 0 \\ -\omega^2 J_5/(1-\omega^2 J_5/k_5) & 1 \end{bmatrix} \begin{bmatrix} \Theta_A \\ T_A \end{bmatrix}_2^L \tag{5-24}$$

This is the point transfer matrix at station 2.

The field transfer matrix is as shown in Eq. (5-16).

$$\begin{bmatrix} \Theta_A \\ T_A \end{bmatrix}_2^L = \begin{bmatrix} 1 & 1/k_2 \\ 0 & 1 \end{bmatrix} \begin{bmatrix} \Theta_A \\ T_A \end{bmatrix}_1^R \tag{5-25}$$

Substituting $\{\Theta_A \quad T_A\}_2^L$ from Eq. (5-25) in (5-24) we obtain the transfer matrix H_2.

$$\begin{bmatrix} \Theta_A \\ T_A \end{bmatrix}_2^R = \begin{bmatrix} 1 & 1/k_2 \\ \dfrac{-\omega^2 J_5}{1-\omega^2 J_5/k_5} & 1-\dfrac{\omega^2 J_5/k_2}{1-\omega^2 J_5/k_5} \end{bmatrix} \begin{bmatrix} \Theta_A \\ T_A \end{bmatrix}_1^R$$

5-6 MYKLESTAD-PROHL METHOD*

Myklestad and Prohl developed a tabular method to find the modes and natural frequencies of structures, such as an airplane wing. It is generally known as the Myklestad method. We shall use the transfer matrix technique for this discussion.

Following the finite element approach illustrated in Fig. 5-10, a structure or a beam can be divided into segments. A typical segment of a beam, as shown in Fig. 5-12, consists of a massless span and a point mass. The flexural properties of the segment is described by the field transfer matrix of the span; the inertial effect of the segment is described by the point transfer matrix of the mass. Hence the procedure is identical to that described in the last section, except for the state variables associated with the problem.

To describe the *field transfer matrix*, consider the free-body sketch of a uniform beam of length L in span n as shown in Fig. 5-12(a). For equilibrium, we require

$$V_n^L = V_{n-1}^R \quad \text{and} \quad M_n^L = M_{n-1}^R - L_n V_{n-1}^R \tag{5-26}$$

where M and V are the moment and the shear force, respectively. The symbols in this section are defined in Fig. 5-9 and the subscripts and superscript notations are the same as the last section. Referring to Fig.

* N. O. Myklestad, "A New Method of Calculating Natural Modes of Uncoupled Bending Vibrations of Airplane Wings and Other Types of Beams," *J. Aeron. Sci.*, (April, 1944), pp. 153–162.
M. A. Prohl, "A General Method for Calculating Critical Speeds of Flexible Rotors," *Trans. ASME*, vol. 66 (1945), p. A–142.

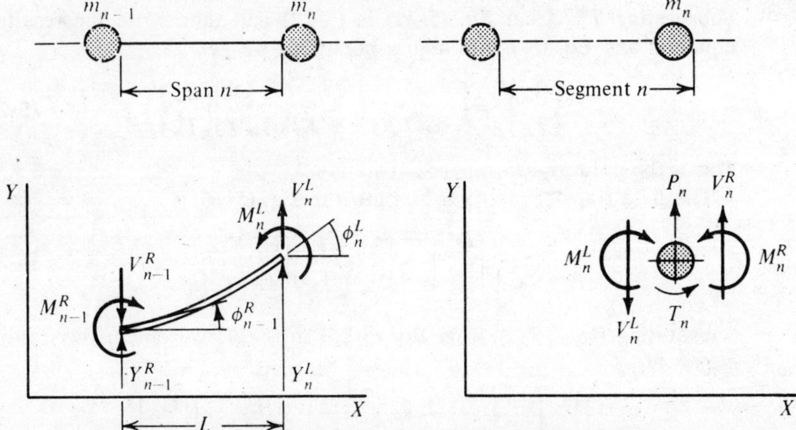

(*a*) Free-body sketch of span (*b*) Free-body sketch of mass

FIG. 5-12. *Derivation of transfer matrix of a beam.*

5-12(*a*), the change in the slope Φ of the span is due to moment M_n^L and the shear V_n^L

$$\Phi_n^L - \Phi_{n-1}^R = \left(\frac{L}{EI}\right)_n M_n^L + \left(\frac{L^2}{2EI}\right)_n V_n^L \tag{5-27}$$

Substituting M_n^L and V_n^L from Eq. (5-26) in (5-27) and rearranging, we get

$$\Phi_n^L = \Phi_{n-1}^R + \left(\frac{L}{EI}\right)_n M_{n-1}^R - \left(\frac{L^2}{2EI}\right)_n V_{n-1}^R \tag{5-28}$$

The change in the deflection Y of the span is

$$Y_n^L - Y_{n-1}^R = L_n \Phi_{n-1}^R + \left(\frac{L^2}{2EI}\right)_n M_n^L + \left(\frac{L^3}{3EI}\right)_n V_n^L \tag{5-29}$$

The first term on the right is the deflection due to the initial slope of the span, the second term due to the moment and the third term the shear force. The shear deformation of the beam is assumed negligible. Substituting M_n^L and V_n^L from Eq. (5-26) in (5-29) and rearranging, we obtain

$$Y_n^L = Y_{n-1}^R + L_n \Phi_{n-1}^R + \left(\frac{L^2}{2EI}\right)_n M_{n-1}^R - \left(\frac{L^3}{6EI}\right)_n V_{n-1}^R \tag{5-30}$$

The field transfer matrix is obtained by writing Eqs. (5-26), (5-28), and

(5-30) in the matrix form.

$$\begin{bmatrix} Y \\ \Phi \\ M \\ V \end{bmatrix}_n^L = \begin{bmatrix} 1 & L & \dfrac{L^2}{2EI} & -\dfrac{L^3}{6EI} \\ 0 & 1 & \dfrac{L}{EI} & -\dfrac{L^2}{2EI} \\ 0 & 0 & 1 & -L \\ 0 & 0 & 0 & 1 \end{bmatrix}_n \begin{bmatrix} Y \\ \Phi \\ M \\ V \end{bmatrix}_{n-1}^R \tag{5-31}$$

To derive the *point transfer matrix*, consider the free-body sketch of m_n in Fig. 5-12(*b*). The D'Alembert's inertia loads are $-\omega^2 m_n Y_n^L$ and $-\omega^2 J_n \Phi_n^L$, where J_n is the mass moment of inertia of m_n about its axis normal to the (x,y) plane. Neglecting the applied force P and the torque T, the equations for the shear and moment are

$$V_n^R = V_n^L - \omega^2 m_n Y_n^L \quad \text{and} \quad M_n^R = M_n^L - \omega^2 J_n \Phi_n^L \tag{5-32}$$

For the rigid body motion of m_n, we have

$$\Phi_n^L = \Phi_n^R \quad \text{and} \quad Y_n^L = Y_n^R \tag{5-33}$$

The point transfer matrix is obtained from Eqs. (5-32) and (5-33).

$$\begin{bmatrix} Y \\ \Phi \\ M \\ V \end{bmatrix}_n^R = \begin{bmatrix} 1 & 0 & 0 & 0 \\ 0 & 1 & 0 & 0 \\ 0 & -\omega^2 J & 1 & 0 \\ -\omega^2 m & 0 & 0 & 1 \end{bmatrix}_n \begin{bmatrix} Y \\ \Phi \\ M \\ V \end{bmatrix}_n^L \tag{5-34}$$

The *transfer matrix* for the segment n is obtained by substituting the state vector $\{Z\}_n^L$ from Eq. (5-31) in (5-34).

$$\begin{bmatrix} Y \\ \Phi \\ M \\ V \end{bmatrix}_n^R = \begin{bmatrix} 1 & 0 & 0 & 0 \\ 0 & 1 & 0 & 0 \\ 0 & -\omega^2 J & 1 & 0 \\ -\omega^2 m & 0 & 0 & 1 \end{bmatrix}_n \begin{bmatrix} 1 & L & \dfrac{L^2}{2EI} & -\dfrac{L^3}{6EI} \\ 0 & 1 & \dfrac{L}{EI} & -\dfrac{L^2}{2EI} \\ 0 & 0 & 1 & -L \\ 0 & 0 & 0 & 1 \end{bmatrix}_n \begin{bmatrix} Y \\ \Phi \\ M \\ V \end{bmatrix}_{n-1}^R$$

$$\begin{bmatrix} Y \\ \Phi \\ M \\ V \end{bmatrix}_n^R = \begin{bmatrix} 1 & L & \dfrac{L^2}{2EI} & -\dfrac{L^3}{6EI} \\ 0 & 1 & \dfrac{L}{EI} & -\dfrac{L^2}{2EI} \\ 0 & -\omega^2 J & 1-\omega^2 J\dfrac{L}{EI} & -L+\omega^2 J\dfrac{L^2}{2EI} \\ -\omega^2 m & -\omega^2 mL & -\omega^2 m\dfrac{L^2}{2EI} & 1+\omega^2 m\dfrac{L^3}{6EI} \end{bmatrix}_n \begin{bmatrix} Y \\ \Phi \\ M \\ V \end{bmatrix}_{n-1}^R \tag{5-35}$$

This is the recurrence formula analogous to Eq. (5-19). Hence the problem can be solved by its repeated application as indicated in Eq. (5-20).

The common boundary conditions for the beam problem are

	Y	Φ	M	V
Simple support	0	Φ	0	V
Free	Y	Φ	0	0
Fixed	0	0	M	V

For example, the deflection Y and the moment M at a simple support must be zero while the slope Φ and the shear force V are unknown and nonzero. At the beginning point or station O of a beam there are two nonzero boundary conditions, dictated by the type of support. Similarly, there are two nonzero boundary conditions at the other end of the beam.

The procedure for a natural frequency calculation is to assume a frequency ω as in the Holzer method and procede with the computation. The ω that satisfies simultaneously the boundary conditions at both ends of the beam is a natural frequency.

Example 8

Find the natural frequencies of the lumped-mass system in Fig. 5-13. Assume $m_1 = m_2 = 20$ kg (44 lb_m) and the flexural stiffness EI of the beam is $3\ kN \cdot m^2$ ($10^6\ lb_f\text{-in}^2$).

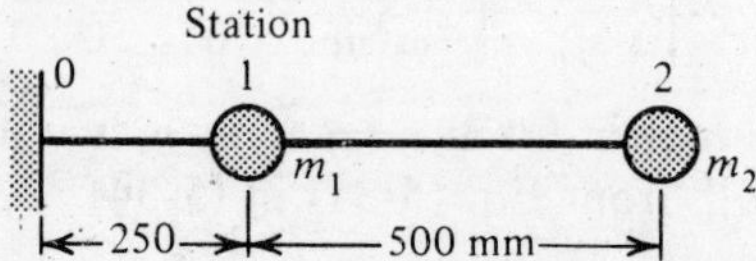

FIG. 5-13. *Lumped-mass representation of a beam.*

Solution:

The recurrence formulas for the computations are

$$\{Z\}_1^R = H_1\{Z\}_o^R \qquad \text{and} \qquad \{Z\}_2^R = H_2\{Z\}_1^R = H_2H_1\{Z\}_o^R$$

Where $\{Z\}_o^R = \{Y \quad \Phi \quad M \quad V\}_o^R = \{0 \quad 0 \quad M_o \quad V_o\}$

and M_o and V_o are the unknown moment and shear at the fixed end. Applying Eq. (5-35) and neglecting J, we get

$$\begin{bmatrix} Y \\ \Phi \\ M \\ V \end{bmatrix}_1^R = \begin{bmatrix} 1 & 0.25 & 10.4(10^{-6}) & -0.87(10^{-6}) \\ 0 & 1 & 83.3(10^{-6}) & -10.42(10^{-6}) \\ 0 & 0 & 1 & -0.25 \\ -20\omega^2 & -5\omega^2 & -208.3(10^{-6})\omega^2 & 1+17.36(10^{-6})\omega^2 \end{bmatrix} \begin{bmatrix} 0 \\ 0 \\ M \\ V \end{bmatrix}_o^R$$

$$\begin{bmatrix} Y \\ \Phi \\ M \\ V \end{bmatrix}_2^R = \begin{bmatrix} 1 & 0.50 & 41.7(10^{-6}) & -6.9(10^{-6}) \\ 0 & 1 & 166.7(10^{-6}) & -41.7(10^{-6}) \\ 0 & 0 & 1 & -0.5 \\ -20\omega^2 & -10\omega^2 & -833.3(10^{-6})\omega^2 & 1+138.9(10^{-6})\omega^2 \end{bmatrix} \begin{bmatrix} Y \\ \Phi \\ M \\ V \end{bmatrix}_1^R$$

From $\{Z\}_2^R = H_2H_1\{Z\}_o^R = [H]\{Z\}_o^R$, we get

$$\begin{bmatrix} Y \\ \Phi \\ M \\ V \end{bmatrix}_2^R = \begin{bmatrix} H_{11} & H_{12} & H_{13} & H_{14} \\ H_{21} & H_{22} & H_{23} & H_{24} \\ H_{31} & H_{32} & H_{33} & H_{34} \\ H_{41} & H_{42} & H_{43} & H_{44} \end{bmatrix} \begin{bmatrix} 0 \\ 0 \\ M_o \\ V_o \end{bmatrix}$$

M_2^R and V_2^R must be zero at the free end of the beam, that is

$$M_2^R = 0 = H_{33}M_o + H_{34}V_o$$
$$V_2^R = 0 = H_{43}M_o + H_{44}V_o$$

For a nontrivial solution of the simultaneous homogeneous equations, the determinant of the coefficients of M_o and V_o must vanish, that is,

$$\begin{vmatrix} H_{33} & H_{34} \\ H_{43} & H_{44} \end{vmatrix} = 0$$

From the given data, the corresponding equation is

$$\begin{vmatrix} 1+104.2(10^{-6})\omega^2 & -0.75-8.68(10^{-6})\omega^2 \\ -2.08(10^{-3})\omega^2-28.9(10^{-9})\omega^4 & 1+486(10^{-6})\omega^2+2.41(10^{-9})\omega^4 \end{vmatrix} = 0$$

or

$$\omega^4 - 73.3(10^3)\omega^2 + 75.4(10^6) = 0$$

The natural frequencies are $\omega_{1,2} = 32.3$ and 269 rad/s or $f_{1,2} = 5.14$ and 42.8 Hz.

Example 9*

Figure 5-14 shows a segment of a turbine blade, which is at a constant angular velocity Ω about the y axis. Derive the transfer matrix of a typical segment of the blade.

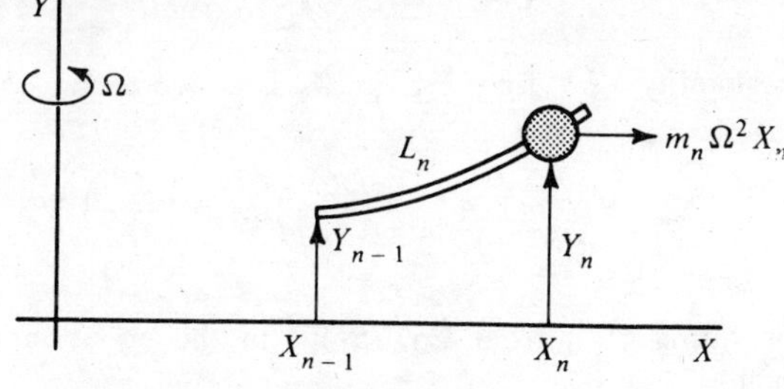

FIG. 5-14. *Centrifugal force on a beam.*

* N. O. Myklestad, *Fundamentals of Vibration Analysis*, McGraw-Hill Book Co., 1956, p. 244.

Solution:

The analysis is similar to that of a beam in the previous example, except that the equations are modified by the centrifugal force. The equations to follow are arranged for the calculations to proceed from right to left, that is, from station N to station o.

The centrifugal force due to m_n is $m_n\Omega^2X_n$ as shown in the figure. The force equation at m_n in the x direction is

$$F_n^L = F_n^R + m_n\Omega^2X_n \tag{5-36}$$

where

$$F_n^L = \sum_{i=n}^{N} m_i\Omega^2X_i \tag{5-37}$$

The component of the force F_n^L normal to the segment is $F_n^L\Phi_n^L$, where Φ_n^L is the slope at station n. Assume the change in slope $\Delta\Phi$ and the change in deflection ΔY due to $F_n^L\Phi_n^L$ are

$$\Delta\Phi = \left(\frac{L^2}{2EI}\right)_n F_n^L\Phi_n^L \quad \text{and} \quad \Delta Y = \left(\frac{L^3}{3EI}\right)_n F_n^L\Phi_n^L \tag{5-38}$$

Consider the span of length L. The moment $-F_n^L(Y_n^L - Y_{n-1}^R)$ is added to the moment equation in Eq. (5-26) to give

$$M_{n-1}^R = M_n^L + L_nV_n^L - F_n^L(Y_n^L - Y_{n-1}^R) \tag{5-39}$$

The quantity $-\Delta\Phi$ from Eq. (5-38) is added to the slope change in Eq. (5-27) to yield

$$\Phi_n^L - \Phi_{n-1}^R = \left(\frac{L}{EI}\right)_n M_n^L + \left(\frac{L^2}{2EI}\right)_n V_n^L - \left(\frac{L^2}{2EI}\right)_n F_n^L\Phi_n^L$$

or

$$\Phi_{n-1}^R = \left[1 + \left(\frac{L^2}{2EI}\right)_n F_n^L\right]\Phi_n^L - \left(\frac{L}{EI}\right)_n M_n^L - \left(\frac{L^2}{2EI}\right)_n V_n^L \tag{5-40}$$

The quantity $-\Delta Y$ from Eq. (5-38) is added to Eq. (5-29) to give

$$Y_n^L - Y_{n-1}^R = L_n\Phi_{n-1}^R + \left(\frac{L^2}{2EI}\right)_n M_n^L + \left(\frac{L^3}{3EI}\right)_n V_n^L - \left(\frac{L^3}{3EI}\right)_n F_n^L\Phi_n^L$$

Substituting Φ_{n-1}^R from Eq. (5-40) in the equation and simplifying, we obtain

$$Y_{n-1}^R = Y_n^L - \left[L_n + \left(\frac{L^3}{6EI}\right)_n F_n^L\right]\Phi_n^L + \left(\frac{L^2}{2EI}\right)_n M_n^L + \left(\frac{L^3}{6EI}\right)_n V_n^L \tag{5-41}$$

The quantity $(Y_n^L - Y_{n-1}^R)$ from Eq. (5-41) is substituted in Eq. (5-39) to obtain the moment equation, that is,

$$M_{n-1}^R = -\left[L_n + \left(\frac{L^3}{6EI}\right)F_n^L\right]F_n^L\Phi_n^L + \left[1+\left(\frac{L^2}{2EI}\right)_n F_n^L\right]M_n^L + \left[L_n + \left(\frac{L^3}{6EI}\right)_n F_n^L\right]V_n^L \quad \textbf{(5-42)}$$

The field transfer matrix is obtained by combining Eqs. (5-40) through (5-42), and $V_{n-1}^R = V_n^L$. Thus,

$$\begin{bmatrix} Y \\ \Phi \\ M \\ V \end{bmatrix}_{n-1}^R = \begin{bmatrix} 1 & -\left(L+\frac{FL^3}{6EI}\right) & \frac{L^2}{2EI} & \frac{L^3}{6EI} \\ 0 & \left(1+\frac{FL^2}{2EI}\right) & -\frac{L}{EI} & -\frac{L^2}{2EI} \\ 0 & -\left(L+\frac{FL^3}{6EI}\right) & \left(1+\frac{FL^2}{2EI}\right) & \left(L+\frac{FL^3}{6EI}\right) \\ 0 & 0 & 0 & 1 \end{bmatrix}_n \begin{bmatrix} Y \\ \Phi \\ M \\ V \end{bmatrix}_n^L \quad \textbf{(5-43)}$$

The point transfer matrix is obtained from Eqs. (5-32) and (5-33) Neglecting the rotary effect due to J, we have

$$\begin{bmatrix} Y \\ \Phi \\ M \\ V \end{bmatrix}_n^L = \begin{bmatrix} 1 & 0 & 0 & 0 \\ 0 & 1 & 0 & 0 \\ 0 & 0 & 1 & 0 \\ \omega^2 m & 0 & 0 & 1 \end{bmatrix}_n \begin{bmatrix} Y \\ \Phi \\ M \\ V \end{bmatrix}_n^R \quad \textbf{(5-4}$$

Substituting the state vector $\{Z\}_n^L$ from Eq. (5-44) in (5-43) and simplifying, we obtain the transfer matrix of the segment.

$$\begin{bmatrix} Y \\ \Phi \\ M \\ V \end{bmatrix}_{n-1}^R = \begin{bmatrix} \left(1+\frac{\omega^2 mL^3}{6EI}\right) & -\left(L+\frac{FL^3}{6EI}\right) & \frac{L^2}{2EI} & \frac{L^3}{6EI} \\ -\frac{\omega^2 mL^2}{2EI} & \left(1+\frac{FL^2}{2EI}\right) & -\frac{L}{EI} & -\frac{L^2}{2EI} \\ \omega^2 m\left(L+\frac{FL^3}{6EI}\right) & -\left(L+\frac{FL^3}{6EI}\right) & \left(1+\frac{FL^2}{2EI}\right) & \left(L+\frac{FL^3}{6EI}\right) \\ \omega^2 m & 0 & 0 & 1 \end{bmatrix}_n \begin{bmatrix} Y \\ \Phi \\ M \\ V \end{bmatrix}_n^R$$

(5-45)

5-7 SUMMARY

Many methods can be used to find the natural frequencies of undamped systems. Simple methods suitable for hand calculations and the transfer matrix technique are presented in this chapter.

Dunkerley's equation and Rayleigh method serve to illustrate the hand methods for estimating the fundamental frequency. Dunkerley's equation assumes that the fundamental frequency is much lower than the harmonics. By neglecting the harmonics, the estimated fundamental frequency is always lower than the actual value.

Rayleigh's method assumes a mode shape for the vibration at a natural frequency. The method is generally used to find the fundamental frequency, because it is more difficult to estimate the mode shapes of the harmonics. If the assumed mode shape is not exact, it is equivalent to having additional constraint in the system. Hence the estimated frequency tends to be higher than the true value.

The Holzer-type problem is discussed in the second half of the chapter. First, the Holzer method is shown as a hand method. A trial frequency is assumed and a solution is achieved if the boundary conditions are satisfied.

The transfer matrix technique may be regarded as an extension of the Holzer method. A state vector shown in Eq. (5-15) is an array of numbers, each of which is the value of a variable at a given station in the system. Hence it describes the state of a system. A transfer matrix relates the state vector of one station to the next, Eq. (5-19). Thus, a recurrence formula is obtained. Example 5 is repeated in Example 6 using the transfer matrix technique. Again, the frequency equation is dictated by the boundary conditions The transfer matrix technique is applied to the Myklestad-Prohl method for beams in Sec. 5-6. More complex problems can be handled in this manner. Computers are inevitably used for the calculations. The computer solutions are illustrated in Chap. 9.

PROBLEMS

5-1 Neglecting the mass of the shafts or beams, use Dunkerley's equation to estimate the fundamental frequencies of the systems shown in each of the following figures:

(**a**) Fig. P4-7(a). Assume $m_1 = 2m_2$ and $L_2 = L_1$.

(**b**) Fig. P4-7(b). Assume $m_1 = 2m_2$ and $L_1 = L_2 = L_3$.

(**c**) Fig. P4-8(a). Assume $L_1 = 150$ mm and $L = 600$ mm.

(**d**) Fig. P4-8(b). Assume $m_1 = m_3 = 2m_2$ and $a = b$.

(**e**) Fig. 5-2. Assume $m_1 = m$, $m_2 = 2m$, $m_3 = 3m$, and the disks and bearings divide the shaft into four equal sections of length L each.

5-2 Repeat Prob. 5-1 by Rayleigh's method. Verify that the estimated frequency tends to be higher than that by the frequency equation.

5-3 A 3,600-rpm turbo-generator set is schematically shown in Fig. P5-1. As a first approximation, assume that the system is symmetrical, friction is negligible, the supports are flexible in the vertical direction but are essentially rigid in the horizontal direction, and the bearings of the turbine and

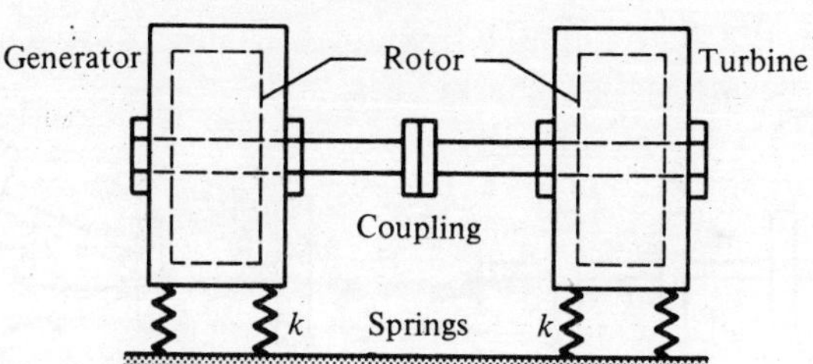

FIG. P5-1.

the generator housings are fairly close to one another. Estimate the critical speeds of the assembly. The data are:

Housing: 200 kg each; springs at each end with $k = 35 \times 10^6$ N/m

Rotor: 45 kg each; $J = 0.7\ \text{m}^2 \cdot \text{kg}$ each

Coupling: 20 kg; $J = 0.1\ \text{m}^2 \cdot \text{kg}$

Shaft: Length between rotors = 1 m; deflection = 0.02 mm under a 200 N force at midpoint

5-4 Use Holzer's method to find the natural frequencies for the first and second modes of the system in Fig. 5-5. Locate the modes along the shaft axis. Assume $J_1 = 1.5\ \text{m}^2 \cdot \text{kg}$, $J_2 = J_1$, $J_3 = 2J_1$, and $k_{t1} = k_{t2} = 100\ \text{N} \cdot \text{m/rad}$.

5-5 Use Holzer's method to find the fundamental frequency of the multidisk system in Fig. P5-2. Data in the SI units are: $J_1 = 300$, $J_2 = 50$, $J_3 = 300$, $J_4 = 20$, $J_5 = 400$, $J_6 = 50$, $J_7 = 4000$, $k_{t1} = 60 \times 10^6$, $k_{t2} = 600 \times 10^6$, $k_{t3} = 60 \times 10^6$, $k_{t4} = 40 \times 10^6$, $k_{t5} = 50 \times 10^6$, and $k_{t6} = 1 \times 10^6$.

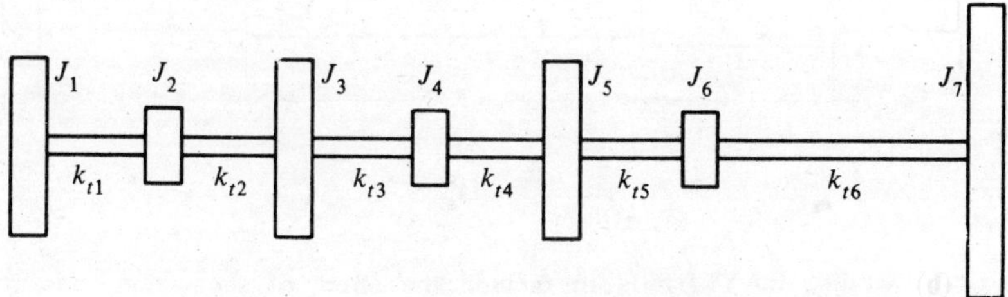

FIG. P5-2.

5-6 A torsional system with one fixed end is shown in Fig. 5-7. Use Holzer's method to find its natural frequencies for the first two modes. Data in the SI units are: $J_1 = 4$, $J_2 = 3$, $J_3 = 2$, $k_{t1} = 2 \times 10^6$, $k_{t2} = 1 \times 10^6$, and $k_{t3} = 1.5 \times 10^6$.

5-7 A torsional system with two fixed ends is shown in Fig. P5-3(a). Use Holzer's method to find its natural frequencies for the first two modes. Data in the SI units are: $J_1 = J_2 = J_3 = 2$ and $k_{t0} = k_{t1} = k_{t2} = k_{t3} = 4 \times 10^6$.

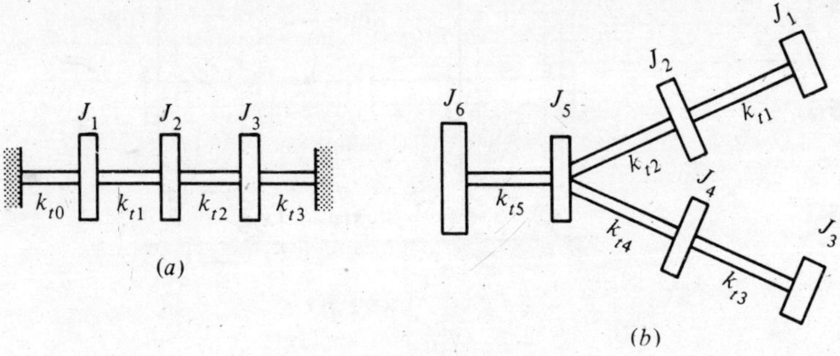

FIG. P5-3.

5-8 A branched-geared system is shown in Fig. P5-3(b). Use Holzer's method to find the fundamental frequency. Data in the SI units are: $J_1 = 5$, $J_2 = 10$, $J_3 = 5$, $J_4 = 10$, $J_5 = 10$, $J_6 = 30$, $k_{t1} = k_{t3} = 100$, $k_{t2} = 200$, $k_{t4} = 300$, and $k_{t5} = 400$.

Computer problems:

5-9 Using the Rayleigh method, the program RAYLEIGH in Fig. 9-10(a) finds the fundamental frequency of undamped multirotor systems.

(**a**) Use the program to find the fundamental frequency of the system in Fig. P5-4. The loads due to the disks are 18 kN and 4 kN. The dimensions are in units of mm.

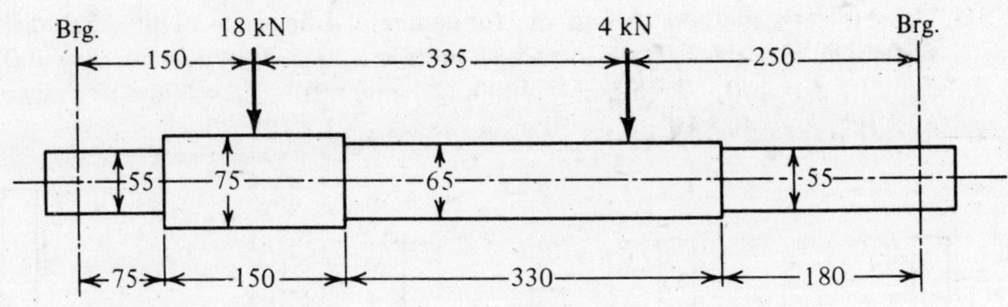

FIG. P5-4.

(**b**) Modify RAYLEIGH to include the effect of shear deformation as shown in Eq. (9-29).

(**c**) Shaft diameters and loads were entered by means of DATA statements in RAYLEIGH. Modify the program such that these values are entered by means of READ statements. Use the data for the shaft assembly in Fig. 9-9 to test the program.

5-10 Using the transfer matrix technique, write a program to find the natural frequencies and the state vectors of semidefinite multirotor systems. Use the program to solve Prob. 5-5.

5-11 A torsional system with one fixed end is shown in Fig. 5-7. Modify the program in Prob. 5-10 for this boundary condition. Use the program to solve Prob. 5-6.

5-12 A torsional system with two fixed ends is shown in Fig. P5-3(*a*). Modify the program in Prob. 5-10 for these boundary conditions. Use the program to solve Prob. 5-7.

5-13 The Myklestad-Prohl method was applied in the program TMXCANTI in Fig. 9-13(*a*) for a cantilever beam. Repeat Example 8 by means of TMXCANTI.

6

Discrete Systems

6-1 INTRODUCTION

Discrete or multidegree-of-freedom systems were introduced in Chap. 4. Additional concepts are presented in this chapter to give a better understanding of the problem. Lagrange's equations are used to examine the properties of the mass matrix M and the stiffness matrix K.

It was shown in Chap. 4 that the values of the elements m_{ij} and k_{ij} in the matrices M and K are dependent on the coordinates selected for the system description. This seems artificial, because m_{ij} and k_{ij} are related to the components of a physical system, which are assumed time invariant. The interpretation is that $[m_{ij}]$ is the quantity in the total kinetic energy of the system due to the generalized velocities. This concept was introduced in Chap. 3 for one-degree-of-freedom systems; an equivalent mass can be defined as that which gives the same kinetic energy. Similarly, $[k_{ij}]$ is the quantity in the total potential energy of the system due to the generalized displacements. The quantities m_{ij} and k_{ij} cannot be fictitious if they are related to the energies of the system, even when their values depend on the coordinates selected to describe the system. By virtue of the energy concept, students should not feel the artificiality in the subsequent matrix manipulations in the chapter.

Solutions by computers are mandatory for the efficient treatment of problems with more than two degrees of freedom and typical computer programs are given in Chap. 9. The general theory in the chapter is essential, because a computer solution of the equations itself would offer little towards the understanding of the problem.

Undamped systems will be treated before systems with damping, because the latter requires addition procedure for the problem solution.

6-2 EQUATIONS OF MOTION—UNDAMPED SYSTEMS

The equations of motion of discrete systems were formulated by Newton's second law and the method of influence coefficients in Chap. 4. Lagrange's equations, developed in App. B, are used to derive the equations of motion in this section.

Lagrange's equations of motion for the free vibration of a conservative system are

$$\frac{d}{dt}\left(\frac{\partial T}{\partial \dot{q}_i}\right)-\frac{\partial T}{\partial q_i}+\frac{\partial U}{\partial q_i}=0 \qquad \textbf{(6-1)}$$

where q_i is a generalized coordinate, $i=1,\ldots,n$, and T and U are the kinetic and potential energy functions, respectively.

The *potential energy* U can be expressed as

$$U=\frac{1}{2}\sum_{i=1}^{n}\sum_{j=1}^{n}k_{ij}q_iq_j \qquad \textbf{(6-2)}$$

where $\{q\}=\{0\}$ is the equilibrium position at which $U_O=0$. As shown in App. B, the *stiffness matrix* $K=[k_{ij}]$ is symmetric, that is, the transpose K^T of K is

$$K^T=K \qquad \text{or} \qquad k_{ij}=k_{ji} \qquad \textbf{(6-3)}$$

The equilibrium state at $\{q\}=\{0\}$ is stable or unstable depending on whether U increases or decreases in value when $\{q\}\neq\{0\}$, respectively. We shall consider only stable systems, regarding semidefinite systems as a special case. If the potential energy U always *increases* for $\{q\}\neq\{0\}$, then U is a *positive definite quadratic function** of $\{q\}$.

The *kinetic energy* T can be expressed as

$$T=\frac{1}{2}\sum_{i=1}^{n}\sum_{j=1}^{n}m_{ij}\dot{q}_i\dot{q}_j \qquad \textbf{(6-4)}$$

We shall use an example to show that the inertia coefficient m_{ij} may be a function of the coordinates $\{q\}$. If this occurs, nonlinear product terms will be introduced in the equations of motion. Since small oscillations about the equilibrium $\{q\}=\{0\}$ are assumed, m_{ij} about the equilibrium are

* A function is positive definite if it is never negative and is zero only if $\{q\}=\{o\}$. A function is positive semidefinite if it is never negative but can be zero for $\{q\}\neq\{0\}$. For example, K is positive semidefinite for a semidefinite system. As shown in Fig. 4-13(a), a semidefinite system could rotate as a rigid body. Thus, U of the system is unchanged while one of the coordinates is not zero.

A matrix is positive definite if every leading principal minor is positive. For example,

$$\begin{vmatrix} 2 & 2 & -1 \\ 2 & 3 & -2 \\ -1 & -2 & 2 \end{vmatrix}, \qquad |2|=2, \qquad \begin{vmatrix} 2 & 2 \\ 2 & 3 \end{vmatrix}=2, \qquad \begin{vmatrix} 2 & 2 & -1 \\ 2 & 3 & -2 \\ -1 & -2 & 2 \end{vmatrix}=1$$

constants, independent of the coordinates $\{q\}$. It follows that T is a function of $\{\dot{q}\}$ only and the term $\partial T/\partial q_i$ in Eq. (6-1) is zero. The *mass matrix* M is symmetric, that is, its transpose M^T is

$$M^T = M \qquad \text{or} \qquad m_{ij} = m_{ji} \tag{6-5}$$

Since the kinetic energy T is *always* positive, T is a positive definite quadratic function of $\{\dot{q}\}$.

The kinetic and potential energy functions in Eqs. (6-4) and (6-2) can be expressed in matrix notations as

$$\begin{aligned} T &= \tfrac{1}{2}\lfloor \dot{q} \rfloor M\{\dot{q}\} \\ U &= \tfrac{1}{2}\lfloor q \rfloor K\{q\} \end{aligned} \tag{6-6}$$

where $\lfloor q \rfloor = \{q\}^T$ is the transpose of $\{q\}$ and $\lfloor \dot{q} \rfloor = \{\dot{q}\}^T$. Note that $\lfloor q \rfloor$ and $\lfloor \dot{q} \rfloor$ are row vectors, since they are the transpose of $\{q\}$ and $\{\dot{q}\}$, respectively.

The equations of motion for the free vibration of a conservative system are obtained by substituting Eq. (6-6) in (6-1). Performing the indicated differentiations and neglecting the $\partial T/\partial q_i$ term, the resulting equation can be expressed in matrix notations as

$$M\{\ddot{q}\} + K\{q\} = \{0\} \tag{6-7}$$

This is a set of simultaneous linear ordinary differential equations as indicated in Eq. (4-36). For the convenience of manipulation in later sections, this is expressed as

$$\{\ddot{q}\} + H\{q\} = \{0\} \tag{6-8}$$

where $H = M^{-1}K$. An alternative form of this equation is

$$G\{\ddot{q}\} + \{q\} = \{0\} \tag{6-9}$$

where $G = K^{-1}M = H^{-1}$. Since K^{-1} is the flexibility (influence coefficient) matrix, Eq. (6-9) may be compared with the influence coefficient formulation of the same type of problem.

Although matrices M and K are symmetric, G and H are not necessarily symmetrical. Symmetry, however, is a convenience as well as a requirement for some computer programs.

A method to obtain a symmetrical H matrix in Eq. (6-8) is to employ a linear transformation

$$\{g\} = [t]\{q\} \qquad \text{or} \qquad \{q\} = [t]^{-1}\{g\} \tag{6-10}$$

where $\{g\}$ is a new set of generalized coordinates and $[t]$ is the transformation matrix. Let $[t]$ satisfy the constraint

$$[t]^T[t] = M \tag{6-11}$$

Substituting Eqs. (6-10) and (6-11) in (6-6) and noting the relation

$[[t]^{-1}]^T[t]^T = [[t][t]^{-1}]^T = I$, the energy functions in Eq. (6-6) become

$$T = \tfrac{1}{2}\lfloor \dot{g} \rfloor [[t]^{-1}]^T[t]^T[t][t]^{-1}\{\dot{g}\}$$
$$U = \tfrac{1}{2}\lfloor g \rfloor [[t]^{-1}]^T K[t]^{-1}\{g\}$$

or

$$\begin{aligned} T &= \tfrac{1}{2}\lfloor \dot{g} \rfloor \{\dot{g}\} \\ U &= \tfrac{1}{2}\lfloor g \rfloor S\{g\} \end{aligned} \tag{6-12}$$

where the matrix S can be shown to be symmetrical. Hence the mass matrix becomes a unit matrix I and the stiffness matrix becomes the symmetric matrix S in the $\{g\}$ coordinates. The equations of motion, corresponding to Eq. (6-8), are

$$\{\ddot{g}\} + S\{g\} = \{0\}$$

The problem can be solved in terms of the $\{g\}$ coordinates and then transformed to $\{q\}$ as shown in Eq. (6-10). Similarly, the matrix G in Eq. (6-9) can be made symmetrical, using the constraint $[t]^T[t] = K$. The matrices G, H, and S are often called the *dynamic matrices.*

A simple method to obtain the matrix $[t]$ in Eq. (6-11) is to factor M into two triangular matrices. For example, if M is 3×3, we assume $M = [t]^T[t]$ to obtain

$$\begin{bmatrix} m_{11} & m_{12} & m_{13} \\ m_{12} & m_{22} & m_{23} \\ m_{13} & m_{23} & m_{33} \end{bmatrix} = \begin{bmatrix} t_{11} & 0 & 0 \\ t_{12} & t_{22} & 0 \\ t_{13} & t_{23} & t_{33} \end{bmatrix} \begin{bmatrix} t_{11} & t_{12} & t_{13} \\ 0 & t_{22} & t_{23} \\ 0 & 0 & t_{33} \end{bmatrix} \tag{6-13}$$

The values of the elements in $[t]$ can be obtained by expanding $[t]^T[t]$ and equating the corresponding elements in Eq. (6-13). This is given as a problem in the chapter.

Example 1. Lagrange's equation—a simple exposition

Let us illustrate Lagrange's equations with a very simple problem. Write the equation of motion for the mass-spring system shown in Fig. 6-1.

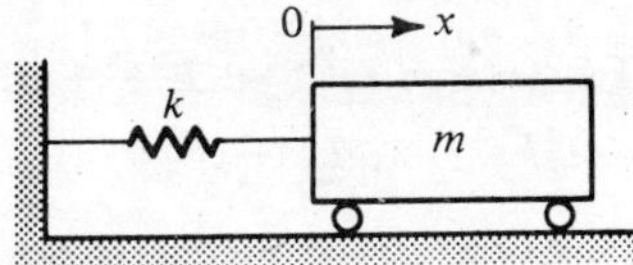

FIG. 6-1. *A mass-spring system.*

Solution:

The kinetic and potential energy functions are

$$T = \tfrac{1}{2}m\dot{x}^2 \quad \text{and} \quad U = \tfrac{1}{2}kx^2$$

Substituting T and U in Eq. (6-1) gives

$$\frac{d}{dt}(m\dot{x}) + kx = 0 \qquad \text{or} \qquad m\ddot{x} + kx = 0$$

Example 2. T as a function of $\{\dot{q}\}$ and $\{q\}$

Derive the equations of motion for the system shown in Fig. 6-2. Assuming small oscillations, find the kinetic and potential energy functions and linearize the equations of motion.

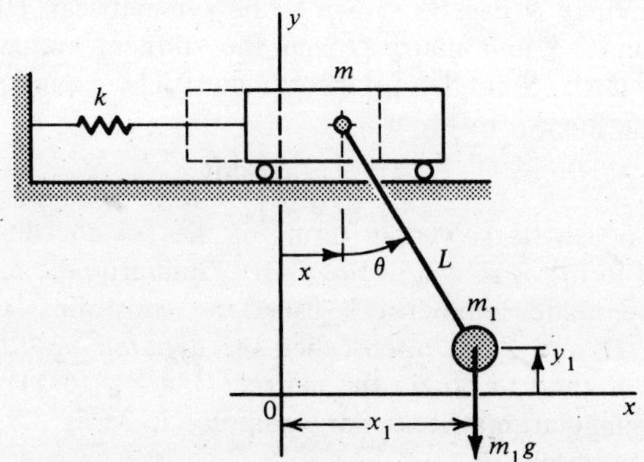

FIG. 6-2. *A two-degree-of-freedom system.*

Solution:

The energy functions can be in any coordinates. Let

$$T = \tfrac{1}{2}m\dot{x}^2 + \tfrac{1}{2}m_1(\dot{x}_1^2 + \dot{y}_1^2)$$
$$U = \tfrac{1}{2}kx^2 + m_1 g y_1$$

Using the generalized coordinates $\{x \quad \theta\}$, we have

$$x_1 = x + L\sin\theta \qquad \text{and} \qquad y_1 = L(1-\cos\theta)$$

Substituting these in T and U above gives

$$\begin{aligned} T &= \tfrac{1}{2}(m+m_1)\dot{x}^2 + \tfrac{1}{2}m_1L^2\dot{\theta}^2 + m_1L\dot{x}\dot{\theta}\cos\theta \\ U &= \tfrac{1}{2}kx^2 + m_1gL(1-\cos\theta) \end{aligned} \tag{6-14}$$

The equations of motion are obtained by the direct substitution of Eq. (6-14) in (6-1). For $q_1 = x$, we get

$$\frac{d}{dt}[(m+m_1)\dot{x} + m_1L\dot{\theta}\cos\theta] + kx = 0$$

$$(m+m_1)\ddot{x} + m_1L(\ddot{\theta}\cos\theta - \dot{\theta}^2\sin\theta) + kx = 0 \tag{6-15}$$

For $q_2 = \theta$, we obtain

$$\frac{d}{dt}(m_1L^2\dot{\theta} + m_1L\dot{x}\cos\theta) + m_1L\dot{x}\dot{\theta}\sin\theta + m_1gL\sin\theta = 0$$

$$m_1L^2\ddot{\theta} + m_1L\ddot{x}\cos\theta + m_1gL\sin\theta = 0 \tag{6-16}$$

The equations of motion in Eqs. (6-15) and (6-16) are nonlinear.

For small oscillations, we assume $\sin\theta \simeq \theta$ and $\cos\theta \simeq (1-\theta^2/2)$. Substituting these in Eq. (6-14) and neglecting the third and higher order terms, T and U can be expressed in matrix notations as

$$T = \tfrac{1}{2}\lfloor \dot{x} \quad \dot{\theta} \rfloor \begin{bmatrix} m+m_1 & m_1L \\ m_1L & m_1L^2 \end{bmatrix} \begin{bmatrix} \dot{x} \\ \dot{\theta} \end{bmatrix} \quad \text{and} \quad U = \tfrac{1}{2}\lfloor x \quad \theta \rfloor \begin{bmatrix} k & 0 \\ 0 & m_1gL \end{bmatrix} \begin{bmatrix} x \\ \theta \end{bmatrix}$$

Thus, the equations of motion are

$$\begin{bmatrix} m+m_1 & m_1L \\ m_1L & m_1L^2 \end{bmatrix} \begin{bmatrix} \ddot{x} \\ \ddot{\theta} \end{bmatrix} + \begin{bmatrix} k & 0 \\ 0 & m_1gL \end{bmatrix} \begin{bmatrix} x \\ \theta \end{bmatrix} = \begin{bmatrix} 0 \\ 0 \end{bmatrix}$$

Alternatively, the equations above can be obtained from Eqs. (6-15) and (6-16) by assuming $\sin\theta \simeq \theta$ and $\cos\theta \simeq 1$ and neglecting the nonlinear product terms.

6-3 UNDAMPED FREE VIBRATION—PRINCIPAL MODES

It was shown in Chap. 4 that, at a principal mode, the entire system executes synchronous harmonic motion at one of its natural frequencies. Matrix technique is used to examine this theory. The natural frequencies are found from the characteristic equation. The adjoint matrix is used as an alternative method to find the modal matrix. (See Example 9, Chap. 4.)

The equations of motion for the free vibration of a conservative system is shown in Eq. (6-8). In seeking a harmonic solution, we assume

$$\{q\} = \{u\}\sin(\omega t + \psi) \triangleq \{u\}p(t) \tag{6-17}$$

where $\{u\}$ is an *eigenvector* or *modal vector*, ω a natural frequency, ψ a constant, and $p(t) \triangleq \sin(\omega t + \psi)$ is a harmonic component of the motion $\{q\}$.

Substituting Eq. (6-17) in (6-8) and factoring out the term $\sin(\omega t + \psi)$, we obtain

$$[\omega^2 I - H]\{u\} = \{0\} \tag{6-18}$$

where I is a unit matrix. For convenience, we define

$$\lambda = \omega^2 \tag{6-19}$$

where λ is called an *eigenvalue.* Thus, Eq. (6-18) becomes

$$[\lambda I - H]\{u\} = \{0\} \tag{6-20}$$

which is a set of simultaneous homogeneous equations. $\Delta(\lambda) \triangleq |\lambda I - H|$ is called the *characteristic determinant.* For a nontrivial solution of Eq. (6-18), we equate $\Delta(\lambda)$ to zero

$$\Delta(\lambda) = |\lambda I - H| = 0 \tag{6-21}$$

This is called the *frequency* or *characteristic equation* and may be compared with Eq. (4-13). The λ's are the eigenvalues of H and $\omega_i = \sqrt{\lambda_i}$, for $i = 1, \ldots, n$. As a consequence of the assumed positive definiteness of M and K, the values of λ are real and positive.* We assume the λ's are distinct.†

The modal vector $\{u\}$ in Eq. (6-20) gives the relative amplitudes of vibration of the discrete masses at a natural frequency ω. For example, at $\omega = \omega_s$ or $\lambda = \lambda_s$, Eq. (6-20) becomes

$$[\lambda_s I - H]\{u\}_S = \{0\} \tag{6-22}$$

where $\{u\}_s$ is the modal vector corresponding to the eigenvalue λ_s. Note that λ_s and H in Eq. (6-22) are known quantities. In solving the homogeneous algebraic equation for $\{u\}_S$, only the relative magnitudes of the elements of $\{u\}_S$ can be determined.

The general solution of the equations of motion in Eq. (6-8) is obtained from the superposition of the harmonic solutions from Eq. (6-17). Considering all modes of vibration, we get

$$\{q\} = \sum_{i=1}^{n} \{u\}_i \sin(\omega_i t + \psi_i) \triangleq \sum_{i=1}^{n} \{u\}_i p_i$$

or

$$\{q\} = [u]\{p\} \quad \text{or} \quad \{p\} = [u]^{-1}\{q\} \tag{6-23}$$

This states that the motion $\{q\}$ of the discrete masses of the system is the superposition of the principal modes $p_i(t)$ as shown in Eqs. (4-20) and (4-21). Each mode is harmonic. The modal matrix $[u]$ is a combination of the modal vectors $\{u\}_i$ as in Eq. (4-23). $[u]$ describes all the modes of vibration of a linear system. It is necessary to prove, however, that the modal vectors, $\{u\}_i$, $i = 1, \ldots, n$, form a *linearly independent set* in the n

* F. B. Hildebrand, *Methods of Applied Mathematics*, Prentice-Hall, Inc., Englewood Cliffs, NJ, 1965, p. 72.

† For the type of problems considered, it rarely happens that two roots of Eq. (6-21) are identical. The roots may be sufficiently closed, however, to pose a computational problem. For a discussion of multiple roots, see, for example, L. A. Pipes, *Matrix Methods for Engineering*, Prentice-Hall, Inc., Englewood Cliffs, NJ, 1963, p. 249; and R. A. Frazer, W. J. Duncan, and A. R. Collar, *Elementary Matrices*, Cambridge University Press, London, 1957, Chap. 3.

space. This will be examined in Sec. 6-6. Furthermore, Eq. (6-23) gives the coordinate transformation between the generalized coordinates $\{q\}$ and the principal coordinates $\{p\}$.

Let us illustrate the discussion above with an example and use the adjoint matrix to find the modal matrix.

Example 3

Repeat Example 5, Chap. 4, and find the modal matrix. Assume $J_1 = J_2 = J_3 = J$ *and* $k_{t1} = k_{t2} = k_{t3} = k$.

Solution:

From Eq. (4-24), the equations of motion are

$$\begin{bmatrix} J & 0 & 0 \\ 0 & J & 0 \\ 0 & 0 & J \end{bmatrix} \begin{bmatrix} \ddot{\theta}_1 \\ \ddot{\theta}_2 \\ \ddot{\theta}_3 \end{bmatrix} + \begin{bmatrix} 2k & -k & 0 \\ -k & 2k & -k \\ 0 & -k & k \end{bmatrix} \begin{bmatrix} \theta_1 \\ \theta_2 \\ \theta_3 \end{bmatrix} = \begin{bmatrix} 0 \\ 0 \\ 0 \end{bmatrix}$$

Defining $b = k/J$ for convenience, the equation above can be expressed in the form of Eq. (6-8) as

$$\{\ddot{\theta}\} + b \begin{bmatrix} 2 & -1 & 0 \\ -1 & 2 & -1 \\ 0 & -1 & 1 \end{bmatrix} \{\theta\} = \{0\}$$

The frequency equation is from the direct application of Eq. (6-21).

$$\Delta(\lambda) = \begin{vmatrix} \lambda - 2b & b & 0 \\ b & \lambda - 2b & b \\ 0 & b & \lambda - b \end{vmatrix} = 0$$

or

$$\Delta(\lambda) = \lambda^3 - 5\lambda^2 b + 6\lambda b^2 - b^3$$

$$= (\lambda - 0.198b)(\lambda - 1.55b)(\lambda - 3.25b) = 0$$

There are three eigenvalues. The corresponding natural frequencies are $\omega_1 = \sqrt{\lambda_1} = \sqrt{0.198k/J}$, $\omega_2 = \sqrt{1.55k/J}$, and $\omega_3 = \sqrt{3.25k/J}$.

The modal vectors can be obtained from the adjoint matrix as illustrated in Example 6, App. B. From the equations of motion, the lambda matrix $[f(\lambda)]$ is

$$[f(\lambda)] = \begin{bmatrix} \lambda - 2b & b & 0 \\ b & \lambda - 2b & b \\ 0 & b & \lambda - b \end{bmatrix}$$

The adjoint $F(\lambda)$ of $[f(\lambda)]$ is

$$F(\lambda) = \begin{bmatrix} (\lambda - b)(\lambda - 2b) - b^2 & -b(\lambda - b) & b^2 \\ -b(\lambda - b) & (\lambda - b)(\lambda - 2b) & -b(\lambda - 2b) \\ b^2 & -b(\lambda - 2b) & (\lambda - 2b)^2 - b^2 \end{bmatrix}$$

A modal vector is obtained from any nonzero column of $F(\lambda)$.

If $\lambda = \lambda_1 = 0.198b$, we have

$$F(\lambda_1) = b^2 \begin{bmatrix} 0.445 & 0.802 & 1.0 \\ 0.802 & 1.445 & 1.802 \\ 1.0 & 1.802 & 2.247 \end{bmatrix}$$

It can be shown that the relative values of the elements in the columns are the same. The modal vector for the first mode is selected as

$$\{u\}_1 = \{1.0 \quad 1.802 \quad 2.247\}$$

In other words, the vibrations of the first mode at $\omega = \omega_1$ are

$$\begin{bmatrix} \theta_1 \\ \theta_2 \\ \theta_3 \end{bmatrix} = \begin{bmatrix} 1.0 \\ 1.802 \\ 2.247 \end{bmatrix} \Theta_1 \sin(\omega_1 t + \psi_1) \triangleq \{u\}_1 p_1$$

where Θ_1 and ψ_1 are arbitrary.

Similarly, the modal vectors for the second and third modes at ω_2 and ω_3 are

$$\{u\}_2 = \{1.0 \quad 0.445 \quad -0.802\}$$

$$\{u\}_3 = \{1.0 \quad -1.247 \quad 0.555\}$$

The system has three natural frequencies and correspondingly three modes of vibration. Each mode of vibration has the form $\{\theta\} = \{u\}_i \Theta_i \sin(\omega_i t + \psi_i) \triangleq \{u\}_i p_i$. By superposition as shown in Eq. (6-23), we get

$$\{\theta\} = [u]\{p\}$$

or

$$\begin{bmatrix} \theta_1 \\ \theta_2 \\ \theta_3 \end{bmatrix} = \begin{bmatrix} 1.0 & 1.0 & 1.0 \\ 1.802 & 0.445 & -1.247 \\ 2.247 & -0.802 & 0.555 \end{bmatrix} \begin{bmatrix} p_1 \\ p_2 \\ p_3 \end{bmatrix}$$

6-4 ORTHOGONALITY AND PRINCIPAL COORDINATES

We shall show that the modal vectors are orthogonal with respect to the mass matrix M and the stiffness matrix K. By virtue of the orthogonality property, the equations of motion can be uncoupled.

Without the coupling terms (see Sec. 4-5), the matrix M becomes a diagonal matrix $[\ulcorner M \lrcorner]$ with m_{ii} as its diagonal elements. Similarly, the matrix K reduces to $[\ulcorner K \lrcorner]$ with k_{ii} as its diagonal elements. The equations of motion, Eq. (6-7), of a conservative system become

$$[\ulcorner M \lrcorner]\{\ddot{p}\} + [\ulcorner K \lrcorner]\{p\} = \{0\} \tag{6-24}$$

or

$$m_{ii}\ddot{p}_i + k_{ii}p_i = 0 \qquad \text{for} \qquad i = 1, 2, \ldots, n \tag{6-25}$$

where $\{p\}$ are the *principal coordinates.* Thus, each mode can be examined as an independent one-degree-of-freedom system.

To uncouple the equations of motion, consider Eq. (6-7)

$$M\{\ddot{q}\} + K\{q\} = \{0\} \tag{6-26}$$

where M and K are symmetric.* At a principal mode, the entire system executes synchronous harmonic motion. For the ith mode, Eq. (6-23) gives $\{q\} = \{u\}_i p_i$. Hence we have for the rth mode

$$\{q\} = \{u\}_r p_r \tag{6-27}$$

Since p_r is harmonic at $\omega = \omega_r$, the corresponding equation of motion is of the form $m_{rr}\ddot{p}_r + k_{rr}p_r = 0$ as shown in Eq. (6-25). This can be expressed as

$$\ddot{p}_r = -\lambda_r p_r \tag{6-28}$$

where $\lambda_r = \omega_r^2 = k_{rr}/m_{rr}$. Similarly, for the sth mode, we get

$$\{q\} = \{u\}_s p_s \tag{6-29}$$

$$\ddot{p}_s = -\lambda_s p_s \tag{6-30}$$

Substituting Eqs. (6-27) and (6-28) in (6-26), factoring out the term p_r, and simplifying, we get

$$\lambda_r M\{u\}_r = K\{u\}_r \tag{6-31}$$

Similarly, substituting Eqs. (6-29) and (6-30) in (6-26) gives

$$\lambda_s M\{u\}_s = K\{u\}_s \tag{6-32}$$

Premultiplying Eq. (6-31) by the transpose $\lfloor u \rfloor_s$ of $\{u\}_s$ and then transposing the resultant equation, we obtain

$$\lambda_r \lfloor u \rfloor_r M\{u\}_s = \lfloor u \rfloor_r K\{u\}_s \tag{6-33}$$

Premultiplying Eq. (6-32) by $\lfloor u \rfloor_r$ yields

$$\lambda_s \lfloor u \rfloor_r M\{u\}_s = \lfloor u \rfloor_r K\{u\}_s \tag{6-34}$$

Upon subtracting Eq. (6-34) from (6-33), we have

$$(\lambda_r - \lambda_s)\lfloor u \rfloor_r M\{u\}_s = 0$$

Since $\lambda_r \neq \lambda_s$, we deduce that

$$\lfloor u \rfloor_r M\{u\}_s = 0 \qquad \text{for} \qquad r \neq s \tag{6-35}$$

* We assume M and K are symmetric and do not consider the general eigenvalue problem in which M and K are nonsymmetric. Since there is inherent symmetry in the problem (see Example 6, Chap. 4), we assume symmetrical M and K matrices can be found.

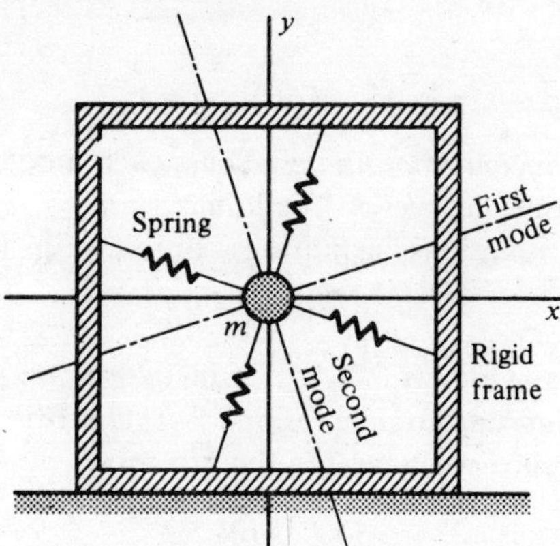

FIG. 6-3. *To illustrate orthogonality of modes.*

Similarly, it can be shown that

$$\lfloor u \rfloor_r K\{u\}_s = 0 \qquad \text{for} \qquad r \neq s \tag{6-36}$$

The last two equations are known as the *orthogonal relations.**

Note that the last two equations are nonzero if $r = s$, because M and K are assumed positive definite. Each of them is a scalar product, having the values m_{rr} and k_{rr}, respectively. Considering all modes of vibration, we get

$$[u]^T M[u] = [\ulcorner M \lrcorner] \qquad \text{and} \qquad [u]^T K[u] = [\ulcorner K \lrcorner] \tag{6-37}$$

where $[u]$ is the modal matrix and $[\ulcorner M \lrcorner]$ and $[\ulcorner K \lrcorner]$ are diagonal matrixes with m_{rr} and k_{rr} as the diagonal elements.

To uncouple the equations of motion in Eq. (6-26), we substitute $\{q\} = [u]\{p\}$ from Eq. (6-23) to get

$$M[u]\{\ddot{p}\} + K[u]\{p\} = \{0\}$$

* Students often ask, "What do you mean by orthogonal?" Let us illustrate with an example. Consider a mass m suspended by springs in a rigid frame as shown in Fig. 6-3. Vibrating in the plane of the paper, the system has two degrees of freedom. It has two natural frequencies and two principal modes of vibration. It is possible to find an axis along which m vibrates at one of its natural frequencies. This is a mode of vibration. It can be shown that the second mode occurs along a perpendicular axis. The verification of this statement is left as an exercise. In this problem, the two modes of vibration are orthogonal and also perpendicular to each other. This is analogous to finding the principal stress in a plane stress problem. Similarly, a mass *particle* in 3-space has three degrees of freedom. The modes of vibration are orthogonal and perpendicular to each other. A rigid body in 3-space, however, has six degrees of freedom. Hence the concept of perpendicularity has to be replaced by a mathematical concept. Mathematically, two vectors are orthogonal to each other if their scalar product is zero. In Eqs. (6-35) and (6-36), the scalar product of the vectors $\{u\}_r$ and $\{u\}_s$ are orthogonal relative to the matrices M and K. For an n-degree-of-freedom system, it is always possible to choose n *linearly independent* modal vectors that are orthogonal relative to M and K. To further illustrate the concept, consider Eq. (3-52) for the Fourier series expansion of a periodic function. In this case, the two functions, sine and cosine, are orthogonal. This will be used in Chap. 7 regarding the orthogonality of modes of continuous systems.

Premultiplying this by the transpose $[u]^T$ of $[u]$ gives

$$[u]^T M[u]\{\ddot{p}\}+[u]^T K[u]\{p\}=\{0\} \tag{6-38}$$

Applying Eq. (6-37), we obtain

$$[\ulcorner M \lrcorner]\{\ddot{p}\}+[\ulcorner K \lrcorner]\{p\}=\{0\}$$

which is identical to Eq. (6-24). Hence the equations of motion are uncoupled. Note that Eq. (6-37) was illustrated in Example 10, Chap. 4.

6-5 NORMAL COORDINATES

For convenience, the principal coordinates $\{p\}$ are normalized to give the *normal coordinates* $\{\eta\}$. In Example 3, we choose the first element of the modal vector $\{u\}_r$ equal to unity. This is one form of normalizing. It is a convenience, such as in plotting the modes.

Another common technique for normalizing is to choose the normal coordinates $\{\eta\}$ such that the diagonal matrix $[\ulcorner M \lrcorner]$ in Eq. (6-38) becomes a unit matrix I. Correspondingly, the matrix $[\ulcorner K \lrcorner]$ is diagonal with $\lambda_i = \omega_i^2$ as its diagonal elements.

Let $[u]=[u_{ij}]$ be the modal matrix relating the generalized coordinates $\{q\}$ and the principal coordinates $\{p\}$. Let $[\mu]=[\mu_{ij}]$ be the *normalized modal matrix* relating $\{q\}$ and the normal coordinates $\{\eta\}$. Thus,

$$\{q\}=[u]\{p\} \qquad \text{and} \qquad \{q\}=[\mu]\{\eta\} \tag{6-39}$$

$[\mu]$ can be obtained from $[u]$ my multiplying each modal vector $\{u\}_s$ in $[u]$ by a constant normalizing factor n_s, that is,

$$\{\mu\}_s=\{u\}_s n_s \qquad \text{or} \qquad [\mu]=[u][\ulcorner n \lrcorner] \tag{6-40}$$

where $[\ulcorner n \lrcorner]$ is diagonal with n_s, $s=1,\ldots,n$, as its diagonal elements.

Using the transformation in Eq. (6-39), the equations of motion in Eq. (6-38) can be expressed in the normal coordinates $\{\eta\}$ as

$$[\mu]^T M[\mu]\{\ddot{\eta}\}+[\mu]^T K[\mu]\{\eta\}=\{0\} \tag{6-41}$$

We require that the mass matrix becomes a unit matrix I. Applying Eq. (6-40) gives

$$\begin{aligned}[\mu]^T M[\mu] &= [\ulcorner n \lrcorner][u]^T M[u][\ulcorner n \lrcorner] \\ &= [\ulcorner n \lrcorner][\ulcorner M \lrcorner][\ulcorner n \lrcorner]=I\end{aligned}$$

Since $[u]^T M[u]=[\ulcorner M \lrcorner]$ is diagonal, the elements in $[\ulcorner n \lrcorner]$ is found by the relation

$$\frac{1}{n_s^2}=\sum_{i=1}^{n}\sum_{j=1}^{n} m_{ij}u_{is}u_{js} \tag{6-42}$$

The kinetic and potential energy functions in terms of the normal coordinates $\{\eta\}$ are

$$
\begin{aligned}
T &= \tfrac{1}{2}\lfloor \dot{q} \rfloor M\{\dot{q}\} = \tfrac{1}{2}\lfloor \dot{\eta} \rfloor [\mu]^T M[\mu]\{\dot{\eta}\} = \tfrac{1}{2}\lfloor \dot{\eta} \rfloor \{\dot{\eta}\} \\
U &= \tfrac{1}{2}\lfloor q \rfloor K\{q\} = \tfrac{1}{2}\lfloor \eta \rfloor [\mu]^T K[\mu]\{\eta\} = \tfrac{1}{2}\lfloor \eta \rfloor \Lambda\{\eta\}
\end{aligned} \tag{6-43}
$$

where Λ is a diagonal matrix with λ's as its diagonal elements. Comparing this with the energy expressions in Eq. (6-12), note that the matrix S in Eq. (6-12) is symmetric but not diagonal, since the generalized coordinates $\{g\}$ are not normal coordinates.

Example 4

Find the normalized modal matrix $[\mu]$ for the system in Example 3.

Solution:

From Example 3, we have

$$
[u] = [u_{ij}] = \begin{bmatrix} 1.0 & 1.0 & 1.0 \\ 1.802 & 0.445 & -1.247 \\ 2.247 & -0.802 & 0.555 \end{bmatrix}
$$

Applying Eq. (6-42) gives

$$
\begin{aligned}
1 &= n_1^2(m_{11}u_{11}^2 + m_{22}u_{21}^2 + m_{33}u_{31}^2) \\
&= n_1^2(1.0^2 + 1.802^2 + 2.247^2)J = n_1^2(9.296J)
\end{aligned}
$$

Thus, $n_1 = 1/3.049\sqrt{J}$. Similarly, $n_2 = 1/1.357\sqrt{J}$ and $n_3 = 1/1.692\sqrt{J}$. From Eq. (6-40), we get

$$
[\mu] = [u][\ulcorner n \lrcorner] = \frac{1}{\sqrt{J}} \begin{bmatrix} 0.328 & 0.737 & 0.591 \\ 0.591 & 0.328 & -0.737 \\ 0.737 & -0.591 & 0.328 \end{bmatrix}
$$

6-6 EXPANSION THEOREM

It was assumed in the previous sections that the motions of the discrete masses are the linear combination of the principal modes. This implies that the modal vectors form a linearly independent set.

The modal vectors are *linearly independent* if one vector cannot be expressed as a linear combination of the others. Mathematically, a set of n vectors $\{u\}_1, \{u\}_2, \ldots, \{u\}_n$ is linearly independent* if no set of constants $c_1, c_2, \ldots, c_n$, at least one of which is not zero, exists such that

$$
c_1\{u\}_1 + c_2\{u\}_2 + \cdots + c_n\{u\}_n = \sum_{s=1}^{n} c_s\{u\}_s = \{0\} \tag{6-44}
$$

Premultiplying this by $\lfloor u \rfloor_r M$ gives

$$
\sum_{s=1}^{n} c_s \lfloor u \rfloor_r M\{u\}_s = 0
$$

* F. B. Hildebrand, *op. cit.*, p. 25.

The orthogonal property in Eq. (6-36) shows that the product $\lfloor u \rfloor_r M\{u\}_s$ is zero only if $r \neq s$, that is, the equation above cannot be satisfied if $r = s$, unless $c_s = 0$ for $s = 1, \ldots, n$. In other words, the c's in Eq. (6-44) are all zero and the modal vectors are linearly independent.

Since the modal vectors are independent, any arbitrary vector $\{V\}$ in the same space can be expressed as a linear combination of $\{u\}_s$, for $s = 1, \ldots, n$, that is,

$$\{V\} = \sum_{s=1}^{n} c_s\{u\}_s \qquad \text{or} \qquad \{V\} = [u]\{c\} \tag{6-45}$$

where the c's are constants, $\{c\}$ is a vector of the constants, and $[u]$ the modal matrix. Each constant c_s determines the degree of participation of $\{u\}_s$ in the expansion of $\{V\}$. This is known as the *expansion theorem.*

A value of c_s is obtained by premultiplying Eq. (6-45) by $\lfloor u \rfloor_s M$ or $\lfloor u \rfloor_s K$. Noting the orthogonal relations, we get

$$\lfloor u \rfloor_s M\{V\} = c_s \lfloor u \rfloor_s M\{u\}_s \qquad \text{or} \qquad \lfloor u \rfloor_s K\{V\} = c_s \lfloor u \rfloor_s K\{u\}_s$$

The equations above can be solved for c_s to give

$$c_s = \frac{\lfloor u \rfloor_s M\{V\}}{\lfloor u \rfloor_s M\{u\}_s} \qquad \text{or} \qquad c_s = \frac{\lfloor u \rfloor_s K\{V\}}{\lfloor u \rfloor_s K\{u\}_s} \tag{6-46}$$

6-7 RAYLEIGH'S QUOTIENT

The Rayleigh method was discussed in Chaps. 2, 3, and 5. If an exact mode shape is assumed, the frequency calculated will be exact. A reasonable mode shape, however, would give satisfactory results. We shall verify this statement.

Consider Eq. (6-34) for which $r = s$. Rearranging the equation gives

$$\lambda_s = \frac{\lfloor u \rfloor_s K\{u\}_s}{\lfloor u \rfloor_s M\{u\}_s} \tag{6-47}$$

where $\omega_s = \sqrt{\lambda_s}$ is the natural frequency. The mode shape for the sth mode is described by the modal vector $\{u\}_s$.

Assume that $\{u\}_s$ is approximated by the vector $\{V\}$. Substituting $\{V\}$ for $\{u\}_s$ in Eq. (6-47) gives

$$\lambda_R = \frac{\lfloor V \rfloor K\{V\}}{\lfloor V \rfloor M\{V\}} \tag{6-48}$$

where λ_R is called the *Rayleigh's quotient.* Comparing Eqs. (6-47) and (6-48), if $\{V\}$ is not an exact modal vector, the frequency calculated from Eq. (6-48) will not be exact.

For convenience, let $\{V\}$ be expanded in terms of the normalized modal vector $\{\mu\}$. Evidently, either $\{u\}$ or $\{\mu\}$ can be used in Eqs. (6-47) and

(6-48). From the expansion in Eq. (6-45), we get

$$\{V\}=[\mu]\{c\} \qquad \text{and} \qquad \lfloor V \rfloor = \lfloor c \rfloor [\mu]^T$$

Substituting these in Eq. (6-48) yields

$$\lambda_R = \frac{\lfloor c \rfloor [\mu]^T K[\mu]\{c\}}{\lfloor c \rfloor [\mu]^T M[\mu]\{c\}} = \frac{\lfloor c \rfloor \Lambda\{c\}}{\lfloor c \rfloor \{c\}}$$

since $[\mu]^T K[\mu]=\Lambda$ and $[\mu]^T M[\mu]=I$ from Eq. (6-43). Alternatively, the expression for λ_R above can be expressed as

$$\lambda_R = \frac{\sum_{i=1}^n c_i^2 \lambda_i}{\sum_{i=1}^n c_i^2} \tag{6-49}$$

The interpretation of Eq. (6-49) is that λ_R is a weighted average of all λ_i, $i=1, \ldots, n$, and the weighting factors of which are c_i^2. Since c_i in Eq. (6-45) defines the degree of participation of each mode, the Rayleigh's quotient will approximate the eigenvalue λ_s if the sth mode is weighted more than the other modes. Thus, the approximate frequency $\omega_s(=\sqrt{\lambda_s})$ can be estimated from λ_R. Furthermore, if $\{V\}$ is a reasonable approximation of $\{\mu\}_s$, then $c_i/c_s<1$ and $(c_i/c_s)^2=\varepsilon_i^2 \ll 1$, where ε_i is a small number.

Assume λ_1 is desired for purpose of illustration. Dividing the numerator and denominator of Eq. (6-49) by c_1^2 gives

$$\lambda_R = \frac{\lambda_1 + \sum_{i=2}^n \varepsilon_i^2 \lambda_i}{1+\sum_{i=2}^n \varepsilon_i^2} = \lambda_1[1+O(\varepsilon^2)] \tag{6-50}$$

where $O(\varepsilon^2)$ denotes an expression in ε of the second order. The equation shows that (1) the estimated natural frequency by Rayleigh's method is greater than or equal to the exact value, and (2) if the error of the assumed mode shape, represented by the values of c's, is of the first order, the error in the estimated λ's is of the second order as indicated by $O(\varepsilon^2)$.

6-8 SEMIDEFINITE SYSTEMS

It was shown in Sec. 4-7 a system is semidefinite if it is capable of a rigid body motion. Since the system is not restrained by a stationary frame, the potential energy function U is a positive semidefinite quadratic function of the generalized coordinates $\{q\}$. The system can be made positive definite by using the orthogonal property to suppress the zero mode.

Let us develop the constraint equation for the zero mode suppression. Let $\{u\}_0$ be the modal vector for the zero mode. Since a rigid body motion requires that $q_1=q_2=\cdots=q_n$, $\{u\}_0$ must consist of elements of equal values. From the orthogonal relation in Eq. (6-35), we get

$$\lfloor u \rfloor_0 M\{u\}_i = 0 \qquad \text{for} \qquad i \neq 0$$

Assume the matrix M is diagonal by the choice of coordinates as illustrated in Sec. 4-4. Neglecting the common constant term in $\lfloor u \rfloor_0$, the constraint equation from the equation above is

$$\sum_{i=1}^{n} m_{ii} q_i = 0 \tag{6-51}$$

Example 5

Determine the modal matrix for the semidefinite system shown in Fig. 6-4(a).

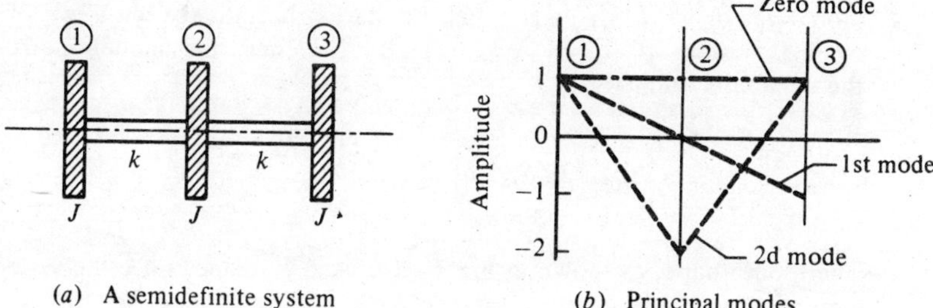

(a) A semidefinite system (b) Principal modes

FIG. 6-4. *Semidefinite system.*

Solution:

Since $m_{ii} = J$, Eq. (6-51) becomes

$$q_1 + q_2 + q_3 = 0$$

Expressing q_3 in terms of q_1 and q_2, we get $q_3 = -q_1 - q_2$, or

$$\begin{bmatrix} q_1 \\ q_2 \\ q_3 \end{bmatrix} = \begin{bmatrix} 1 & 0 \\ 0 & 1 \\ -1 & -1 \end{bmatrix} \begin{bmatrix} q_1 \\ q_2 \end{bmatrix}$$

This 3×2 matrix describes the constraint for the coordinates and can be called the *constraint matrix.* Thus, the kinetic energy T and the potential energy U become

$$T = \frac{1}{2} [\dot{q}_1 \quad \dot{q}_2] \begin{bmatrix} 1 & 0 & -1 \\ 0 & 1 & -1 \end{bmatrix} \begin{bmatrix} J & 0 & 0 \\ 0 & J & 0 \\ 0 & 0 & J \end{bmatrix} \begin{bmatrix} 1 & 0 \\ 0 & 1 \\ -1 & -1 \end{bmatrix} \begin{bmatrix} \dot{q}_1 \\ \dot{q}_2 \end{bmatrix}$$

$$= \frac{J}{2} [\dot{q}_1 \quad \dot{q}_2] \begin{bmatrix} 2 & 1 \\ 1 & 2 \end{bmatrix} \begin{bmatrix} \dot{q}_1 \\ \dot{q}_2 \end{bmatrix}$$

$$U = \frac{1}{2} [q_1 \quad q_2] \begin{bmatrix} 1 & 0 & -1 \\ 0 & 1 & -1 \end{bmatrix} \begin{bmatrix} k & -k & 0 \\ -k & 2k & -k \\ 0 & -k & k \end{bmatrix} \begin{bmatrix} 1 & 0 \\ 0 & 1 \\ -1 & -1 \end{bmatrix} \begin{bmatrix} q_1 \\ q_2 \end{bmatrix}$$

$$= \frac{k}{2} [q_1 \quad q_2] \begin{bmatrix} 2 & 1 \\ 1 & 5 \end{bmatrix} \begin{bmatrix} q_1 \\ q_2 \end{bmatrix}$$

The system is positive definite with respect to the *new* M and K matrices. Defining $b = k/J$, the corresponding equations of motion are

$$\begin{bmatrix} 2 & 1 \\ 1 & 2 \end{bmatrix} \begin{bmatrix} \ddot{q}_1 \\ \ddot{q}_2 \end{bmatrix} + b \begin{bmatrix} 2 & 1 \\ 1 & 5 \end{bmatrix} \begin{bmatrix} q_1 \\ q_2 \end{bmatrix} = \begin{bmatrix} 0 \\ 0 \end{bmatrix}$$

It can be shown that the frequency equation is

$$\Delta(\lambda) = \begin{vmatrix} 2\lambda - 2b & \lambda - b \\ \lambda - b & 2\lambda - 5b \end{vmatrix} = 3(\lambda - b)(\lambda - 3b) = 0$$

For $\lambda = b$, the modal vector is $\{q_1 \quad q_2\} = \{1 \quad 0\}$. Since the constraint is $q_3 = -q_1 - q_2$, the corresponding value of q_3 is -1. Hence the modal vector $\{u\}_1$ of the original system is $\{q_1 \quad q_2 \quad q_3\} = \{1 \quad 0 \quad -1\}$. For $\lambda = 3b$, the modal vector is $\{q_1 \quad q_2\} = \{1 \quad -2\}$. Similarly, the modal vector $\{u\}_2$ of the original system is $\{q_1 \quad q_2 \quad q_3\} = \{1 - 2 \quad 1\}$. Thus, the modal matrix of the original system is

$$[u] = [\{u\}_0 \ \{u\}_1 \ \{u\}_2] = \begin{bmatrix} 1 & 1 & 1 \\ \text{[illegible]} & \text{[illegible]} & -2 \\ \text{[illegible]} & \text{[illegible]} & \text{[illegible]} \end{bmatrix}$$

The mode shape, as shown in Fig. 6-4(b), is a [illegible]ited, since the system is symmetrical.

6-9 MATRIX ITERATION

The method of Duncan and Collar,* described in this section, is generally called matrix iteration, although there are other iterative methods. The method yields one frequency and modal vector at a time. It first iterates for the lowest, or the highest, frequency and mode. Then this mode is suppressed and the process repeated to obtain the next mode.

Let $p(t)$ represent a principal mode at a natural frequency ω and $\{u\}$ the corresponding modal vector. Hence the displacement vector is $\{q\} = \{u\}p$ and $\{\ddot{q}\} = -\omega^2\{u\}p = -\lambda\{u\}p$. Substituting these quantities in Eqs. (6-8) and (6-9) and simplifying, we get

$$\lambda\{u\} - H\{u\} \tag{6-52}$$

$$\frac{1}{\lambda}\{u\} = G\{u\} \tag{6-53}$$

Either equation can be used to start the iteration. Iterating on Eq. (6-52) will first yield the highest frequency and mode, and Eq. (6-53) the lowest frequency and mode.

* W. J. Duncan and A. R. Collar, "A Method for the Solution of Oscillation Problems by Matrices," *Phil. Mag.*, 1934 ser. 7, vol. 17, p. 865.

Let us use Eq. (6-52) for illustration. Assume an arbitrary vector $\{V\}_0$. Expanding $\{V\}_0$ in terms of the modal vectors as shown in Eq. (6-45) gives

$$\{V\}_0 = \sum_{i=1}^{n} c_i\{u\}_i \tag{6-54}$$

where the c's are constants. Next, form a sequence of vectors as follows:

$$\{V\}_0, \quad \{V\}_1 = H\{V\}_0, \quad \{V\}_2 = H\{V\}_1 = H^2\{V\}_0, \ldots \tag{6-55}$$

Substituting Eq. (6-55) in (6-54) and noting the relation $\lambda\{u\} = H\{u\}$ in Eq. (6-52), we obtain

$$\{V\}_1 = H\{V\}_0 = \sum_{i=1}^{n} c_i H\{u\}_i = \sum_{i=1}^{n} c_i\lambda_i\{u\}_i$$

$$\{V\}_2 = H^2\{V\}_0 = \sum_{i=1}^{n} c_i H^2\{u\}_i = \sum_{i=1}^{n} c_i\lambda_i H\{u\}_i = \sum_{i=1}^{n} c_i\lambda_i^2\{u\}_i$$

$$\cdots\cdots$$

$$\{V\}_s = H^s\{V\}_0 = \sum_{i=1}^{n} c_i H^s\{u\}_i = \sum_{i=1}^{n} c_i\lambda_i^s\{u\}_i$$

We assume that the λ's are distinct and $\lambda_1 > \lambda_2 > \cdots > \lambda_n$. If s is sufficiently large, $\lambda_1^s \gg \lambda_i^s$, for $i = 2, \ldots, n$. Thus, we obtain

$$\{V\}_s = H^s\{V\}_0 \simeq c_1\lambda_1^s\{u\}_1 \tag{6-56}$$

If one more iteration is performed, it can be seen that the result is

$$\{V\}_{s+1} = c_1\lambda_1^{s+1}\{u\}_1 = \lambda_1\{V\}_s \tag{6-57}$$

Thus, the first eigenvalue λ_1 and the corresponding modal vector $\{u\}_1$ can be obtained.

The second mode is found by suppressing the first mode p_1, that is, by introducing the constraint $p_1 = 0$. Then the equations of motion are modified accordingly and the iteration procedure repeated.

To suppress the first mode, consider the transformation

$$\{q\} = [u]\{p\} \qquad \text{and} \qquad \{p\} = [u]^{-1}\{q\} \triangleq [v]\{q\}$$

Using the constraint equation $p_1 = 0$ gives

$$p_1 = v_{11}q_1 + v_{12}q_2 + \cdots + v_{1n}q_n = 0 \tag{6-58}$$

The rest of the coordinates remain unchanged, that is,

$$q_i = q_i \qquad \text{for} \qquad i = 2, 3, \ldots, n$$

The last two equations can be conveniently expressed in matrix notations.

Using a 3×3 matrix to illustrate the last two equations, we have

$$\begin{bmatrix} v_{11} & v_{12} & v_{13} \\ 0 & 1 & 0 \\ 0 & 0 & 1 \end{bmatrix} \begin{bmatrix} q_1 \\ q_2 \\ q_3 \end{bmatrix} = \begin{bmatrix} 0 & 0 & 0 \\ 0 & 1 & 0 \\ 0 & 0 & 1 \end{bmatrix} \begin{bmatrix} q_1 \\ q_2 \\ q_3 \end{bmatrix}$$

$$\begin{bmatrix} q_1 \\ q_2 \\ q_3 \end{bmatrix} = \begin{bmatrix} v_{11} & v_{12} & v_{13} \\ 0 & 1 & 0 \\ 0 & 0 & 1 \end{bmatrix}^{-1} \begin{bmatrix} 0 & 0 & 0 \\ 0 & 1 & 0 \\ 0 & 0 & 1 \end{bmatrix} \begin{bmatrix} q_1 \\ q_2 \\ q_3 \end{bmatrix}$$

or

$$\{q\} = S_1\{q\} \tag{6-59}$$

where S_1 as defined above is called the *sweeping matrix.*

The matrix H in Eq. (6-52) is modified by S_1 to give a new matrix H_1.

$$H_1 = HS_1 \tag{6-60}$$

Now, H_1 can be used in Eq. (6-52) for the iteration of the second mode, since the first mode has been suppressed. Equations (6-55) to (6-57) can be used for the iterations as before.

The v's in Eq. (6-58), however, cannot be determined directly, because only $\{u\}_1$ is found and $[u]^{-1} = [v]$ is as yet unknown. From the orthogonal property in Eq. (6-35), we have

$$[u]^T M [u] = [^\frown M_\smile]$$

where $[^\frown M_\smile]$ is diagonal. Premultiplying both sides of the equation by $[^\frown M_\smile]^{-1}$ and postmultiplying by $[u]^{-1}$, we get

$$[^\frown M_\smile]^{-1}[u]^T M = [u]^{-1} = [v]$$

Since $[^\frown M_\smile]$ is diagonal, the first row of this equation gives

$$[v_{11} \quad v_{12} \cdots v_{1n}] = (\text{constant}) \lfloor u \rfloor_1 M \tag{6-61}$$

Thus, the sweeping matrix in Eq. (6-59) can be constructed.

Example 6

Determine the natural frequencies and the modal vectors of the system in Fig. 6-5.

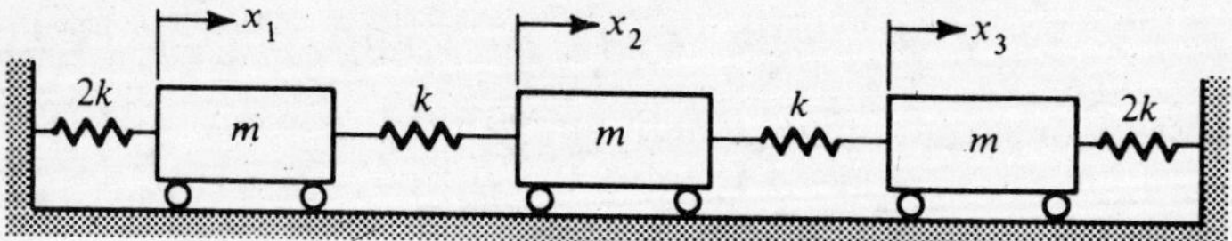

FIG. 6-5. *A three-degree-of-freedom system.*

Solution:

The equations of motion are

$$\begin{bmatrix} m & 0 & 0 \\ 0 & m & 0 \\ 0 & 0 & m \end{bmatrix}\begin{bmatrix} \ddot{x}_1 \\ \ddot{x}_2 \\ \ddot{x}_3 \end{bmatrix} + \begin{bmatrix} 3k & -k & 0 \\ -k & 2k & -k \\ 0 & -k & 3k \end{bmatrix}\begin{bmatrix} x_1 \\ x_2 \\ x_3 \end{bmatrix} = 0$$

Premultiplying by K^{-1}, the equation can be rearranged in the form of Eq. (6-53) as

$$\begin{bmatrix} 5 & 3 & 1 \\ 3 & 9 & 3 \\ 1 & 3 & 5 \end{bmatrix}\{u\} = \frac{1}{\lambda^*}\{u\} \qquad \text{where } \lambda^* = \frac{m\lambda}{12k}$$

or

$$G\{u\} = \frac{1}{\lambda^*}\{u\}$$

Let $\{V\}_0 = \{1 \quad 1 \quad 1\}$ arbitrarily to start the iteration. From Eq. (6-55), we have

$$\{V\}_1 = G\{V\}_0 = \begin{bmatrix} 5 & 3 & 1 \\ 3 & 9 & 3 \\ 1 & 3 & 5 \end{bmatrix}\begin{bmatrix} 1 \\ 1 \\ 1 \end{bmatrix} = \begin{bmatrix} 9 \\ 15 \\ 9 \end{bmatrix} = 9\begin{bmatrix} 1.0 \\ 1.67 \\ 1.0 \end{bmatrix}$$

The constant 9 is discarded. Continuing the iteration gives

$$\{V\}_2 = G\{V\}_1 = \begin{bmatrix} 5 & 3 & 1 \\ 3 & 9 & 3 \\ 1 & 3 & 5 \end{bmatrix}\begin{bmatrix} 1.0 \\ 1.67 \\ 1.0 \end{bmatrix} = \begin{bmatrix} 11 \\ 21 \\ 11 \end{bmatrix} = 11\begin{bmatrix} 1.0 \\ 1.91 \\ 1.0 \end{bmatrix}$$

Again, the constant 11 is discarded. The fourth and fifth iterations give

$$\{V\}_4 = 11.98\begin{bmatrix} 1.0 \\ 2.0 \\ 1.0 \end{bmatrix} \qquad \text{and} \qquad \{V\}_5 = 12.0\begin{bmatrix} 1.0 \\ 2.0 \\ 1.0 \end{bmatrix}$$

Hence, we choose $\{u\}_1 = \{1 \quad 2 \quad 1\}$ and $1/\lambda_1^* = 12 = 12k/m\lambda_1$, or

$$\omega_1^2 = \lambda_1 = \frac{12k}{12m} = \frac{k}{m}$$

The first mode is suppressed by the constraint $p_1 = 0$ as shown in Eq. (6-58). The v's are found from Eq. (6-61) as

$$\lfloor v_{11} \quad v_{12} \quad v_{13} \rfloor = \lfloor 1 \quad 2 \quad 1 \rfloor \begin{bmatrix} m & 0 & 0 \\ 0 & m & 0 \\ 0 & 0 & m \end{bmatrix}$$

The common constant terms in the vector $\lfloor v \rfloor$ can be discarded. The sweeping matrix from Eq. (6-59) is

$$S_1 = \begin{bmatrix} 1 & 2 & 1 \\ 0 & 1 & 0 \\ 0 & 0 & 1 \end{bmatrix}^{-1}\begin{bmatrix} 0 & 0 & 0 \\ 0 & 1 & 0 \\ 0 & 0 & 1 \end{bmatrix} = \begin{bmatrix} 0 & -2 & -1 \\ 0 & 1 & 0 \\ 0 & 0 & 1 \end{bmatrix}$$

From Eq. (6-60), the new matrix G_1 is

$$G_1 = GS_1 = \begin{bmatrix} 5 & 3 & 1 \\ 3 & 9 & 3 \\ 1 & 3 & 5 \end{bmatrix} \begin{bmatrix} 0 & -2 & -1 \\ 0 & 1 & 0 \\ 0 & 0 & 1 \end{bmatrix} = \begin{bmatrix} 0 & -7 & -4 \\ 0 & 3 & 0 \\ 0 & 1 & 4 \end{bmatrix}$$

The second mode is obtained by assuming an arbitrary vector $\{V\}_0 = \{1 \quad 1 \quad 1\}$ and iterating on G_1. After 14 iterations we have $\{u\}_2 = \{-1 \quad 0 \quad 1\}$ and $1/\lambda_2^* = 4 = 12k/m\lambda_2$. Hence

$$\omega_2^2 = \lambda_2 = \frac{12k}{4m} = \frac{3k}{m}$$

The first and the second modes are suppressed by the constraints $p_1 = p_2 = 0$. The vectors $\lfloor v \rfloor_1$ and $\lfloor v \rfloor_2$ can be found from Eq. (6-61) as before. Thus, the sweeping matrix S_2 from Eq. (6-59) is

$$S_2 = \begin{bmatrix} 1 & 2 & 1 \\ -1 & 0 & 1 \\ 0 & 0 & 1 \end{bmatrix}^{-1} \begin{bmatrix} 0 & 0 & 0 \\ 0 & 0 & 0 \\ 0 & 0 & 1 \end{bmatrix} = \begin{bmatrix} 0 & 0 & 1 \\ 0 & 0 & -1 \\ 0 & 0 & 1 \end{bmatrix}$$

The new matrix G_2 from Eq. (6-60) is

$$G_2 = G_1 S_2 = \begin{bmatrix} 0 & -7 & -4 \\ 0 & 3 & 0 \\ 0 & 1 & 4 \end{bmatrix} \begin{bmatrix} 0 & 0 & 1 \\ 0 & 0 & -1 \\ 0 & 0 & 1 \end{bmatrix} = \begin{bmatrix} 0 & 0 & 3 \\ 0 & 0 & -3 \\ 0 & 0 & 3 \end{bmatrix}$$

The third mode is obtained by assuming an arbitrary vector $\{V\}_0 = \{1 \quad 1 \quad 1\}$ and iterating with G_2. After two iterations we obtain $\{u\}_3 = \{1 \quad -1 \quad 1\}$ and $1/\lambda_3^* = 3 = 12k/m\lambda_3$. Hence

$$\omega_3^2 = \lambda_3 = \frac{12k}{3m} = \frac{4k}{m}$$

Summarizing the calculations, the natural frequencies are $\omega_1 = \sqrt{k/m}$, $\omega_2 = \sqrt{3k/m}$, $\omega_3 = \sqrt{4k/m}$. The modal matrix is

$$[u] = \begin{bmatrix} 1 & 1 & 1 \\ 2 & 0 & -1 \\ 1 & -1 & 1 \end{bmatrix}$$

6-10 UNDAMPED FORCED VIBRATION—MODAL ANALYSIS

The method of *modal analysis* transforms the simultaneous equations of motion of a discrete system into a set of independent second-order differential equations. Each of the uncoupled equations in the principal coordinate $p_i(t)$ describes a mode of vibration. The resultant motion is obtained from the superposition of the motions of the modes.

The procedure consists of (1) using the orthogonal relation in Eq. (6-37) to uncouple the equations of motion in order to describe the equations in the principal coordinates $\{p\}$, (2) employing the transformation $\{p(0)\}=[u]^{-1}\{q(0)\}$ in Eq. (6-23) to find the contribution of the initial conditions towards the excitation of each mode p_i, (3) superposing the solutions due to the initial conditions and the excitation for each mode p_i as shown in Eq. (2-74), and (4) finding the response in the original coordinates $\{q\}$ by the transformation $\{q\}=[u]\{p\}$ as shown in Eq. (6-23). The method is conceptually simple. Computers, however, are necessary for the numerical solution of the problem.

We shall first discuss the particular solution due to the transient excitation and then the complementary solution due to the initial conditions, since the two can be handled separately.

Consider the equations of motion of a conservative system

$$M\{\ddot{q}\}+K\{q\}=\{Q(t)\} \tag{6-62}$$

where M is the mass matrix, K the stiffness matrix, $\{q\}$ the generalized coordinates, and $\{Q(t)\}$ the generalized excitation force. For convenience, we use the transformation $\{q\}=[\mu]\{\eta\}$ from Eq. (6-39) to relate $\{q\}$ and the normal coordinates $\{\eta\}$. Thus, Eq. (6-62) yields

$$M[\mu]\{\ddot{\eta}\}+K[\mu]\{\eta\}=\{Q(t)\}$$

Premultiply this by $[\mu]^T$. Note that from Eq. (6-43) $[\mu]^T M[\mu]=I$, a unit matrix, and $[\mu]^T K[\mu]=\Lambda$, a diagonal matrix. Hence the equations of motion become

$$\{\ddot{\eta}\}+\Lambda\{\eta\}=[\mu]^T\{Q(t)\}\triangleq\{N(t)\} \tag{6-63}$$

where $\{N(t)\}$ is a column vector representing the normalized excitation force. In other words, $\{N(t)\}$ represents the contribution of the transient force $\{Q(t)\}$ toward the excitation of the normal modes. Since Λ is diagonal and $\lambda_i=\omega_i^2$, each of the uncoupled equations in Eq. (6-63) can be written as

$$\ddot{\eta}_i+\omega_i^2\eta_i=N_i(t) \qquad \text{for } i=1,2,\ldots,n \tag{6-64}$$

The equation above can be treated as an independent one-degree-of-freedom system. From Eq. (2-71), the particular solution due to $N_i(t)$ is

$$\eta_i(t)=\int_0^t h_i(t-\tau)N_i(\tau)\,d\tau$$

where $h_i(t)$ is the impulse response of the system in Eq. (6-64). From Eq. (2-69), the impulse response of an undamped system is

$$h_i(t)=\frac{1}{\omega_i}\sin\omega_i t$$

Combining the last two equations gives the transient response for the ith mode for zero initial conditions; that is,

$$\eta_i(t)=\frac{1}{\omega_i}\int_0^t N_i(\tau)\sin\omega_i(t-\tau)\,d\tau \qquad \text{for } i=1,2,\ldots,n \qquad \textbf{(6-65)}$$

For the complementary solution, we first find the contribution of the initial conditions towards the excitation of the normal modes. The original initial conditions are $\{q(0)\}$ and $\{\dot{q}(0)\}$. By the transformation $\{\eta\}=[\mu]^{-1}\{q\}$ in Eq. (6-39), we get

$$\{\eta(0)\}=[\mu]^{-1}\{q(0)\} \qquad \text{and} \qquad \{\dot{\eta}(0)\}=[\mu]^{-1}\{\dot{q}(0)\} \qquad \textbf{(6-66)}$$

Let η_{io} and $\dot{\eta}_{io}$ be the initial conditions in the normal coordinates for the ith mode. Hence the complementary solution of Eq. (6-64) due to the initial conditions is

$$\eta_i(t)=\eta_{io}\cos\omega_i t+\frac{1}{\omega_i}\dot{\eta}_{io}\sin\omega_i t \qquad \textbf{(6-67)}$$

The complete solution for the ith mode is the sum of the responses due to (1) the excitation in Eq. (6-65) and (2) the initial condition in Eq. (6-67). As indicated in Eq. (2-74), these responses are added directly to give the complete response.

The complete solution in the generalized coordinates $\{q\}$ is the superposition of the responses of the modes as indicated in Eq. (6-39). Let $\eta_i(t)$ be the sum of the responses from Eqs. (6-65) and (6-67). Thus,

$$\{q(t)\}=\sum_{i=1}^{n}\{\mu\}_i\eta_i(t) \qquad \text{or} \qquad \{q(t)\}=[\mu]\{\eta(t)\} \qquad (6\text{-}68)$$

Since the transient excitation $\{Q(t)\}$ is arbitrary, only the formal solution, as shown in Eq. (6-68), can be presented. Computer methods for the numerical solutions will be discussed in Chap. 9. The equations above can be expressed in alternative forms. Since these do not introduce additional concepts, we shall not further reduce the equations.

6-11 SYSTEMS WITH PROPORTIONAL DAMPING

If a system possesses damping, the modal analysis described in the last section generally does not apply. The equations of motion cannot be uncoupled by the modal matrix of undamped systems except for systems with proportional damping.

Consider the equations of motion of a system with damping.

$$M\{\ddot{q}\}+D\{\dot{q}\}+K\{q\}=\{Q(t)\} \qquad \textbf{(6-69)}$$

where the mass matrix M, the damping matrix D, and the stiffness matrix K are symmetric. $\{Q(t)\}$ denotes the generalized force and $\{q\}$ the generalized coordinates. *Proportional damping** occurs if the damping matrix D can be expressed as

$$D = \alpha M + \beta K \tag{6-70}$$

where α and β are constants.

To uncouple the equations, we substitute Eq. (6-70) and the transformation $\{q\} = [\mu]\{\eta\}$ from Eq. (6-39) in (6-69).

$$M[\mu]\{\ddot{\eta}\} + [\alpha M + \beta K][\mu]\{\dot{\eta}\} + K[\mu]\{\eta\} = \{Q(t)\}$$

where $\{\eta\}$ are the normal coordinates of the corresponding conservative system. Premultiplying this by $[\mu]^T$ and recalling that $[\mu]^T M[\mu] = I$ and $[\mu]^T K[\mu] = \Lambda$ from Eq. (6-43), we get

$$\{\ddot{\eta}\} + [\alpha I + \beta \Lambda]\{\dot{\eta}\} + \Lambda\{\eta\} = [\mu]^T\{Q(t)\} \triangleq \{N(t)\} \tag{6-71}$$

or

$$\ddot{\eta}_i + (\alpha + \beta\omega_i^2)\dot{\eta}_i + \omega_i^2\eta_i = N_i(t) \qquad \text{for } i = 1, \ldots, n \tag{6-72}$$

Hence the equations are uncoupled. Equation (6-71) can be treated as a set of one-degree-of-freedom systems.

6-12 ORTHOGONALITY OF MODES OF DAMPED SYSTEMS

The equations of motion of systems with viscous damping can be uncoupled by reducing the equations to a set of first-order differential equations.† Operating on the reduced equations, the analyses are essentially the same as those for undamped systems. For example, the same techniques are used to find the characteristic equation and to derive the orthogonality of the modes. The eigenvalues and the modal vectors, however, are complex quantities. Computer solutions, as illustrated in Chap. 9, are necessary for the numerical solution of the problem.

Consider the equations of motion for the free vibration of systems with viscous damping, in which the matrices M, C, and K are symmetric of order n.

$$M\{\ddot{q}\} + C\{\dot{q}\} + K\{q\} = \{0\} \tag{6-73}$$

The reduced equations are formed by introducing a $2n \times 1$ *state vector* $\{y\}$.

$$\{y\} = \begin{bmatrix} \dot{q} \\ q \end{bmatrix} \tag{6-74}$$

It can be shown that Eq. (6-73) can be expressed as

$$\underset{2n\times 2n}{\begin{bmatrix} 0 & M \\ M & C \end{bmatrix}} \underset{2n\times 1}{\begin{bmatrix} \ddot{q} \\ \dot{q} \end{bmatrix}} + \underset{2n\times 2n}{\begin{bmatrix} -M & 0 \\ 0 & K \end{bmatrix}} \underset{2n\times 1}{\begin{bmatrix} \dot{q} \\ q \end{bmatrix}} = \underset{2n\times 1}{\begin{bmatrix} 0 \\ 0 \end{bmatrix}} \tag{6-75}$$

* Lord Rayleigh, *The Theory of Sound*, vol. 1, Dover, New York, 1945, p. 130.
† Frazer, Duncan, and Collar, *op. cit.*, p. 289

or

$$A\{\dot{y}\}+B\{y\}=\{0\} \tag{6-76}$$

where A and B are defined by comparing the last two equations. The matrices M, C, and K become submatrices of A and B. Hence Eq. (6-73) is reduced to a set of $2n$ first-order equations shown in Eq. (6-76). It can be expressed in a more convenient form by premultiplying by A^{-1} and defining $H=-A^{-1}B$. Thus,

$$\{\dot{y}\}-H\{y\}=\{0\} \tag{6-77}$$

The eigenvalues are obtained by assuming the solution of Eq. (6-77) is of the form

$$\{y\}=\{\Psi\}e^{\gamma t} \tag{6-78}$$

where γ is a complex number and $\{\Psi\}$ a $2n\times 1$ *modal vector* with complex elements. Substituting Eq. (6-78) in (6-77) yields

$$[\gamma I-H]\{\Psi\}=\{0\} \tag{6-79}$$

Hence the characteristic equation is

$$\Delta(\gamma)=|\gamma I-H|=0 \tag{6-80}$$

This may be compared with Eq. (6-21) for undamped systems. Since H is a square matrix of order $2n$, there are $2n$ eigenvalues, which are necessarily complex conjugates. We assume the eigenvalues are distinct.

A modal vector $\{\Psi\}$ is found by substituting an eigenvalue γ in Eq. (6-79) and solving the corresponding homogeneous algebraic equations for $\{\Psi\}$. The process was illustrated in Example 9, Chap. 4, and it is conceptually simple. The subroutine $CHOMO, listed in App. C, can be used to solve the complex homogeneous algebraic equations. Since the matrix H is of order 2n, there are $2n$ modal vectors, which are necessarily complex conjugates. The *modal matrix* $[\Psi]$ is a linear combination of the eigenvectors and is of order $2n$.

$$[\Psi]=[\{\Psi\}_1\ \{\Psi\}_2\ \cdots\ \{\Psi\}_{2n}] \tag{6-81}$$

Let us show that the modal vectors $\{\Psi\}$ are orthogonal relative to the matrices A and B. Substituting Eq. (6-78) in (6-76) and factoring out the $e^{\gamma t}$ term, we have

$$\gamma A\{\Psi\}+B\{\Psi\}=\{0\}$$

Substituting γ_r and γ_s for the rth and sth modes in the equation above gives

$$\gamma_r A\{\Psi\}_r+B\{\Psi\}_r=\{0\}$$

$$\gamma_s A\{\Psi\}_s+B\{\Psi\}_s=\{0\}$$

Premultiplying the first of these equations by the transpose $\lfloor\Psi\rfloor_s$ of $\{\Psi\}_s$ and transposing the resulting equations, and premultiplying the second equation by $\lfloor\Psi\rfloor_r$, we get

$$\begin{aligned}\gamma_r\lfloor\Psi\rfloor_r A\{\Psi\}_s+\lfloor\Psi\rfloor_r B\{\Psi\}_s&=0\\ \gamma_s\lfloor\Psi\rfloor_r A\{\Psi\}_s+\lfloor\Psi\rfloor_r B\{\Psi\}_s&=0\end{aligned} \qquad \textbf{(6-82)}$$

The difference of these equations gives

$$(\gamma_r-\gamma_s)\lfloor\Psi\rfloor_r A\{\Psi\}_s=0$$

Since $\gamma_r\neq\gamma_s$, we deduce that

$$\lfloor\Psi\rfloor_r A\{\Psi\}_s=0 \qquad \text{for} \qquad r\neq s \qquad \textbf{(6-83)}$$

Similarly, it can be shown that

$$\lfloor\Psi\rfloor_r B\{\Psi\}_s=0 \qquad \text{for} \qquad r\neq s \qquad \textbf{(6-84)}$$

The last two equations are the *orthogonal relations* for systems with viscous damping. They may be compared with Eqs. (6-35) and (6-36). It is implicit that Eqs. (6-83) and (6-84) are not zero for $r=s$.

By virtue of the orthogonal property, using the modal matrix $[\Psi]$ to include all the modal vectors we obtain

$$[\Psi]^T A[\Psi]=[\ulcorner A\lrcorner] \qquad \text{and} \qquad [\Psi]^T B[\Psi]=[\ulcorner B\lrcorner] \qquad \textbf{(6-85)}$$

where $[\ulcorner A\lrcorner]$ and $[\ulcorner B\lrcorner]$ are diagonal. This may be compared with Eq. (6-37) for undamped systems. If the matrices A and B are diagonalized as shown above, then the equations of motion in Eq. (6-76) are uncoupled correspondingly.

6-13 DAMPED FORCED VIBRATION—MODAL ANALYSIS

The technique for the modal analysis of systems with viscous damping is similar to that for undamped systems, except that the complex modal matrix $[\Psi]$ is applied to the reduced equations. We shall first uncouple the equations of motion, find the particular solution due to the excitation, and then the complementary solution due to the initial conditions.

When the excitation $\{Q(t)\}$ is applied to the system, the equations of motion in Eq. (6-73) becomes

$$M\{\ddot{q}\}+C\{\dot{q}\}+K\{q\}=\{Q(t)\}$$

As shown in Eqs. (6-75) and (6-76), this can be reduced to a set of $2n$

simultaneous first-order equations as

$$\underset{2n\times 2n}{\begin{bmatrix} 0 & M \\ M & C \end{bmatrix}} \underset{2n\times 1}{\begin{bmatrix} \ddot{q} \\ \dot{q} \end{bmatrix}} + \underset{2n\times 2n}{\begin{bmatrix} -M & 0 \\ 0 & K \end{bmatrix}} \underset{2n\times 1}{\begin{bmatrix} \dot{q} \\ q \end{bmatrix}} = \underset{2n\times 1}{\begin{bmatrix} 0 \\ Q(t) \end{bmatrix}} \tag{6-86}$$

$$A\{\dot{y}\}+B\{y\}=\{E(t)\} \tag{6-87}$$

where $\{E(t)\}$ is a $2n\times 1$ excitation vector representing the generalized force as defined above.

The reduced equations in Eq. (6-87) can be uncoupled by means of the modal matrix $[\Psi]$. Introducing a new state vector $\{z\}$ as defined by the transformation

$$\{y\}=[\Psi]\{z\} \qquad \text{or} \qquad \{z\}=[\Psi]^{-1}\{y\} \tag{6-88}$$

Eq. (6-87) becomes

$$A[\Psi]\{\dot{z}\}+B[\Psi]\{z\}=\{E(t)\}$$

Note that the transformation in Eq. (6-88) may be compared with Eq. (6-23), which relates the generalized coordinates and the principal coordinates of an undamped system. Premultiplying the equation above by the transpose $[\Psi]^T$ of the modal matrix $[\Psi]$ and noting the orthogonal relations in Eq. (6-85), we get

$$[\ulcorner A \lrcorner]\{\dot{z}\}+[\ulcorner B \lrcorner]\{z\}=[\Psi]^T\{E(t)\}\triangleq\{N(t)\} \tag{6-89}$$

or

$$a_{ii}\dot{z}_i+b_{ii}z_i=N_i(t) \qquad \text{for} \qquad i=1,\ldots,2n \tag{6-90}$$

Hence the equations of motion are uncoupled. Furthermore, from Eq. (6-82) when $r=s=i$, we get

$$b_{ii}=-\gamma_i a_{ii} \tag{6-91}$$

Thus, Eq. (6-90) can be further simplified to become

$$\dot{z}_i-\gamma_i z_i=\frac{1}{a_{ii}}N_i(t) \qquad \text{for} \qquad i=1,\ldots,2n \tag{6-92}$$

The particular solution of Eq. (6-92) is found from the direct application of the convolution integral in Eq. (2-71).

$$z_i=\frac{1}{a_{ii}}\int_0^t e^{\gamma_i(t-\tau)}N_i(\tau)\,d\tau \qquad \text{for} \qquad i=1,\ldots,2n \tag{6-93}$$

where $e^{\gamma_i t}/a_{ii}$ is the impulse response of Eq. (6-92) with zero initial conditions.

The initial conditions in the $\{z\}$ coordinates are found by the transformation $\{z\}=[\Psi]^{-1}\{y\}=[\Psi]^{-1}\{\dot{q} \quad q\}$ in Eq. (6-88), that is,

$$\{z(0)\}=[\Psi]^{-1}\{y(0)\}=[\Psi]^{-1}\begin{bmatrix}\dot{q}(0)\\ q(0)\end{bmatrix} \tag{6-94}$$

Let z_{io} be the initial condition for the ith mode. Hence the complementary solution of Eq. (6-92) is of the form

$$z_i = z_{io}e^{\gamma_i t} \qquad \text{for} \qquad i=1,\ldots,2n \tag{6-95}$$

The complete solution z_i for the ith mode is the sum of the solutions from Eqs. (6-93) and (6-95). Note that the particular solution and the complementary solution are evaluated separately. The complete solution in the $\{q\}$ coordinates is determined by means of the transformation in Eq. (6-88).

$$\begin{bmatrix}\dot{q}(t)\\ q(t)\end{bmatrix}=\{y(t)\}=[\Psi]\{z(t)\} \tag{6-96}$$

We have presented the formal solutions in this discussion. Computer methods for the numerical solutions will be discussed in Chap. 9. The equations above can be expressed in alternative forms. Since these do not introduce additional concepts, we shall not further reduce the equations.

6-14 SUMMARY

Lagrange's equations are used to derive the equations of motion in Sec. 6-2. In Eq. (6-6), the mass matrix M denotes the total kinetic energy T of the system due to the generalized velocities; the stiffness matrix K denotes the total potential energy U of the system due to the generalized displacements. Since energies can be expressed in any coordinates, the reader should not feel the artificiality in the numerous coordinate transformations in the chapter. T is a positive definite quadratic function of $\{\dot{q}\}$; U is a positive definite quadratic function of $\{q\}$, except when the system is semidefinite.

Undamped free vibration is discussed in Secs. 6-3 to 6-9. The eigenvalues are determined from the characteristic equation in Eq. (6-21). The modal vectors are found by substituting the respective eigenvalues in Eq. (6-22) and solving the set of homogeneous algebraic equations. A modal vector represents a mode of vibration.

By virtue of the orthogonal property, the matrices M and K can be diagonalized by means of the modal matrix $[u]$ as shown in Eq. (6-37). This property is used for the modal analysis in later sections.

The modal vectors are linearly independent as shown in Sec. 6-6. Hence the motions of the discrete masses of a system is the superposition of the principal modes of vibration. Since the modal vectors are independent, an arbitrary vector $\{V\}$ can be expressed in terms of the modal vectors by means of the expansion theorem. An assumed modal vector $\{V\}$ is employed in the Rayleigh's quotient in Sec. 6-7 to find a natural frequency of the system. If $\{V\}$ is not an exact modal vector, the estimated natural frequency tends to be greater than the exact value.

Since the principal modes are independent, a mode can be suppressed or eliminated from consideration by means of a coordinate transformation. Thus, a semidefinite system can be made positive definite by suppressing the zero mode as shown in Sec. 6-8. The matrix iteration in Sec. 6-9 finds a natural frequency and the corresponding modal vector by an iterative procedure. Then the known modes are suppressed in order to iterate for the next mode.

The modal analysis of undamped systems in Sec. 6-10 first uses the orthogonal property in Eq. (6-37) to uncouple the equations of motion. For each mode of vibration, (1) the particular solution due to the excitation is found by the convolution integral as shown in Eq. (6-65), and (2) the complementary solution due to the initial conditions is found independently from Eq. (6-67). The particular and the complementary solutions are added to give the complete solution. The complete solution in the generalized coordinates $\{q\}$ is the superposition of the responses of the modes as shown in Eq. (6.68).

Similar procedures are used for the modal analysis of systems with viscous damping. The n simultaneous second-order differential equations of motion are first reduced to a set of $2n$ first-order equations in Eq. (6-76). The modal vectors are complex conjugates and are orthogonal relative to the matrices A and B of the reduced equations as shown in Eq. (6-85). By virtue of the orthogonal property, the equations in Eq. (6-89) are uncoupled. Hence except for the reduced equations and complex numbers, the same technique is used for systems with and without damping.

PROBLEMS

6-1 Lagrange's equations as shown in Eq. (6-1) was applied in Example 2. Use the Lagrangian method to derive the equations of motion for each of the systems in Fig. P4 1. Obtain the linearized equations by assuming small oscillations.

6-2 Use the Lagrangian method to derive Eqs. (4-27) to (4-29), which described the coordinate coupling of the system in Fig. 4-9. In each case, start the problem by expressing the energy quantities in the (x_1, θ) coordinates.

6-3 Use the Lagrangian method to derive the equations of motion for small oscillations for each of the systems shown in Fig. P6-1.

(a) A uniform bar hinged at 0 and rotated with constant velocity Ω.

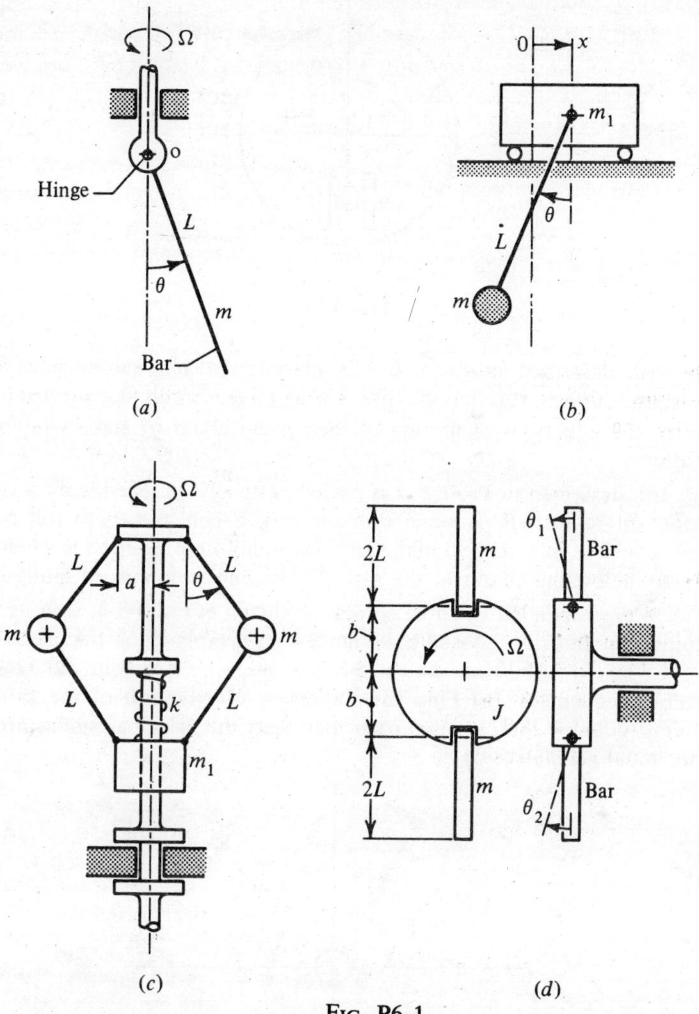

FIG. P6-1.

(b) A pendulum suspended from the mass m_1.

(c) A flyball governor with constant angular velocity Ω. Assume the spring k is unstressed when the arms L are vertical.

(d) Bars hinged to disk J. Assume the constant angular velocity Ω is sufficiently high that the gravitational effect may be neglected.

6-4 A horizontal rotating disk of mass m_1 and mass moment of inertia J is shown in Fig. P6-2. The mass m is constrained to move radially. Assume that the free length of the spring k is ℓ, there is no friction between m and its guides, and a torque $T_o \sin \omega t$ is applied to the disk. Derive the equations of motion of the system.

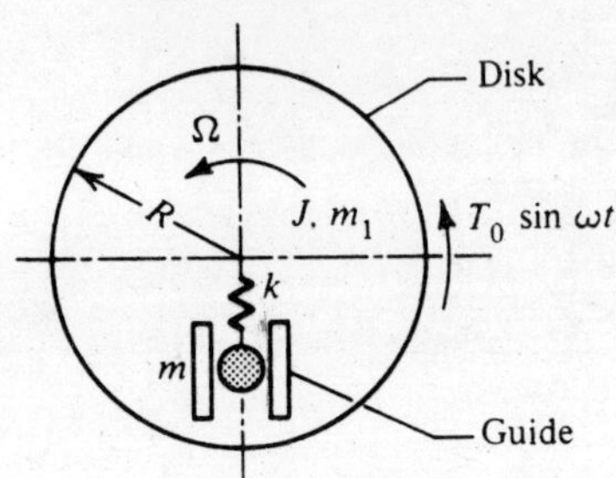

FIG. P6-2.

6-5 The disk described in Prob. 6-4 is placed vertically on its edge on a horizontal surface. Assume the disk is constrained to roll in a vertical plane. Derive the equations of motion of the system about its static equilibrium position.

6-6 The disk described in Prob. 6-4 is placed vertically on its edge on a curved surface of radius $4R$. Assume that the disk is constrained to roll on the curved surface in a vertical plane and the equilibrium position is when m is directly below the center of the disk. Derive the equations of motion.

6-7 A mass m, suspended by three springs as shown in Fig. P6-3, is in its static equilibrium position. Assume m moves in the plane of the paper. Let $m = 1.3$ kg, $k_1 = 350$ N/m, $k_2 = 875$ N/m, and $k_3 = 1225$ N/m. (**a**) Find the natural frequencies. (**b**) Find the directions of vibration of the principal modes. (**c**) Show that, for this particular case, the principal modes are also orthogonal geometrically.

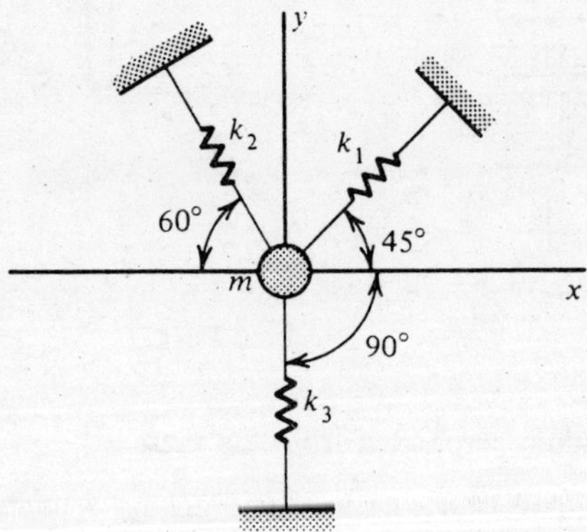

FIG. P6-3.

6-8 Consider the undamped system $M\{\ddot{x}\}+K\{x\}=\{Q(t)\}$.

$$\begin{bmatrix} 4 & 0 \\ 0 & 1 \end{bmatrix}\begin{bmatrix} \ddot{x}_1 \\ \ddot{x}_2 \end{bmatrix}+\begin{bmatrix} 24 & -4 \\ -4 & 6 \end{bmatrix}=\begin{bmatrix} 8 \\ 0 \end{bmatrix}$$

(**a**) Find the eigenvalues λ of the characteristic equation $|\lambda I - H| = 0$, where $H = M^{-1}K$.

(b) Evaluate the modal matrix $[u]$.

(c) Find the uncoupled equations expressed in the principal coordinates.

(d) Verify that $[u]^{-1}H[u]=\Lambda$, where Λ is a diagonal matrix with λ's as the diagonal elements.

(e) Normalize $[u]$ such that the matrix M becomes a unit matrix as shown in Eq. (6-43).

(f) Express the uncoupled equations in the normal coordinates.

6-9 Repeat Prob. 6-8 for the system shown in Fig. 4-11, that is,

$$\begin{bmatrix} 2 & 0 & 0 \\ 0 & 1 & 0 \\ 0 & 0 & 2 \end{bmatrix}\begin{bmatrix} \ddot{x}_1 \\ \ddot{x}_2 \\ \ddot{x}_3 \end{bmatrix}+\begin{bmatrix} 4 & -1 & 0 \\ -1 & 2 & -1 \\ 0 & -1 & 4 \end{bmatrix}\begin{bmatrix} x_1 \\ x_2 \\ x_3 \end{bmatrix}=\begin{bmatrix} F_1(t) \\ F_2(t) \\ F_3(t) \end{bmatrix}$$

6-10 Determine the modal matrix of the torsional system with fixed ends as shown in Fig. P5-3(a). Assume $J_1=J_2=6$, $J_3=9$, $k_{t0}=6$, and $k_{t1}=k_{t2}=k_{t3}=18$.

6-11 A semidefinite system can be made positive definite by suppressing the zero mode as illustrated in Example 5. Consider the free vibration of each of the following semidefinite systems:

System	J_1	J_2	J_3	k_{t1}	k_{t2}
(1)	2	5	8	54	36
(2)	4	4	16	24	32
(3)	8	2	8	8	8

(a) Find the modal matrix $[u]$. Show that $[u]^{-1}H[u]=\Lambda$.

(b) Write the uncoupled equations.

(c) Normalize $[u]$ such that the mass matrix becomes a unit matrix.

(d) Express the uncoupled equations in the normal coordinates.

6-12 Find the natural frequencies and the modal matrices of each of the systems in the following figures by matrix iteration:

(a) A two-mass system in Fig. 4-2(a). Assume $m_1=4$ kg $m_2=1$ kg, $k_1=20$ kN/m, $k_2=2$ kN/m, and $k=4$ kN/m.

(b) A double pendulum shown in Fig. P4-1(a). Assume $m_1=m_2$, and $L_1=L_2$.

(c) The torsional system in Fig. P5-3(a). Assume $J_1=J_2=6$, $J_3=9$, $k_{t0}=6$, and $k_{t1}=k_{t2}=k_{t3}=18$.

(d) Each of the configurations of the system shown in Fig. 4-9. Assume $m=1.0$ kg, $J_{cg}=0.6\ \text{m}^2\cdot\text{kg}$, $k_1=60$ N/m, $k_2=40$ N/m, $L_1=1.5$ m, and $L_2=1.0$ m.

6-13 Referring to the equations of motion in Prob. P6-8, find the transient response for each of the following initial conditions:

(**a**) $\{x(0)\}=\{0\}$, $\{\dot{x}(0)\}=\{0\}$;

(**b**) $\{x(0)\}=\{1 \quad 0\}$, $\{\dot{x}(0)\}=\{2 \quad 1\}$.

Check the solutions by the classical method.

6-14 Consider each of the following systems:

(**a**) $$\begin{bmatrix}5 & 0\\0 & 3\end{bmatrix}\begin{bmatrix}\ddot{x}_1\\\ddot{x}_2\end{bmatrix}+\begin{bmatrix}3 & -1\\-1 & 2\end{bmatrix}\begin{bmatrix}\dot{x}_1\\\dot{x}_2\end{bmatrix}+\begin{bmatrix}15 & -6\\-6 & 8\end{bmatrix}\begin{bmatrix}x_1\\x_2\end{bmatrix}=\begin{bmatrix}0\\0\end{bmatrix}$$

(**b**) $$\begin{bmatrix}1 & 0\\0 & 2\end{bmatrix}\begin{bmatrix}\ddot{x}_1\\\ddot{x}_2\end{bmatrix}+\begin{bmatrix}0.4 & -0.2\\-0.2 & 0.2\end{bmatrix}\begin{bmatrix}\dot{x}_1\\\dot{x}_2\end{bmatrix}+\begin{bmatrix}5 & -4\\-4 & 4\end{bmatrix}\begin{bmatrix}x_1\\x_3\end{bmatrix}=\begin{bmatrix}0\\0\end{bmatrix}$$

Find the characteristic equation by means of the reduced equation as shown in Eq. (6-86). Check the characteristic equation by substituting $D(=d/dt)$ in the system equations above and expanding the determinant of the coefficient matrix of $\{x\}$.

Computer problems:

6-15 Given the equations of motion $M\{\ddot{q}\}+K\{q\}=\{0\}$, where M and K are symmetric. By means of the transformation $\{g\}=[t]\{q\}$ in Eq. (6-10), the equations become $\{\ddot{g}\}+S\{g\}=\{0\}$, where S is symmetric. The form of $[t]$ is as shown in Eq. (6-13).

(**a**) Write a subroutine for the transformation matrix $[t]$.

(**b**) Write a program to use this subroutine and to show that $[t]^T[t]=M$, $[t]^TK[t]=S$, and $[[t]^T]^{-1}M[t]^{-1}=I$.

(**c**) Use the following data to verify your program:

$$\begin{bmatrix}6 & 5 & 3\\5 & 5 & 3\\3 & 3 & 3\end{bmatrix}\begin{bmatrix}\ddot{q}_1\\\ddot{q}_2\\\ddot{q}_3\end{bmatrix}+\begin{bmatrix}20 & -7 & -1\\-7 & 13 & -2\\-1 & -2 & 11\end{bmatrix}\begin{bmatrix}q_1\\q_2\\q_3\end{bmatrix}=\begin{bmatrix}0\\0\\0\end{bmatrix}$$

6-16 Use the program ITERATE listed in Fig. 9-14(a) to find the natural frequencies and the modal matrices of each of the systems described in Prob. 4-35.

6-17 Rewrite the program ITERATE listed in Fig. 9-14(a) as a subroutine. Repeat Prob. 6-16, using this subroutine.

6-18 Given the semidefinite system (see Example 5, Chap. 5)

$$\begin{bmatrix}4 & 0 & 0 & 0\\0 & 1 & 0 & 0\\0 & 0 & 4 & 0\\0 & 0 & 0 & 1\end{bmatrix}\begin{bmatrix}\ddot{\theta}_1\\\ddot{\theta}_2\\\ddot{\theta}_3\\\ddot{\theta}_4\end{bmatrix}+10^5\begin{bmatrix}1 & -1 & 0 & 0\\-1 & 3 & -2 & 0\\0 & -2 & 3 & -1\\0 & 0 & -1 & 1\end{bmatrix}\begin{bmatrix}\theta_1\\\theta_2\\\theta_3\\\theta_4\end{bmatrix}=\begin{bmatrix}0\\0\\0\\0\end{bmatrix}$$

Write a program (**a**) to transform the system to become positive definite, (**b**) to use the subroutine in Prob. 6-17 to find the natural frequencies and the modal matrix, and (**c**) to express the modal matrix in the original coordinates.

6-19 Referring to the cantilever shown in Fig. 9-12, write a program (**a**) to formulate the problem by the method of influence coefficients, and (**b**) to use the subroutine in Prob. 6-17 to find the natural frequencies and the modal matrix.

6-20 Rewrite the subroutine $CMODL listed in Fig. C-11 for the modal matrix of discrete systems with viscous damping as a main program. Find the modal matrix of the system shown below but do not list the complex conjugates.

$$\begin{bmatrix} 3 & 0 & 0 \\ 0 & 1 & 0 \\ 0 & 0 & 2 \end{bmatrix}\begin{bmatrix} \ddot{x}_1 \\ \ddot{x}_2 \\ \ddot{x}_3 \end{bmatrix} + \begin{bmatrix} 10 & -3 & -2 \\ -3 & 8 & -4 \\ -2 & -4 & 7 \end{bmatrix}\begin{bmatrix} \dot{x}_1 \\ \dot{x}_2 \\ \dot{x}_3 \end{bmatrix} + \begin{bmatrix} 140 & -60 & -20 \\ -60 & 220 & -80 \\ -20 & -80 & 200 \end{bmatrix}\begin{bmatrix} x_1 \\ x_2 \\ x_3 \end{bmatrix} = \begin{bmatrix} 0 \\ 0 \\ 0 \end{bmatrix}$$

6-21 Modify the program in Prob. 6-20 to obtain the print-outs as follows:

(**a**) The complex conjugate roots γ of the characteristic equation.

(**b**) The modal matrix $[\Psi]$.

(**c**) The product $[\Psi]^T A[\Psi] = [\ulcorner A \lrcorner]$ and $[\Psi]^T B[\Psi] = [\ulcorner B \lrcorner]$ to show the orthogonal relations as shown in Eq. (6-85).

(**d**) The product $[\Psi]^{-1} H[\Psi] = [\ulcorner \gamma \lrcorner]$, where $H = A^{-1}B$ and $[\ulcorner \gamma \lrcorner]$ is a diagonal matrix with γ's as the diagonal elements.

Note that the modal matrix $[\Psi]$ of order $2n$ can be partitioned into two submatrices.

$$\underset{2n \times n}{[\Psi]} = \begin{bmatrix} \Psi_2 \\ \Psi_1 \end{bmatrix}$$

where

$$\underset{n \times 2n}{[\Psi_2]} = \underset{n \times 2n}{[\Psi_1]} \; \underset{2n \times 2n}{[\ulcorner \gamma \lrcorner]}$$

Verify the equations above from the print-out.

6-22 Use the program TRESPDAM in Fig. 9-15(*a*) to find the transient response of a discrete system with viscous damping. Choose the appropriate initial conditions, $\{x(0)\}$ and $\{\dot{x}(0)\}$, and excitations $\{F(t)\}$. Consider the problem in three parts as follows:

(**a**) $\{F(t)\} \neq \{0\}$, $\{x(0)\} \neq \{0\}$, and $\{\dot{x}(0)\} \neq \{0\}$.

(**b**) $\{F(t)\} \neq \{0\}$, $\{x(0)\} = \{0\}$, and $\{\dot{x}(0)\} = \{0\}$.

(**c**) $\{F(t)\} = \{0\}$, $\{x(0)\} \neq \{0\}$, and $\{\dot{x}(0)\} \neq \{0\}$.

Verify from the computer print-out that the values of $\{x(t)\}$ and $\{\dot{x}(t)\}$ from part **a** is the sum of that of parts **b** and **c**.

6-23 Modify the program TRESPDAM in Fig. 9-15(a) such that the values of $\{x(t)\}$ are stored in one file and that of $\{\dot{x}(t)\}$ in another. Choose the appropriate initial conditions and/or excitations. Execute the program. Use the program PLOTFILE in Fig. 9-5(a) to plot the results.

6-24 Modify the program TRESPDAM in Fig. 9-15(a) such that only every nth computed value of $\{x(t)\}$ is entered into a data file and every nth value of $\{\dot{x}(t)\}$ into another. Choose the appropriate initial conditions and/or excitations. Execute the program. Use the program PLOTFILE in Fig. 9-5(a) to plot the results.

7

Continuous Systems

7-1 INTRODUCTION

A beam is an example of a continuous system. Its mass is distributed, inseparable from the elasticity of the beam. A continuous system has its mass, elasticity, and damping distributed. By contrast, a discrete system consists of distinct, separate, and idealized masses, springs, and dampers. Although complex continuous systems are often modeled as discrete systems, a basic understanding of the continuous system is helpful in formulating the equivalent discrete system. A continuous model, however, may be justified for many applications.

The analysis of a continuous system is more involved. Its vibratory motion is described in terms of the space and time coordinates, and the equation of motion is a partial differential equation. Furthermore, the elasticity of a beam and the manner of its support must be considered. Thus, two additional aspects are introduced in the analysis, namely, elasticity and boundary-value problem. These two topics are separate studies in themselves. Fortunately, the conceptual aspect of its vibration theory closely resembles that of discrete systems.

This chapter is an introductory study of continuous systems. The continuous model is assumed linear and the material in the system homogeneous and isotropic. The topics selected are similar to those for discrete systems. One-dimensional continuous systems are treated in some detail.

7-2 CONTINUOUS SYSTEMS—A SIMPLE EXPOSITION

To illustrate a continuous system, we shall first derive the equation of motion for the lateral vibration of a taut string and then describe the solution in terms of the principal modes as for discrete systems.

A flexible string of mass m per unit length under the tension T is shown

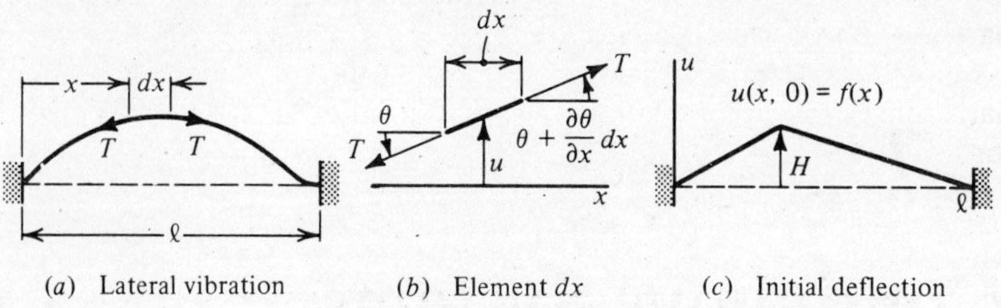

(a) Lateral vibration (b) Element dx (c) Initial deflection

FIG. 7-1. *Lateral vibration of a flexible taut string.*

in Fig. 7-1(*a*). The lateral deflection u along the string is a function of the space variable x and time t.

$$u = u(x,t) \tag{7-1}$$

The free-body sketch of an element dx of the string is shown in Fig. 7-1(*b*). If the lateral deflection is small, the change in T due to the deflection is negligible. Applying Newton's second law and assuming small deflections u and θ, the equation of motion is

$$m dx \frac{\partial^2 u}{\partial t^2} = T\left(\theta + \frac{\partial \theta}{\partial x}\, dx\right) - T\theta \tag{7-2}$$

Substituting $\theta = \partial u/\partial x$ and simplifying, the equation becomes

$$\frac{\partial^2 u}{\partial t^2} = c^2 \frac{\partial^2 u}{\partial x^2} \tag{7-3}$$

where $c^2 = T/m$. This is a one-dimensional *wave equation.*

The solution of Eq. (7-3) has four arbitrary constants. The problem must be compatible with the *boundary conditions* at the fixed ends, which are

$$u(0,t) = 0 \qquad \text{and} \qquad u(\ell,t) = 0 \tag{7-4}$$

Two of the constants are specified by Eq. (7-4); the other two by the *initial conditions*

$$u(x,0) = f(x) \qquad \text{and} \qquad \frac{\partial u}{\partial t}(x,0) = g(x) \tag{7-5}$$

For example, if the string is plucked at some intermediate point as shown in Fig. 7-1(*c*) and released with zero initial velocity, then $f(x)$ defines the initial deflection and $g(x) = 0$.

Let us pause for a moment to consider the problem formulation. (1) The equation of motion is derived from Newton's second law, Eq. (7-2). (2) The resulting equation is a partial differential equation, Eq. (7-3). (3) The string problem has four arbitrary constants to be evaluated by the boundary conditions, Eq. (7-4), and the initial conditions, Eq. (7-5). (4) It may be anticipated that the boundary conditions specify the type or the mode of vibration, that is, the possible ways that the system will oscillate.

(5) The initial conditions give the actual motion, that is, the degree of participation of each mode.

Since the system is undamped, we assume that a mode of vibration is harmonic as for discrete systems. Thus, the solution of Eq. (7-3) is of the form

$$u(x,t) = \phi(x)\sin(\omega t + \psi) \tag{7-6}$$

where ω is a natural frequency, ψ a constant, and $\phi(x)$ an *eigenfunction*, which describes the mode shape of the string at the frequency ω. Substituting Eq. (7-6) in (7-3) and factoring out the sine term, we get

$$\frac{d^2\phi(x)}{dx^2} + \frac{\omega^2}{c^2}\phi(x) = 0 \tag{7-7}$$

The solution of this equation is

$$\phi(x) = A\sin\frac{\omega x}{c} + B\cos\frac{\omega x}{c} \tag{7-8}$$

Combining this with Eq. (7-6) gives the formal solution

$$u(x,t) = \left(A\sin\frac{\omega x}{c} + B\cos\frac{\omega x}{c}\right)\sin(\omega t + \psi) \tag{7-9}$$

The boundary condition $u(0,t) = 0$ requires that $B = 0$. The condition $u(\ell,t) = 0$ gives

$$\sin\frac{\omega\ell}{c} = 0 \qquad \text{or} \qquad \frac{\omega\ell}{c} = i\pi \qquad \text{for} \qquad i = 1, 2, \ldots \tag{7-10}$$

This is called the *frequency* or the *characteristic equation.* For example, the frequency for the ith mode is $\omega_i = i\pi c/\ell$ rad/s. A continuous system has an infinite number of natural frequencies as indicated in Eq. (7-10).*

Since the system is linear, the general solution is the superposition of the principal modes, that is,

$$u(x,t) = \sum_{i=1}^{\infty} A_i \sin\frac{i\pi x}{\ell}\sin(\omega_i t + \psi_i) \tag{7-11}$$

where $\omega_i = i\pi c/\ell$. Since $\phi(x) = \sin i\pi x/\ell$ was determined by the boundary conditions, the modes are stipulated by the boundary conditions.

The constants A_i and ψ_i in Eq. (7-11) are evaluated by the initial conditions. Let the equation be expressed in a more convenient form

$$u(x,t) = \sum_{i=1}^{\infty}\left(\sin\frac{i\pi x}{\ell}\right)(C_i\sin\omega_i t + D_i\cos\omega_i t) \tag{7-12}$$

* A continuous system is said to have an infinite number of degrees of freedom. Then it is necessary to explain that the natural frequencies are countable instead of being "distributed." An alternative viewpoint is that the mass and elasticity of continuous systems are distributed. If principal modes exist, then the natural frequencies are countable, but there are an infinite number of them.

Applying the initial conditions from Eq. (7-5) yields

$$u(x,0) = \sum_{i=1}^{\infty} D_i \sin \frac{\omega_i x}{c} = f(x)$$
$$\frac{\partial u}{\partial t}(x,0) = \sum_{i=1}^{\infty} C_i \omega_i \sin \frac{\omega_i x}{c} = g(x) \qquad \textbf{(7-13)}$$

Consider the Fourier series expansion of $f(x)$ and $g(x)$.

$$f(x) = \sum_{i=1}^{\infty} f_i \sin \frac{\omega_i x}{c}$$
$$g(x) = \sum_{i=1}^{\infty} g_i \sin \frac{\omega_i x}{c} \qquad \textbf{(7-14)}$$

The constants C_i and D_i are found by equating the coefficients in Eqs. (7-13) and (7-14), that is, $C_i \omega_i = g_i$ and $D_i = f_i$. Since C_i and D_i are associated with the vibratory motion at ω_i, in Eq. (7-12) the initial conditions determine the contribution of each principal mode to the vibration of the system.

7-3 SEPARATION OF THE TIME AND SPACE VARIABLES

Many methods can be used to solve the wave equation shown in Eq. 7-3). The D'Alembert's solution expresses the wave motion as the superposition of two travelling waves in opposite directions, that is,

$$u(x,t) = F_1(x - ct) + F_2(x + ct)$$

where F_1 and F_2 are arbitrary functions. Laplace transform can be used to transform Eq. (7-3) with respect to either x or t. We shall discuss the method of the *separation of variables*, which has general applications.

Let the partial differential equation in Eq. (7-3) be satisfied by the functions of the form

$$u(x,t) = \phi(x) q(t) \qquad \textbf{(7-15)}$$

where $\phi(x)$ is a function of the space variable x alone and $q(t)$ a function of t alone. Hence, they are written as ϕ and q in subsequent equations; their differentiations give $d\phi/dx$ and dq/dt instead of partial derivatives. Substituting Eq. (7-15) in (7-3) and simplifying, we obtain

$$c^2 \frac{1}{\phi} \frac{d^2\phi}{dx^2} = \frac{1}{q} \frac{d^2 q}{dt^2} \qquad \textbf{(7-16)}$$

Since the left side of the equation is independent of t and the right side is independent of x and the equality holds for all values of t and x, it follows that each side must be a constant. Let the constant be $-\omega^2$. Thus, we obtain two ordinary differential equations

$$\begin{aligned} \frac{d^2\phi}{dx^2} + \frac{\omega^2}{c^2}\phi = 0 \\ \frac{d^2q}{dt^2} + \omega^2 q = 0 \end{aligned} \tag{7-17}$$

Note that the reader should assume constants, positive, real, or complex, and deduce the conclusions.

The solutions of Eq. (7-17) are

$$\begin{aligned} \phi(x) = A \sin\frac{\omega x}{c} + B \cos\frac{\omega x}{c} \\ q(t) = C \sin \omega t + D \cos \omega t \end{aligned} \tag{7-18}$$

Thus, the formal solution from Eq. (7-15) is

$$u(x,t) = \left(A \sin\frac{\omega x}{c} + B \cos\frac{\omega x}{c}\right)(C \sin \omega t + D \cos \omega t) \tag{7-19}$$

he coefficients A, B, C, and D are to be evaluated by the boundary and ne initial conditions. Note that $\phi(x)$ and $q(t)$ are related only by the frequency ω. For a given ω, $\phi(x)$ defines the mode shape and $q(t)$ determines the vibration as in Eq. (7-6).

For purpose of illustration, let the boundary conditions be as stated in Eq. (7-4). The condition $u(0,t) = 0$ requires that $B = 0$ in Eq. (7-19). For $u(\ell,t) = 0$, we get

$$A \sin\frac{\omega\ell}{c} = 0 \qquad \text{or} \qquad \frac{\omega\ell}{c} = i\pi \qquad \text{for} \qquad i = 1, 2, 3, \ldots$$

This equation is identical to Eq. (7-10). Incorporating A into C and D, Eq. (7-19) becomes

$$u(x,t) = \sum_{i=1}^{\infty}\left(\sin\frac{i\pi x}{\ell}\right)(C_i \sin \omega_i t + D_i \cos \omega_i t) \tag{7-20}$$

This is identical to Eq. (7-12). Hence the method of separation of variables and the method of assumed principal modes give identical solutions.

Example 1. Vibrating String

Assume the string in Fig. 7-1(c) is plucked at the midpoint and released with zero initial velocity. Find the motion $u(x,t)$.

Solution:

The solution of the string problem is given in Eq. (7-20), in which $\omega_i = i\pi c/\ell$. Let H be the initial displacement at the midpoint of the string. The initial displacement is

$$u(x,0) = \begin{cases} 2Hx/\ell & \text{for} \quad 0 \le x \le \ell/2 \\ 2H(1-x/\ell) & \text{for} \quad \ell/2 \le x \le \ell \end{cases}$$

For zero initial velocity, $C_i = 0$ in Eq. (7-20). A value of D_i is obtained by multiplying Eq. (7-13) by $\sin(\omega_i x/c)$ and integrating for $0 \le x \le \ell$. For $f(x) = u(x,0)$ above, we get

$$D_i \int_0^\ell \sin^2 \frac{i\pi x}{\ell}\, dx = \int_0^\ell f(x) \sin \frac{i\pi x}{\ell}\, dx$$

or

$$D_i = \frac{2}{\ell}\left[\int_0^{\ell/2} \frac{2Hx}{\ell} \sin \frac{i\pi x}{\ell}\, dx + \int_{\ell/2}^{\ell} 2H\left(1-\frac{x}{\ell}\right)\sin \frac{i\pi x}{\ell}\, dx\right]$$

$$= 4H\left[\int_0^{\frac{1}{2}} \frac{x}{\ell} \sin \frac{i\pi x}{\ell}\, d\frac{x}{\ell} + \int_{\frac{1}{2}}^{1} \left(1-\frac{x}{\ell}\right)\sin \frac{i\pi x}{\ell}\, d\frac{x}{\ell}\right]$$

$$= \begin{cases} 0 & \text{for } i \text{ even} \\ \dfrac{8H}{i^2\pi^2}(-1)^{(i-1)/2} & \text{for } i \text{ odd} \end{cases}$$

Hence Eq. (7-20) becomes

$$u(x,t) = \frac{8H}{\pi^2}\left(\sin \frac{\pi x}{\ell} \cos \omega_1 t - \frac{1}{9} \sin \frac{3\pi x}{\ell} \cos \omega_3 t + \frac{1}{25} \sin \frac{5\pi x}{\ell} \cos \omega_5 t + \cdots\right)$$

Defining $i = 2n+1$, the equation can be expressed as

$$u(x,t) = \frac{8H}{\pi^2} \sum_{n=0}^{\infty} \frac{(-1)^n}{(2n+1)^2} \sin \frac{(2n+1)\pi x}{\ell} \cos \frac{(2n+1)\pi ct}{\ell}$$

7-4 PROBLEMS GOVERNED BY THE WAVE EQUATION

Many problems, such as the propogation of sound, are governed by the wave equation in Eq. (7-3). We shall consider the longitudinal vibration of rods and the torsional oscillation of shafts as examples of the wave equation in vibration. Either the assumption of principal modes or the separation of variables can be used to solve the equation.

Longitudinal Vibration of Rods

A long rod is shown in Fig. 7-2(a). Let $u(x,t)$ be the axial displacement of an element dx of the rod as shown in Fig. 7-2(b). From Hooke's law,

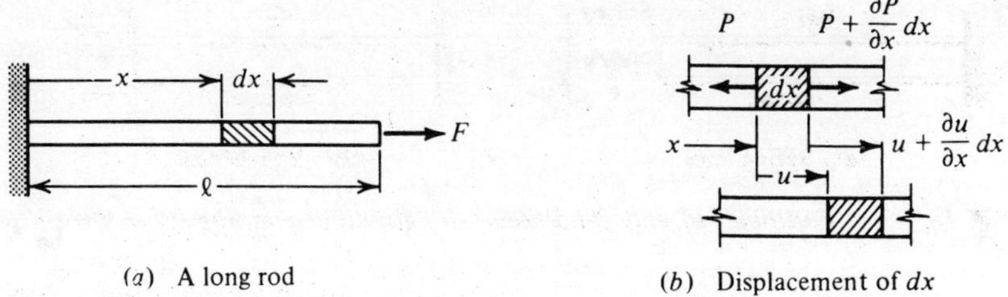

(a) A long rod (b) Displacement of dx

FIG. 7-2. *Longitudinal vibration of a rod.*

the stress-strain relation is

$$\frac{P}{A} = E\frac{\partial u}{\partial x} \tag{7-21}$$

where P is the force at x, A the cross-sectional area, and E the Young's modulus. Applying Newton's second law gives

$$dm\frac{\partial^2 u}{\partial t^2} = \left(P + \frac{\partial P}{\partial x}\,dx\right) - P \tag{7-22}$$

where $dm = \rho A dx$ is the mass of the element dx and ρ = mass per unit volume. Substituting Eq. (7-21) into (7-22) and simplifying, we obtain the equation of motion

$$\rho A\frac{\partial^2 u}{\partial t^2} = \frac{\partial}{\partial x}\left(AE\frac{\partial u}{\partial x}\right) \tag{7-23}$$

If AE is constant, we get

$$\frac{\partial^2 u}{\partial t^2} = c^2\frac{\partial^2 u}{\partial x^2} \tag{7-24}$$

where $c^2 = E/\rho$. This is identical to the wave equation in Eq. (7-3).

Boundary conditions are usually difficult to determine in applications. For example, if the end of the rod is fixed, the boundary condition is $u = 0$. We imply that the end is rigidly clamped. It is difficult, however, to ascertain that an end is absolutely fixed in an experiment. If an end is free, the condition is $\partial u/\partial x = 0$, since there is no stress at the free end. Typical boundary conditions for the longitudinal vibration of rods are

END CONDITION	BC LEFT	BC RIGHT	
Fixed	$u(0,t) = 0$	$u(\ell,t) = 0$	
Free	$\frac{\partial u}{\partial x} = 0$	$\frac{\partial u}{\partial x} = 0$	
Spring load	$ku = EA\frac{\partial u}{\partial x}$	$ku = -EA\frac{\partial u}{\partial x}$	**(7-25)**
Inertia load	$m\frac{\partial^2 u}{\partial t^2} = EA\frac{\partial u}{\partial x}$	$m\frac{\partial^2 u}{\partial t^2} = -EA\frac{\partial u}{\partial x}$	

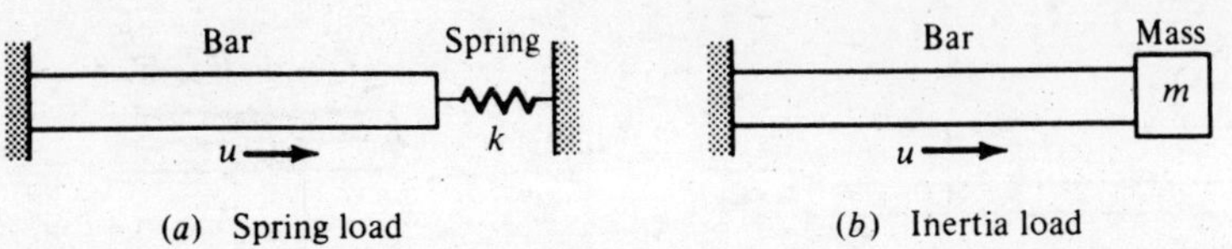

(a) Spring load (b) Inertia load

FIG. 7-3. *Examples of end conditions—longitudinal vibration of a bar.*

The spring and inertia loads are illustrated in Fig. 7-3. For example, with inertia load, the inertia force is equal to the force due to the strain in the rod.

Example 2

A long uniform rod with one end fixed and the other end free is shown in Fig. 7-2(*a*) Assume an axial force F is applied at the free end and released at time $t=0$. Find the motion $u(x,t)$ of the rod.

Solution:

By the method of separation of variables, the solution $u(x,t)=\phi(x)q(t)$ can be expressed as shown in Eq. (7-19).

The boundary conditions at the fixed end at $x=0$ and the free end at $x=\ell$ are

$$u(0,t)=0 \qquad \text{and} \qquad \frac{\partial u}{\partial x}(\ell,t)=0$$

or

$$\phi(0)=0 \qquad \text{and} \qquad \frac{d\phi}{dx}(\ell)=0$$

For the given boundary conditions, it follows from Eq. (7-19) that

$$B=0 \qquad \text{and} \qquad \frac{\omega}{c}A\cos\frac{\omega\ell}{c}=0$$

Thus, the frequency equation $\cos \omega\ell/c=0$ gives

$$\frac{\omega\ell}{c}=i\frac{\pi}{2} \qquad \text{for} \qquad i=1,3,5,\ldots$$

For $\omega_i = i\pi c/2\ell$, from Eq. (7-19), we get

$$u(x,t)=\sum_{i=1,3,5,\ldots}^{\infty}\sin\frac{i\pi x}{2\ell}\left(C_i\sin\frac{i\pi ct}{2\ell}+D_i\cos\frac{i\pi ct}{2\ell}\right)$$

The constants C_i and D_i are determined from the initial conditions

$$u(x,0)=\frac{Fx}{AE} \qquad \text{and} \qquad \frac{\partial u}{\partial t}(x,0)=0$$

The first condition gives the uniform initial strain at $t=0$. The second co[illegible] that the initial velocity is zero, since the rod is released at

$t=0$. Thus, $C_i=0$. A value of D_i is obtained by multiplying Eq. (7-13) by $\sin(\omega_i x/c)$ and integrating for $0 \le x \le \ell$. For $f(x)=u(x,0)$ above, we have

$$D_i \int_0^\ell \sin^2 \frac{i\pi x}{2\ell}\,dx = \int_0^\ell \frac{Fx}{AE} \sin \frac{i\pi x}{2\ell}\,dx$$

For $i=$ odd from the frequency equation above, we get

$$D_i = \frac{8F\ell}{i^2\pi^2 AE}(-1)^{(i-1)/2} \qquad \text{for} \qquad i=1,3,5,\ldots$$

Defining $i=2n+1$, the longitudinal motion of the rod is

$$u(x,t) = \frac{8F\ell}{\pi^2 AE} \sum_{n=0}^{\infty} \frac{(-1)^n}{(2n+1)^2} \sin \frac{(2n+1)\pi x}{2\ell} \cos \frac{(2n+1)\pi ct}{2\ell}$$

Torsional Vibration of Shafts

Let $\theta(x,t)$ be the angular displacement of an element dx of the shaft shown in Fig. 7-4(a). From Hooke's law, the torque T at x is

$$T = I_p G \frac{\partial \theta}{\partial x} \tag{7-26}$$

where I_pG is the torsional stiffness of the shaft, I_p the polar moment of inertia of the cross-section, and G the shear modulus of the material. Applying Newton's second law to an element dx shown in Fig. 7-4(b), we get

$$\rho I_p \frac{\partial^2 \theta}{\partial t^2}\,dx = \left(T + \frac{\partial T}{\partial x}\,dx\right) - T \tag{7-27}$$

where ρ is mass per unit volume. The equation of motion is obtained by substituting Eq. (7-26) in (7-27).

$$\rho I_p \frac{\partial^2 \theta}{\partial t^2} = \frac{\partial}{\partial x}\left(I_p G \frac{\partial \theta}{\partial x}\right) \tag{7-28}$$

If I_pG is constant, we get

$$\frac{\partial^2 \theta}{\partial t^2} = c^2 \frac{\partial^2 \theta}{\partial x^2} \tag{7-29}$$

where $c^2 = G/\rho$. Again, this is identical to Eq. (7-3).

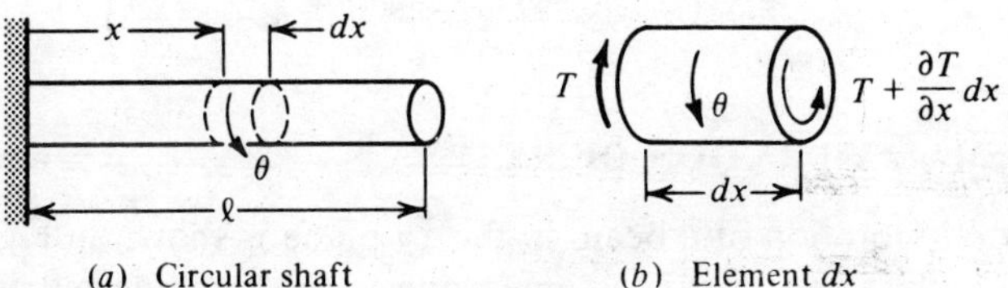

(a) Circular shaft (b) Element dx

FIG. 7-4. *Torsional vibration of long shaft.*

The boundary conditions are analogous to those for the longitudinal vibration of rods. Thus, from Eq. (7-25) we obtain

END CONDITION	BC LEFT	BC RIGHT
Fixed	$\theta(0,t)=0$	$\theta(\ell,t)=0$
Free	$\dfrac{\partial\theta}{\partial x}=0$	$\dfrac{\partial\theta}{\partial x}=0$
Spring load	$k_t\theta=I_pG\dfrac{\partial\theta}{\partial x}$	$k_t\theta=-I_pG\dfrac{\partial\theta}{\partial x}$
Inertia load	$J_0\dfrac{\partial^2\theta}{\partial t^2}=I_pG\dfrac{\partial\theta}{\partial x}$	$J_0\dfrac{\partial^2\theta}{\partial t^2}=-I_pG\dfrac{\partial\theta}{\partial x}$

(7-30)

where k_t is the stiffness of a torsional spring and J_0 the mass moment of inertia of a disk.

Example 3

A drill pipe of an oil well is modeled as a long shaft, fixed at one end and carrying a cutting bit of effective mass moment of inertia J_0 at the other end. Find the frequency equation of the system.

Solution:

The general solution of the wave equation from Eq. (7-19) is

$$\theta(x,t)=\left(A\sin\frac{\omega x}{c}+B\cos\frac{\omega x}{c}\right)(C\sin\omega t+D\cos\omega t)$$

The boundary conditions at the ends $x=0$ and $x=\ell$ are

$$\theta(0,t)=0 \quad\text{and}\quad J_0\frac{\partial^2\theta}{\partial t^2}(\ell,t)=-I_pG\frac{\partial\theta}{\partial x}(\ell,t)$$

The first condition requires $B=0$. Substituting the second condition in the solution above and factoring out the terms $A(C\sin\omega t+D\cos\omega t)$, we obtain the frequency equation

$$-J_0\omega^2\sin\frac{\omega\ell}{c}=-I_pG\frac{\omega}{c}\cos\frac{\omega\ell}{c}$$

Noting $c^2=G/\rho$ and $J_{\text{shaft}}=I_p\ell\rho$, by simple algebraic operations, the equation can be expressed more conveniently as

$$\frac{\omega\ell}{c}\tan\frac{\omega\ell}{c}=J_{\text{shaft}}/J_0$$

7-5 LATERAL VIBRATION OF BEAMS

The lateral vibration of a beam in the xy plane is shown in Fig. 7-5(a). When the deflection $u(x,t)$ is assumed due to the bending moment only, the model is often called the *Euler-Bernouli beam*.

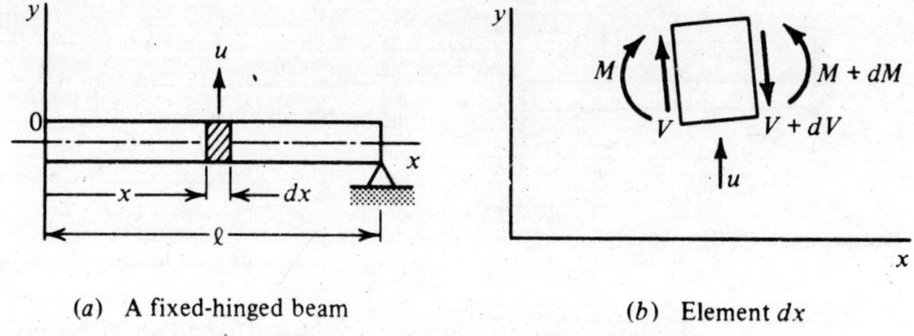

(a) A fixed-hinged beam (b) Element dx

FIG. 7-5. *Lateral vibration of beam.*

Consider the free-body sketch of an element of the beam shown in Fig. 7-5(b). From Newton's second law, the dynamic force equation in the lateral direction is

$$m\frac{\partial^2 u}{\partial t^2}\,dx = -\left(V + \frac{\partial V}{\partial x}\,dx\right) + V$$

or

$$m\frac{\partial^2 u}{\partial t^2} = -\frac{\partial V}{\partial x} \tag{7-31}$$

where m is mass/length and V the shear force. Summing the moments M about any point on the right face of the element yields

$$\frac{\partial M}{\partial x}\,dx - V\,dx \simeq 0$$

or

$$\frac{\partial M}{\partial x} = V \tag{7-32}$$

From elementary strength of materials, the beam curvature and the moment M are related by

$$EI\frac{\partial^2 u}{\partial x^2} = M \tag{7-33}$$

where EI is the flexural stiffness of the beam. Combining Eqs. (7-31) to (7-33), the beam equation for its lateral vibration is

$$m\frac{\partial^2 u}{\partial t^2} = -\frac{\partial^2}{\partial x^2}\left(EI\frac{\partial^2 u}{\partial x^2}\right) \tag{7-34}$$

If EI is constant, defining $a^2 = EI/m$, the beam equation reduces to

$$\frac{\partial^2 u}{\partial t^2} + a^2\frac{\partial^4 u}{\partial x^4} = 0 \tag{7-35}$$

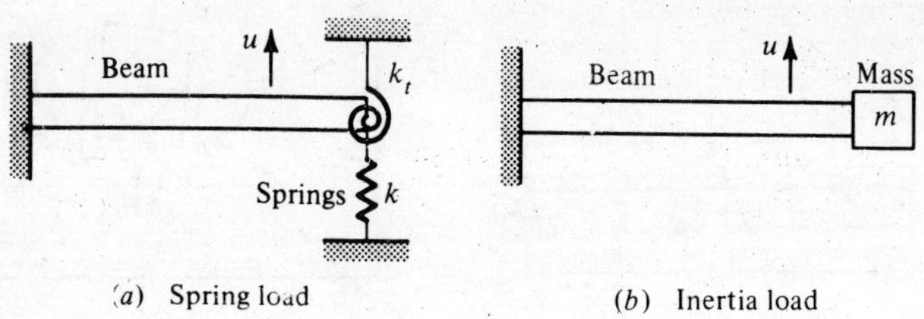

(a) Spring load (b) Inertia load

FIG. 7-6. *Examples of end conditions for lateral vibration of beam.*

The common boundary conditions are as follows:

END CONDITION	DEFLECTION	SLOPE	MOMENT*	SHEAR*	
	u	$\theta=\dfrac{\partial u}{\partial x}$	$M=EI\dfrac{\partial^2 u}{\partial x^2}$	$V=EI\dfrac{\partial^3 u}{\partial x^3}$	
Fixed	$u=0$	$\theta=0$			
Free			$M=0$	$V=0$	**(7-36)**
Hinged	$u=0$		$M=0$		
Spring load			$M=-k_t\dfrac{\partial u}{\partial x}$	$V=ku$	
Inertia load			$M=0$	$V=m\dfrac{\partial^2 u}{\partial t^2}$	

* BC's are for the right end of beam. Change the signs of M and V for BC's for the left end of beam.

The conditions of spring load and inertia load are shown in Fig. 7-6. Four quantities, namely, deflection, slope, moment, and shear, are considered for the end conditions. Each end condition is associated with two of these quantities. For example, a fixed end will require both the deflection and slope be zero. Conceivably, conditions other than those listed can be devised.

Example 4

Derive the frequency equation for the lateral vibration of a uniform beam with one end fixed and the other simply supported as shown in Fig. 7-5(a).

Solution:

By the method of separation of variables, the formal solution of the beam equation from Eq. (7-15) is

$$u(x,t)=\phi(x)q(t)$$

Substituting this in Eq. (7-35) and simplifying, we get

$$-a^2 \frac{1}{\phi}\frac{d^4\phi}{dx^4} = \frac{1}{q}\frac{d^2q}{dt^2}$$

Letting each side of the equation equal to $-\omega^2$, where ω is a constant, we obtain

$$\frac{d^4\phi}{dx^4} - \beta^4\phi = 0 \qquad \text{and} \qquad \frac{d^2q}{dt^2} + \omega^2 q = 0$$

where $\beta^4 = \omega^2/a^2 = m\omega^2/EI$. It can be shown from theory of linear differential equations that the solutions are

$$\phi(x) = C_1 \sin \beta x + C_2 \cos \beta x + C_3 \sinh \beta x + C_4 \cosh \beta x$$
$$q(t) = C_5 \sin \omega t + C_6 \cos \omega t$$

The boundary conditions from Eq. (7-36) for the fixed end are

$$(1) \quad u(0,t) = 0 \qquad \text{and} \qquad (2) \quad \frac{\partial u}{\partial x}(0,t) = 0$$

and those for the simply supported end are

$$(3) \quad u(\ell,t) = 0 \qquad \text{and} \qquad (4) \quad \frac{\partial^2 u}{\partial x^2}(\ell,t) = 0$$

Condition (1) requires that $C_2 + C_4 = 0$. Condition (2) gives $C_1 + C_3 = 0$. Thus,

$$\phi(x) = C_1(\sin \beta x - \sinh \beta x) + C_2(\cos \beta x - \cosh \beta x)$$

Applying conditions (3) and (4) yields

$$(\sin \beta\ell - \sinh \beta\ell)C_1 + (\cos \beta\ell - \cosh \beta\ell)C_2 = 0$$
$$-(\sin \beta\ell + \sinh \beta\ell)C_1 - (\cos \beta\ell + \cosh \beta\ell)C_2 = 0$$

The nontrivial solution for C_1 and C_2 requires that the determinant of their coefficients be zero, that is,

$$\begin{vmatrix} (\sin \beta\ell - \sinh \beta\ell) & (\cos \beta\ell - \cosh \beta\ell) \\ -(\sin \beta\ell + \sinh \beta\ell) & -(\cos \beta\ell + \cosh \beta\ell) \end{vmatrix} = 0$$

Expanding the determinant yields the frequency equation

$$\cos \beta\ell \sinh \beta\ell - \sin \beta\ell \cosh \beta\ell = 0$$

or

$$\tan \beta\ell = \tanh \beta\ell$$

7-6 ROTARY INERTIA AND OTHER EFFECTS

The natural frequency of a system is determined by its mass and stiffness. An effect that increases the effective mass and/or decreases the effective stiffness of a system will tend to lower its natural frequency and vice versa. The rotary inertia and other effects on the beam vibration can be interpreted in like manner.

Shear Deformation and Rotary Inertia Effects

The beam deformation was assumed due to the bending moment in the last section. When the effects of shear deformation and rotary inertia are considered, the beam is sometimes called the *Timoshenko beam.*

A beam in its lateral vibration will assume a deflection curve. Consider an element of a beam in the (x,y) plane in Fig. 7-7. The slope of the deflection curve is $\partial u/\partial x$. Hence the rotation of the element is $\partial u/\partial x$. The *rotary inertia effect* is due to this "rocking" motion $\partial u/\partial x$ of dx in the (x,y) plane. The rotary inertia is equivalent to an increase in mass and therefore will cause a decrease in natural frequency. Furthermore, the effect is more pronounced at the higher frequencies and more influencial on the higher modes. It is well to be aware of the problem. An accurate modeling of a system for the higher modes, however, may be difficult.

To include the rotary inertia effect, a dynamic moment equation is introduced in addition to the dynamic force equation in Eq. (7-31); that is,

$$m\frac{\partial^2 u}{\partial t^2} = -\frac{\partial V}{\partial x} \quad \text{and} \quad J\frac{\partial^2 \theta}{\partial t^2} = \frac{\partial M}{\partial x} - V \tag{7-37}$$

where m is mass per unit length and J the rotaty inertia per unit length of the beam in the (x,y) plane.

An element dx of the beam is deformed by the shear force and the moment. When the shear force is zero, the center line of dx is normal to the face of the cross-section as shown in Fig. 7-7. Let θ be the slope due to the moment M. Neglecting the interaction between the shear and the moment, the shear force V will cause a rectangular element to become a

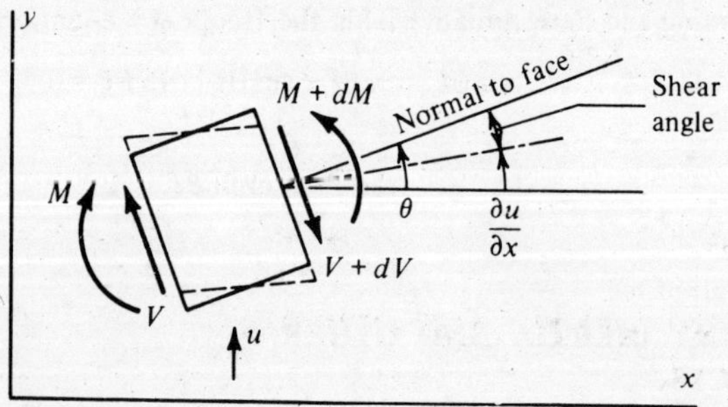

FIG. 7-7. *Lateral vibration of beam with rotary inertia and shear deformation.*

parallelogram without a rotation of the faces. Thus, the slope of the deflection curve is decreased by the shear angle. The slope $\partial u/\partial x$ is due to the moment M and the shear V.

$$\frac{\partial u}{\partial x} = \theta - (\text{shear angle})$$

$$\text{Shear angle} = \frac{V}{KAG} \tag{7-38}$$

where A is the cross-sectional area, G the shear modulus, and K a shape factor of the cross-section. The beam curvature in Eq. (7-33) can be expressed as

$$EI\frac{\partial \theta}{\partial x} = M \tag{7-39}$$

For a uniform beam, Eqs. (7-37) to (7-39) are combined to give

$$EI\frac{\partial^4 u}{\partial x^4} + m\frac{\partial^2 u}{\partial t^2} - \left(J + \frac{EIm}{KAG}\right)\frac{\partial^4 u}{\partial x^2 \partial t^2} + \frac{Jm}{KAG}\frac{\partial^4 u}{\partial t^4} = 0 \tag{7-40}$$

The algebraic manipulation is left as an exercise. This is the equation of motion for the lateral vibration of a uniform beam including the shear deformation and rotary inertia effects.

Example 5

A shaft between bearings may be considered as a hinged-hinged beam. Determine the rotary-inertia effect on the natural frequency of a uniform shaft between bearings.

Solution:

Let us first find the natural frequency of the beam without the rotary-inertia effect. Assuming the existence of principal modes as in Eq. (7-6), the lateral motion of the beam is

$$u(x,t) = \phi(x)\sin(\omega t + \psi)$$

substituting this into the beam equation shown in Eq. (7-35) and simplifying, we get

$$a^2\frac{d^4\phi}{dx^4} - \omega^2\phi = 0 \qquad \text{or} \qquad \frac{d^4\phi}{dx^4} - \beta^4\phi = 0$$

where $a^2 = EI/m$, $\beta^4 = \omega^2/a^2$, and ω is a natural frequency. The solution of the equation is

$$\phi = C_1 \sin \beta x + C_2 \cos \beta x + C_3 \sinh \beta x + C_4 \cosh \beta x$$

The boundary conditions for a hinged-hinged beam from Eq. (7-36) are

$$\phi|_{x=0} = \left.\frac{d^2\phi}{dx^2}\right|_{x=0} = \phi|_{x=\ell} = \left.\frac{d^2\phi}{dx^2}\right|_{x=\ell} = 0$$

Applying the boundary conditions, it can be shown that $C_2 = C_3 = C_4 = 0$ and

$$C_1 \sin \beta\ell = 0 \qquad \text{or} \qquad \beta\ell = n\pi \qquad \text{for} \qquad n = 1, 2, 3, \ldots$$

For convenience, let $m = \rho A$, where $\rho =$ mass/volume and $A =$ cross-sectional area. Hence the natural frequency is

$$\omega_n = \beta^2 a = \left(\frac{n\pi}{\ell}\right)^2 \sqrt{\frac{EI}{\rho A}}$$

where ω_n is a natural frequency without the rotary-inertia effect.

The equation of motion with rotary-inertia effect is deduced from Eq. (7 40) by neglecting the shear effect. Letting $m = \rho A$ as before and $J = \rho I$, the equation becomes

$$EI\frac{\partial^4 u}{\partial x^4} + \rho A\frac{\partial^2 u}{\partial t^2} - \rho I\frac{\partial^4 u}{\partial x^2 \partial t^2} = 0$$

Since we have a hinged-hinged beam, the mode shapes must be sine functions as before. The solution of the equation above is of the form

$$u(x,t) = C \sin \frac{n\pi x}{\ell} \sin(\omega_c t + \psi)$$

where ω_c is a natural frequency corrected for the rotary-inertia effect. Substituting $u(x,t)$ into the equation of motion above and simplifying, we obtain

$$\omega_c^2\left[1 + \frac{I}{A}\left(\frac{n\pi}{\ell}\right)^2\right] = \frac{EI}{\rho A}\left(\frac{n\pi}{\ell}\right)^4$$

The right side of this equation is ω_n^2. Thus,

$$\omega_c \simeq \omega_n\left[1 - \frac{1}{2}\frac{I}{A}\left(\frac{n\pi}{\ell}\right)^2\right]$$

Note that (1) the effect of rotary inertia is to decrease the natural frequency, and (2) the effect is more pronounced for the higher modes.

Effect of Axial Loading

A beam in lateral vibration may be subjected to an axial load. For example, in studying a fixed-fixed beam, the tension or compression in the beam due to the clamped ends may be difficult to determine.

Consider the free-body sketch of an element dx of a long beam in lateral vibration as shown in Fig. 7-8. The tension T is assumed constant for small deflections of the beam and the other effects are negligible. The forces involved are (1) that of the string problem in Eq. (7-2) and (2) the beam problem in Eq. (7-31). Let $m =$ mass/length, $\theta = \partial u/\partial x =$ slope, $V =$ shear force, and the deflection $u = u(x,t)$. From Newton's second law, the

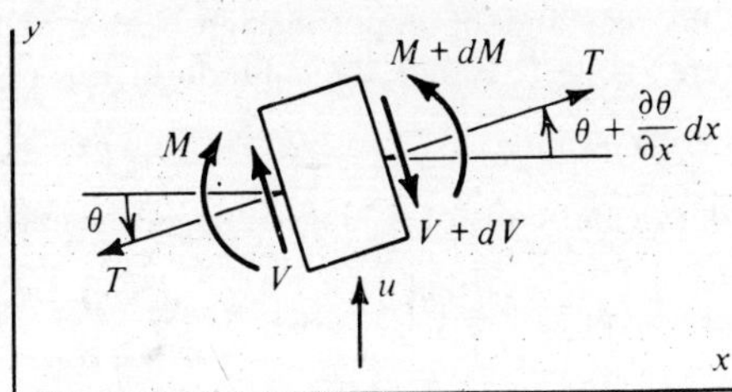

FIG. 7-8. *Lateral vibration of beam with axial tension.*

dynamic force equation is

$$m\frac{\partial^2 u}{\partial t^2}\,dx = -\left(V+\frac{\partial V}{\partial x}\,dx\right)+V+T\left(\theta+\frac{\partial \theta}{\partial x}\,dx\right)-T\theta$$

or

$$m\frac{\partial^2 u}{\partial t^2} = -\frac{\partial V}{\partial x}+T\frac{\partial^2 u}{\partial x^2} \tag{7-41}$$

Substituting Eqs. (7-32) and (7-33) in (7-41), we obtain

$$m\frac{\partial^2 u}{\partial t^2} = -\frac{\partial^2}{\partial x^2}\left(EI\frac{\partial^2 u}{\partial x^2}\right)+T\frac{\partial^2 u}{\partial x^2} \tag{7-42}$$

which is the beam equation including the effect of axial tension. The sign of T is reversed if the beam is under compression.

Example 6

Find the natural frequency of a uniform hinged-hinged beam subjected to an axial tension T.

Solution:

Assume the existence of principal modes as shown in Eq. (7-6).

$$u(x,t) = \phi(x)\sin(\omega t+\psi)$$

Substituting this in Eq. (7-42) and simplifying, we get

$$EI\frac{d^4\phi}{dx^4}-T\frac{d^2\phi}{dx^2}-m\omega^2\phi = 0 \tag{7-43}$$

Let the solution be $\phi = Ce^{sx}$, where C and s are constants. It can be shown the corresponding characteristic equation is

$$EIs^4 - Ts^2 - m\omega^2 = 0$$

The quadratic roots are

$$s_{1,2}^2 = \frac{T}{2EI}[1\pm(1+4m\omega^2 EI/T^2)^{\frac{1}{2}}]$$

which must be real and of opposite sign. Let $s_1^2 = a^2$ and $s_2^2 = -b^2$, or $s_2 = \pm jb$, where $j = \sqrt{-1}$. Hence the solution of Eq. (7-43) is

$$\phi = C_1 \sinh ax + C_2 \cosh ax + C_3 \sin bx + C_4 \cos bx$$

The boundary conditions of a hinged-hinged beam from Eq. (7-36) are

$$\phi|_{x=0} = \left.\frac{d^2\phi}{dx^2}\right|_{x=0} = \phi|_{x=\ell} = \left.\frac{d^2\phi}{dx^2}\right|_{x=\ell} = 0$$

It can be shown from these conditions that $C_1 = C_2 = C_4 = 0$ and

$$C_3 \sin b\ell = 0 \qquad \text{or} \qquad b\ell = n\pi \qquad \text{for} \qquad n = 1, 2, 3, \ldots$$

The natural frequency is obtained by equating

$$b^2 = \left(\frac{n\pi}{\ell}\right)^2 = -\frac{T}{2EI}[1 - (1 + 4m\omega^2 EI/T^2)^{\frac{1}{2}}]$$

or

$$\omega_n^2 = \left(\frac{n\pi}{\ell}\right)^2 \frac{T}{m} + \left(\frac{n\pi}{\ell}\right)^4 \frac{EI}{m}$$

Note that if $T = 0$ the natural frequency is that of a simply supported beam as shown in Example 5. If $EI = 0$, we have a flexible taut string as shown in Example 1. The tension will stiffen the beam and thereby increase its natural frequency.

7-7 THE EIGENVALUE PROBLEM

The wave equation and the beam equation can be generalized as eigenvalue problems. This gives a perspective of the problem and shows the similarity between discrete and continuous systems. We assume the rotary inertia and other effects are negligible.

Let the partial differential equations of the wave and the beam equations be generalized as

$$M(\ddot{u}) + L(u) = 0 \tag{7-44}$$

where L and M are linear differential operators* involving only the space variable x. The displacement u is as defined in Eq. (7-15), that is,

$$u(x,t) = \phi(x)q(t) \tag{7-45}$$

* A linear differential operator of the space variable x is of the form

$$L = a_0 + a_1\frac{\partial}{\partial x} + a_2\frac{\partial^2}{\partial x^2} + \cdots$$

$$L(u) = a_0 u + a_1\frac{\partial u}{\partial x} + a_2\frac{\partial^2 u}{\partial x^2} + \cdots$$

For simplicity, the symbol $(\dot{\ })$ denotes the partial differentiation with respect to time and $(')$ with respect to the space variable x. The symbol M is also used to denote a moment and L a length in the previous sections.

Substituting this in Eq. (7-44) and simplifying, we obtain

$$\frac{L(\phi)}{M(\phi)} = -\frac{\ddot{q}}{q} = \lambda \tag{7-46}$$

where λ is a constant, since $L(\phi)/M(\phi)$ is a function of x and $\ddot{q}/q$ is a function of t. Thus,

$$\ddot{q} + \lambda q = 0 \tag{7-47}$$

$$L(\phi) = \lambda M(\phi) \tag{7-48}$$

The eigenvalue problem formulation of continuous systems is as shown in Eq. (7-48), where λ is an *eigenvalue* and $\phi(x)$ an *eigenfunction.* The problem is to find the values of λ for which there are functions $\phi(x)$ satisfying Eq. (7-48) and the boundary conditions. Note that Eqs. (7-47) and (7-48) give the generalized form of Eq. (7-17).

The forms of L and M depend on the problem. For most problems, M is merely a weighting factor. M becomes a constant for a uniform beam. Examples of the forms of L were illustrated in the previous problems. For the wave equation, we have

$$L = -\frac{\partial}{\partial x}\left(k\frac{\partial}{\partial x}\right) \tag{7-49}$$

where $k = T$ in the vibrating string problem, $k = AE$ in Eq. (7-23), and $k = I_pG$ in Eq. (7-28). For the beam equation,

$$L = \frac{\partial^2}{\partial x^2}\left(k\frac{\partial^2}{\partial x^2}\right) \tag{7-50}$$

where $k = EI$ in Eq. (7-34).

The boundary conditions in Eqs. (7-25), (7-30), and (7-36) are of the form

$$B(\phi) = \lambda C(\phi) \tag{7-51}$$

where B and C are linear differential operators involving the space variable x only. The right side of the equation is from the inertia term by replacing $\partial^2 u/\partial t^2$ *by* $-\lambda u$. Hence if the boundary does not have an inertia load, we get

$$B(\phi) = 0 \tag{7-52}$$

In other words, the eigenvalue λ will appear in the boundary condition if the system has an inertia load as shown in Figs. 7-3 and 7-6.

Comparing with discrete systems, M in Eq. (7-48) is analogous to the mass matrix, L the stiffness matrix, and $\phi(x)$ the modal vector. There is no "exact" analogy for boundary conditions. It is anticipated that the eigenfunctions of continuous systems are orthogonal.

7-8 ORTHOGONALITY

We shall follow the steps outlined for discrete systems in Sec. 6-4 to show that the eigenfunctions $\phi(x)$ of continuous systems are orthogonal with respect to the operators L and M. It is assumed that the eigenvalues are distinct and the eigenfunctions are the solutions of a selfadjoint problem.*

From Eq. (7-48), for the rth and sth modes, we have

$$L(\phi_r) = \lambda_r M(\phi_r) \quad \text{or} \quad \phi_s L(\phi_r) = \lambda_r \phi_s M(\phi_r)$$

$$L(\phi_s) = \lambda_s M(\phi_s) \quad \text{or} \quad \phi_r L(\phi_s) = \lambda_s \phi_r M(\phi_s)$$

Subtracting one equation from the other and integrating over the domain Σ of the elastic body, we obtain

$$\int_\Sigma [\phi_s L(\phi_r) - \phi_r L(\phi_s)]\, d\sigma = \int_\Sigma [\lambda_r \phi_s M(\phi_r) - \lambda_s \phi_r M(\phi_s)]\, d\sigma$$

By virtue of the selfadjoint property, we get

$$\int_\Sigma [\phi_s L(\phi_r) - \phi_r L(\phi_s)]\, d\sigma = 0 \tag{7-53}$$

$$(\lambda_r - \lambda_s) \int_\Sigma \phi_r M(\phi_s)\, d\sigma = 0 \tag{7-54}$$

Since $\lambda_r \neq \lambda_s$, we deduce that

$$\int_\Sigma \phi_r M(\phi_s)\, d\sigma = 0 \qquad \text{for} \qquad r \neq s \tag{7-55}$$

Similarly, it can be shown that

$$\int_\Sigma \phi_r L(\phi_s)\, d\sigma = 0 \qquad \text{for} \qquad r \neq s \tag{7-56}$$

The *orthogonal relations* are as shown in Eqs. (7-55) and (7-56). We shall apply these relations to the wave and the beam equation.

* The eigenfunctions are selfadjoint if

$$\int_\Sigma \phi_r L(\phi_s)\, d\sigma = \int_\Sigma \phi_s L(\phi_r)\, d\sigma$$

$$\int_\Sigma \phi_r M(\phi_s)\, d\sigma = \int_\Sigma \phi_s M(\phi_r)\, d\sigma$$

This is analogous to the symmetry of the stiffness matrix K and the mass matrix M of discrete systems, that is, $k_{ij} = k_{ji}$ and $m_{ij} = m_{ji}$.

Case 1. Boundary Conditions Independent of λ

(1) Consider the *wave equation*. Substituting Eq. (7-49) in (7-53) and then integrating by parts, we obtain

$$\int_0^\ell [\phi_r(k\phi_s')' - \phi_s(k\phi_r')']\,dx$$

$$= \left[\phi_r k\phi_s' \big|_0^\ell - \int_0^\ell k\phi_s'\phi_r'\,dx\right] - \left[\phi_s k\phi_r' \big|_0^\ell - \int_0^\ell k\phi_s'\phi_r'\,dx\right]$$

$$= [\phi_r k\phi_s' - \phi_s k\phi_r']_0^\ell \tag{7-57}$$

The problem is selfadjoint and the eigenfunctions orthogonal if this equation is equal to zero for $r \neq s$. The equation is satisfied if the boundary conditions in Eqs. (7-25) and (7-30) are of the form.*

$$a\phi + b\phi' = 0 \tag{7-58}$$

where a and b may depend on the boundary points.

From Eqs. (7-55) and (7-56), the *orthogonal relations* of the *wave equation* are

$$\int_0^\ell m(x)\phi_r\phi_s\,dx = 0 \qquad \text{for} \qquad r \neq s \tag{7-59}$$

$$\int_0^\ell k\phi_r'\phi_s'\,dx - [k\phi_r\phi_s']_0^\ell = 0 \qquad \text{for} \qquad r \neq s \tag{7-60}$$

Neglecting the spring load, Eq. (7-60) becomes

$$\int_0^\ell k\phi_r'\phi_s'\,dx = 0 \qquad \text{for} \qquad r \neq s \tag{7-61}$$

Note that $k = T$ for the vibrating string, $k = AE$ in Eq. (7-23) for the longitudinal vibration of a bar, and $k = I_pG$ in Eq. (7-28) for the torsional oscillation of a rod.

(2) Consider the *beam equation*. Substituting Eq. (7-50) in (7-53) and then integrating by parts, we obtain

$$\int_0^\ell [\phi_r(k\phi_s'')'' - \phi_s(k\phi_r'')'']\,dx$$

$$= \left\{[\phi_r(k\phi_s'')']_0^\ell - \int_0^\ell \phi_r'(k\phi_s'')'\,dx\right\}$$

$$- \left\{[\phi_s(k\phi_r'')']_0^\ell - \int_0^\ell \phi_s'(k\phi_r'')'\,dx\right\}$$

$$= \left\{[\phi_r(k\phi_s'')' - \phi_r'(k\phi_s'')]_0^\ell + \int_0^\ell k\phi_r''\phi_s''\,dx\right\}$$

$$- \left\{[\phi_s(k\phi_r'')' - \phi_s'(k\phi_r'')]_0^\ell + \int_0^\ell k\phi_s''\phi_r''\,dx\right\}$$

$$= [\phi_r(k\phi_s'')' - \phi_s(k\phi_r'')' - \phi_r'k\phi_s'' + \phi_s'k\phi_r'']_0^\ell \tag{7-62}$$

* K. N. Tong, *Theory of Mechanical Vibrations*, John Wiley & Sons, Inc., New York, 1960, p. 264.

The problem is selfadjoint and Eq. (7-62) is identically zero if the boundary conditions at each end can be expressed as*

$$a\phi + b(k\phi'')' = 0$$
$$c\phi' + d(k\phi'') = 0 \tag{7-63}$$

The constants a, b, c, and d depend on the boundary conditions. Except for the inertia load, the two boundary conditions at each end of a beam in Eq. (7-36) are described by Eq. (7-63).

The *orthogonal relations* for the *beam equation* from Eqs. (7-55) and (7-56) are

$$\int_0^\ell m(x)\phi_r\phi_s\,dx = 0 \qquad \text{for} \qquad r \neq s \tag{7-64}$$

$$\int_0^\ell k\phi_r''\phi_s''\,dx + [\phi_r(k\phi_s'')' - k\phi_r'\phi_s'']_0^\ell = 0 \qquad \text{for} \qquad r \neq s \tag{7-65}$$

Neglecting the spring load, Eq. (7-65) becomes

$$\int_0^\ell k\phi_r''\phi_s''\,dx = 0 \qquad \text{for} \qquad r \neq s \tag{7-66}$$

where k is the flexural stiffness EI of the beam.

Example 7. Uniform Free-free Beam

Components of machines are often tested under free-free conditions, because boundary conditions, such as fixed or hinged, are difficult to achieve. Determine the frequency equation and the eigenfunctions for the lateral vibration of a uniform free-free beam.

Solution:

Let the lateral deflection be $u(x,t) = \phi(x)q(t)$. Comparing the beam equation in Eq. (7-34) and the generalized equation in Eq. (7-44), we have $M = m$ and $L = EI\partial^4/\partial x^4$. Thus, the generalized equations in Eqs. (7-47) and (7-48) reduce to

$$\frac{d^2q}{dt^2} + \omega^2 q = 0 \qquad \text{and} \qquad \frac{d^4\phi}{dx^4} - \beta^4\phi = 0$$

where $\omega^2 = \lambda$ and $\beta^4 = \omega^2/a^2 = \omega^2 m/EI$. The solution of the second equation is

$$\phi(x) = C_1 \sin\beta x + C_2\cos\beta x + C_3\sinh\beta x + C_4\cosh\beta x$$

The boundary conditions from Eq. (7-36) are

$$(1)\quad \phi''|_{x=0} = 0 \qquad (2)\quad \phi'''|_{x=0} = 0$$
$$(3)\quad \phi''|_{x=\ell} = 0 \qquad (4)\quad \phi'''|_{x=\ell} = 0$$

* Tong, *ibid*, p. 269.

Substituting conditions (1) and (2) in $\phi(x)$, we get

$$-C_2+C_4=0 \qquad \text{and} \qquad -C_1+C_3=0$$

Conditions (3) and (4) require that

$$(\sin \beta\ell - \sinh \beta\ell)C_3 + (\cos \beta\ell - \cosh \beta\ell)C_4 = 0$$
$$(\cos \beta\ell - \cosh \beta\ell)C_3 - (\sin \beta\ell + \sinh \beta\ell)C_4 = 0$$

A nontrivial solution for C_3 and C_4 requires that the determinant of the coefficients of C_3 and C_4 be zero. Thus, the frequency equation is

$$\cos \beta\ell \cosh \beta\ell = 1$$

When $\beta = 0$, the solutions of the differential equations are

$$q(t) = A_1 + A_2 t$$
$$\phi(x) = B_0 + B_1 x + B_2 x^2 + B_3 x^3$$

The function $q(t)$ indicates that the free-free beam is semidefinite. The boundary conditions (3) and (4) require that $B_2 = B_3 = 0$. Let the corresponding eigenfunctions be

$$\phi_o(x) = B_o \qquad \text{and} \qquad \phi_1(x) = B_1(x - \ell/2)$$

where B_o and B_1 are arbitrary. This shows that, corresponding to $\beta = 0$ or $\omega = 0$, the rigid body motion can be a translation and/or a rotation about the mass center of the system.

When $\beta \neq 0$, we obtain from the equations above

$$\frac{C_3}{C_4} = -\frac{(\cos \beta\ell - \cosh \beta\ell)}{(\sin \beta\ell - \sinh \beta\ell)} = \frac{(\sin \beta\ell + \sinh \beta\ell)}{(\cos \beta\ell - \cosh \beta\ell)}$$

$$\phi_r(x) = C_r[(\sin \beta_r\ell + \sinh \beta_r\ell)(\sin \beta_r x + \sinh \beta_r x) + (\cos \beta_r\ell - \cosh \beta_r\ell)(\cos \beta_r x + \cosh \beta_r x)]$$

where $r = 2, 3, \ldots$

The modes of single-span uniform beam for different end conditions are shown in Fig. 7-9. The corresponding equations are summarized in Table 7-1. The roots $\beta\ell$ of the frequency equations are listed in Table 7-2.

Case 2. Boundary Conditions Dependent on λ

An eigenvalue λ will appear in the boundary condition if the system has an inertia load. Under such conditions, the eigenfunctions in Eq. (7-59) will not be orthogonal. We shall derive the orthogonal relation by redefining the mass distribution in the system.

Consider the longitudinal vibration of a long uniform rod of length ℓ and mass/length m. Let a mass m_a be attached at the free end as shown in Fig. 7-3(b). The boundary conditions are as given in Eq. (7-25). Let m_a

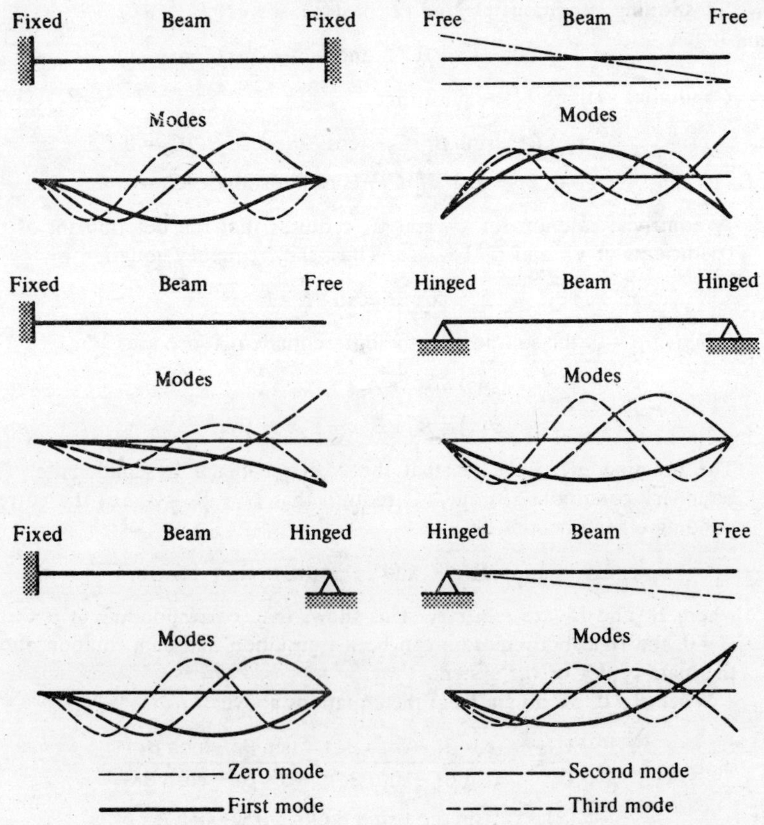

FIG. 7-9. *Modes of vibration of single-span uniform beam for different end conditions.*

be an integral part of the rod. Hence

$$\text{Total mass} = m\ell + m_a$$

or

$$\text{Mass/length} = m + m_a\delta(x-\ell) \tag{7-67}$$

Since m_a is a concentrated mass, its mass/length becomes a delta function* $m_a\delta(x-\ell)$. If m_a is part of the rod, then the boundary condition beyond m_a is that of a free end.

* Dirac's delta function was described in Chap. 2. From Eq. (2-64) we have

$$\delta(x-\ell)=0 \quad \text{for} \quad x \neq \ell \quad \text{and} \quad \int_{-\infty}^{\infty} \delta(x-\ell)\,dx = 1$$

For any arbitrary function $f(x)$, this gives

$$\int_{-\infty}^{\infty} \delta(x-\ell)f(x)\,dx = f(\ell)$$

TABLE 7-1. *Summary of Equations for Lateral Vibration of Single-span Uniform Beam.*

END CONDITIONS	EQUATIONS
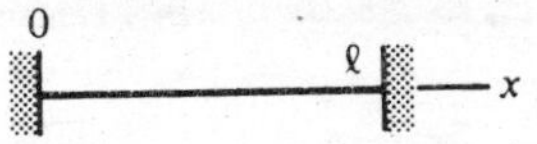 0 ℓ x	(1) $u(0,t)=u'(0,t)=u(\ell,t)=u'(\ell,t)=0$ (2) $\cos\beta\ell\cosh\beta\ell=1$ (3) $\phi(x)=A(\cos\beta x-\cosh\beta x)+(\sin\beta x-\sinh\beta x)$ $A=-\dfrac{\sin\beta\ell-\sinh\beta\ell}{\cos\beta\ell-\cosh\beta\ell}=\dfrac{\cos\beta\ell-\cosh\beta\ell}{\sin\beta\ell+\sinh\beta\ell}$
0 ℓ x	(1) $u''(0,t)=u'''(0,t)=u''(\ell,t)=u'''(\ell,t)=0$ (2) $\cos\beta\ell\cosh\beta\ell=1$* (3) $\phi(x)=A(\cos\beta x+\cosh\beta x)+(\sin\beta x+\sinh\beta x)$ $A=-\dfrac{\sin\beta\ell-\sinh\beta\ell}{\cos\beta\ell-\cosh\beta\ell}=\dfrac{\cos\beta\ell-\cosh\beta\ell}{\sin\beta\ell+\sinh\beta\ell}$
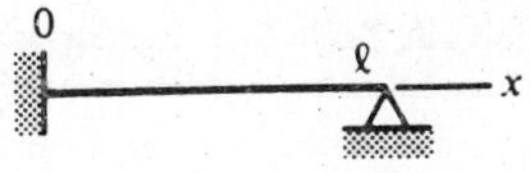 0 ℓ x	(1) $u(0,t)=u'(0,t)=u(\ell,t)=u''(\ell,t)=0$ (2) $\tan\beta\ell=\tanh\beta\ell$ (3) $\phi(x)=A(\cos\beta x-\cosh\beta x)+(\sin\beta x-\sinh\beta x)$ $A=-\dfrac{\sin\beta\ell-\sinh\beta\ell}{\cos\beta\ell-\cosh\beta\ell}=-\dfrac{\sin\beta\ell+\sinh\beta\ell}{\cos\beta\ell+\cosh\beta\ell}$
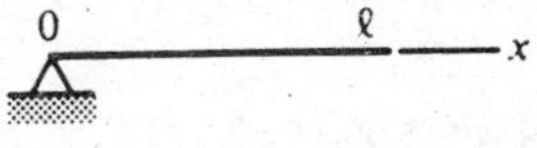 0 ℓ x	(1) $u(0,t)=u''(0,t)=u''(\ell,t)=u'''(\ell,t)=0$ (2) $\tan\beta\ell=\tanh\beta\ell$* (3) $\phi(x)=A\sin\beta x+\sinh\beta x$ $A=\dfrac{\sinh\beta\ell}{\sin\beta\ell}=\dfrac{\cosh\beta\ell}{\cos\beta\ell}$
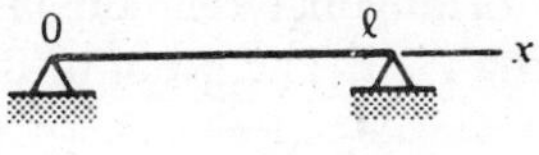 0 ℓ x	(1) $u(0,t)=u''(0,t)=u(\ell,t)=u''(\ell,t)=0$ (2) $\sin\beta\ell=0$ (3) $\phi(x)=A\sin\beta x$
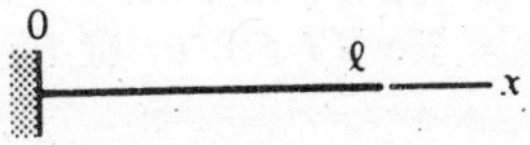 0 ℓ x	(1) $u(0,t)=u'(0,t)=u''(\ell,t)=u'''(\ell,t)=0$ (2) $\cos\beta\ell\cosh\beta\ell=-1$ (3) $\phi(x)=A(\cos\beta x-\cosh\beta x)+(\sin\beta x-\sinh\beta x)$ $A=-\dfrac{\sin\beta\ell+\sinh\beta\ell}{\cos\beta\ell+\cosh\beta\ell}=\dfrac{\cos\beta\ell+\cosh\beta\ell}{\sin\beta\ell-\sinh\beta\ell}$

Lateral deflection $=\sum\phi(x)q(t)$ $m=$ mass/length $\omega=$ freq., rad/s

Beam equation: $\dfrac{\partial^2 u}{\partial t^2}+a^2\dfrac{\partial^4 u}{\partial x^4}=0$ $\dfrac{d^4\phi}{dx^4}-\beta^4\phi=0$ $\dfrac{d^2 q}{dt^2}+\omega^2 q=0$

$$\beta^4=\omega^2/a^2=m\omega^2/EI$$

(1) End condition   (2) Frequency equation   (3) Eigenfunction

* Semidefinite systems

TABLE 7-2. *Roots $\beta\ell$ of Frequency Equation for Single-span Uniform Beam*†

END CONDITIONS	$\beta_1\ell$	$\beta_2\ell$	$\beta_3\ell$	$\beta_4\ell$
clamped–clamped	4.730041	7.853205	10.995608	14.137166
 free–free*	4.730041	7.853205	10.995608	14.137166
 clamped–hinged	3.926602	7.068583	10.210176	13.351768
hinged–free*	3.926602	7.068583	10.210176	13.351768
hinged–hinged	3.141593	6.283185	9.424778	12.566370
 clamped–free	1.875104	4.694091	7.854757	10.995541

* Semidefinite system. Zero modes not included.

† The roots $\beta\ell$ are computed from the frequency equations in Table 7-1. Similar tables are found in the literature. See, for example, R. Bishop and D. Johnson, *Vibration Analysis Tables*, Cambridge University Press, New York, 1955.

Using $m(x) = m + m_a\delta(x - \ell)$, the orthogonal relation in Eq. (7-59) is

$$\int_0^\ell [m + m_a\delta(x-\ell)]\phi_r\phi_s\,dx = \int_0^\ell m\phi_r\phi_s\,dx + m_a\phi_r(\ell)\phi_s(\ell)$$

$$= 0 \qquad \text{for} \qquad r \neq s \tag{7-68}$$

The same technique can be used to find the orthogonal relation of a torsional shaft or a vibrating beam with an inertia load. The proof of this statement is left as an exercise.

Since the operator L in Eq. (7-56) is not affected by m_a, the corresponding orthogonal relation remains as shown in Eq. (7-60) or (7-65).

Example 8

A long uniform rod with an attached mass m_a at $x = \ell$ is shown in Fig. 7-3(*b*). For the longitudinal vibration, find (**a**) the frequency equation, and (**b**) the frequencies of the first four modes. (**c**) Illustrate that the eigenfunctions are orthogonal and sketch the mode shapes of the first four modes.

Solution:

(**a**) Let the longitudinal displacement be $u(x,t) = \phi(x)q(t)$ as defined in Eq. (7-15). From Eq. (7-18), we get

$$\phi(x) = C_1 \sin\frac{\omega}{c}x + C_2\cos\frac{\omega}{c}x$$

$$q(t) = C_3 \sin\omega t + C_4 \cos\omega t$$

where $c^2 = E/\rho$, ρ = mass/volume of the rod, and C's are constants. From Eq. (7-25), the boundary conditions are

$$(1) \quad u(0,t)=0 \qquad (2) \quad m_a\ddot{u}(\ell,t) = -EAu'(\ell,t)$$

Condition (1) gives $C_2 = 0$. Condition (2) yields

$$m_a\omega^2 \sin \omega\ell/c = EA\omega/c \cos \omega\ell/c$$

This can be rearranged to give the frequency equation

$$\frac{\omega\ell}{c}\tan\frac{\omega\ell}{c} = \frac{\rho A\ell}{m_a} = \frac{\text{mass of rod}}{\text{attached mass}}$$

This may be compared with the frequency equation in Example 3 for a torsional shaft with an inertia load.

(b) The values of $\omega\ell/c$ in the frequency equation are obtained by numerical solution or from tables.* For numerical answers, we assume:

Rod: mass/length, $m = 20$ kg/m, $\ell = 0.25$ m, $\rho = 8000$ kg/m^3

$$E = 200 \text{ GPa}, \quad c = \sqrt{E/\rho} = 5 \text{ km/s}$$

attached mass: $m_a = 5$ kg, mass ratio $= 20 \times 0.25/5 = 1.0$ The resuits of the first four modes are:

Mode	1	2	3	4
$\omega\ell/c$	0.8603	3.4256	6.4373	9.5293
Frequency, Hz	2 740	10 900	20 500	30 300

(c) Let us use $\phi_1(x)$ and $\phi_3(x)$ to illustrate the orthogonal relation $\int_0^\ell m\phi_r\phi_s\,dx + m_a\phi_r(\ell)\phi_s(\ell) = 0$.

$$\phi_1(x) = A_1 \sin 3.44x \qquad \phi_3(x) = A_3 \sin 25.75x$$

Substituting the ϕ's in Eq. (7-68), we obtain†

$$A_1A_3\left[\int_0^{0.25} 20(\sin 3.44x \sin 25.75x)\,dx + 5\sin 0.8603 \sin 6.4373\right]$$
$$= A_1A_3[20(-0.0145 - 0.0145) + 5(0.7580)(0.1535)]$$
$$= 0$$

Substituting $\phi_1(x)$ and $\phi_3(x)$ into $\int_0^\ell AE\phi_r'\phi_s'\,dx$ in Eq. (7-61) gives

$$AE\int_0^{0.25} [3.44(\cos 3.44x)][25.75(\cos 25.75x)]\,dx$$
$$= AE(3.44)(25.75)(-0.0145 + 0.0145) = 0$$

* See, for example, Jahnke and Emde, *Tables of Functions*, 4th ed., Dover Publications, Inc., 1945.

† The identities used for the integration are

$$\int \sin ax \sin bx\,dx = \frac{\sin(a-b)x}{2(a-b)} - \frac{\sin(a+b)x}{2(a+b)}$$

$$\int \cos ax \cos bx\,dx = \frac{\sin(a-b)x}{2(a-b)} + \frac{\sin(a+b)x}{2(a+b)}$$

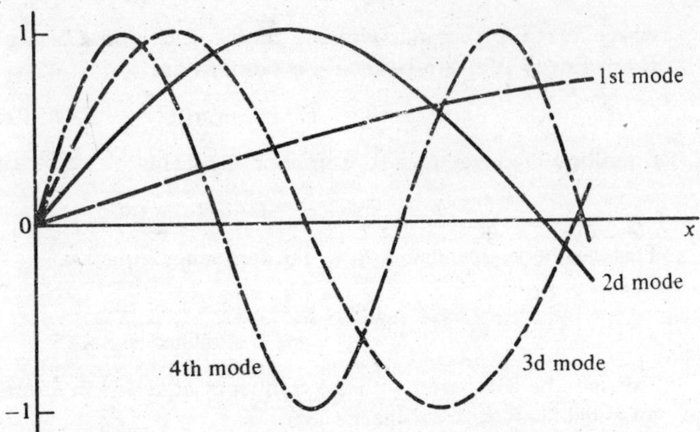

FIG. 7-10. *Mode shape for system with inertia load as shown in Fig. 7-3(b): mass ratio* = 1.

The mode shapes are illustrated in Fig. 7-10. The displacement of the inertia mass m_a decreases with increasing frequency ω, because the inertia load of m_a is proportional to ω^2. Hence m_a is almost stationary at the high frequencies.

7-9 LAGRANGE'S EQUATIONS

We shall relate the linear operators M and L of continuous systems in Eq. (7-44) to the kinetic and potential energy functions of the system. For discrete systems, the mass matrix M and the stiffness matrix K are related to the energy functions in like manner.

Consider the deflection $u(x,t)$ of a continuous system. In terms of the principal modes, it can be expressed as

$$u(x,t) = \sum_{i=1}^{\infty} \phi_i(x) q_i(t) \tag{7-69}$$

Since the inertia force is described by the operator M in Eq. (7-44), the kinetic energy function T is

$$T = \frac{1}{2} \int_0^e \dot{u} M(\dot{u})\, dx \tag{7-70}$$

Let M be a weighting function, that is, $M = m(x)$ mass/length. Substituting Eq. (7-69) in (7-70) and noting the orthogonal relation in Eq. (7-55), we obtain

$$\begin{aligned} T &= \frac{1}{2} \int_0^e m(x) \sum_{i=1}^{\infty} \phi_i \dot{q}_i \sum_{j=1}^{\infty} \phi_j \dot{q}_j \, dx \\ &= \frac{1}{2} \sum_{i=1}^{\infty} \sum_{j=1}^{\infty} \dot{q}_i \dot{q}_j \int_0^e m(x) \phi_i \phi_j \, dx \\ &= \frac{1}{2} \sum_{i=1}^{\infty} \dot{q}_i^2 m_{ii} \end{aligned} \tag{7-71}$$

where

$$m_{ii} = \int_0^{\ell} m(x)\phi_i^2 \, dx \tag{7-72}$$

and m_{ii} is the *generalized mass* of the continuous system for the ith mode. This may be compared with the orthogonal relations in Eqs. (7-55), (7-59), and (7-64). If the system has a mass m_a attached at $x=\ell$, from Eq. (7-68), the generalized mass m_{ii} in Eq. (7-71) is

$$m_{ii} = \int_0^{\ell} m(x)\phi_i^2 \, dx + m_a\phi_i^2(\ell) \tag{7-73}$$

Similarly, the operator L describes the elastic properties of the system. Hence the potential energy function U is

$$U = \frac{1}{2}\int_0^{\ell} uL(u) \, dx \tag{7-74}$$

Substituting $u(x,t)$ from Eq. (7-69) and noting the orthogonal property in Eq. (7-56), we have

$$\begin{aligned} U &= \frac{1}{2}\int_0^{\ell} \sum_{i=1}^{\infty} \phi_i q_i L\left(\sum_{j=1}^{\infty} \phi_j q_j\right) dx \\ &= \frac{1}{2}\sum_{i=1}^{\infty}\sum_{j=1}^{\infty} q_i q_j \int_0^{\ell} \phi_i L(\phi_j) \, dx \\ &= \frac{1}{2}\sum_{i=1}^{\infty} q_i^2 k_{ii} \end{aligned} \tag{7-75}$$

where

$$k_{ii} = \int_0^{\ell} \phi_i L(\phi_i) \, dx \tag{7-76}$$

and k_{ii} is the *generalized stiffness* for the ith mode. This may be compared with the orthogonal relations in Eqs. (7-56), (7-60), and (7-65). Furthermore, the expressions for the kinetic energy T and the potential energy U in Eqs. (7-71) and (7-75) may be compared with the corresponding expressions in Eq. (6-43) for discrete systems. T is always positive definite. U can be positive definite or semidefinite.

The Lagrange's equations of motion for the small oscillations of a conservative system are

$$\frac{d}{dt}\left(\frac{\partial T}{\partial \dot{q}_i}\right) + \frac{\partial U}{\partial q_i} = 0 \tag{7-77}$$

Substituting Eqs. (7-71) and (7-75) into (7-77), we obtain

$$m_{ii}\ddot{q}_i + k_{ii}q_i = 0 \tag{7-78}$$

or

$$\ddot{q}_i + \omega_i^2 q_i = 0 \tag{7-79}$$

where $\omega_i^2 = k_{ii}/m_{ii}$. This was anticipated, since T and U in Eqs. (7-71) and (7-75) are expressed in the eigenfunctions, or normal modes. There is no coupling in the equations of motion.

Let us relate the *initial conditions*, as shown in Eqs. (7-13) and (7-14), to our present discussion. From Eq. (7-69) the initial conditions can be expressed as

$$u(x,0) = \sum_{i=1}^{\infty} \phi_i(x)q_i(0) = f(x) \tag{7-80}$$

$$\dot{u}(x,0) = \sum_{i=1}^{\infty} \phi_i(x)\dot{q}_i(0) = g(x) \tag{7-81}$$

Using the orthogonal relation in Eq. (7-68) with m_a at $x = \ell$, the corresponding initial conditions in the *normal coordinates* are

$$q_i(0) = \frac{1}{m_{ii}}\left[\int_0^{\ell} m(x)f(x)\phi_i\,dx + m_a f(\ell)\phi_i(\ell)\right] \tag{7-82}$$

$$\dot{q}_i(0) = \frac{1}{m_{ii}}\left[\int_0^{\ell} m(x)g(x)\phi_i\,dx + m_a g(\ell)\phi_i(\ell)\right] \tag{7-83}$$

where m_{ii} is defined in Eq. (7-73). The terms due to m_a are neglected if no mass is attached at $x = \ell$.

Similarly, $q_i(0)$ and $\dot{q}_i(0)$ can be obtained from the generalized stiffness in Eq. (7-76). Except for the zero mode, neglecting the spring load as in Eq. (7-61) for simplicity, for the wave equation, we get

$$\begin{aligned} q_j(0) &= \frac{1}{k_{jj}}\int_0^{\ell} kf'(x)\phi_j'\,dx \\ \dot{q}_j(0) &= \frac{1}{k_{jj}}\int_0^{\ell} kg'(x)\phi_j'\,dx \end{aligned} \tag{7-84}$$

where k is as defined in Eq. (7-49) for a general problem. Similarly, neglecting the spring load as in Eq. (7-66), the beam equation gives

$$\begin{aligned} q_j(0) &= \frac{1}{k_{jj}}\int_0^{\ell} EIf''(x)\phi_j''\,dx \\ \dot{q}_j(0) &= \frac{1}{k_{ii}}\int_0^{\ell} EIg''(x)\phi_i''\,dx \end{aligned} \tag{7-85}$$

Example 9

Show that $k_{ii}/m_{ii} = \omega_i^2$ for the first three modes of vibration of a uniform cantilever beam of m mass/length.

Solution:

Let the lateral deflection of the beam be $u(x,t)=\phi(x)q(t)$. From Example 4, the formal solutions of the beam equation are

$$\phi(x)=C_1 \sin \beta x + C_2 \cos \beta x + C_3 \sinh \beta x + C_4 \cosh \beta x$$

$$q(t)=C_5 \sin \omega t + C_6 \cos \omega t$$

where $\beta^4=\omega^2/a^2=m\omega^2/EI$ and $m=$ mass/length.

The boundary conditions for a cantilever beam from Eq. (7-36) are

$$(1)\quad u(0,t)=0 \qquad (2)\quad u'(0,t)=0$$

$$(3)\quad u''(\ell,t)=0 \qquad (4)\quad u'''(\ell,t)=0$$

Conditions (1) and (2) give $C_2+C_4=0$ and $C_1+C_3=0$. Conditions (3) and (4) yield

$$(\sin \beta\ell + \sinh \beta\ell)C_1 + (\cos \beta\ell + \cosh \beta\ell)C_2 = 0$$

$$(\cos \beta\ell + \cosh \beta\ell)C_1 - (\sin \beta\ell - \sinh \beta\ell)C_2 = 0$$

By letting the determinant of the coefficients of C_1 and C_2 equal zero, we obtain the frequency equation

$$\cos \beta\ell \cosh \beta\ell = -1$$

Let us normalize m_{ii} in order to show that $\omega_i^2 = k_{ii}/m_{ii}$. Let $\int \phi_i^2\, dx = \alpha_i$, a constant. Hence, from Eq. (7-72) $m_{ii} = \int m\phi_i^2\, dx = \alpha_i m$, a normalized mass.

For a uniform beam, k_{ii} can be calculated from Eq. (7-66) instead of (7-76), that is, $k_{ii} = \int EI(\phi_i'')^2\, dx$. Note that (1) $\int (\phi_i'''')\phi_i\, dx = \int (\phi_i'')^2\, dx$ from Eqs. (7-65) and (7-66), and (2) $d^4\phi/dx^4 = \beta^4\phi$. Thus, we have

$$k_{ii} = EI\int_0^\ell (\phi_i'')^2\, dx = EI\int_0^\ell (\phi_i'''')\phi_i\, dx = EI\beta_i^4 \int_0^\ell \phi_i^2\, dx$$

$$= EI\beta_i^4\alpha_i$$

$$\therefore \frac{k_{ii}}{m_{ii}} = \frac{EI\beta_i^4\alpha_i}{m\alpha_i} = \frac{EI}{m}\beta_i^4 = \frac{EI}{m}\left(\frac{m\omega_i^2}{EI}\right) = \omega_i^2 \qquad \text{for all modes}$$

Alternatively, let us calculate m_{ii} and k_{ii} in order to find ω_i^2. From Table 7-1, the expression for $\phi(x)$ is

$$\phi_i = A_i(\cos \beta_i x - \cosh \beta_i x) + (\sin \beta_i x - \sinh \beta_i x)$$

where $A_i = (\cos \beta_i\ell + \cosh \beta_i\ell)/(\sin \beta_i\ell - \sinh \beta_i\ell)$. The calculated values of A_i for the first three modes are -1.3622, -0.9818, and -1.0008. Thus, $\phi_i(x)$ can be substituted into Eq. (7-72) to evaluate m_{ii}.

For $i=1$, we get

$$m_{11} = \int_0^\ell m\phi_1^2\, dx = \left(\frac{m}{\beta_1}\right)\int_0^{\beta_1\ell} \phi_1^2\, d\beta_1 x$$

$$= \left(\frac{m}{\beta_1}\right)\int_0^{\beta_1\ell} [A_1(\cos \beta_1 x - \cosh \beta_1 x) + (\sin \beta_1 x - \sinh \beta_1 x)]^2\, d\beta_1 x$$

$$= \left(\frac{m}{\beta_1}\right)[A_1^2(4.8155) + 2A_1(2.4865) + (1.3181)]$$

$$= 3.4795 m/\beta_1$$

Obtaining $\phi_1''(x)$ from $\phi_1(x)$ and substituting $\phi_1''(x)$ into Eq. (7-66), we get

$$k_{11} = EI\beta_1^3(17.2120 - 23.3269 + 9.5944) = 3.4795EI\beta_1^3$$

$$\therefore k_{11}/m_{11} = EI\beta_1^4/m = (EI/m)(m\omega_1^2/EI) = \omega_1^2$$

The results for the first three modes to show that $\omega_i^2 = k_{ii}/m_{ii}$ are as follows:

	$i=1$	$i=2$	$i=3$
$\beta_i\ell$	1.8751	4.6941	7.8548
A_i	−1.3622	−0.9819	−1.0008
m_{ii}	$3.4795m/\beta_1$	$4.5254m/\beta_2$	$7.8418m/\beta_3$
k_{ii}	$3.4795EI\beta_1^3$	$4.5254EI\beta_2^3$	$7.8418EI\beta_3^3$

Example 10

Repeat Example 2, using Eqs. (7-82) and (7-83) to evaluate the constants of integration from the initial conditions.

Solution:

From Example 2, the formal solution in the form of $u(x,t) = \sum \phi(x)q(t)$ is

$$u(x,t) = \sum_{i=1,3,5,\ldots}^{\infty} \sin\frac{i\pi x}{2\ell}\left(C_i \sin\frac{i\pi ct}{2\ell} + D_i \cos\frac{i\pi ct}{2\ell}\right)$$

where $c^2 = E/\rho$, E = Young's modulus and ρ = mass/volume. The initial conditions are

$$u(x,0) = \frac{Fx}{AE} = f(x) \qquad \text{and} \qquad \dot{u}(x,0) = g(x) = 0$$

To evaluate $q_i(0)$, we use Eq. (7-72) to find m_{ii} and then (7-82) to find $q_i(0)$. For $\phi_i(x) = \sin(i\pi x/2\ell)$, we have

$$m_{ii} = \int_0 \rho A\phi_i^2\, dx = \int_0^\ell \rho A \sin^2\frac{i\pi}{2\ell}x\, dx = \frac{\rho A\ell}{2}$$

$$q_i(0) = \frac{2}{\rho A\ell}\int_0^\ell \rho A\frac{Fx}{AE}\sin\frac{i\pi}{2\ell}x\, dx = \frac{8F\ell}{\pi^2 AE}\frac{(-1)^{(i-1)/2}}{i^2}$$

for $i = 1, 3, 5, \ldots$. Defining $i = (2n+1)$, we get

$$q_n(0) = \frac{8F\ell}{\pi^2 AE}\frac{(-1)^n}{(2n+1)^2} \qquad \text{for} \qquad n = 0, 1, 2, 3, \ldots$$

This gives D_i in $u(x,t)$. From Eq. (7-83), $C_i = 0$. Thus,

$$u(x,t) = \frac{8F\ell}{\pi^2 AE}\sum_{n=0}^{\infty}\frac{(-1)^n}{(2n+1)^2}\sin\frac{(2n+1)\pi x}{2\ell}\cos\frac{(2n+1)\pi ct}{2\ell}$$

7-10 UNDAMPED FORCED VIBRATION—MODAL ANALYSIS

The partial differential equation for the undamped forced vibration of a continuous system from Eq. (7-44) is

$$M(\ddot{u})+L(u)=F(x,t) \tag{7-86}$$

where M and L are linear differential operators and $F(x,t)$ is an excitation force. Let the deflection $u(x,t)$ be

$$u(x,t)=\sum_{i=1}^{\infty}\phi_i(x)q_i(t) \tag{7-87}$$

Analogous to the modal analysis of discrete systems in Sec. 6-11, Eq. (7-86) can be uncoupled by substituting Eq. (7-87) in (7-86) and using the orthogonal properties in Eqs. (7-55) and (7-56). It can be shown readily that

$$m_{ii}\ddot{q}_i+k_{ii}q_i=\int_0^{\ell}F(x,t)\phi_i(x)\,dx\triangleq Q_i(t) \tag{7-88}$$

where $Q_i(t)$ as defined above is the generalized force for the ith mode and m_{ii} and k_{ii} are evaluated as shown in Eqs. (7-73) and (7-76). The complementary and the particular solutions of Eq. (7-88) can be obtained as described in Sec. 2-7. The solutions $q_i(t)$ are then substituted in Eq. (7-87) to obtain the deflection $u(x,t)$.

Example 11. Concentrated Force

Find the deflection $u(x,t)$ of a continuous system due to a concentrated force $F(t)$ applied at $x=\xi$.

Solution:

The concentrated force at $x=\xi$ can be expressed as $F(t)\delta(x-\xi)$, where $\delta(x-\xi)$ is a delta function occurring at $x=\xi$. The generalized force $Q_i(t)$ for the ith mode from Eq. (7-88) is

$$Q_i=\int_0^{\ell}[F(t)\delta(x-\xi)]\phi_i(x)\,dx \tag{7-89}$$

$$Q_i=F(t)\phi_i(\xi) \tag{7-90}$$

Since Eq. (7-88) is of the same form as (6-64), its particular solution can be found by the convolution integral as shown in Eq. (6-65). Defining $\omega_i^2=k_{ii}/m_{ii}$ and the force $Q_i(t)$ as shown above, the particular solution for the ith mode is

$$q_i=\frac{\phi_i(\xi)}{m_{ii}\omega_i}\int_0^{t}F(\tau)\sin\omega_i(t-\tau)\,d\tau \tag{7-91}$$

The initial conditions $q_i(0)$ and $\dot{q}_i(0)$ are found from $u(x,0)$ and $\dot{u}(x,0)$ by Eqs. (7-82) and (7-83) or as shown in Example 10. Analogous to Eq. (6-67), the complementary solution of Eq. (7-88) is

$$q_i = q_i(0)\cos\omega_i t + \frac{1}{\omega_i}\dot{q}_i(0)\sin\omega_i t \tag{7-92}$$

The complete solution for the ith mode is the superposition of Eqs. (7-91) and (7-92).

$$q_i = q_i(0)\cos\omega_i t + \frac{1}{\omega_i}\dot{q}_i(0)\sin\omega_i t + \frac{\phi_i(\xi)}{m_{ii}\omega_i}\int_0^t F(\tau)\sin\omega_i(t-\tau)\,d\tau \tag{7-93}$$

This is substituted into Eq. (7-87) to obtain $u(x,t) = \sum \phi_i(x)q_i(t)$.

Note that if the system is positive definite, the complementary solution due to the initial conditions is always in the form shown in Eq. (7-92). Hence, we shall examine only the particular solutions due to typical excitations in the examples to follow.

Example 12. Distributed Force

A distributed force $F(x,t)$ shown in Fig. 7-11(a) is suddenly applied to a simply-supported uniform beam. Find the generalized force and the motion of the beam. Assume zero initial conditions.

Solution:

The eigenfunction $\phi_i(x)$ and the frequency equation to obtain ω_i are given in Table 7-1.

$$\phi_i(x) = A_i \sin\frac{i\pi x}{\ell} \quad \text{and} \quad \omega_i^2 = \frac{EI(i\pi)^4}{m\ell^4} \quad \text{for} \quad i = 1, 2, 3, \ldots$$

The generalized mass m_{ii} from Eq. (7-72) is

$$m_{ii} = m\int_0^\ell \phi_i^2\,dx = mA_i^2\frac{\ell}{2}$$

where m is mass/length. Let $A_i = \sqrt{2}$. Thus, the normal mode and the

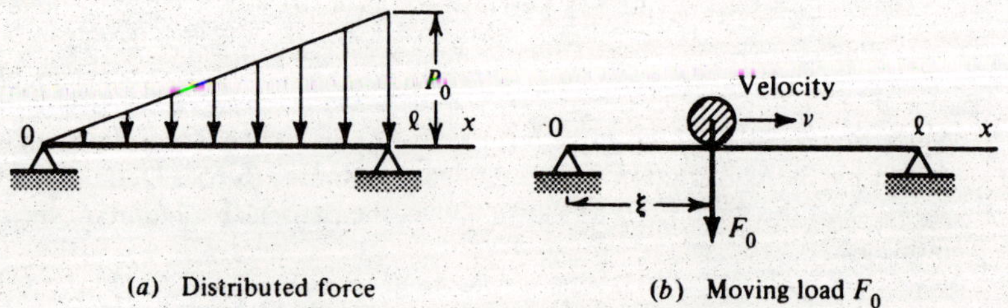

(a) Distributed force $p(x) = P_0x/\ell$

(b) Moving load F_0

FIG. 7-11. *Forced vibration of beam.*

normalized m_{ii} become

$$\phi_i(x)=\sqrt{2}\sin\frac{i\pi x}{\ell} \qquad \text{and} \qquad m_{ii}=m\ell$$

The distributed force $F(x,t)$ is

$$F(x,t)=p(x)=P_0 x/\ell$$

which is suddenly applied at $t=0$. Substituting this in Eq. (7-88) gives

$$Q_i=P_0\int_0^\ell \frac{x}{\ell}\sqrt{2}\sin\frac{i\pi x}{\ell}\,dx$$

$$=(-1)^{i+1}P_0\frac{\sqrt{2}\ell}{i\pi}$$

From Eq. (7-88), the equation of motion is

$$\ddot{q}_i+\omega_i^2 q_i=Q_i/m\ell$$

where Q_i is as defined above. From the convolution integral in Eq. (6-65), the particular solution for the ith mode is

$$q_i(t)=(-1)^{i+1}P_0\frac{\sqrt{2}\ell}{m\ell i\pi\omega_i}\int_0^t \sin\omega_i(t-\tau)\,d\tau$$

$$=(-1)^{i+1}P_0\frac{\sqrt{2}}{mi\pi\omega_i^2}(1-\cos\omega_i t) \qquad \text{for} \qquad i=1,2,3,\ldots$$

For zero initial conditions, the complementary solution becomes zero. Thus, the motion of the beam is

$$u(x,t)=\sum_{i=1}^{\infty}\phi_i(x)q_i(t)=\sum_{i=1}^{\infty}\sqrt{2}\sin\frac{i\pi x}{\ell}q_i(t)$$

Example 13. Concentrated Moving Force

A moving load on a bridge with constant velocity v is shown in Fig. 7-11(b). Find the deflection $u(x,t)$ of the bridge. Assume zero initial conditions.

Solution:

Let us approximate the bridge as a simply-supported uniform beam of m mass/length. The concentrated force F_0 is applied at $x=\xi=vt$ for $0\le vt\le \ell$. The generalized force Q_i from Eq. (7-88) is

$$Q_i=\int_0^\ell [F_0\delta(x-vt)]\phi_i(vt)\,dx$$

$$=\begin{cases}F_0\phi_i(vt) & \text{for} \quad 0\le t\le \ell/v\\ 0 & \text{for} \quad t>\ell/v\end{cases}$$

From the last example, we have $\phi_i(vt)=\sqrt{2}\sin(i\pi vt/\ell)$, $m_{ii}=m\ell$, and $\omega_i^2=EI(i\pi)^4/(m\ell^4)$. The particular solution is obtained by means of the convolution integral as shown in Eq. (6-65). Thus, for $0 \le t \le l/v$, $q_i(t)$ for the ith mode is

$$q_i(t) = \frac{1}{m_{ii}\omega_i} \int_0^t \left(F_0\sqrt{2} \sin \frac{i\pi v\tau}{\ell} \right) \sin \omega_i (t-\tau)\, d\tau$$

$$= \frac{\sqrt{2}F_0}{m\ell\omega_i} \frac{1}{a_i^2 - \omega_i^2} (a_i \sin \omega_i t - \omega_i \sin a_i t)$$

where $a_i = i\pi v/\ell$. The complementary solution is zero if the bridge was originally at rest. Substituting $q_i(t)$ in $u(x,t)$ in Eq. (7-87), we obtain

$$u(x,t) = \sum_{i=1}^{\infty} \frac{2F_0}{m\ell\omega_i} \frac{1}{a_i^2 - \omega_i^2} (a_i \sin \omega_i t - \omega_i \sin a_i t) \left(\sin \frac{i\pi x}{\ell} \right)$$

For $t > \ell/v$, the bridge is in free vibration. The initial conditions are $q_i(\ell/v)$ and $\dot{q}_i(\ell/v)$, which are the end conditions for the interval $0 \ge t \ge \ell/v$. The free vibration can be obtained from Eq. (7-92).

7-11 RAYLEIGH'S QUOTIENT

The Rayleigh method is perhaps the basis of a majority of approximate methods for vibration analysis. The technique used for continuous systems is similar to that for discrete systems in Sec. 6-7. This section prepares the background for the Rayleigh-Ritz method in the next section.

The Rayleigh's quotient is formed from the kinetic and the potential energy functions as for the discrete system shown in Eq. (6-48). Let the deflection $u(x,t)$ of a continuous system be approximated by

$$u(x,t) = \psi(x)q(t) \tag{7-94}$$

where $\psi(x)$ is an assumed mode shape. Note that we defined $u(x,t) = \phi(x)q(t)$ in the previous sections, where $\phi(x)$ is an eigenfunction or the exact mode shape. Substituting Eq. (7-94) in (7-70) and (7-74), the Rayleigh's quotient is

$$\lambda_R[\psi(x)] = \frac{\int_0^\ell \psi L(\psi)\, dx}{\int_0^\ell \psi M(\psi)\, dx} \tag{7-95}$$

where L and M are linear differential operators of the variable x as defined in Eq. (7-44). The quotient is a function of the function $\psi(x)$ and is called a *functional.* There are restrictions, however, on the assumed mode $\psi(x)$. Essentially, the restrictions are that (1) the functions $L(\psi)$ and $M(\psi)$ must exist and the integration possible, and (2) $\psi(x)$ must satisfy all the boundary conditions. It will be shown in the next section that the boundary conditions can be relaxed.

Example 14

Find the fundamental frequency of a uniform cantilever beam. Assume the mode shape $\psi(x)$ is that of a static curve, when the beam is deflected by its own weight.

Solution:

From elementary strength of materials, the static curve of a cantilever from the fixed end is

$$\psi(x) = \frac{mg}{24EI}(x^4 - 4\ell x^3 + 6\ell^2 x^2)$$

where m = mass/length. Substituting $\psi(x)$ in Eq. (7-95), noting from Eq. (7-34) that $L = EI\partial^4/\partial x^4$ and $M = m$ for a uniform beam, and performing the integration, we obtain $\lambda_R = 12.48EI/m\ell^4$ or $\omega = 3.530\sqrt{EI/m\ell^4}$. The exact value from Table 7-2 is $\omega = (\beta\ell)^2\sqrt{EI/m\ell^4} = 3.516\sqrt{EI/m\ell^4}$

Let us write Rayleigh's quotient in terms of the maximum potential and kinetic energies. Let the beam deflection be $u(x,t) = \psi(x)q(t)$, where $q(t)$ is harmonic. It can be shown from strength of materials that the potential energy U of a beam due to a bending moment is

$$U = \frac{1}{2}q^2(t)\int_0^\ell EI(x)\left(\frac{d^2\psi}{dx^2}\right)^2 dx$$

Assuming $q(t)_{max} = 1$, the maximum potential energy is

$$U_{max} = \frac{1}{2}\int_0^\ell EI(x)\left(\frac{d^2\psi}{dx}\right)^2 dx \tag{7-96}$$

The kinetic energy of the beam can be expressed as

$$T = \frac{1}{2}\int_0^\ell m(x)\left(\frac{\partial u}{\partial t}\right)^2 dx = \frac{1}{2}\dot{q}^2(t)\int_0^\ell m(x)\psi^2\, dx$$

Assuming $q(t)_{max} = 1$, the maximum kinetic energy is

$$T_{max} = \frac{1}{2}\omega^2\int_0^\ell m(x)\psi^2\, dx \triangleq \omega^2 T^*_{max} \tag{7-97}$$

Equating the maximum potential and kinetic energies, we get

$$\omega^2 = \frac{U_{max}}{T^*_{max}} \tag{7-98}$$

where $T^*_{max} = \frac{1}{2}\int_0^\ell m(x)\psi^2\, dx$ is called the *reference kinetic energy*. The quotient in Eq. (7-98) is equivalent to λ_R in Eq. (7-95).

7-12 RAYLEIGH-RITZ METHOD

The Rayleigh-Ritz method assumes the deflection curve $\psi(x)$ in Eq. (7-94) can be approximated by a finite series.

$$\psi(x) = \sum_{i=1}^{n} a_i \gamma_i(x) \tag{7-99}$$

where the a's are parameters and $\gamma_i(x)$ form a *generating set*, consisting of linearly independent functions satisfying all the boundary conditions. In limiting the generating set to a finite number of functions, the analysis can be interpreted as approximating a continuous system by means of a n-degree-of-freedom discrete system. Since (1) the assumed mode in Eq. (7-99) is a finite series instead of infinite, and (2) the assumed mode is not exact, the results can be interpreted as geometric constraints, both tending to raise the estimated natural frequency. The *Rayleigh-Ritz method* gives a procedure to minimize the estimated frequencies.

Substituting Eq. (7-99) into the Rayleigh's quotient in Eq. (7-95), we obtain

$$\lambda_R = \frac{\int_0^\ell \sum_{i=1}^n a_i \gamma_i L(\sum_{j=1}^n a_j \gamma_j)\, dx}{\int_0^\ell \sum_{i=1}^n a_i \gamma_i M(\sum_{j=1}^n a_j \gamma_j)\, dx}$$

$$= \frac{\sum_{i=1}^n \sum_{j=1}^n a_i a_j \int_0^\ell \gamma_i L(\gamma_j)\, dx}{\sum_{i=1}^n \sum_{j=1}^n a_i a_j \int_0^\ell \gamma_i M(\gamma_j)\, dx} \triangleq \frac{\sum_{i=1}^n \sum_{j=1}^n a_i a_j k_{ij}}{\sum_{i=1}^n \sum_{j=1}^n a_i a_j m_{ij}}$$

which can be expressed in matrix notations as

$$\lambda_R(a_1, a_2, \ldots, a_n) = \frac{\lfloor a \rfloor [k_{ij}] \{a\}}{\lfloor a \rfloor [m_{ij}] \{a\}} \triangleq \frac{N_R}{D_R} \tag{7-100}$$

where $\{a\}$ is a vector of the parameter a's and its transpose is $\lfloor a \rfloor$.

$$k_{ij} = \int_0^\ell \gamma_i L(\gamma_j)\, dx \quad \text{and} \quad m_{ij} = \int_0^\ell \gamma_i M(\gamma_j)\, dx \tag{7-101}$$

Note that k_{ij} and m_{ij} are known values for the assumed values of γ's in Eq. (7-99). The problem is selfadjoint and

$$k_{ij} = k_{ji} \quad \text{and} \quad m_{ij} = m_{ji} \tag{7-102}$$

The problem is to minimize λ_R above with respect to the a's. Let $\partial\lambda_R/\partial a_i = 0$ for minimum. Thus, Eq. (7-100) yields

$$\frac{\partial \lambda_R}{\partial a_i} = \frac{D_R(\partial N_R/\partial a_i) - N_R(\partial D_R/\partial a_i)}{D_R^2} = 0 \tag{7-103}$$

If ω^2 is the minimum of λ_R, from Eq. (7-100) we get

$$\omega^2 = \left(\frac{N_R}{D_R}\right)_{min} \tag{7-104}$$

Taking the partial derivatives with respect to a_i and recalling $k_{ij} = k_{ji}$ and $m_{ij} = m_{ji}$, we have

$$\begin{aligned} \frac{\partial N_R}{\partial a_i} &= \frac{\partial}{\partial a_i} \lfloor a \rfloor [k_{ij}]\{a\} = 2\sum_{j=1}^{n} k_{ij}a_j \qquad \text{for} \qquad i = 1, 2, \ldots, n \\ \frac{\partial D_R}{\partial a_i} &= \frac{\partial}{\partial a_i} \lfloor a \rfloor [m_{ij}]\{a\} = 2\sum_{j=1}^{n} m_{ij}a_j \qquad \text{for} \qquad i = 1, 2, \ldots, n \end{aligned} \tag{7-105}$$

Substituting Eqs. (7-104) and (7-105) in (7-103) and equating the numerator to zero for minimum, we obtain

$$\sum_{j=1}^{n} (k_{ij} - \omega^2 m_{ij}) a_j = 0 \qquad \text{for} \qquad i = 1, 2, \ldots, n \tag{7-106}$$

This is a set of n homogeneous equations with a's as the unknowns. For a nontrivial solution, we set the determinant of the coefficients to zero.

$$|k_{ij} - \omega^2 m_{ij}| = 0 \tag{7-107}$$

This gives the frequency equation, resembling that of a discrete system.

Example 15

Find the fundamental frequency of a uniform cantilever beam and compare the answer with that given in Example 14. Assume $\psi(x) = a_1x^2 + a_2x^3$

Solution:

The maximum potential energy from Eq. (7-96) is

$$\begin{aligned} U_{max} &= \frac{1}{2} EI \int_0^{\ell} (2a_1 + 6a_2x)^2 \, dx \\ &= \frac{1}{2} EI(4a_1^2\ell + 12a_1a_2\ell^2 + 12a_2^2\ell^3) \end{aligned}$$

$$\frac{\partial U_{max}}{\partial a_1} = EI(4a_1\ell + 6a_2\ell^2) \triangleq \sum_{j=1}^{2} k_{1j}a_j$$

$$\frac{\partial U_{max}}{\partial a_2} = EI(6a_1\ell^2 + 12a_2\ell^3) \triangleq \sum_{j=1}^{2} k_{2j}a_j$$

From the derivatives above and Eq. (7-105), we obtain

$$k_{11} = 4EI\ell \qquad k_{12} = k_{21} = 6EI\ell^2 \qquad k_{22} = 12EI\ell^3$$

The maximum reference kinetic energy from Eq. (7-97) is

$$T^*_{\max} = \frac{m}{2}\int_0^\ell \psi^2\,dx = \frac{m}{2}\int_0^\ell (a_1x^2 + a_2x^3)^2\,dx$$

$$= \frac{m}{2}\left(a_1\frac{\ell^5}{5} + 2a_1a_2\frac{\ell^6}{6} + a_2\frac{\ell^7}{7}\right)$$

Similarly, m_{ij} from Eq. (7-105) are

$$m_{11} = m\ell^5/5 \qquad m_{12} = m_{21} = m\ell^6/6 \qquad m_{22} = m\ell^7/7$$

Substituting k_{ij} and m_{ij} in Eq. (7-107) gives the frequency equation

$$\begin{vmatrix} 4EI\ell - \omega^2 m\ell^5/5 & 6EI\ell^2 - \omega^2 m\ell^6/6 \\ 6EI\ell^2 - \omega^2 m\ell^6/6 & 12EI\ell^3 - \omega^2 m\ell^7/7 \end{vmatrix} = 0$$

or

$$\omega^4 - 1{,}224(EI/m\ell^4)\omega^2 + 15{,}120(EI/m\ell^4)^2 = 0$$

The estimated frequencies are

$$\omega = \frac{3.533}{\ell^2}\sqrt{\frac{EI}{m}} \quad \text{and} \quad \omega = \frac{34.81}{\ell^2}\sqrt{\frac{EI}{m}}$$

The exact frequencies for the first two modes from Table 7-2 are

$$\omega = \frac{3.516}{\ell^2}\sqrt{\frac{EI}{m}} \quad \text{and} \quad \omega = \frac{22.03}{\ell^2}\sqrt{\frac{EI}{m}}$$

Using a simple power series for the problem above, the method gives good results for the fundamental frequency, but the correlation for the second mode is poor. Greater accuracy in the second mode can be achieved if the mode $\psi(x)$ is approximated by three or more terms.

It may be difficult to form a function $\psi(x)$ in Eq. (7-99) to meet all the boundary conditions. In fact, the $\psi(x)$ in Example 15 meets only the boundary conditions at the fixed end of the cantilever. The types of boundary conditions are (1) *geometric*, such as deflection and slope, and (2) *natural*, such as moment and shear. It can be shown* the boundary conditions can be relaxed and the quotient in Eq. (7-98) needs to satisfy only the geometric boundary conditions.

Example 16

Find the fundamental frequency of the wedge-shape cantilever beam shown in Fig. 7-12. Assume the beam is of constant unit width.

* See, for example, L. Meirovitch, *Analytical Methods in Vibrations*, The Macmillan Company, New York, 1967, p. 227.

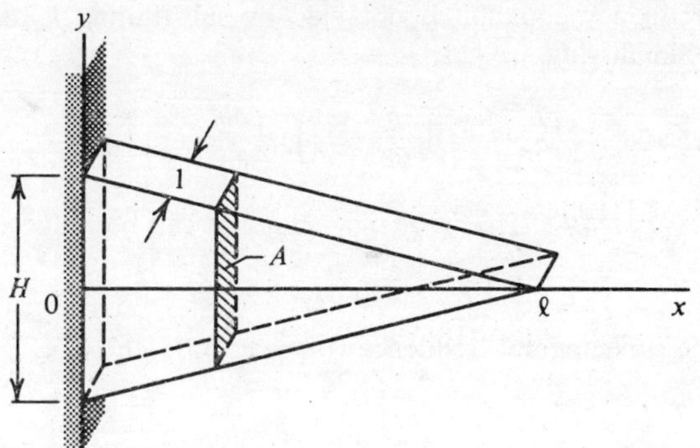

FIG. 7-12. *Lateral vibration of wedge-shape cantilever.*

Solution:

Since the height H is linear with x, we have

$$m(x) = \rho A(x) = \rho H(1 - x/\ell)$$
$$EI(x) = E[H(1 - x/\ell)]^3/12$$

where $m(x)$ = mass/length, ρ = mass/volume, and A = area of section. Let $\psi = a_1x^2 + a_2x^3$. From Eq. (7-96), the maximum potential energy U_{max} is

$$U_{max} = \frac{EH^3}{24\ell^3}\int_0^\ell (\ell - x)^3(2a_1 + 6a_2x)^2\, dx$$
$$= \frac{EH^3}{24\ell^3}\left(a_1^2\ell^4 + \frac{6}{5}a_1a_2\ell^5 + \frac{3}{5}a_2^2\ell^6\right)$$

Applying Eq. (7-105) yields

$$\frac{\partial U_{max}}{\partial a_1} = \frac{EH^3}{12\ell^3}\left(a_1\ell^4 + \frac{3}{5}a_2\ell^5\right) = \sum_{j=1}^{2} k_{1j}a_j$$
$$\frac{\partial U_{max}}{\partial a_2} = \frac{EH^3}{12\ell^3}\left(\frac{3}{5}a_1\ell^5 + \frac{3}{5}a_2\ell^6\right) = \sum_{j=1}^{2} k_{2j}a_j$$

Thus,

$$k_{11} = \frac{EH^3}{12\ell^3}\ell^4 \qquad k_{12} = k_{21} = \frac{EH^3}{12\ell^3}\frac{3}{5}\ell^5 \qquad k_{22} = \frac{EH^3}{12\ell^3}\frac{3}{5}\ell^6$$

The maximum reference kinetic energy from Eq. (7-97) is

$$T^*_{max} = \frac{\rho H}{2\ell}\int_0^\ell (\ell - x)(a_1x^2 + a_2x^3)^2\, dx$$
$$= \frac{\rho H}{2\ell}\left(a_1^2\frac{\ell^6}{30} + 2a_1a_2\frac{\ell^7}{42} + a_2^2\frac{\ell^8}{56}\right)$$

The partial derivatives are evaluated by Eq. (7-105) to give

$$m_{11} = \frac{\rho H\ell^5}{30}, \qquad m_{12} = m_{21} = \frac{\rho H\ell^6}{42}, \qquad \text{and} \qquad m_{22} = \frac{\rho H\ell^7}{56}$$

The frequency equation is obtained by substituting k_{ij} and m_{ij} in Eq. (7-107). Simplifying, we get

$$\left(\frac{b^2}{12}-\frac{\omega^2}{30}\right)\left(\frac{3b^2}{60}-\frac{\omega^2}{56}\right)-\left(\frac{3b^2}{60}-\frac{\omega^2}{42}\right)^2=0$$

where $b^2 = EH^2/\rho\ell^4$. The frequency equation can be reduced to

$$10\omega^4 - 273b^2\omega^2 + 588b^4 = 0$$

Hence the fundamental frequency is $\omega = 1.535H\sqrt{E/\rho\ell^4}$

7-13 SUMMARY

The one-dimensional wave equation and beam equation are treated in this chapter. The vibration of continuous systems involves (1) the elasticity of the material, (2) the boundary value problem, and (3) vibration theory. The chapter shows the similarity of the vibrational aspect of continuous and discrete systems.

The methods of problem formulation and solution of continuous systems are illustrated in Sec. 7-2. (1) The motion is a function of the space and time variables. The partial differential equation in Eq. (7-3) is derived from Newton's law of motion. (2) The problem is governed by the boundary conditions and the initial conditions. The boundary conditions stipulate the frequency equation and the modes of vibration. The initial conditions define the degree of participation of each mode. (3) Similar to discrete systems, the motion of continuous systems is obtained from the superposition of the modes. The procedure above is formalized by the classical method of separation of variables in Sec. 7-3.

Examples of problems governed by the wave equation are shown in Sec. 7-4. The Euler-Bernouli beam equation is discussed in Sec. 7-5. The methods of problem formulation and solution are essentially as outlined above.

Rotary inertia and other effects will influence the effective inertia and stiffness of a beam in vibration. A decrease in stiffness and/or an increase in inertia will decrease the natural frequency and vice versa. Some of the effects on beam vibration are discussed in Sec. 7-6. Again, the general method is the same as described above, but the procedure can be quite involved.

The continuous problem is generalized in Sec. 7-7, using the linear differential operators L and M. These may be compared with the matrices K and M of discrete systems. The forms of the operator L are as shown in Eqs. (7-49) and (7-50). The boundary conditions are generalized in Eqs. (7-51) and (7-52).

The proof of orthogonality of the eigenfunctions in Sec. 7-8 essentially follows the procedure for discrete systems. The orthogonal relations for the wave equation are shown in Eqs. (7-59) to (7-61) and for the beam equation in Eqs. (7-64) to (7-66). If the system has an inertia load, the orthogonal relation is modified as shown in Eq. (7-68).

The modal analysis in Sec. 7-10 closely resembles that for discrete systems. The equation of motion is first uncoupled by means of the eigenfunctions. The uncoupled equations in Eq. (7-88) are expressed in terms of the generalized mass m_{ii}, stiffness k_{ii}, and force $Q_i(t)$. The initial conditions for the modes are found from Eqs. (7-80) to (7-85). The complete solution for each mode is shown in Eq. (7-93). The general solution is the superposition of the modes $u(x,t)=\sum \phi_i(x)q_i(t)$.

The Rayleigh's quotient in Eq. (7-95) is an extension of Eq. (6-47) for discrete systems. The Rayleigh-Ritz method approximates a deflection curve $\psi(x)$ by a finite series in Eq. (7-99), where the a's are arbitrary. The estimated natural frequency by the Rayleigh's method tends to be higher than the actual value. The Rayleigh-Ritz method gives a procedure to minimize the Rayleigh's quotient and to obtain a frequency equation.

PROBLEMS

7-1 Find the motion $u(x,t)$ for the lateral vibration of a taut string shown in Fig. 7-1 for each of the following sets of initial conditions:

(a) $u(x,0)=0$ and $\dfrac{\partial u}{\partial t}(x,0)=\begin{cases} 2Vx/\ell & \text{for} \quad 0\le x\le \ell/2 \\ 2V(1-x/\ell) & \text{for} \quad \ell/2\le x\le \ell \end{cases}$

(b) $u(x,0)=f(x)=Hx(\ell-x)/\ell^2$ and $\dfrac{\partial u}{\partial t}=0$ for $0\le x\le \ell$

where H and V are constants.

7-2 A heavy but perfectly flexible cable of mass m per unit length and length L hangs freely from a fixed point A as shown in Fig. P7-1. (**a**) Assuming its

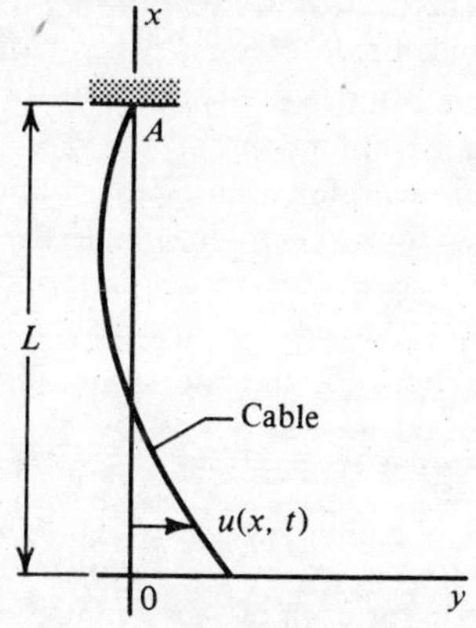

FIG. P7-1.

vibratory motion $u(x,t)$ is in the plane of the paper, derive the equation of motion. (**b**) By the method of separation of variables, $u(x,t)=\phi(x)q(t)$, show that the equation of motion can be separated into two ordinary differential equations

$$x\phi''+\phi'+\lambda\phi=0 \qquad \text{and} \qquad \ddot{q}+g\lambda q=0$$

where $\lambda=$ constant and $g=$ gravitational constant.

7-3 A uniform bar, free at both ends, is initially compressed by equal forces at the ends. Find the longitudinal vibration $u(x,t)$ if the forces are suddenly removed.

7-4 A uniform bar of length ℓ is fixed, or built-in, at both ends. The bar, initially at rest, is given a constant velocity V in the longitudinal x direction. Find the motion $u(x,t)$.

7-5 A spring loaded uniform bar is shown in Fig. 7-3(a). Derive its frequency equation.

7-6 Four end conditions for the longitudinal vibration of rods are shown in Eq. (7-25). (**a**) Tabulate the types of boundary conditions from the combinations of these end conditions. (**b**) Derive the frequency equations for five of these boundary conditions. (Note that the same statements can be made for the torsional vibration of shafts for the end conditions given in Eq. (7-30). Furthermore, various initial conditions can be applied to problems with different boundary conditions.)

7-7 The drill-pipe problem was discussed in Example 3. To examine the effect of the inertial load, we define $J_{\text{shaft}}/J_0=\alpha$ and $\omega\ell/c=\beta$. The frequency equation becomes $\beta \tan \beta=\alpha$. Find the first three roots β for $\alpha=0.1$, 0.5, 1.0, and 2.0.

7-8 (**a**) Derive the frequency equation for the longitudinal vibration of a uniform rod with an attached inertial mass as shown in Fig. 7-3(a). (**b**) Assuming that $(\text{mass})_{\text{rod}}/(\text{attached mass})$ is small, find the fundamental frequency. This is equivalent to assuming that $\tan \beta_1 \simeq \beta_1$. (**c**) Repeat part **b** by assuming that $\tan \beta_1 \simeq \beta_1+\beta_1^3/3$.

7-9 The semidefinite system shown in Fig. 4-13(a) has two disks J_1 and J_2 at the ends of a uniform circular shaft. (**a**) Derive the frequency equation. (**b**) Show that the fundamental frequency can be approximated as shown in Eq. (4-46) if J_1 and J_2 are much greater than the inertia of the shaft.

7-10 Derive the frequency equation and the eigenfunction for the lateral vibration of a uniform beam for each of the following boundary conditions: (**a**) fixed at both ends; (**b**) one end hinged and the other end free. (Note that the corresponding equations for common types of boundary conditions are derived in various examples and the results tabulated in Table 7-1.)

7-11 Typical end conditions for the lateral vibration of beams are shown in Eq. (7-36). With the aid of a sketch, show the types of boundary conditions from the combination of these end conditions.

7-12 A uniform beam of length ℓ and m mass/length has a mass m_a attached at the end as shown in Fig. P7-2(b). As shown, m_a is at $x=0$ or $x=\ell$. Find the frequency equation for each of these configurations. Assume the mass ratio $m_a/m\ell=1$. Compare the fundamental frequency with that from Prob. 2-6.

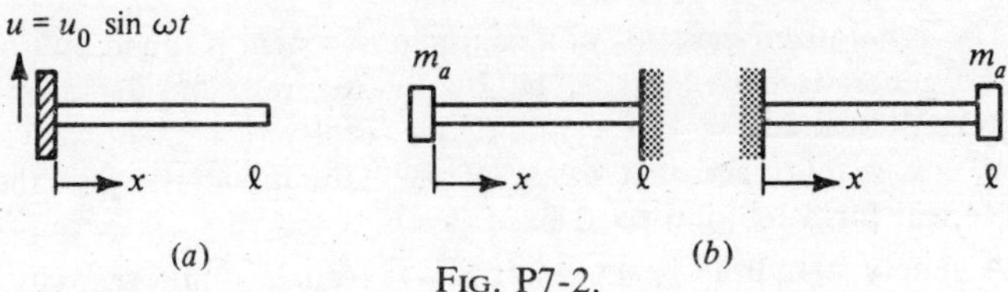

FIG. P7-2.

7-13 A turbine blade is modeled as a cantilever beam as shown in Fig. P7-2(*a*). The end support of the cantilever has a motion $u(0,t) = u_0 \sin \omega t$. (**a**) Find the frequency equation. (**b**) Evaluate the coefficients of the eigenfunction.

7-14 Data for a simply supported I beam are as follows: length = 9 m, $I_{AA} = 1.25 \times 10^{-3}\ \mathrm{m}^4$, mass/length $m = 150$ kg/m, $E = 200$ GPa. Assume a mass of 500 kg is placed at a distance $a = \ell/4$ from one end and this mass is removed at $t = 0$. Find the subsequent vibration of the beam.

7-15 Find the frequency equation for the lateral vibration of each of the uniform beams shown in Fig. P7-3.

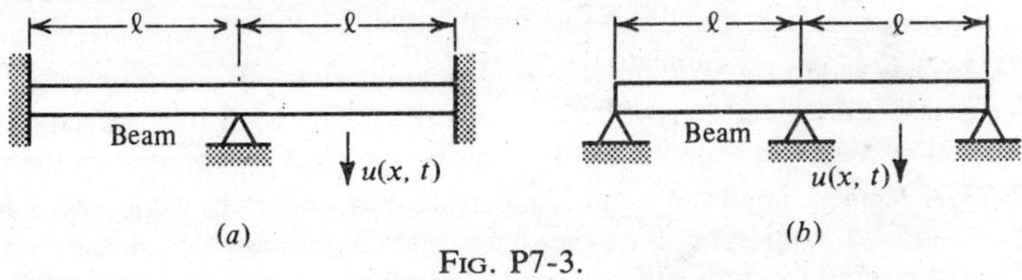

FIG. P7-3.

7-16 A rectangular steel bar is used as a fixed-fixed beam for vibration testing in a student project. The dimensions are width = 10 mm, height = 50 mm, and length = 1.2 m. It is found that the clamping at the ends introduces a tensile stress of 1.4 MPa in the beam. Estimate the error in the fundamental frequency of the lateral vibration due to the clamping.

7-17 Show that the eigenfunctions for the lateral vibrations $u(x,t)$ of each of the beams in Fig. P7-4 are orthogonal.

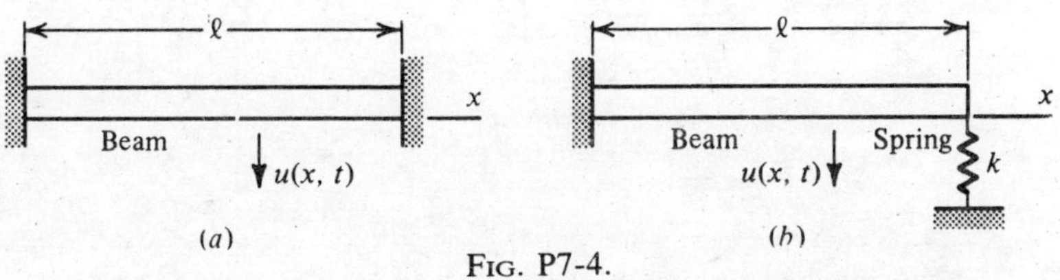

FIG. P7-4.

7-18 Find the steady-state response of the longitudinal vibration $u(x,t)$ for each of the uniform bars shown in Fig. P7-5.

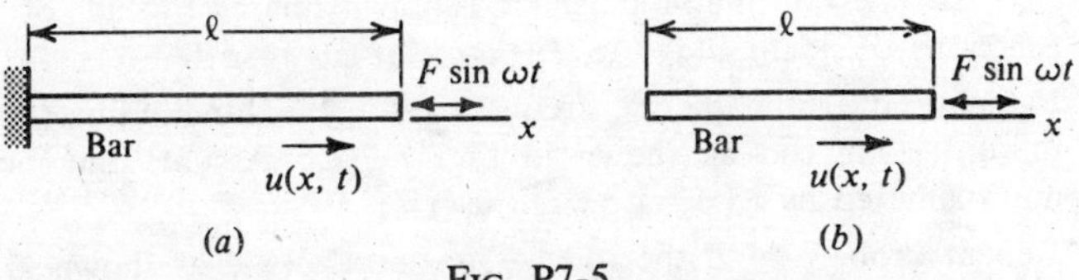

FIG. P7-5.

7-19 The generalized mass m_{ii} of a continuous system is found by Eq. (7-72) and the generalized stiffness k_{ii} by Eq. (7-76). (**a**) Consider a uniform fixed-hinged beam. Use the values from Tables 7-1 and 7-2 to show that $\omega_i^2 = k_{ii}/m_{ii}$ for the first mode of the lateral vibration of the beam. (**b**) Repeat for a uniform fixed-fixed beam.

7-20 A simply supported beam is initially at rest. If a uniform velocity V in the lateral direction is imparted on the beam, find the motion induced in the beam.

7-21 Find the response $u(x,t)$ for the lateral vibration of each of the following uniform beams described below. Assume zero initial conditions and the excitation force is $F_0 \sin \omega t$.

(**a**) A simply supported beam with the force applied at a distance ξ from one of its ends.

(**b**) A simply supported beam with the force applied uniformly along its length.

7-22 The blade of a large turbine can be approximated as a cantilever. Assume that, during the start-up, the base of the blade is given a constant acceleration. Write an expression for the motion of the blade relative to the base.

7-23 A simply supported beam deflects 10 mm under the action of a force F_0 applied at midspan. Find the steady-state amplitude if $F_0 \sin \omega t$ is applied at the same location with $\omega/\omega_1 = 1/2$, where ω_1 is the fundamental frequency of the beam.

7-24 Find the uncoupled equations for the lateral vibration $u(x,t)$ of each of the simply supported beams with the constraints as shown in Fig. P7-6.

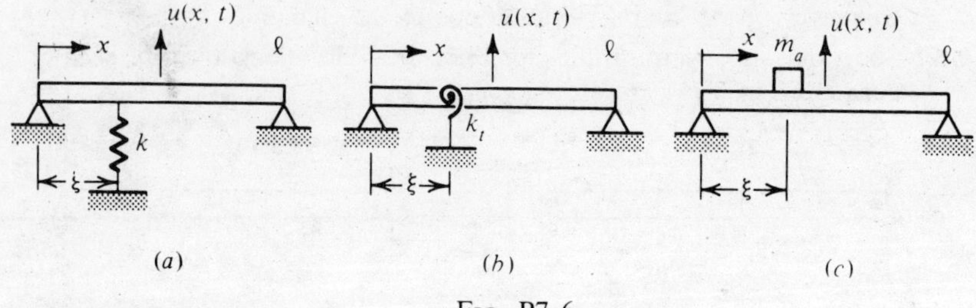

FIG. P7-6.

7-25 A moving load across a bridge was shown in Fig. 7-11(b). Find the lateral vibration of the bridge if the moving load gives a sinusoidal force $F_0 \sin \omega t$.

7-26 Estimate the fundamental frequency of the vibrating string shown in Fig. 7-1. Assume that the mode shape is approximated by each of the following functions $\psi(x)$: (**a**) $\psi(x) = a_1(\ell - x)x$; (**b**) $\psi(x) = a_1(\ell - x)x + a_2(\ell - x)^2x^2$.

7-27 Estimate the fundamental frequency for the longitudinal vibration of a uniform long rod as shown in Fig. 7-2(a). Assume the mode shape is approximated by $\psi(x) = a_1(x/\ell) + a_2(x/\ell)^2$.

7-28 Repeat Prob. 7-27 if the rod has an inertia load as shown in Fig. 7-3(a). Assume the attached mass is equal to the mass of the rod.

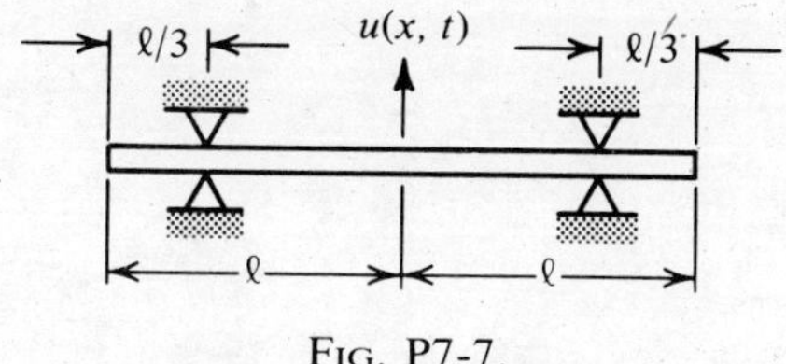

FIG. P7-7.

7-29 A uniform shaft between bearings is shown in Fig. P7-7. Assume an appropriate mode shape and estimate the fundamental frequency.

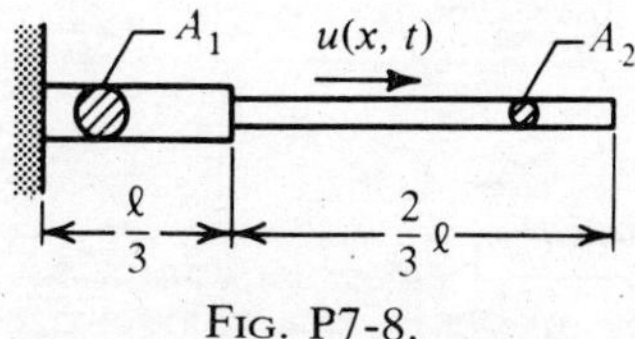

FIG. P7-8.

7-30 Estimate the fundamental frequency for the longitudinal vibration of the rod shown in Fig. P7-8. Assume the area $A_2/A_1 = 1/2$ and

$$\psi(x) = a_1 \sin \frac{\pi}{2\ell} x + a_2 \sin \frac{3\pi}{2\ell} x$$

7-31 Hysteretic damping was discussed in Example 28, Chap. 3. Consider the longitudinal vibration of the long rod shown in Fig. 7-2(a) with hysteretic damping. (**a**) Derive the equation of motion. (**b**) Find the formal solution by the method of separation of variables.

8

Nonlinear Systems

8-1 INTRODUCTION

Most engineering problems can be linearized. Some phenomena, however, cannot be predicted by linear theory. The shimmy of automobile wheels is an example. Engineers should be alert to the nonlinear phenomena around them. It is necessary to have some means to estimate the effect of the linearization and to know when a nonlinear analysis is required.

In comparing linear and nonlinear systems, the latter has two distinct attributes. (1) The superposition principle is not valid for nonlinear systems. For example, if the force applied to a system is doubled, its response is not necessarily doubled. In other words, the response of a nonlinear system may be dependent on both the frequency and the amplitude of the excitation. (2) A linear system has one and only one position of equilibrium. A nonlinear system could have more than one equilibrium, depending on the conditions of operation.

This chapter gives a simple general analysis of one-degree-of-freedom nonlinear systems in order to provide a perspective of the problem. The general analysis is necessary, although computers can be used to solve nonlinear initial value problems routinely. Computer solutions are specific. For example, when the parameters of a nonlinear equation are modified, the specific answer may be totally unexpected or vastly different from the previous answer.

The chapter consists of two main parts, conforming to the two approaches to nonlinear analysis. The beginning sections cover the geometric analysis and the graphical methods. The latter sections give some analytical methods to estimate the desired answer.*

* In nonlinear analysis, the technique employed depends on the information desired. No one method will give a general answer, such as the solution of linear ordinary differential equations with constant coefficients. Hence the sentence is phrased in the manner. We also avoid the statement that there is no general method to solve nonlinear differential equations and there are only few known exact solutions.

8-2 STABILITY AND EXAMPLES OF NONLINEAR SYSTEMS

The examples in Fig. 8-1 serve to illustrate the concept of stability of nonlinear systems. A system is *stable*† about an equilibrium position if its motion, due to an arbitrary small initial disturbance, does not increase with time. The vertical hanging pendulum in Fig. 8-1(a) is stable. Given a small initial displacement and/or velocity, the resulting motion will remain in the neighborhood of its equilibrium. In addition, if the resulting

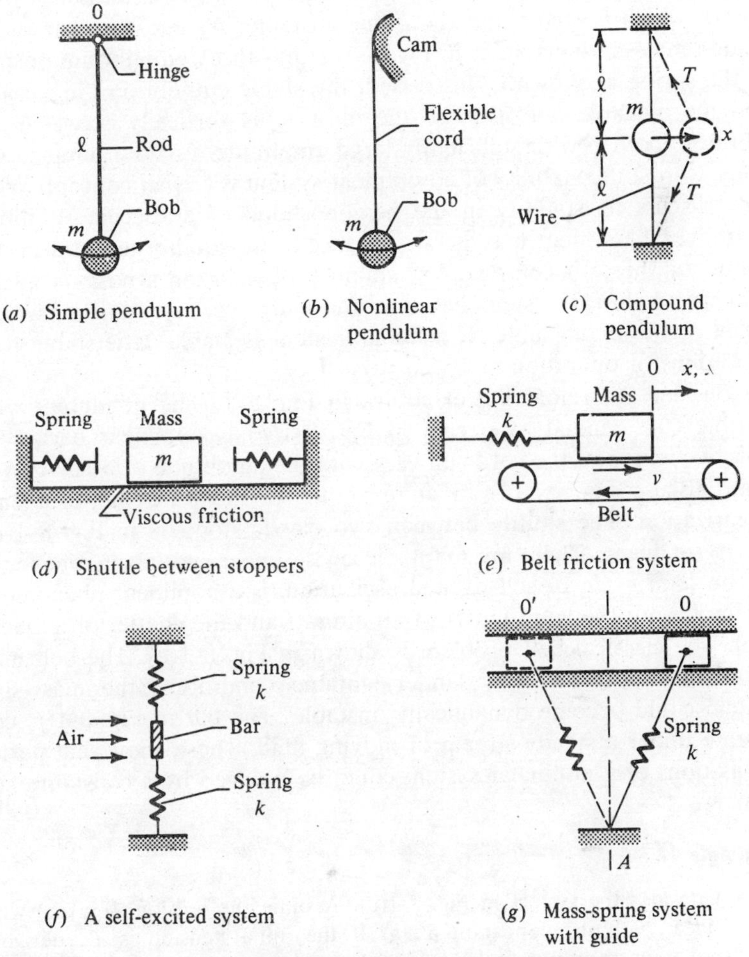

FIG. 8-1. *Examples of simple nonlinear systems.*

† There are many definitions of stability for nonlinear systems. See, for example, J. LaSalle and S. Lefschetz, *Stability by Liapunov's Direct Method*, Academic Press, New York, 1961.

motion tends towards the equilibrium position with increasing time, the system is *asymptotically stable*. The hanging pendulum is asymptotically stable if it is viscously damped.

A nonlinear system can have more than one equilibrium position. For small oscillations, the hanging pendulum is linear and has one and only one equilibrium. For large oscillations, the pendulum becomes nonlinear as illustrated in Eq. (2-14). For very large oscillations, a second equilibrium is when the bob m is vertically above the hinge o. This equilibrium is *unstable*, since a small disturbance will cause the motion to increase with time. Hence, the pendulum in Fig. 8-1(a) has two equilibrium positions, one of which is stable and the other unstable. As another example, the guided mass-spring system in Fig. 8-1(g) has three equilibrium positions. If the spring as shown is unstressed, the stable equilibriums are o and o' and the unstable one is when the mass m is vertically above A. If m vibrates about o with sufficiently large amplitude, it could jump to o'. In other words, the stability of a nonlinear system is a *local* concept. We can only discuss its stability in the neighborhood of a specific equilibrium position. The system may be stable about one equilibrium and unstable about another. In contrast, the stability of a linear system is a *global* concept. A linear system has one and only one equilibrium, which is either stable or unstable. If a linear system is stable, it is stable for all conditions of operation and vice versa.

Consider the remaining examples in Fig. 8-1. The pendulum with a flexible cord guided by a cam in Fig. 8-1(b) is nonlinear because the effective length of the cord will vary with the amplitude of oscillation. We shall discuss Fig. 8-1(c) in an example to illustrate the problem of linearization. The shuttle between two elastic stoppers in Fig. 8-1(d) is *piecewise* linear. There are many piecewise linear systems in engineering.

The *limit cycle*, or self-excited oscillation, is a nonlinear phenomenon. Examples are the wing-flutter of an aircraft and the chatter of a machine tool. The classical belt problem is shown in Fig. 8-1(e). The belt moves with constant velocity v. Under suitable conditions, the mass-spring system could become dynamically unstable. The bar in Fig. 8-1(f) could vibrate under a steady stream of moving fluid. These show that periodic oscillations of a nonlinear system could be induced by a constant excitation.

Example 1.

Consider the system in Fig. 8-1(c). Assume the tension T in the wire is linear with its elongation δ. (**a**) If the initial tension is T_o, derive the equation of motion and linearize the problem. (**b**) Repeat for $T_o = 0$.

Solution:

(**a**) The equation of motion of the mass m is

$$m\ddot{x} + F_x = 0$$

where F_x is the force in the x-direction. The length of the elongated wire is

$$(\ell+\delta)=\sqrt{\ell^2+x^2}$$

or

$$\frac{\delta}{\ell}=[\sqrt{1+(x/\ell)^2}-1]\simeq\frac{x^2}{2\ell^2}$$

The tension T in the wire is

$$T=T_o+AE\frac{\delta}{\ell}\simeq T_o+\frac{AE}{2\ell^2}x^2$$

where A = area and E = Young's modulus of the wire. The force F_x due to the displacement x is

$$F_x=2T\frac{x}{\ell+\delta}\simeq\frac{2Tx}{\ell}\simeq\frac{2}{\ell}\left(T_o+\frac{AE}{2\ell^2}x^2\right)x$$

Substituting this in the equation of motion gives

$$m\ddot{x}+\frac{2}{\ell}\left(T_o+\frac{AE}{2\ell^2}x^2\right)x=0$$

We have already used two approximations and the equation of motion is still nonlinear. Neglecting the x^3 term gives

$$m\ddot{x}+\frac{2}{\ell}T_o x=0$$

(**b**) If $T_o=0$, the equation of motion becomes

$$m\ddot{x}+\frac{AE}{\ell^3}x^3=0$$

If the x^3 term is neglected, the linearized model is that of an unrestrained rigid body, which is unrealistic. In conclusion, part **a** of the problem can be linearized if T_o is large, and part **b** of the problem cannot be linearized.

8-3 THE PHASE PLANE

The solution of a differential equation can be presented as a curve in a phase plane. A family of such curves gives a perspective of the general solution of the equation. We shall discuss the characteristics of a phase plane in this section.

The differential equation of a one-degree-of-freedom system has the general form

$$\ddot{x}+f(x,\dot{x},t)=0$$

If the independent variable t does not appear explicitly in the equation and only as a differential dt, the system is called *autonomous* and the equation becomes

$$\ddot{x}+f(x,\dot{x})=0 \tag{8-1}$$

In other words, the elastic and the damping properties of the system as described in $f(x,\dot{x})$ are time independent. It is then allowable to shift the time origin or to change the time scale without affecting the system behavior. We shall discuss only autonomous systems.

The *state* of the system at a given time t can be specified by the values of its displacement $x(t)$ and velocity $\dot{x}(t)$, which are called *state variables.* The vector $\{x \quad \dot{x}\}$ consisting of the state variables is called the *state vector.* Thus, the state of a system at a specific time is defined by its state vector. Instead of the state vector $\{x \quad \dot{x}\}$, the concept can be generalized* and the system described by the state vector $\{x_1 \quad x_2\}$, that is,

$$x_1=x \qquad \text{and} \qquad x_2=\dot{x} \tag{8-2}$$

Thus, Eq. (8-1) can be expressed as two simultaneous first-order differential equations

$$\begin{aligned}\dot{x}_1&=x_2\\ \dot{x}_2&=-f(x_1,x_2)\end{aligned} \tag{8-3}$$

or, as a single equation in which time is implicit.

$$\frac{dx_2}{dx_1}=\frac{-f(x_1,x_2)}{x_2} \tag{8-4}$$

A solution of the equation above can be portrayed in a *phase plane,* such as shown in Fig. 8-2. The coordinates are $(x,\dot{x})$ or (x_1,x_2). The solution curve is called a *phase trajectory.* Starting from the initial conditions $(x_o,\dot{x}_o)$, the trajectory in this illustration spirals inward towards the origin 0. Hence this trajectory shows that $(x,\dot{x})$ will decrease with increasing time and the system is asymptotically stable Another trajectory (in dash line) for the same system but with different initial conditions is shown in the figure to show the pattern of the trajectories. In other words, a family of trajectories in the phase plane gives a perspective of the general solution of the equation.

* State variables may be compared with generalized coordinates of discrete systems, although the latter specifies only the spatial coordinates. As a system can be described by more than one set of generalized coordinates, its state can also be described by more than one set of state variables. The choice of $\{x \quad \dot{x}\}$ here is obvious. Other choices, however, may be more convienient. There is no fixed and fast rules for choosing state variables. Furthermore, the discussion on generalized coordinates, such as coordinate transformation and principal coordinates, has its counterpart in state variables. The term *phase variables* usually denotes the set in which each variable is defined as the derivative of the preceding variable.

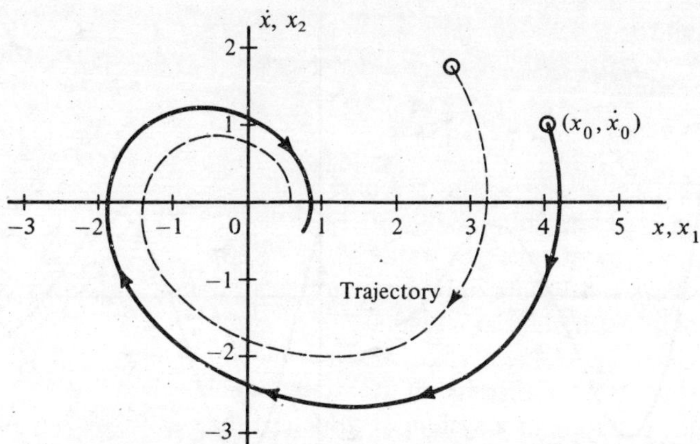

FIG. 8-2. *Phase plane and phase trajectories.*

Since time is implicit in the phase plane, the average velocity $\dot{x}_{av}$ is

$$\dot{x}_{av} = \Delta x/\Delta t \qquad \text{or} \qquad \Delta t = \Delta x/\dot{x}_{av} \tag{8-5}$$

where Δx and Δt are the corresponding displacement and time intervals. Thus, a time scale along the trajectory can be found, although the procedure is laborious.

Let us illustrate the phase trajectory with a familiar linear system. Techniques for the graphical construction of trajectories will be discussed in Sec. 8-5.

Example 2

Consider the free vibration of the system

$$\ddot{x} + 2\zeta\omega_n\dot{x} + \omega_n^2 x = 0$$

where $\zeta = 0.2$, $\omega_n = 0.8$, and $(x_o, \dot{x}_o) = (4,1)$. (**a**) Plot the solutions x vs t and $\dot{x}$ vs t. (**b**) Plot the corresponding values of $(x, \dot{x})$ in a phase plane.

Solution:

(**a**) From Eq. (2-34), the solution is

$$x = e^{-\zeta\omega_n t}(A_1 \cos \omega_d t + A_2 \sin \omega_d t)$$

where $\zeta\omega_n = 0.16$ and $\omega_d = \sqrt{1-\zeta^2}\omega_n = 0.78$. From the initial conditions, we obtain $(A_1, A_2) = (4, 2.09)$. It can be shown from Eq. (2-35) that

$$x = 4.51e^{-0.16t} \sin(0.78t + 62.4°)$$

$$\dot{x} = 3.61e^{-0.16t} \sin(0.78t + 163.9°)$$

The plots of x vs t and $\dot{x}$ vs t are shown in Fig. 8-3.

(**b**) The corresponding values of $(x, \dot{x})$ can be obtained from Fig. 8-3 and plotted in the phase plane in Fig. 8-2.

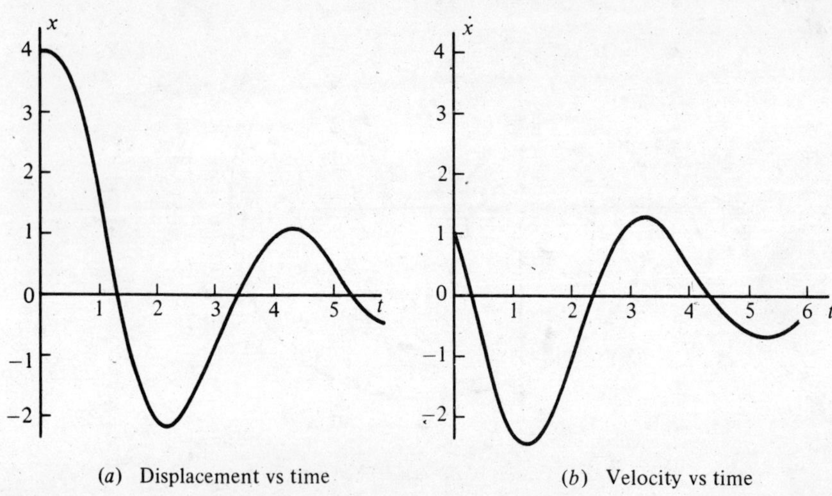

(*a*) Displacement vs time (*b*) Velocity vs time

FIG. 8-3. *Free vibration: Example 2.*

The *characteristics of phase trajectories* can be observed in Fig. 8-2. (1) The trajectory is always in a clockwise direction. If the position of $(x,\dot{x})$ is above the x axis, the velocity $\dot{x}$ is positive and the trajectory points towards increasing x. The converse is true if $(x,\dot{x})$ is below the x axis. (2) The trajectory always crosses the x axis perpendicularly. Along the x axis, we have $(x,\dot{x})=(x,0)$, that is, $\dot{x}=0$ or $\Delta x/\Delta t=0$. With $\Delta x=0$, the trajectory is normal to the x axis. (3) The trajectory crosses the $\dot{x}$ axis with any slope, which can be positive, negative, or zero. Along the $\dot{x}$ axis, we have $(x,\dot{x})=(0,\dot{x})$. Since $\dot{x}$ dictates the direction of the trajectory, only its direction is defined. (4) The trajectories do not intersect in the phase plane. A trajectory is a particular solution of a differential equation. If the solutions are unique, there is only one trajectory through each point in the phase plane. The trajectories will meet at a singular point. This will be discussed in the next section.

8-4 STABILITY OF EQUILIBRIUM

As illustrated in Fig. 8-1, a nonlinear system can have more than one equilibrium position, some of which may be stable while others unstable. In nonlinear analysis, it is mandatory (1) to define the equilibrium positions, and (2) to determine the stability of the system about each equilibrium. The condition $\{\dot{x} \quad \ddot{x}\}=\{\dot{x}_1 \quad \dot{x}_2\}=\{0 \quad 0\}$ is an equilibrium. From Eq. (8-4), an equilibrium can also be expressed as $dx_2/dx_1=0/0$. Hence, an equilibrium position is called a *singular point.* The other points in the phase plane are called *regular points.*

To find the equilibrium positions, let a one-degree-of-freedom system be described by two first-order differential equations

$$\dot{x}_1 = X_1(x_1,x_2) \qquad \text{and} \qquad \dot{x}_2 = X_2(x_1,x_2) \tag{8-6}$$

where X_1 and X_2 are functions of (x_1,x_2). Note that this is a generalization of Eq. (8-3). Let (s_1,s_2) be an equilibrium. It follows that

$$\dot{x}_1 = X_1(s_1,s_2) = 0 \qquad \text{and} \qquad \dot{x}_2 = X_2(s_1,s_2) = 0 \tag{8-7}$$

The values of s_1 and s_2 can be found from Eq. (8-7).

Let the origin of the equations be transferred to (s_1,s_2) by a coordinate translation as shown in Fig. 8-4.

$$x_1 = y_1 + s_1 \qquad \text{and} \qquad x_2 = y_2 + s_2 \tag{8-8}$$

Thus, Eq. (8-6) becomes

$$\dot{x}_1 = \dot{y}_1 = X_1(s_1 + y_1,\ s_2 + y_2)$$
$$\dot{x}_2 = \dot{y}_2 = X_2(s_1 + y_1,\ s_2 + y_2)$$

Expanding X_1 and X_2 about (s_1,s_2) by Taylor's series gives

$$\dot{y}_1 = a_{11}y_1 + a_{12}y_2 + \phi_1(y_1,y_2)$$
$$\dot{y}_2 = a_{21}y_1 + a_{22}y_2 + \phi_2(y_1,y_2)$$

$$\begin{bmatrix}\dot{y}_1\\ \dot{y}_2\end{bmatrix} = \begin{bmatrix}a_{11} & a_{12}\\ a_{21} & a_{22}\end{bmatrix}\begin{bmatrix}y_1\\ y_2\end{bmatrix} + \begin{bmatrix}\phi_1\\ \phi_2\end{bmatrix} \tag{8-9}$$

or

$$\{\dot{y}\} = A\{y\} + \{\phi\} \tag{8-10}$$

where (y_1,y_2) are the state variables referred to the origin at (s_1,s_2), the a's in the matrix A are constants, and the ϕ's are the nonlinear terms. Thus, the *linearized equations* of the system about the equilibrium at (s_1,s_2) are

$$\{\dot{y}\} = A\{y\} \tag{8-11}$$

The stability of the system *about* (s_1,s_2) can be examined by the trajectory of Eq. (8-11) in a phase plane. Applying linear theory, the

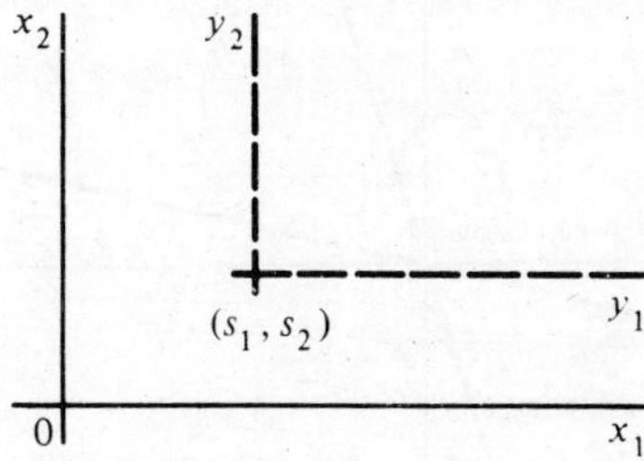

FIG. 8-4. *Coordinate translation.*

solution of Eq. (8-11) is

$$\{y\}=\{y_o\}e^{\lambda t} \tag{8-12}$$

where λ is an eigenvalue and $\{y_o\}$ a vector of constants. Substituting Eq. (8-12) in (8-11), factoring out the $e^{\lambda t}$ term, and simplifying, we get

$$[\lambda I - A]\{y_o\}=\{0\} \tag{8-13}$$

where I is a 2×2-unit matrix. The values of λ are found from the characteristic equation

$$\Delta(\lambda)=|\lambda I - A|=0 \tag{8-14}$$

Since A is a 2×2 matrix, there are two eigenvalues, which can be real, imaginary, or complex. From the theory of linear differential equations, the system is stable about the equilibrium (s_1,s_2) only if the roots λ_1 and λ_2 from Eq. (8-14) have zero or negative real parts.

Let us uncouple Eq. (8-11) in order to have a simple presentation of the phase trajectories about the equilibrium at (s_1,s_2). Assume (1) a similarity transformation

$$\{y\}=B\{z\} \qquad \text{or} \qquad \begin{bmatrix} y_1 \\ y_2 \end{bmatrix}=\begin{bmatrix} b_{11} & b_{12} \\ b_{21} & b_{22} \end{bmatrix}\begin{bmatrix} z_1 \\ z_2 \end{bmatrix} \tag{8-15}$$

as illustrated in Fig. 8-5, where B is a nonsingular matrix, and (2) the equations in the variables (z_1,z_2) are uncoupled. Substituting Eq. (8-15) in (8-11) and premultiplying the resulting equation by B^{-1}, the uncoupled equations are

$$\{\dot{z}\}=B^{-1}AB\{z\} \tag{8-16}$$

or

$$\begin{bmatrix} \dot{z}_1 \\ \dot{z}_2 \end{bmatrix}=\begin{bmatrix} \lambda_1 & 0 \\ 0 & \lambda_2 \end{bmatrix}\begin{bmatrix} z_1 \\ z_2 \end{bmatrix} \tag{8-17}$$

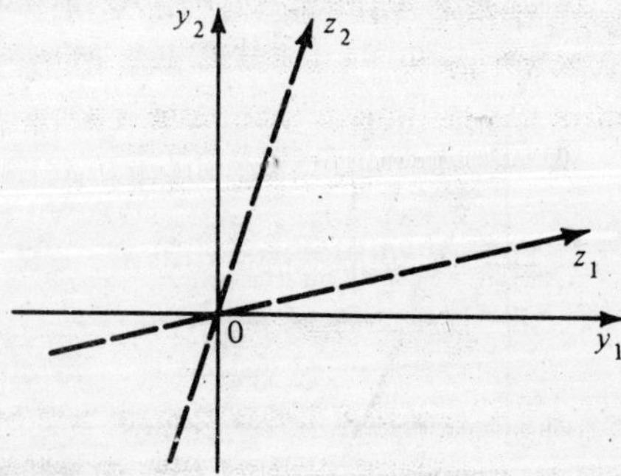

FIG. 8-5. *Similarity transformation.*

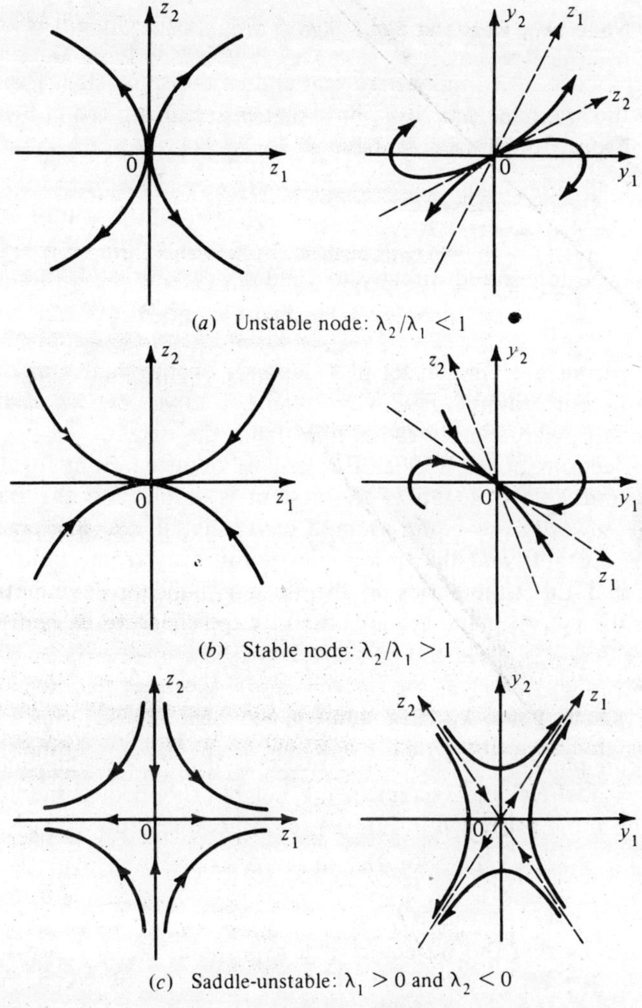

(*a*) Unstable node: $\lambda_2/\lambda_1 < 1$

(*b*) Stable node: $\lambda_2/\lambda_1 > 1$

(*c*) Saddle-unstable: $\lambda_1 > 0$ and $\lambda_2 < 0$

FIG. 8-6. *Trajectories for $\lambda_{1,2}$ real.*

or

$$\dot{z}_1 = \lambda_1 z_1 \qquad \text{and} \qquad \dot{z}_2 = \lambda_2 z_2 \tag{8-18}$$

The stability of the system about an equilibrium (s_1, s_2) can be examined in the (y_1, y_2) or (z_1, z_2) phase plane,* such as shown in Fig. 8-6. In other words, the trajectories are mapped from one plane to the other. We shall consider the stability of the four cases to follow.

* The eigenvalues are independent of the similarity transformation and the system stability can be examined in either phase plane. The eigenvalues of Eq. (8-11) and (8-16) are identical because the matrices A and $B^{-1}AB$ possess the same determinant. Recalling $|B^{-1}| = |B|^{-1}$, we have

$$|B^{-1}AB| = |B^{-1}||A||B| = |B|^{-1}|A||B| = |A|$$

Case 1. Node: $\lambda_{1,2}$ Real and Same Sign

A *node* occurs if λ_1 and λ_2 are real and of the same sign. Examples of nodes in the (z_1,z_2) and (y_1,y_2) phase planes are illustrated in Figs. 8-6(a) and (b). From Eq. (8-18), we have

$$\frac{dz_2}{dz_1}=\frac{\lambda_2 z_2}{\lambda_1 z_1} \qquad \text{or} \qquad \frac{dz_2}{z_2}=\frac{\lambda_2}{\lambda_1}\frac{dz_1}{z_1}$$

which can be integrated directly to yield

$$z_2 = Cz_1^{\lambda_2/\lambda_1} \tag{8-19}$$

where C = constant. The exact plot depends on the values of C and the ratio λ_2/λ_1. For example, if $\lambda_2/\lambda_1=2$ and $C=1$, we get $z_2=z_1^2$. If $\lambda_2=\lambda_1$, the trajectories are simply radial lines from the origin.

The directions of the trajectories can be deduced from Eq. (8-18). If both λ_1 and λ_2 are negative, the system is stable and the trajectories converge towards the equilibrium. Conversely, if $\lambda_{1,2}$ are positive, the system is unstable and the trajectories point away from equilibrium.

Note that the trajectories in the (z_1,z_2) plane are symmetrical, but those in the (y_1,y_2) plane are governed by the characteristics discussed in Sec. 8-3.*

Case 2. Saddle Point: $\lambda_{1,2}$ Real and Opposite Sign

A *saddle point* occurs if $\lambda_{1,2}$ are real and of opposite sign. Since one of the λ's is positive, the system in the neighborhood of a saddle point is always unstable as shown in Fig. 8-6(c). Since the ratio λ_1/λ_1 is negative, Eq. (8-19) can be expressed as

$$z_2 = Cz_1^{-|\lambda_2/\lambda_1|} \quad \text{or} \quad z_2 z_1^{|\lambda_2/\lambda_1|}=C \tag{8-20}$$

This indicates that the trajectory is a hyperbola.

Case 3. Vortex: $\lambda_{1,2}$ Imaginary

A *vortex*, or *center*, occurs if $\lambda_{1,2}=\pm j\beta$ are imaginary, where $j=\sqrt{-1}$ and β = constant. From Eq. (8-18) we get

$$\dot{z}_1 = j\beta z_1 \qquad \text{and} \qquad \dot{z}_2 = -j\beta z_2$$

or

$$z_1 = z_{10}e^{j\beta t} \qquad \text{and} \qquad z_2 = z_{20}e^{-j\beta t}$$

* Since (y_1,y_2) are obtained from (x_1,x_2) by a coordinate *translation*, the x's and y's have the same physical interpretation. For example, if x_1 denotes a displacement, and so does y_1. Similarly, if x_2 denotes a velocity, and so does y_2. Analogous to principal coordinates, (z_1,z_2) do not necessarily have such physical interpretations. Thus, the characteristics of phase trajectories, as discussed in the last section, do not apply to the plots in the (z_1,z_2) phase plane.

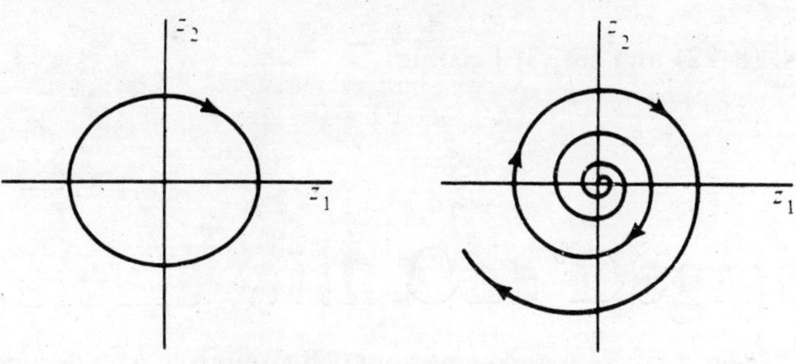

(a) Vortex: $\lambda_{1,2}$ imaginary (b) Unstable focus: $\lambda_{1,2}$ complex

FIG. 8-7. *Phase trajectories for $\lambda_{1,2}$ imaginary and complex.*

where (z_{10}, z_{20}) = constants. The factor $e^{\pm j\beta t}$ represents a harmonic motion of unit magnitude with circular frequency β as discussed in Sec. 1-5. Hence, the resulting motion in the (z_1, z_2) plane is a combination of two harmonic motions, which is an ellipse as shown in Fig. 8-7(a). The motion in the neighborhood of equilibrium in the (z_1, z_2) or (y_1, y_2) plane forms a closed curve. By definition, the system is stable.

Case 4. Focus: $\lambda_{1,2}$ Complex Conjugates

A *focus* occurs if $\lambda_{1,2} = \alpha \pm j\beta$ are complex conjugates, where $j = \sqrt{-1}$ and α and β are constants. From Eq. (8-18), we get

$$\dot{z}_1 = (\alpha + j\beta)z_1 \quad \text{and} \quad \dot{z}_2 = (\alpha - j\beta)z_2 \tag{8-21}$$

or

$$z_1 = (z_{10}e^{\alpha t})e^{j\beta t} \quad \text{and} \quad z_2 = (z_{20}e^{\alpha t})e^{-j\beta t}$$

where (z_{10}, z_{20}) = constants. The factor $e^{\pm j\beta t}$ represents a harmonic motion of unit magnitude with circular frequency β as before. If $\alpha > 0$, $e^{\alpha t}$ increases exponentially with time t and the trajectory in the (z_1, z_2) plane is a divergent logarithmic spiral as shown in Fig. 8-7(b). Hence for $\alpha > 0$, the equilibrium is unstable. If $\alpha < 0$, the trajectory is a convergent logarithmic spiral and the system is asymptotically stable.

In summary, the stability about an equilibrium at (s_1, s_2) can be examined from the roots of the characteristic equation. Substituting $A = [a_{ij}]$ from Eq. (8-9) in (8-14), the characteristic equation and the roots $\lambda_{1,2}$ are

$$\lambda^2 - (a_{11} + a_{22})\lambda + (a_{11}a_{22} - a_{12}a_{21}) = 0 \tag{8-22}$$

$$\lambda_{1,2} = \tfrac{1}{2}\{(a_{11} + a_{22}) \pm [(a_{11} + a_{22})^2 - 4(a_{11}a_{22} - a_{12}a_{21})]^{1/2}\} \tag{8-23}$$

For convenience, we introduce the parameters u and v as

$$u = (a_{11} + a_{22}) \quad \text{and} \quad v = (a_{11}a_{22} - a_{12}a_{21}) \tag{8-24}$$

Thus, Eqs. (8-22) and (8-23) become

$$\lambda^2 - u\lambda + v = 0 \tag{8-25}$$

$$\lambda_{1,2} = \tfrac{1}{2}(u \pm \sqrt{u^2 - 4v}) \tag{8-26}$$

The four cases above are summarized as follows:

Case 1. Node: $\lambda_{1,2}$ real and same sign. This requires $u^2 > 4v$ and $v > 0$, with $u^2 = 4v$ as a limiting case. The node is stable if $u < 0$ and unstable if $u > 0$.

Case 2. Saddle point: $\lambda_{1,2}$ real and opposite sign. This requires $u^2 > 4v$ and $v < 0$. This is always unstable.

Case 3. Center: $\lambda_{1,2}$ imaginary. This requires $u^2 < 4v$ and $u = 0$. By definition, the system is stable.

Case 4. Focus: $\lambda_{1,2}$ complex conjugates. This requires $u^2 < 4v$ and $u \gtrless 0$. The system is stable if $u < 0$ and unstable if $u > 0$.

The stability of the cases enumerated are summarized and mapped in the (u,v) plane as shown in Fig. 8-8. The demarcation between the nodes and foci is given by

$$u^2 = 4v \tag{8-27}$$

The center is mapped in the positive real axis. The stable regions are given in the fourth quadrant of the plot.

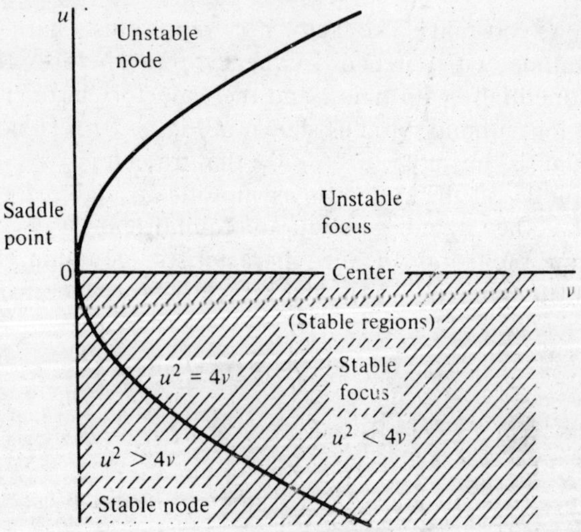

FIG. 8-8. *Stability of equilibrium: (u,v) defined in Eq. (8-24).*

Example 3

Find the equilibrium positions of the simple pendulum in Fig. 8-1(a) and the stability about each equilibrium.

Solution:

The equation of motion of the pendulum is

$$J\ddot{\theta} + mg\ell \sin\theta = 0$$

where J is the mass moment of inertia about the hinge o. Let $x_1 = \theta$. From Eq. (8-3) we get

$$\dot{x}_1 = x_2 \qquad \text{and} \qquad \dot{x}_2 = -(mg\ell/J)\sin x_1$$

From Eq. (8-7), the equilibrium positions are

$$\dot{x}_1 = x_2 = 0 \qquad \text{and} \qquad \dot{x}_2 = -(mg\ell/J)\sin x_1 = 0$$

This gives $x_2 = 0$ and $x_1 = n\pi$, for $n = 1,2,\ldots$. The equilibrium positions are $(s_1,s_2) = (n\pi,0)$. We obtain from Eq. (8-8)

$$x_1 = n\pi + y_1 \qquad \text{and} \qquad x_2 = 0 + y_2 = y_2$$

Hence the system equations from Eq. (8-9) are

$$\dot{y}_1 = y_2 \qquad \text{and} \qquad \dot{y}_2 = -(mg\ell/J)\sin(n\pi + y_1)$$

Note that (1) $\sin(n\pi + y_1) = \sin y_1$ for $n =$ even, (2) $\sin(n\pi + y_1) = -\sin y_1$ for $n =$ odd, and (3) $\sin y_1 \simeq (y_1 - y_1^3/3!)$.

For $n =$ even, the system equations from Eq. (8-9) are

$$\begin{bmatrix}\dot{y}_1\\ \dot{y}_2\end{bmatrix} = \begin{bmatrix}0 & 1\\ -mg\ell/J & 0\end{bmatrix}\begin{bmatrix}y_1\\ y_2\end{bmatrix} + \begin{bmatrix}0\\ y_1^3/6\end{bmatrix}\frac{mg\ell}{J}$$

Thus, the elements in the square matrix are $a_{11} = 0$, $a_{12} = 1$, $a_{21} = -mg\ell/J$, and $a_{22} = 0$. From Eq. (8-24) we get

$$u = 0 \qquad \text{and} \qquad v = mg\ell/J$$

It can be shown that the equilibrium at $(n\pi,0)$ for n even is a center as shown in Case 3.

Similarly, for $n =$ odd, we obtain from Eq. (8-9)

$$\begin{bmatrix}\dot{y}_1\\ \dot{y}_2\end{bmatrix} = \begin{bmatrix}0 & 1\\ mg\ell/J & 0\end{bmatrix}\begin{bmatrix}y_1\\ y_2\end{bmatrix} - \begin{bmatrix}0\\ y_1^3/6\end{bmatrix}\frac{mg\ell}{J}$$

Hence the elements in the square matrix are $a_{11} = 0$, $a_{12} = 1$, $a_{21} = mg\ell/J$, and $a_{22} = 0$. From Eq. (8-24) we get

$$u = 0 \qquad \text{and} \qquad v = -mg\ell/J$$

It can be shown that the equilibrium at $(n\pi,0)$ for $n =$ odd is a saddle point, which is always unstable, as shown in Case 2.

Example 4

A hydraulic machine is modeled by the equation

$$\ddot{x}-(0.1-3\dot{x}^2)\dot{x}+x+x^2=0$$

(**a**) Find the equilibrium positions. (**b**) Determine the types of stability. (**c**) Sketch the trajectories in the neighborhood of each equilibrium.

Solution:

(**a**) Let $x=x_1$. The equation can be expressed as

$$\dot{x}_1=x_2 \qquad \text{and} \qquad \dot{x}_2=(0.1-3x_2^2)x_2-x_1-x_1^2$$

The equilibrium positions are determined from Eq. (8-7) by equating $\dot{x}_1=\dot{x}_2=0$, that is,

$$x_2=0 \qquad \text{and} \qquad x_1(1+x_1)=0$$

Hence the equilibriums are at (0, 0) and (−1, 0).

(**b**) For the equilibrium at (0,0), we have from Eq. (8-8)

$$x_1=0+y_1=y_1 \qquad \text{and} \qquad x_2=0+y_2=y_2$$

The system equations in the form of Eq. (8-9) are

$$\begin{bmatrix}\dot{y}_1\\ \dot{y}_2\end{bmatrix}=\begin{bmatrix}0 & 1\\ -1 & 0.1\end{bmatrix}\begin{bmatrix}y_1\\ y_2\end{bmatrix}-\begin{bmatrix}0\\ y_1^2+3y_2^3\end{bmatrix}$$

The elements of the square matrix above are $a_{11}=0$, $a_{12}=1$, $a_{21}=-1$, and $a_{22}=0.1$. From Eq. (8-24) we get

$$u=0.1 \qquad \text{and} \qquad v=1.0$$

It can be shown that the equilibrium at (0,0) is an unstable focus as shown in Case 4. The trajectory is shown in Fig. 8-9.

For the equilibrium at (−1,0), Eq. (8-8) gives

$$x_1=-1+y_1 \qquad \text{and} \qquad x_2=y_2$$

The system equations from Eq. (8-9) are

$$\dot{y}_1=y_2 \qquad \text{and} \qquad \dot{y}_2=(0.1-3y_2^2)y_2-(-1+y_1)-(-1+y_1)^2$$

or,

$$\begin{bmatrix}\dot{y}_1\\ \dot{y}_2\end{bmatrix}=\begin{bmatrix}0 & 1\\ 1 & 0.1\end{bmatrix}\begin{bmatrix}y_1\\ y_2\end{bmatrix}-\begin{bmatrix}0\\ y_1^2+3y_2^3\end{bmatrix}$$

The elements in the square matrix above are $a_{11}=0$, $a_{12}=1$, $a_{21}=1$, $a_{22}=0.1$. From Eq. (8-24) we get

$$u=0.1 \qquad \text{and} \qquad v=-1$$

It can be shown that the equilibrium at (−1,0) is a saddle point, which is unstable, as shown in Case 2.

(**c**) To plot the trajectories in the (y_1,y_2) plane in Fig. 8-9, it is necessary to map the $z_{1,2}$ axes in the (y_1,y_2) plane. From Eq. (8-26) we have

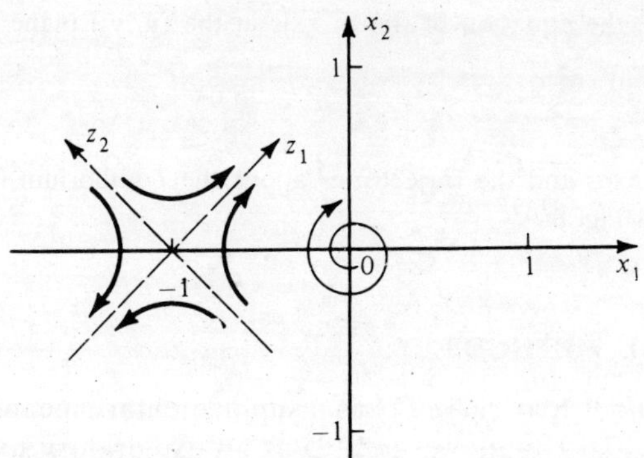

FIG. 8-9. *Stability of equilibrium; Example 4.*

$\lambda_{1,2} = 1.05$ and -0.95. The matrix B and the similarity transformation $\{y\} = B\{z\}$ in Eq. (8-15) can be obtained by the method shown in Example 6, App. A. The lambda matrix $[f(\lambda)]$ of matrix A in Eq. (8-11) is

$$[f(\lambda)] = \begin{bmatrix} \lambda - a_{11} & -a_{12} \\ -a_{21} & \lambda - a_{22} \end{bmatrix}$$

The adjoint matrix $F(\lambda)$ of $[f(\lambda)]$ is

$$F(\lambda) = \begin{bmatrix} \lambda - a_{22} & a_{12} \\ a_{21} & \lambda - a_{11} \end{bmatrix}$$

Any nonzero column of $F(\lambda_i)$ can be used for a column $\{b_{1i} \quad b_{2i}\}$ in the matrix B, that is,

$$\frac{b_{2i}}{b_{1i}} = \frac{a_{21}}{\lambda_i - a_{22}} = \frac{\lambda_i - a_{11}}{a_{12}} \triangleq \mu_i \qquad \text{for } i = 1,2 \tag{8-28}$$

Hence the similarity transformation is*

$$\begin{bmatrix} y_1 \\ y_2 \end{bmatrix} = \begin{bmatrix} b_{11} & b_{12} \\ b_{21} & b_{22} \end{bmatrix} \begin{bmatrix} z_1 \\ z_2 \end{bmatrix} = \begin{bmatrix} 1 & 1 \\ \mu_1 & \mu_2 \end{bmatrix} \begin{bmatrix} z_1 \\ z_2 \end{bmatrix} \tag{8-29}$$

Substituting $\lambda_{1,2} = 1.05$ and -0.95 in Eq. (9-28) and for the values $a_{11} = 0$, $a_{12} = 1$, $a_{21} = 1$, and $a_{22} = 0.1$, we get

$$\begin{bmatrix} y_1 \\ y_2 \end{bmatrix} = \begin{bmatrix} 1 & 1 \\ 1.05 & -0.95 \end{bmatrix} \begin{bmatrix} z_1 \\ z_2 \end{bmatrix}$$

The equation of the z_1 axis in the (y_1, y_2) plane is $z_2 = 0$. Thus, from the equation above we get

$$\frac{y_2}{y_1} = \mu_1 = 1.05$$

* Note the similarity between Eqs. (4-16) and (8-28), and Eqs. (4-20) and (8-29)

Similarly, the equation of the z_2 axis in the (y_1,y_2) plane is

$$\frac{y_2}{y_1} = \mu_2 = -0.95$$

The $z_{1,2}$ axes and the trajectories about the equilibrium $(-1,0)$ are as shown in Fig. 8-9.

8-5 GRAPHICAL METHODS

Graphical methods may serve (1) as a supplementary means for solving nonlinear differential equations, and (2) as an exploratory tool to obtain the perspective of a problem. We shall discuss the isocline and the Pell's method.

Let us first describe the basis of the graphical methods by plotting the trajectories from Example 2. The equation of motion

$$\ddot{x} + 2\zeta\omega_n\dot{x} + \omega_n^2 x = 0$$

can be expressed in the form of Eq. (8-4) as

$$\frac{d\dot{x}}{dx} = \frac{-2\zeta\omega_n\dot{x} - \omega_n^2 x}{\dot{x}} = F(x,\dot{x}) \tag{8-30}$$

The slope of the trajectory in the $(x,\dot{x})$ phase plane is $F(x,\dot{x})$, that is, $F(x,\dot{x})$ is tangential to the trajectory at the point $(x,\dot{x})$. Substituting the values of $(x_0,\dot{x}_0)$ in $F(x,\dot{x})$ gives the slope at $(x_0,\dot{x}_0)$. A small tangential line segment through $(x_0,\dot{x}_0)$ can be drawn as shown in Fig. 8-10. In other words, the trajectory is extrapolated from $(x_0,\dot{x}_0)$ to $(x,\dot{x})$ by following this line segment. Using the end point of the previous step as the beginning point of the next step, a new slope can be calculated for additional

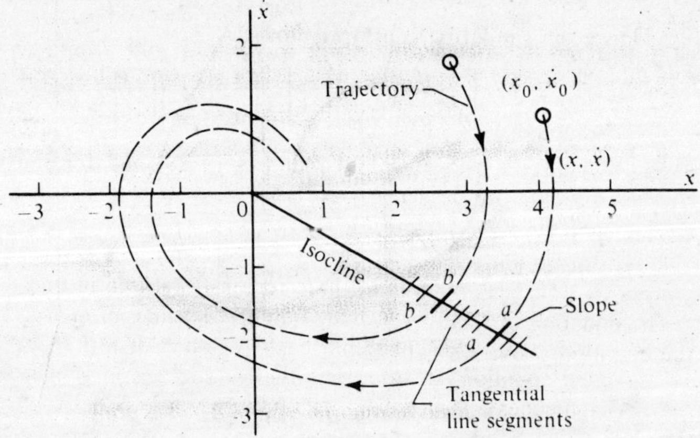

Fig. 8-10: *Isocline method; linear system in Example 2.*

extrapolations. The process is repeated to obtain a trajectory. The phase plane can be filled with a family of trajectories to show the pattern of the solution of the differential equation.

Isocline Method

An *isocline* in a phase plane is the locus joining the points of trajectories having the same slope. For example, the tangential line segments a-a and b-b of the trajectories in Fig. 8-10 have the same slope. A locus joining these segments is called an iso ine. All the line segments drawn on the isocline have the same slope as shown in the figure.

Consider an autonomous system

$$\frac{d\dot{x}}{dx} = F(x,\dot{x}) \tag{8-31}$$

where $F(x,\dot{x})$ is the slope of the phase trajectory at $(x,\dot{x})$. An isocline is defined by the equation

$$F(x,\dot{x}) = C \tag{8-32}$$

where C = constant. This can be plotted in the $(x,\dot{x})$ phase plane to obtain one isocline. A different value of C gives another isocline. Thus, a family of isoclines can be plotted.

The procedure is (1) to plot the isoclines in the phase plane to cover the area of interest, (2) to draw the line segments on each isocline to indicate the respective slopes, and (3) to use the slopes provided to guide the extrapolation in order to sketch the trajectories.

Example 5

A piecewise linear system consisting of a mass m shunting between two stoppers is shown in Fig. 8-11(a). The springs k and dampers c are linear, the clearance between the stoppers is 2Δ, and $\Delta = 1$. When m is in contact with the right stopper, the equation of motion is

$$\ddot{x} + 0.2\dot{x} + (x - \Delta) = 0$$

Plot the trajectory for the initial conditions $(x_0,\dot{x}_0) = (2.5,3.2)$.

Solution:

Let us divide the phase plane into three regions as shown in Fig. 8-11(b). Due to symmetry, we need to calculate only the isoclines of regions (1) and (2). Since the system is linear within each region, Eq. (8-32) must be a linear algebraic equation. In other words, the isoclines in a linear region must be straight lines*.

* It should be noted that if Eq. (8-32) is a nonlinear algebraic equation, the corresponding isocline is not a straight line.

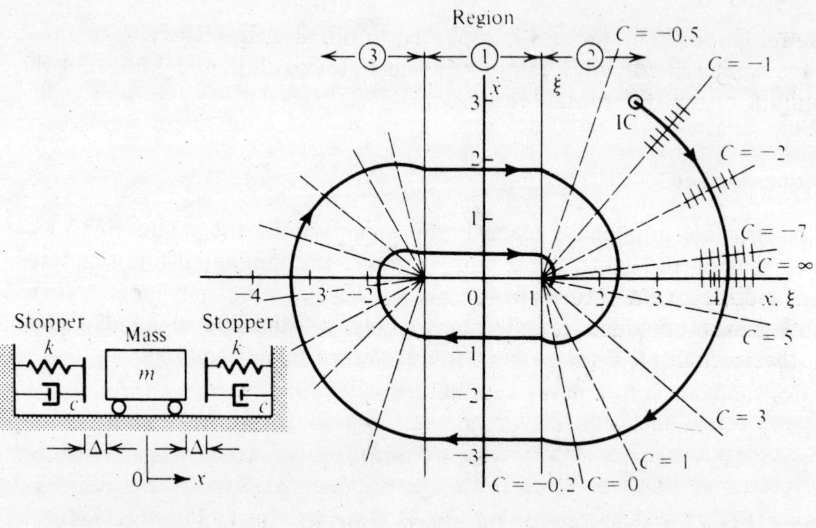

FIG. 8-11. *Isocline method: Example 5.*

In region (1) for $-\Delta < x < \Delta$, the equation of motion is

$$\ddot{x} = 0 \qquad \text{or} \qquad \frac{d\dot{x}}{dx} = 0$$

Thus, $F(x,\dot{x}) = 0$ and the slope of the trajectory is always zero. The mass moves across region (1) with constant velocity.

In region (2) for $x > \Delta$, the equation of motion is

$$\ddot{\xi} + 0.2\dot{\xi} + \xi = 0$$

where $\xi = (x - \Delta)$. From Eq. (8-32) we get $F(\xi,\dot{\xi}) = -(0.2\dot{\xi} + \xi)/\dot{\xi} = C$, or

$$(C + 0.2)\dot{\xi} = -\xi$$

The isoclines for different values of C are plotted as dash lines in Fig. 8-11(b). The short line segments along an isocline indicate the slope of the trajectories as specified by the value of C.

Starting from the initial conditions IC, a trajectory can be drawn by following the slopes provided by the isoclines. Note that the line segments are not necessarily normal to the isoclines. In fact, the isoclines for the zero slope $C = 0$ are not perpendicular to the ξ axis.

Pell's Method

Many graphical methods can be used to construct trajectories in a phase plane.* Pell's method considers a second-order autonomous equation of the form

$$\ddot{x} + \phi(\dot{x}) + \psi(x) = 0 \tag{8-33}$$

or

$$\frac{d\dot{x}}{dx} = \frac{-[\phi(\dot{x}) + \psi(x)]}{\dot{x}} \tag{8-34}$$

where $\phi(\dot{x})$ and $\psi(x)$ are known nonlinear functions. Note that the slope of the trajectory is $d\dot{x}/dx$. The Pell's method gives a procedure (1) to construct the slope of a trajectory, and (2) to extrapolate from a point at $(x,\dot{x})$ according to the slope to obtain the trajectory.

Pell's method first constructs graphically the quantity $\dot{x}/[\phi(\dot{x}) + \psi(\dot{x})]$ for the given point at $(x,\dot{x})$. Since the slope of the trajectory from Eq. (8-34) is $-[\phi(\dot{x}) + \psi(x)]/\dot{x}$, from simple geometry, a line segment perpendicular to $\dot{x}/[\phi(\dot{x}) + \psi(x)]$ is the slope of the trajectory.

Referring to Fig. 8-12, the detail construction is as follows: (1) Plot the given $\psi(x)$ versus x as shown in the fourth quadrant. (2) Plot the given

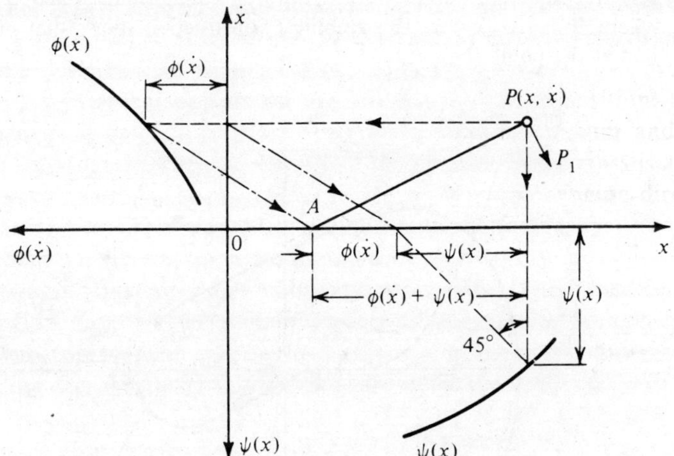

FIG. 8-12. *Pell's method; graphical construction of trajectory.*

* The isocline method is a general method but, nonetheless, rather tedious. The delta method due to Jacobsen is widely used. The Lienard method may be considered as a special case of Pell's method. To our experience, Pell's method seems easier and faster.

L. S. Jacobsen, "On a General Method of Solving Second-order Differential Equations by Phase Plane Displacements," *J. Appl. Math.*, vol. 19, 1952, pp. 543–553.

A. Lienard, "Etudes des Oscillations Entretenues," *Rev. Gen. de l'Elec.*, 23, 1928, pp. 901–12, 946–54.

W. H. Pell, "Graphical Solution of Single-degree-of-freedom Vibration Problems with Arbitrary Damping and Restoring forces," *J. Appl. Mech.*, vol. 24, 1957, pp. 311–312.

$\phi(\dot{x})$ versus $\dot{x}$ as shown in the second quadrant. (3) For the given point P at $(x,\dot{x})$, follow the dash line to obtain $\psi(x)$ and the dotted line for $\phi(\dot{x})$ as indicated in the figure. (4) $\phi(\dot{x})$ is added to $\psi(x)$ on the x axis graphically by means of the parallel dash-dot lines. This gives $[\phi(\dot{x})+\psi(x)]$. The slope of the line PA is $\dot{x}/[\phi(\dot{x})+\psi(x)]$. (5) From P draw a small line segment PP_1 perpendicular to the line PA. This is a small line segment of the trajectory. (6) Using P_1 as a new initial point, the process is repeated to obtain the trajectory.

Example 6

Referring to Fig. 8-1(d), when the mass m is in contact with the right spring, the equation of motion is

$$\ddot{x}+0.2\dot{x}+(x-\Delta)^2=0$$

where x is measured from the center line of the system and the total clearance is 2Δ. Let $\Delta=1$. Find the trajectory for the initial conditions $(x,\dot{x})=(3,-2)$.

Solution:

The function $\phi(\dot{x})=0.2\dot{x}$ is plotted as a dash-dot line and $\psi(x)=(x-\Delta)^2$ as a dash line in Fig. 8-13. The trajectory is as shown. The graphical construction is relatively simple and is omitted. Starting from the initial conditions IC, the trajectory spirals inward towards equilibrium, which can be anywhere for $-\Delta<x<\Delta$. The graphical construction shows that m comes to rest at the point B.

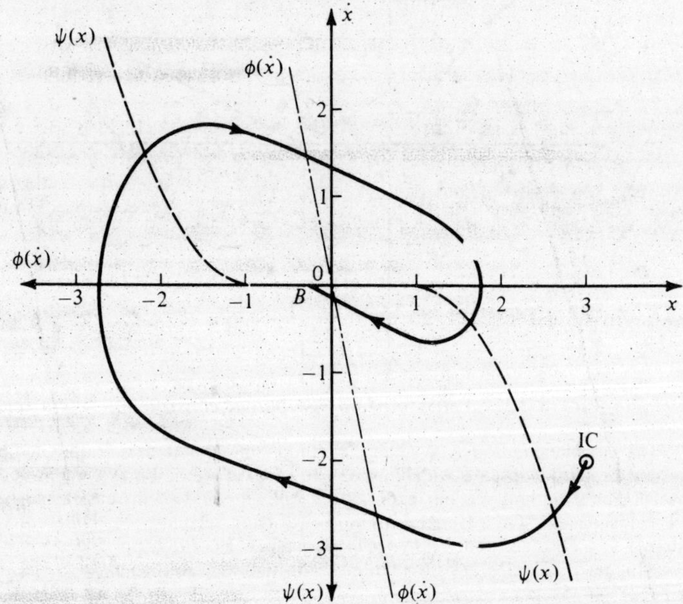

FIG. 8-13. *Phase trajectory by Pell's method: Example 6.*

8-6 SELF-EXCITED OSCILLATIONS

Shimmy of automobile wheels was mentioned as an example of self-excited oscillation or limit cycle.* It differs from (1) the harmonic response of a linear system, and (2) the free vibration of a conservative system, both of which are also periodic.

Consider the harmonic response of a linear system. From Eqs. (2-18) and (2-38), we have

$$m\ddot{x} + c\dot{x} + kx = F \sin \omega t$$
$$x = X \sin(\omega t - \phi) \tag{8-35}$$

The frequency ω of the harmonic response $x(t)$ is the excitation frequency, regardless of the natural frequency of the system. The amplitude X is proportional to F and is dependent on ω as shown in Eq. (2-39). Hence both the frequency and the amplitude of $x(t)$ are dictated by the excitation $F \sin \omega t$.

Consider the free vibration of a conservative system, such as a pendulum. It vibrates at its natural frequency but the amplitude is determined by the initial conditions. In other words, the trajectories due to different initial conditions form a family of concentric closed curves in a $(x,\dot{x})$ phase plane, one of which is illustrated in Fig. 8-7(a).

A stable limit cycle is shown in Fig. 8-14. In contrast with the forced vibration in Eq. (8-35), the nonlinear system is autonomous as shown in

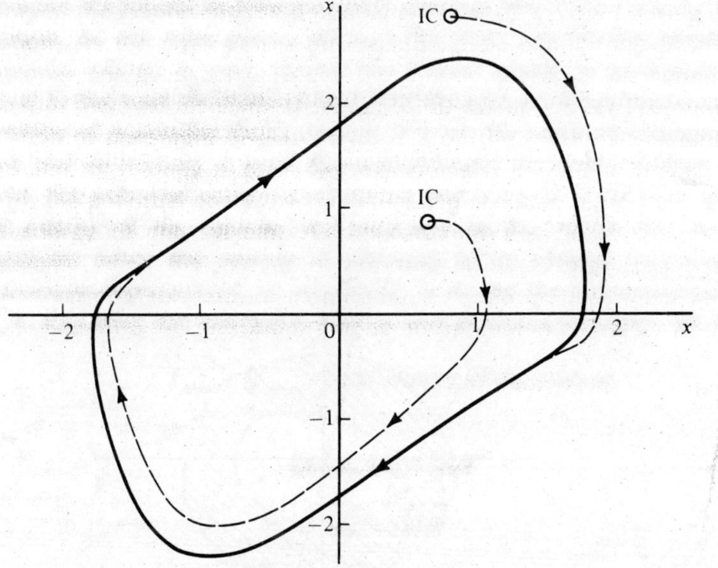

FIG. 8-14. *Limit cycle:* $\ddot{x} + (x^2 - 1)\dot{x} + x = 0$.

*A limit cycle may be stable, unstable, or semistable and a system may possess more than one limit cycles. We shall limit our attention to systems with one stable limit cycle.

Eq. (8-1). In contrast with the conservative linear system, a limit cycle is a nonconservative and nonlinear phenomenon. As illustrated in Fig. 8-14, there is only one closed trajectory, which is independent of the initial conditions. The initial conditions may be inside or outside of the closed trajectory. Once started, the system will eventually "lock" itself into a limit cycle with constant amplitude and frequency. A *limit cycle* occurs if, over one cycle, the net energy input from the excitation is equal to the energy dissipation within the system.

Self-excitation is due to an excitation that is a function of the motion itself, such as displacement and/or velocity. If the excitation force in Eq. (8-35) is $F\dot{x}$, we get

$$m\ddot{x} + c\dot{x} + kx = F\dot{x} \tag{8-36}$$

or

$$m\ddot{x} + (c - F)\dot{x} + kx = 0$$

The equation is autonomous. If $(c-F)>0$, the system is stable. If $(c-F)<0$, we have *negative damping* and the amplitude increases with each oscillation. If $(c-F)=0$, sustained oscillation is possible. In any case, the oscillations occur near the natural frequency of the system.

To illustrate the conditions for a stable limit cycle, assume the damping term in Eq. (8-36) is $(cx^2)\dot{x}$. Hence the equation of motion is

$$m\ddot{x} + (cx^2)\dot{x} + kx = F\dot{x}$$

or

$$m\ddot{x} + c(x^2 - F/c)\dot{x} + kx = 0 \tag{8-37}$$

The damping force is governed by two tendencies, each dominating a range of operation. If $x^2 < F/c$, the term F/c dominates and the damping is negative. Hence the system near equilibrium is unstable as illustrated in Fig. 8-7(*b*). If $x^2 > F/c$, the damping is positive. Thus, for large oscillations, the phase trajectory spirals inward as illustrated in Fig. 8-2.

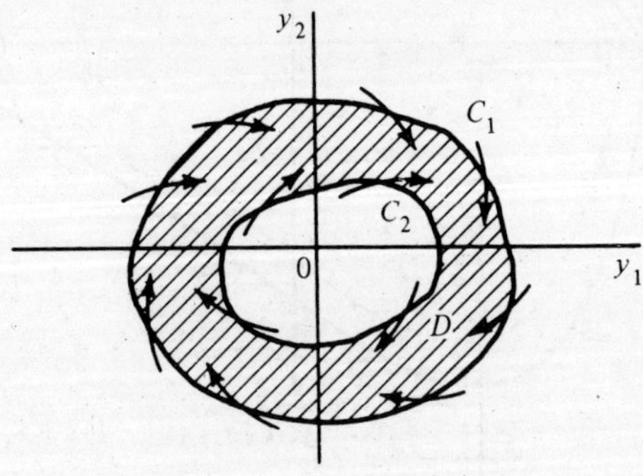

FIG. 8-15. *Illustration of Bendixson's second theorem.*

Intermediate between the stable and unstable operations, the phase trajectory converges to a limit cycle as illustrated in Fig. 8-14.

It is difficult to establish the existence of a limit cycle for a given system. Bendixson* has shown a sufficient condition for the *absence* of limit cycles. Bendixson's second theorem may be helpful to visualize a limit cycle in the phase plane. Consider two closed curves C_1 and C_2 enclosing a region D in Fig. 8-15. Assume C_2 contains a singular point but D contains only regular points. If the phase trajectories enter and remain within D indefinitely, then there is a limit cycle in region D.

8-7 ANALYTICAL METHODS

To complement the phase-plane technique, the rest of the chapter is devoted to analytical methods for the analysis of nonlinear systems. A limited number of methods are selected in order to restrict the scope of the chapter. We shall discuss free and forced vibrations.

Most methods are approximations of a nonlinear problem. The general approach is (1) to assume that the "small" nonlinearity is separable from the linear part of the equation, (2) to neglect the nonlinear terms, as a first approximation, to obtain a *generating solution*, and (3) to use the generating solution with the original equation to get the *corrective terms* due to the nonlinearity.

It is well to note that our purpose is to explain the nonlinear phenomena, because (1) computers can be used to solve nonlinear problems, and (2) greater accuracy is not our objective. Several methods may be used to investigate a problem, since each method gives the information characteristic of the technique.

8-8 FREE VIBRATION

Three methods will be presented in this section, namely, the method of perturbation, variation of parameters, and harmonic balance. Each method may serve a different purpose, although they may be used to obtain the same answer for certain problems.

Perturbation Method

The *perturbation method* assumes the solution of a nonlinear equation can be expressed as a power series, consisting of a generating solution and the added corrective terms. We shall discuss this method with an example.

*T. Bendixson, "Sur les courbes définies par les équations différentielles", *Acta Mathematica*, vol. 24, (1901), pp. 1–88.

Example 7

Consider the equation of motion from Example 1.

$$m\ddot{x}\cdot\frac{2}{\ell}\left(T_0+\frac{AE}{2\ell^2}x^2\right)x=0$$

or

$$\ddot{x}+\omega_0^2x+hx^3=0 \tag{8-38}$$

where $\omega_0=\sqrt{2T_0/(\ell m)}$ is the natural frequency of the linearized system and $h=AE/(\ell^3m)$ is the constant associated with the nonlinearity. Find the first-order corrective term and the first-order corrected solution.

Solution:

Let the solution be expressed as a power series

$$x(t)=x_0(t)+hx_1(t)+h^2x_2(t)+\cdots \tag{8-39}$$

where $x_0(t)$ is the generating solution, h is a small perturbation parameter, and $x_1(t)$, etc., are the added corrective terms. If h is sufficiently "small", the series will converge to the solution. Since the frequency is influenced by the amplitude A of the oscillation, we perturb the frequency ω by letting

$$\omega^2=\omega_0^2+hb_1(A)+h^2b_2(A)+\cdots$$

or

$$\omega_0^2=\omega^2-hb_1(A)-h^2b_2(A)-\cdots \tag{8-40}$$

where $b_1(A)$, etc., are the corrective terms for the frequency.

Substituting Eqs. (8-39) and (8-40) into (8-38) gives

$$(\ddot{x}_0+h\ddot{x}_1+h^2\ddot{x}_2+\cdots)+(\omega^2-hb_1-h^2b_2-\cdots)(x_0+hx_1+h^2x_2+\cdots)$$
$$+h(x_0+hx_1+h^2x_2+\cdots)^3=0$$

Expanding the equation and collecting the coefficients of h, we get

$$h^0:\quad \ddot{x}_0+\omega^2x_0=0 \tag{8-41}$$
$$h^1:\quad \ddot{x}_1+\omega^2x_1=b_1x_0-x_0^3 \tag{8-42}$$
$$h^2:\quad \ldots \qquad \ldots$$

The *generating solution* from the solution of Eq. (8-41) is

$$x_0=A\cos\omega t+B\sin\omega t. \tag{8-43}$$

where A and B are determined by the initial conditions. Let $x_0=A_0$ and $\dot{x}_0=0$ at $t=0$. Thus,

$$x_0=A_0\cos\omega t \tag{8-44}$$

The first *corrective term* $x_1(t)$ is obtained from Eq. (8-42). Since x_0 is known, we have*

$$\ddot{x}_1+\omega^2x_1=b_1A_0\cos\omega t-(A_0\cos\omega t)^3$$
$$=(b_1A_0-\tfrac{3}{4}A_0^3)\cos\omega t-\tfrac{1}{4}A_0^3\cos 3\omega t$$

* We use the identity: $\cos^3\theta=\frac{3}{4}\cos\theta+\frac{1}{4}\cos 3\theta$

The solution $x_1(t)$ is the sum of the complementary function and the particular integral, that is,

$$x_1 = A_1 \cos \omega t + B_1 \sin \omega t + \frac{t}{2\omega}(b_1 A_0 - \tfrac{3}{4}A_0^3)\sin \omega t + \frac{A_0^3}{32\omega^2}\cos 3\omega t \tag{8-45}$$

where A_1 and B_1 are constants. The third term on the right side corresponds to a resonance condition for the free vibration of a conservative system, which is contrary to the physical problem. It is called a *secular term.* It is removed by requiring $(b_1 A_0 - 3A_0^3/4) = 0$. Since $A_0 \neq 0$, we have

$$b_1 = \tfrac{3}{4}A_0^2 \tag{8-46}$$

The initial conditions for the problem are already accounted for in the generating solution. Assuming zero initial conditions and suppressing the secular term in Eq. (8-45), we get

$$x_1 = -\frac{A_0^3}{32\omega^2}(\cos \omega t - \cos 3\omega t) \tag{8-47}$$

If only the first-order correction is desired, the solution of Eq. (8-38) is

$$x(t) = x_0(t) + hx_1(t)$$

or

$$x(t) = A_0 \cos \omega t - \frac{h}{32\omega^2}A_0^3(\cos \omega t - \cos 3\omega t)$$

$$\omega^2 = \omega_0^2 + \tfrac{3}{4}hA_0^2 \tag{8-48}$$

The solution in Eq. (8-48) is as anticipated. The spring force is of the form $k(x+ax^3)$, where α is a positive constant. It is an example of a *hard spring,* in which the spring force increases with the spring deformation. Hence the natural frequency increases with the amplitude of oscillation. On the other hand, we may anticipate a decrease in frequency with increasing amplitude for a *soft spring,* where the spring force is $k(x-\alpha x^3)$. The simple pendulum in Example 3 is a case of a soft spring, because the equation of motion can be approximated as

$$J\ddot{\theta} + mg\ell(\theta - \theta^3/6) \simeq 0$$

Note also that the effect of the nonlinear spring is to distort the oscillation by introducing higher harmonic terms in the solution.

Variation of Parameter Method

The variation of parameter method allows small changes in amplitude and phase angle in the time solution. In contrast, the perturbation method gives a steady-state solution.

Consider the equation

$$\ddot{x} + \omega_0^2 x + \mu\phi(x,\dot{x},t) = 0 \tag{8-49}$$

where $\phi(x,\dot{x},t)$ is the nonlinearity and μ a constant. Neglecting $\phi(x,\dot{x},t)$, the generating solution from the corresponding linear equation $\ddot{x}+\omega_0^2x=0$ is

$$x = A\cos(\omega_0 t+\theta) \tag{8-50}$$

$$\dot{x} = -\omega_0 A\sin(\omega_0 t+\theta) \tag{8-51}$$

For convenience, we denote

$$\psi=\omega_0 t+\theta \qquad \text{and} \qquad \dot{\psi}=\omega_0+\dot{\theta}$$

Differentiating Eq. (8-50) with respect to time t gives

$$\dot{x}=\dot{A}\cos\psi-(A\sin\psi)(\omega_0+\dot{\theta})$$

Subtracting Eq. (8-51) from this, we get

$$\dot{A}\cos\psi-\dot{\theta}A\sin\psi=0 \tag{8-52}$$

Differentiating Eq. (8-51) with respect to t gives

$$\ddot{x}=-\omega_0\dot{A}\sin\psi-(\omega_0 A\cos\psi)(\omega_0+\dot{\theta}) \tag{8-53}$$

Substituting Eqs. (8-50) and (8-51) into (8-49) and subtracting the resulting equation from (8-53), we obtain

$$-\dot{A}\omega_0\sin\psi-\dot{\theta}\omega_0 A\cos\psi=-\mu\phi(A\cos\psi,-\omega_0 A\sin\psi,t) \tag{8-54}$$

Equations (8-52) and (8-54) can be solved for $\dot{A}$ and $\dot{\theta}$, that is,

$$\dot{A}=\frac{\mu}{\omega_0}\sin\psi\;\phi(A\cos\psi,-\omega_0 A\sin\psi,t)$$

$$\dot{\theta}=\frac{\mu}{\omega_0 A}\cos\psi\;\phi(A\cos\psi,-\omega_0 A\sin\psi,t) \tag{8-55}$$

It seems that we have traded one nonlinear differential equation for two simultaneous nonlinear differential equations.

The approximate solution of Eq. (8-55) is found by assuming μ is sufficiently small such that A and θ do not change rapidly. If the changes in A and θ are small *over one cycle*, we may consider the average values of $\dot{A}$ and $\dot{\theta}$ rather than the instantaneous values. Thus,

$$\begin{aligned}\dot{A}_{\text{av}}&=\frac{\mu}{2\pi\omega_0}\int_0^{2\pi}\sin\psi\;\phi(A\cos\psi,-\omega_0 A\sin\psi,t)\,d\psi\\ \dot{\theta}_{\text{av}}&=\frac{\mu}{2\pi\omega_0 A}\int_0^{2\pi}\cos\psi\;\phi(A\cos\psi,-\omega_0 A\sin\psi,t)\,d\psi\end{aligned} \tag{8-56}$$

where A is assumed constant in the integrand.

Example 8

Repeat Example 7 using the method of variation of parameters.

Solution:

From example 7 and Eq. (8-49), we have

$$\ddot{x}+\omega_0^2 x+hx^3=0$$

$$\ddot{x}+\omega_0^2 x+\mu\phi(x,\dot{x},t)=0$$

The generating solution from Eq. (8-50) is

$$x=A\cos(\omega_0 t+\theta)=A\cos\psi$$

Since $\mu\phi(x,\dot{x},t)=hx^3$, from Eq. (8-56) we get

$$\dot{A}_{av}=\frac{h}{2\pi\omega_0}\int_0^{2\pi}A^3\sin\psi\cos^3\psi\,d\psi=0$$

$$\dot{\theta}_{av}=\frac{h}{2\pi\omega_0 A}\int_0^{2\pi}A^3\cos^4\psi\,d\psi=\frac{3hA^2}{8\omega_0}$$

or

$$A_{av}=A=C_1$$

$$\theta_{av}=\frac{3hA^2}{8\omega_0}t+C_2$$

where C_1 and C_2 are constants. For the initial conditions $x=A_0$ and $\dot{x}=0$, we get $C_1=A_0$ and $C_2=0$. Thus,

$$x=A_0\cos\left(\omega_0+\frac{3h}{8\omega_0}A_0^2\right)t$$

Harmonic Balance

The method of *harmonic balance* seeks a steady-state periodic solution, expressed as a Fourier series. We assume only the fundamental component is of interest. Substituting the displacement and its derivatives into the system equation, the solution is adjusted to satisfy all terms of the fundamental frequency. A steady-state oscillation exists if the assumed periodic solution is verified. The method neglects the transient as well as the stability of the oscillation. Let us illustrate this with an example.

Example 9

Solve Van der Pol's equation shown in Fig. 8-14 by the method of harmonic balance.

Solution:

The system equation is

$$\ddot{x}+\varepsilon(x^2-1)\dot{x}+x=0$$

Let

$$x=A\sin(\omega t+\theta)$$

$$\dot{x}=\omega A\cos(\omega t+\theta)$$

$$\ddot{x}=-\omega^2 A\sin(\omega t+\theta)$$

For convenience, let $\varepsilon = 1$ and $\psi = (\omega t + \theta)$. Substituting x, $\dot{x}$, and $\ddot{x}$ into the system equation gives

$$-\omega^2 A \sin \psi + (A^2 \sin^2 \psi - 1)(\omega A \cos \psi) + A \sin \psi = 0$$

Expanding the equation and simplifying, we obtain

$$A(1-\omega^2)\sin \psi - \omega A\left(1 - \frac{A^2}{4}\right)\cos \psi - \omega A^3 \cos 3\psi = 0$$

The $\cos 3\psi$ term is a third harmonic and therefore neglected. From the coefficients of the sine and cosine terms we get

$$\omega = 1 \qquad \text{and} \qquad A = 2$$

Hence the solution is

$$x = 2 \sin(t + \theta)$$

where θ is arbitrary. The method seeks a steady-state solution and does not satisfy the initial conditions.

8-9 FORCED VIBRATIONS

This section describes certain nonlinear behavior that do not exist in linear systems. We shall examine the periodic response due to harmonic excitations.

Jump Phenomenon

Consider a system described by Duffing's equation.

$$m\ddot{x} + c\dot{x} + k(x + \alpha x^3) = F \cos \omega t \tag{8-57}$$

where $k(x + \alpha x^3)$ is a nonlinear spring force. Let us use the hard spring, where α is positive, to illustrate the jump phenomenon.

By the method of harmonic balance and considering only the fundamental component, the steady-state solution is

$$x = A \cos(\omega t - \theta)$$

It is more convenient to associate the phase angle θ with the excitation. The system equation and the response become

$$m\ddot{x} + c\dot{x} + k(x + \alpha x^3) = F \cos(\omega t + \theta)$$

or

$$\ddot{x} + 2\beta\dot{x} + \omega_0^2 x + hx^3 = F_c \cos \omega t + F_s \sin \omega t \tag{8-58}$$

$$x = A \cos \omega t \tag{8-59}$$

where F_c and F_s have units of force per unit mass, and hx^3 is the nonlinear term. We first treat the simpler case of undamped vibrations.

Case 1. Undamped systems

If a system is undamped, the steady-state response is either in-phase or 180° out-of-phase with the excitation. The phase angle can be accounted for by the sign change in A in Eq. (8-59). Hence the second force $F_s \sin \omega t$ can be neglected. The system equation becomes

$$\ddot{x} + \omega_0^2 x + hx^3 = F_c \cos \omega t \tag{8-60}$$

Substituting Eq. (8-59) in (8-60) and simplifying,* we get

$$-\omega^2 A + \omega_0^2 A + \tfrac{3}{4} hA^3 = F_c \tag{8-61}$$

The values of A versus ω in the equation can be solved for the given values of ω_0, h, and F_c.

To solve Eq. (8-61) graphically, let

$$\begin{aligned} y_1 &= \tfrac{3}{4} hA^3 \\ y_2 &= (\omega^2 - \omega_0^2)A + F_c \end{aligned} \tag{8-62}$$

where ω_0, h, and F_c are known as illustrated in Example 10 to follow. The y_1 curve is plotted for a given value of h as shown in Fig. 8-16. For an assumed value of ω, the y_2 curve is a straight line. The intersection of the y_1 and y_2 curves gives the value of A for the assumed value of ω. Note that (1) the value of A can be positive or negative, corresponding to the

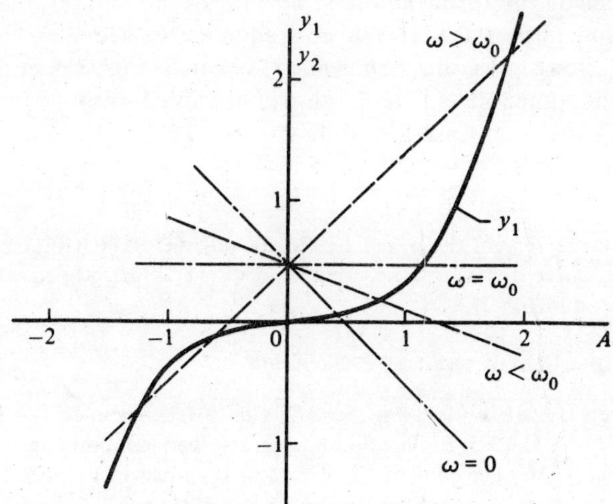

FIG. 8-16. *Graphical solution of cubic equation: Example 10.*

* The identity $\cos^3 \omega t = \frac{3}{4}\cos \omega t + \frac{1}{4}\cos 3\omega t$ is used for the manipulation. Since $\cos 3\omega t$ is a third harmonic and we consider only the fundamental components, the $\cos 3\omega t$ term is neglected.

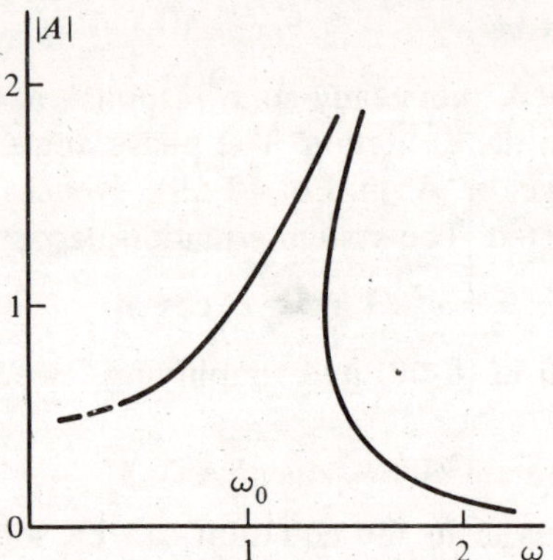

FIG. 8-17. *Harmonic response: hard spring, Example 10(a).*

phase angle of 0 or 180° of the steady-state response relative to the excitation, and (2) there can be either one or three values of A. The particular plots of y_2 for $\omega = 0$ and $\omega = \omega_0$ are shown as dash-dot lines. The general plots of y_2 for $\omega \gtrless \omega_0$ are shown as dash lines.

The values of $|A|$ versus ω are plotted in Fig. 8-17. Comparing with the harmonic response of a linear system in Fig. 2-9, the nonlinear response skews towards higher frequencies for the case of a hard spring. Physically, as the driving frequency increases towards resonance, $|A|$ increases and thereby increasing the spring stiffness. This raises the natural frequency of the system and pushes the resonance frequency to a higher value. The graphical solution shows the response $|A|$ versus ω curve is multivalued. The jump phenomenon is due to the multivalued response. We shall discuss this behavior in Case 2 to follow.

Example 10

Consider the system described by Eq. (8-60). (**a**) Sketch the steady-state response curve for $\omega_0 = 1$, $h = 1/2$, and $F_c = 1/2$. (**b**) Repeat for $\omega_0 = 1$, $h = -1/2$, and $F_c = 1/2$.

Solution:

(**a**) Given $\omega_0 = 1$, $h = 1/2$, and $F_c = 1/2$, Eq. (8-62) becomes $y_1 = 3A^3/8$ and $y_2 = (\omega^2 - 1)A + 1/2$. The equations are solved simultaneously as shown in Fig. 8-16. The plot of $|A|$ versus ω is shown in Fig. 8-17.

(**b**) For the case of a soft spring, Eq. (8-61) becomes

$$-\omega^2 A + \omega_0^2 A - \tfrac{3}{4}|h|A^3 = F_c$$

This can be solved graphically as before. The plot of $|A|$ versus ω for $\omega_0 = 1$, $h = -1/2$, and $F_c = 1/2$ is shown in Fig. 8-18.

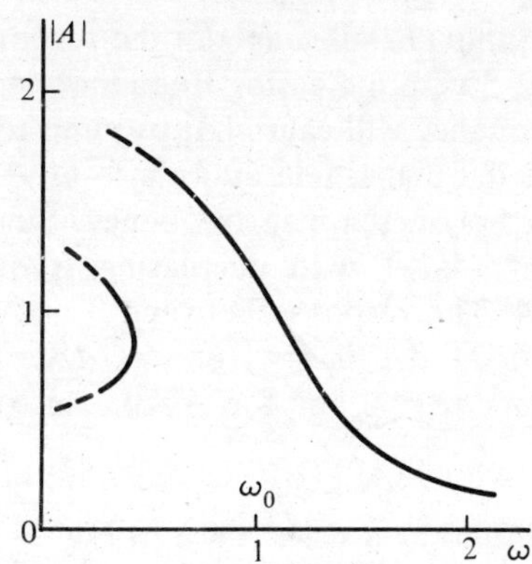

FIG. 8-18. *Harmonic response: soft spring, Example 10(b).*

Case 2. Damped systems

Systems with damping can be analyzed by the method of harmonic balance as before. Substituting Eq. (8-59) in (8-58), neglecting the third harmonic term, and collecting the coefficients of the cosine and sine terms, we obtain

$$\cos \omega t: \qquad -\omega^2 A + \omega_0^2 A + \tfrac{3}{4} h A^3 = F_c$$

$$\sin \omega t: \qquad -2\beta\omega A = F_s$$

The values of A and F can be related by the last two equations. Thus,

$$[(\omega_0^2 - \omega^2)A + \tfrac{3}{4} h A^3]^2 + (2\beta\omega A)^2 = F^2 \tag{8-63}$$

where $F^2 = F_c^2 + F_s^2$. The equation can be solved to yield $|A|$ versus ω for the given parameters ω_0, h, β, and F. A graphical solution, however, is not feasible. If the damping is light, the response curve will resemble that of an undamped system in Example 10. A typical plot of $|A|$ versus ω for a lightly damped system is shown in Fig. 8-19.

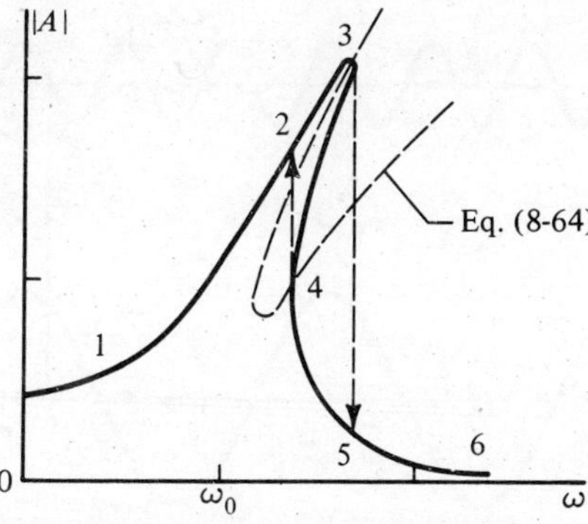

FIG. 8-19. *Jump phenomenon: system with hard spring.*

To describe the *jump phenomenon*, let the response $|A|$ in Fig. 8-19 be along the curve 1-2-3 with increasing frequency. At Point 3, an infinitesimal increase in frequency will cause $|A|$ to jump to Point 5, that is, there is a step change in the magnitude and phase of A. The response will be along 5-6 with further increase in frequency. Similarly, the response is along the curve 6-5-4-2-1 with decreasing frequency, with the jump occuring at Point 4. The response between 3 and 4 is unstable.

The jump occurs at $dA/d\omega = \infty$, or $d\omega^2/dA = 0$. Differentiating Eq. (8-63) to obtain $d\omega^2/dA = 0$ and for $A \neq 0$, the condition is

$$[(\omega_0^2 - \omega^2) + \tfrac{3}{4}hA^2][(\omega_0^2 - \omega^2) + \tfrac{9}{4}hA^2] + (2\beta\omega)^2 = 0 \tag{8-64}$$

The equation is plotted as a dash curve in Fig. 8-19. The jump may not occur if the system is heavily damped.

The jump phenomenon is due to the multivalued response. It occurs in the steady-state response with an excitation of constant amplitude at varying frequency as shown above. Similarly, a jump could also occur with an excitation of varying amplitude at constant frequency. In other words, a jump could occur if the response curve of $|A|$ versus amplitude of the excitation is multivalued.

Subharmonic Oscillations

The input and the output wave forms of a system undergoing subharmonic oscillations are illustrated in Fig. 8-20. The frequency of the output is an integral submultiple of the input or the driving frequency. The figure shows that the output frequency is 1/3 of that of the input. In general, if ω_1 is the driving frequency, then ω_1/n is the frequency of the subharmonic oscillation, where n is an integer. We shall illustrate the 1/3 subharmonic by the perturbation method.

Consider Duffing's equation from Eq. (8-60).

$$\ddot{x} + \omega_0^2 x + \mu h x^3 = F \cos \omega_1 t \tag{8-65}$$

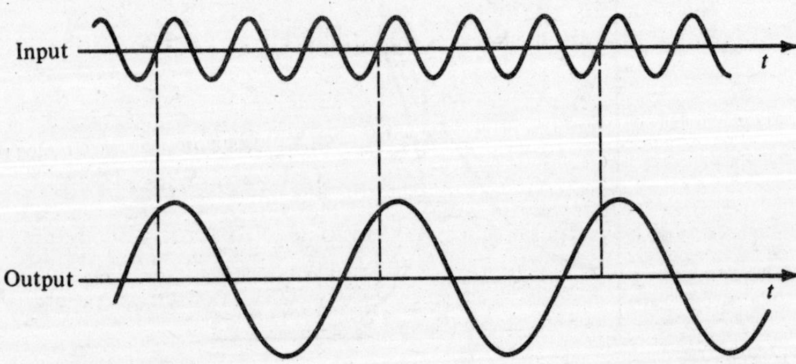

FIG. 8-20. *Input and output wave forms: 1/3 subharmonic oscillations.*

where ω_1 is the driving frequency and μ a constant. Since the 1/3 subharmonic is of interest, we assume

$$\omega_1 = 3\omega \tag{8-66}$$

where ω is the output frequency.

Applying the perturbation method, let

$$x(t) = x_0(t) + \mu x_1(t) + \cdots \tag{8-67}$$

$$\omega^2 = \omega_0^2 + \mu b_1(A) + \cdots \tag{8-68}$$

as shown in Eqs. (8-39) and (8-40). Substituting Eqs. (8-66) through (8-68) in (8-65) and retaining only the first-order correction terms, we obtain

$$\ddot{x}_0 + \mu\ddot{x}_1 + \omega^2 x_0 + \mu\omega^2 x_1 - \mu b_1 x_0 + \mu h x_0^3 = F\cos 3\omega t$$

Collecting the coefficients of μ gives

$$\mu^0: \qquad \ddot{x}_0 + \omega^2 x_0 = F\cos 3\omega t \tag{8-69}$$

$$\mu^1: \qquad \ddot{x}_1 + \omega^2 x_1 = b_1 x_0 - h x_0^3 \tag{8-70}$$

From Eq. (8-69), the generating solution for zero initial velocity is

$$x_0 = A\cos\omega t + X\cos 3\omega t \tag{8-71}$$

where A is as yet undefined. A $\cos\omega t$ is the subharmonic and $X\cos 3\omega t$ is the particular integral. From linear theory, the amplitude X is

$$X = \frac{F}{\omega^2 - 9\omega^2} = \frac{F}{-8\omega^2} \tag{8-72}$$

We shall show that the amplitude and the frequency of the subharmonic and the particular integral are related, since the system is nonlinear.

To consider the first-order correction term, substituting Eq. (8-71) in (8-70) gives

$$\ddot{x}_1 + \omega^2 x_1 = b_1(A\cos\omega t + X\cos 3\omega t) - h(A\cos\omega t + X\cos 3\omega t)^3$$

The right side of this equation can be expanded and simplified to yield

$$\begin{aligned} &A(b_1 - \tfrac{3}{4}hA^2 - \tfrac{3}{4}hAX - \tfrac{3}{2}hX^2)\cos\omega t \\ &\qquad + (b_1 X - \tfrac{1}{4}hA^3 - \tfrac{3}{2}hA^2X - \tfrac{3}{4}hX^3)\cos 3\omega t \\ &\qquad - \tfrac{3}{4}hAX(A+X)\cos 5\omega t - \tfrac{3}{4}hAX^2\cos 7\omega t - \tfrac{1}{4}hX^3\cos 9\omega t \end{aligned}$$

The first term in the expression is equated to zero to avoid a secular term. Since $A \neq 0$, we get

$$b_1 = \tfrac{3}{4}h(A^2 + AX + 2X^2)$$

Substituting b_1 in Eq. (8-68) and assuming $\mu = 1$, we obtain the relation for the amplitude and frequency of the subharmonic and the particular integral, that is,

$$\omega^2 = \omega_0^2 + \tfrac{3}{4}h(A^2 + AX + 2X^2) \tag{8-73}$$

Let us relate the output frequency ω and ω_0 of the system. When the amplitude A of the subharmonic becomes zero, Eq. (8-73) gives

$$(\omega^2 - \omega_0^2) = \tfrac{3}{2}X^2$$

Substituting X from Eq. (8-72), we get

$$(\omega^2 - \omega_0^2)\omega^4 = \frac{3hF^2}{128} \tag{8-74}$$

To relate ω and ω_0 for $A \neq 0$, Eq. (8-73) can be solved for A. Since the equation is quadratic in A, the solution for real values of A requires that

$$\frac{16}{3h}(\omega^2 - \omega_0^2) \geq 7X^2$$

Substituting X from Eq. (8-72) yields

$$\frac{16}{3h}(\omega^2 - \omega_0^2)\omega^4 \geq \frac{7F^2}{64} \tag{8-75}$$

The inequality must be satisfied for A to be real. The minimum value of ω is when Eq. (8-75) becomes an equality. Let $\omega/\omega_0 = (1+\varepsilon)$, where $\varepsilon \ll 1$. Substituting this in Eq. (8-75) and simplifying, we obtain

$$2\varepsilon\omega_0^6 \geq \frac{21hF^2}{1024}$$

Hence the minimum value for ω is

$$\omega_{\min} = \omega_0(1+\varepsilon) = \omega_0\left(1 + \frac{21hF^2}{2048\omega_0^6}\right) \tag{8-76}$$

8-10 SUMMARY

Most engineering problems can be linearized. There are common problems, however, that cannot be explained by linear theory. The chapter describes the nonlinear phenomena of one-degree-of-freedom systems.

Stability and examples of nonlinear systems are illustrated in Sec. 8-2. A nonlinear system could have more than one equilibrium, some of which may be stable and others unstable.

A trajectory in a phase plane represents the solution of a differential equation. The pattern of the trajectories in a phase plane gives a perspective of the general solution of the equation. The characteristics of trajectories are discussed in Sec. 8-3.

A one-degree-of-freedom system is at equilibrium when $(\dot{x},\ddot{x}) = (0,0)$. The equilibrium positions of a system are found from Eq. (8-7). If the system equation is linearized about an equilibrium as in Eq. (8-11), the stability about each equilibrium can be determined by linear theory. The system is stable only if the roots of its characteristic equation in Eq. (8-14) has zero or negative real parts. The types of equilibrium are summarized in Fig. 8-8.

The isocline and Pell's method for the graphical construction of trajectories in the phase plane are described in Sect. 8-5. (1) The slope of the trajectory is calculated from the system equation. (2) Using the slope for extrapolation, a trajectory is constructed.

Most analytical methods are approximations of a nonlinear problem. The general approach is (1) to separate the small nonlinearity from the linear part of the equation, (2) to solve the linear equation, as a first approximation, and then (3) apply the corrections due to the nonlinearity. Generally, each method yields the information characteristic of that technique.

Three common analytical methods are presented in Sec. 8-8 in connection with free vibration. The perturbation method gives the steady-state solution. The variation of parameter method allows for slow changes in the amplitude and phase angle in the solution. The harmonic balance method seeks a steady-state preiodic solution as in a Fourier series. Harmonic distortions and limit cycles are illustrated in Examples 8 and 9.

Force vibrations with harmonic excitation is treated in Sect. 8-9. A jump phenomenon describes an abrupt change in the amplitude and phase of the steady-state response due to a small change in the excitation. A jump could occur when the response curve is multivalued as shown in Fig. 8-19. Subharmonic oscillations are steady-state vibrations, in which the frequency of the response is a submultiple of that of the excitation as illustrated in Fig. 8-20. A plausible explanation is that there is harmonic distortion in the response and the amplitude and frequency of the subharmonic and the particular integral are related in Eq. (8-73). It is possible for the excitation to supply energy to the system at one of the harmonics to sustain its steady-state oscillation, which is at a frequency close to the natural frequency of the system.

PROBLEMS

8-1 Derive the equation of motion for each of the systems shown in Fig. P8-1. Assume (1) the springs k are linear, and (2) the systems as shown are in their static equilibrium positions.

(a) A simple pendulum with the flexible cord wrapped around a cylinder.

(b) A simple pendulum with an attached spring k.

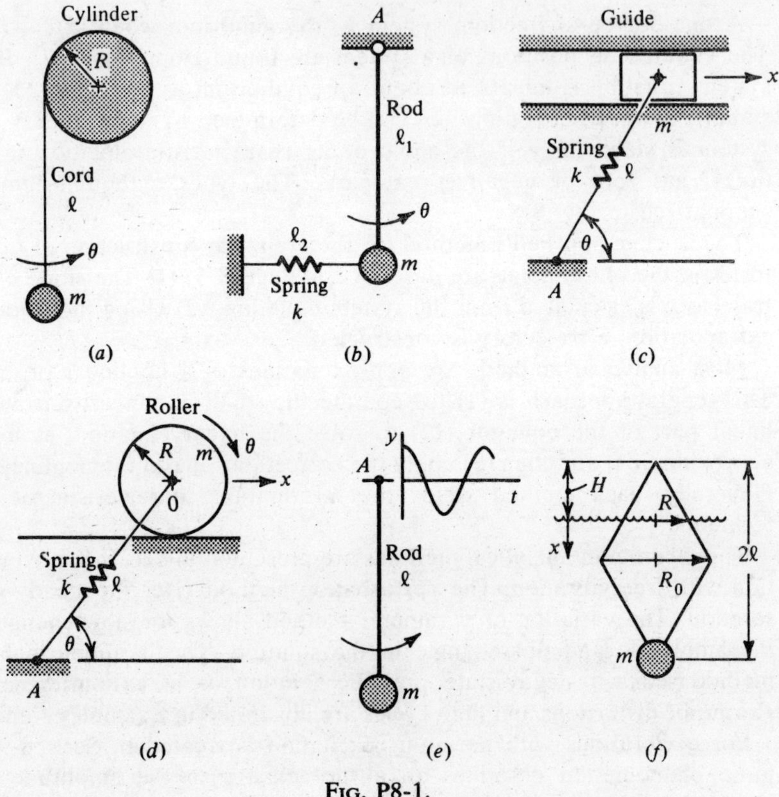

Fig. P8-1.

(c) A guided spring-mass system. Assume friction is negligible.

(d) A roller of mass m on a horizontal surface without slippage. Assume the roller is always in contact with the surface.

(e) A simple pendulum, the pivot A of which has a vertical motion $y = Y \cos \omega t$.

(f) A buoy consisting of two cones.

8-2 Consider the equation of motion

$$\ddot{x} + 0.6\dot{x} + 2.25x = 0$$

Assume the initial conditions are $x(0) = -2$ and $\dot{x}(0) = 3$.

(a) Plot $x(t)$ versus t and $\dot{x}(t)$ versus t for $0 \le t \le 5$, and

(b) obtain a phase trajectory in a $(x, \dot{x})$ phase plane from corresponding values of $x(t)$ and $\dot{x}(t)$ in part **a**.

8-3 Show that the trajectories of the system $m\ddot{x} + kx = 0$ form a family of concentric ellipses. Let $m = 1$ kg and $k = 100$ N/m. Sketch the trajectories in a $(x,\dot{x})$ phase plane and indicate their directions for each of the following initial conditions:

(a) $(x,\dot{x})=(1,5)$; **(b)** $(x,\dot{x})=(1,10)$.

8-4 Find the equilibrium position(s), determine the type(s) of singularity, and sketch the trajectories in the neighborhood of equilibrium for each of the following systems:

(a) $m\ddot{x}+c\dot{x}+k(1+\alpha x^2)x=0$, where $\alpha>0$

(b) $\ddot{x}+0.2(x^2-1)\dot{x}+x=0$

(c) $\ddot{x}+0.1\dot{x}+f(x)=0$,

where $f(x)=0$ for $-\Delta\leq x\leq\Delta$

$f(x)=\psi(x)|\psi(x)|$ and $\psi(x)=[x-\Delta\,\mathrm{sgn}(x)]$ for $|x|>\Delta$

8-5 Sketch the trajectories in the neighborhood of equilibrium for each of the following systems. The state variables x_1 and x_2 do not necessarily represent displacement and velocity.

(a) $\dot{x}_1=3x_1-4x_1x_2$ and $\dot{x}_2=2x_2-5x_1x_2$ for $x_1>0$ and $x_2>0$

(b) $\dot{x}_1=-3x_1+5x_2$ and $\dot{x}_2=-5x_1+3x_2$

(c) $\dot{x}_1=-41x_1-12x_2+70$ and $\dot{x}_2=-12x_1-34x_2-10$

8-6 Use the *isocline method* to obtain the phase trajectories for each of the systems described below:

(a) $\ddot{x}+0.32\dot{x}+0.64x=0$ for the initial conditions $(x_0,\dot{x}_0)=(4,1)$

(b) The system shown in Fig. 8-1(*d*), for the initial conditions $x(0)=4$ and $\dot{x}(0)=0$. Assume $m=1$ kg, $k=2$ N/m, $c=0.25$ N·s/m, and the total gap between m and the springs $=2$ m.

(c) Repeat part **b** except that the spring force is proportional to the square of the spring deformation.

8-7 Repeat Prob. 8-6 using the *Pell's method.*

8-8 Find the graphical solution of each of the following equations:

(a) $\ddot{x}+0.3(x^2-1)\dot{x}+x=0$ Assume initial conditions $x(0)=3$ and $\dot{x}(0)=2$.

(b) $\ddot{x}+(\dot{x}-1)\dot{x}+x=0$ Assume initial conditions $x(0)=-2$ and $\dot{x}(0)=0$. Compare the given equation with Eq. (8-37).

(c) $m\ddot{x}+C\,\mathrm{sgn}(\dot{x})+kx=0$ Assume $m=1$, $C=0.6$, $k=1$, and $x(0)=3.2$ and $\dot{x}(0)=0$. See Sec. 3-8 on systems with Coulomb damping.

(d) $m\ddot{x}+C\dot{x}^2\,\mathrm{sgn}(\dot{x})+kx=0$ Assume $m=1$, $C=0.6$, $k=0.5$, and $x(0)=3$ and $\dot{x}(0)=0$. See Sec. 3-8 on systems with velocity squared damping.

(**e**) $\ddot{x}+f(\dot{x})+0.5x=0$ where $f(\dot{x})$ is as shown in Fig. P8-2(a). Assume $x(0)=2$ and $\dot{x}(0)=1.5$.

(**f**) $\ddot{x}+\dot{x}+f(x)=0$ where $f(x)$ is as shown in Fig. P8-2(b). Assume $x(0)=2$ and $\dot{x}(0)=1.5$.

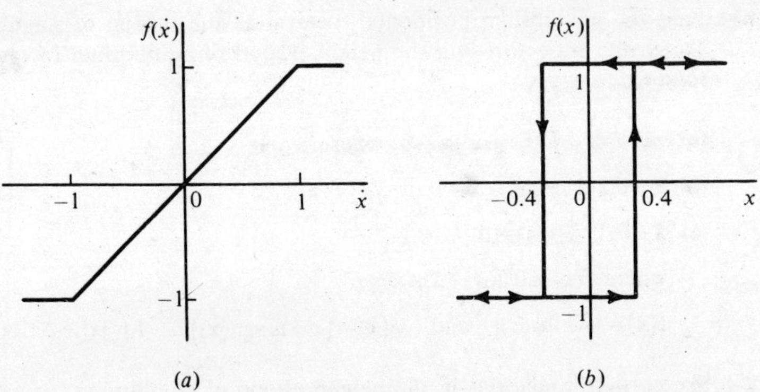

FIG. P8-2.

8-9 Consider the simple pendulum shown in Fig. 8-1(a). Assume that $\sin\theta \simeq \theta-\theta^3/6$, and $\theta(0)=A_0$ and $\dot{\theta}(0)=0$. Use the perturbation method to find the approximate period of oscillation.

Computer problems:

Remarks: Two types of nonlinear initial value problems are considered. (**1**) The nonlinearity is an expression as in Prob. 8-10. (**2**) The nonlinearity has discontinuities as in Prob. 8-11. In the first case, the method described in Sec. 9-2 can be applied directly. In the second case, the method can be modified by IF statements to allow for the discontinuities.

8-10 Given: $\dot{x}=2.5x-xy$ and $\dot{y}=-2.5y+2xy$

(**a**) Estimate the solution by means of a graphical method. (**b**) Write a program to solve the equation. Assume $x(0)=0.4$ and $y(0)=1.0$.

8-11 Write a program to solve Prob. 8-8(f). Assume $x(0)=3$ and $\dot{x}(0)=0$.

9

Solutions by Digital Computers

9-1 INTRODUCTION

The chapter introduces the use of digital computers to solve typical problems in vibration. (1) Computers are inevitably used for the numerical solution of such problems in industry. (2) The engineer should be aware of the assumptions and techniques when using a canned program. (3) This knowledge is helpful in communicating with programmers. The programs are written in FORTRAN* for time sharing.

The sections in the chapter are arranged to conform to the sequence of the topics presented in the text. The sections and the previous chapters can be used concurrently. In fact, some problems in the prior chapters require computer solutions. The advantage in grouping the programs in one place is that it would be unnecessary to repeat the explanations.

The programs, as listed in Table 9-1, are sufficient for the compution and the plotting of the results of discrete systems, damped or undamped. The first three programs employ the one-degree-of-freedom system to explain (1) the organization of the programs, (2) the use of subroutines, and (3) the application of the plotting subroutine. The reader not familiar with time sharing is encouraged to try the first three programs before the

*The choice of language for the programs is a matter of preference. The programs were originally written in PL/1. At the suggestion of the reviewers and publisher, the programs are rewritten in FORTRAN for time sharing.

There are many versions of FORTRAN. The IBM CALL/360-OS, Manual GH20-0710-2, is used for reference. Some statements may differ slightly from other versions of FORTRAN. If this occurs, the computer center should be consulted.

Many digital simulation or modeling programs can be used to solve similar problems. We choose FORTRAN because (1) it is widely known and used, and (2) the programs can be written to run parallel to the theory developed in the text.

TABLE 9-1. *Typical Vibration Analysis Programs*

PROGRAM	TITLE AND DESCRIPTION
1. TRESP1	Transient response of one-degree-of-freedom linear systems. 4th-order Runge-Kutta method. Assume constant excitation.
2. TRESPSUB	Transient response of one-degree-of-freedom linear systems. Use subroutine for computations. Transient excitation.
3. TRESPF1	Transient response of one-degree-of-freedom linear systems. Write output in file for plotting. Transient excitation.
4. PLOTFILE	Program to plot data from a file.
5. FRESP1	Frequency response of one-degree-of-freedom systems Mechanical impedance method.
6. FRESPN	Frequency response of n-degree-of-freedom systems with viscous damping. Mechanical impedance method.
7. TRESPUND	Transient response of undamped discrete systems. Modal analysis. Transient excitation.
RAYLEIGH	Rayleigh's method for undamped multi-rotor systems.
9. TMXCANT1	Transfer matrix analysis of uniform cantilever. Mykelstad method.
10. ITERATE	Matrix iteration of undamped discrete systems. Duncan and Collar method.
11. TRESPDAM	Transient response of discrete systems with viscous damping. Modal analysis. Transient excitation.

others. Each program is preceded by a brief description and a recapitulation of the equations. For pedagogical reasons, the organization of the programs and typical terminal commands for the program execution and editing are described with the program.

The reader is assumed to have some working knowledge of FORTRAN. Hence programming per se will not be discussed. Numerical methods, not directly related to vibration, are also omitted. Subroutines are used extensively in order to simplify the main programs. The subroutines commonly used are listed and discussed in App. C.

9-2 ONE-DEGREE-OF-FREEDOM SYSTEMS—TRANSIENT RESPONSE

A one-degree-of-freedom system is described by a second-order differential equation.

$$\ddot{x} = f(x,\dot{x},t) \tag{9-1}$$

where $f(x,\dot{x},t)$ is a function of x, $\dot{x}$, and t. The equation can be linear or nonlinear. For vibration problems, we assume that x is the displacement, $\dot{x}$ the velocity, and the time t the independent variable. For purpose of computation, Eq. (9-1) is expressed as two simultaneous first-order equations.

$$\begin{aligned} \dot{x} &= y \\ \dot{y} &= f(x,y,t) \end{aligned} \tag{9-2}$$

Given the initial conditions x_0 and y_0 at $t=0$, Eq. (9-2) is solved by finding the increment Δx and Δy for each time increment Δt. Let t_i be a typical time for the ith time interval and x_i and y_i the corresponding values of x and y. Thus, the recurrence formulas are

$$\begin{aligned} t_{i+1} &= t_i + \Delta t \\ x_{i+1} &= x_i + \Delta x \\ y_{i+1} &= y_i + \Delta y \end{aligned} \tag{9-3}$$

An initial value problem is solved by the repeated application of the recurrence formulas above.

Initial value problems can be solved by a number of methods of integration.* For example, consider the equation of motion from Eq. (2-18).

$$m\ddot{x} + c\dot{x} + kx = F(t) \tag{9-4}$$

or

$$\ddot{x} + 2\zeta\omega_n\dot{x} + \omega_n^2 x = \frac{1}{m}F(t) \tag{9-5}$$

where $F(t)$ is an arbitrary excitation, $\omega_n^2 = k/m$, and $2\zeta\omega_n = c/m$ as defined in Eq. (2-27). The equation can be expressed as

$$\begin{aligned} \dot{x} &= y \\ \dot{y} &= -\left[2\zeta\omega_n y + \omega_n^2 x - \frac{1}{m}F(t)\right] \end{aligned} \tag{9-6}$$

* See, for example, R. W. Hamming, *Numerical Methods for Scientists and Engineers*, McGraw-Hill Book Company, New York, 1962.

as shown in Eq. (9-2). Applying the fourth-order Runge-Kutta method* gives

$$\begin{aligned} \Delta x &= (K_1+2K_2+2K_3+K_4)/6 \\ \Delta y &= (L_1+2L_2+2L_3+L_4)/6 \end{aligned} \tag{9-7}$$

where

$$\begin{aligned} K_1 &= y\Delta t & L_1 &= [-2\zeta\omega_n y - \omega_n^2 x + F]\Delta t \\ K_2 &= \left(y+\frac{L_1}{2}\right)\Delta t & L_2 &= \left[-2\zeta\omega_n\left(y+\frac{L_1}{2}\right)-\omega_n^2\left(x+\frac{K_1}{2}\right)+F\right]\Delta t \\ K_3 &= \left(y+\frac{L_2}{2}\right)\Delta t & L_3 &= \left[-2\zeta\omega_n\left(y+\frac{L_2}{2}\right)-\omega_n^2\left(x+\frac{K_2}{2}\right)+F\right]\Delta t \\ K_4 &= (y+L_3)\Delta t & L_4 &= [-2\zeta\omega_n(y+L_3)-\omega_n^2(x+K_3)+F]\Delta t \end{aligned} \tag{9-8}$$

and F is the value of $F(t)/m$ evaluated at $t = t_i$, the ith time interval under consideration. Using Eqs. (9-7) and (9-8), the recurrence formulas in (9-3) can be applied repeatedly to solve an initial value problem.

9-3 PROGRAM—TRESP1

The program TRESP1 gives the *t*ransient *resp*onse of a one-degree-of-freedom system described in Eq. (9-6). For simplicity, $F(t)$ is assumed constant. The case of an arbitrary $F(t)$ will be discussed in the next section. This program serves to illustrate the arrangement of the programs in the chapter.

The program, as listed in Fig. 9-1(a), is divided into paragraphs. For example, Par. I in TRESP1 consists of the format and input statements, and Par. II the calculations and output. For ease of reading, most of the documentary comments are given in the text rather than in the program itself.

Each line in a program is identified by a *line number.* The first line in TRESP1 is line 100; the character quote (") denotes a *comment* statement, giving the title of the program. The character percent (%) in line 140 indicates that the FORTRAN statement number (80) is to continue in the next line.

Groups of number are assigned to each type of statements. The *format* statements begin with the statement number (80). They are grouped at the beginning of a program for ease of reference. Ordinary statement numbers begin with (10) as illustrated in line 310. Numbers (40) to (79) are assigned to the DO loops. Furthermore, the statements within a DO loop are indented to facilitate the tracing of the loop. For example, DO

*J. B. Scarborough, *Numerical Mathematical Analysis*, Johns Hopkins Press, 1958, 4th ed., p. 302.

TRESP1

```
"   *** ONE-DEGREE-OF-FREEDOM SYSTEMS - RUNGE-KUTTA METHOD ***
"
"   *** #I. FORMAT AND INPUT. ***
    REAL K1, K2, K3, K4, L1, L2, L3, L4
 80 FORMAT (' TRANSIENT RESPONSE - ONE-DEGREE-OF-FREEDOM SYSTEMS.'%
    ,/,' ASSUME FORCE = CONSTANT.   METHOD: RUNGE-KUTTA.',//, %
    ' DIFFERENTIAL EQUATION:',/,'     D2X/DT2 + 2*Z*WN*DX/DT + WN↑2'%
    '*X = FORCE/M',/,'    WN = FREQ, RAD/S     Z = DAMPING FACTOR',/,%
    '    X  = DISPLACEMENT    Y = VELOCITY = DX/DT',/)
 81 FORMAT (' ENTER: (1) WN = NATURAL FREQUENCY, RAD/S',/,8X, %
    '(2) Z  = DAMPING FACTOR',/,8X,'(3) F  = FORCE/M, A CONSTANT')
 82 FORMAT (' WN =',F8.4,'    Z =',F8.4,'    F =',F8.4,/, %
    ' *** IS THIS CORRECT?    1 = YES;  2 = NO.')
 83 FORMAT (' ENTER DATA:  X0,  Y0,  DT,  TIME')
 84 FORMAT (' X0 = ',F8.5,'   Y0 = ',F8.5,'   DT = ',F8.5,%
    '    TIME = ',F8.5,/,' *** IS THIS CORRECT?   1 = YES;  2 = NO.')
 85 FORMAT (T12, 'TIME',T24, 'DISPLACEMENT',T41,'VELOCITY',/)
 86 FORMAT (3F16.4)
 87 FORMAT (/,' *** RUN AGAIN?    1 = YES;   2 = NO.')
    WRITE (6,80)
 10 WRITE (6,81)
    READ (5,*) WN, Z, F
    WRITE (6,82) WN, Z, F
    READ (5,*) IANS
    IF (IANS = 2) GOTO 10
 11 WRITE (6,83)
    READ (5,*) X, Y, DT, TIME
    WRITE (6,84) X, Y, DT, TIME
    READ (5,*) IANS
    IF (IANS = 2) GOTO 11
    WRITE (6,85)
    T = 0
    WRITE (6,86) T, X, Y
"
"   *** #II. CALCULATIONS AND OUTPUT. ***
    A = WN↑2
    B = 2.*Z*WN
 12 DO 40 I=1,4
      K1 = DT*Y
      L1 = DT*(-B*Y - A*X + F)
      K2 = DT*(Y + L1/2.)
      L2 = DT*(-B*(Y+L1/2.) - A*(X+K1/2.) + F)
      K3 = DT*(Y + L2/2.)
      L3 = DT*(-B*(Y+L2/2.) - A*(X+K2/2.) + F)
      K4 = DT*(Y + L3)
      L4 = DT*(-B*(Y+L3) - A*(X+K3) + F)
      DX = (K1 + 2.*K2 + 2.*K3 + K4)/6.
      DY = (L1 + 2.*L2 + 2.*L3 + L4)/6.
      T = T + DT
      X = X + DX
 40   Y = Y + DY
"   *OUTPUT.
    WRITE (6,86) T, X, Y
    IF (T - TIME) 12, 12, 13
 13 WRITE (6,87)
    READ (5,*) IRUN
    IF (IRUN = 1) GOTO 11
    STOP
    END
```

(*a*) Program listing

FIG. 9-1. *Transient response of one-degree-of-freedom systems with constant excitation—TRESP1.*

```
LOAD TRESP1
READY

RUN

TRESP1


TRANSIENT RESPONSE - ONE-DEGREE-OF-FREEDOM SYSTEMS.
ASSUME FORCE = CONSTANT.    METHOD: RUNGE-KUTTA.

DIFFERENTIAL EQUATION:
  D2X/DT2 + 2*Z*WN*DX/DT + WN↑2*X = FORCE/M
  WN = FREQ, RAD/S          Z = DAMPING FACTOR
  X  = DISPLACEMENT         Y = VELOCITY = DX/DT

ENTER: (1) WN = NATURAL FREQUENCY, RAD/S
       (2) Z  = DAMPING FACTOR
       (3) F  = FORCE/M, A CONSTANT
?3   .2   10

WN =   3.0000      Z =   0.2000      F = 10.0000
*** IS THIS CORRECT?      1 = YES;   2 = NO.
?1

ENTER DATA:  X0,  Y0,   DT,   TIME
?0  , 0   .025   .6

X0 =   0.0          Y0 =   0.0          DT =   0.02500    TIME =   0.60000
*** IS THIS CORRECT?      1 = YES;   2 = NO.
?1

           TIME            DISPLACEMENT          VELOCITY

          0.0                 0.0                 0.0
          0.1000              0.0477              0.9283
          0.2000              0.1795              1.6734
          0.3000              0.3749              2.1934
          0.4000              0.6101              2.4700
          0.5000              0.8609              2.5074
          0.6000              1.1044              2.3295
          0.7000              1.3209              1.9757

*** RUN AGAIN?     1 = YES;   2 = NO.
?2

STOP
```

(*b*) Print-out

FIG. 9-1. (*Continued*)

40 begins at line 480 and ends at line 610. It will be shown in later programs that the amount of indention can also be used to indicate the level of a DO loop.

The program TRESP1 finds the solution of Eq. (9-6) by means of the 4th-order Runge-Kutta method in Eqs. (9-7) and (9-8) and the recurrence formulas in (9-3). Paragraph I consists of the format and input statements. Line 130 declares that the K's and L's are real variables. The format statements are self-explanatory. The data input and their verifications are given in lines 300 to 410. For example, the natural frequency ω_n (WN), the damping factor ζ (Z), and the force F/m (F) are entered by the READ(5,*) statement in line 320. The verification is by the WRITE(6,82) statement in line 330.

In Par. II for the calculations, Eq. (9-8) can be identified readily in lines 490 through 560, Eq. (9-7) in lines 570 and 580, and Eq. (9-3) in lines 590 to 610. The DO 40 loop provides a print-out for every $4\Delta t$. The IF statement in line 640 controls the duration of the run, where TIME is the specified duration. Lines 650 to 670 allow the program to be repeated conveniently. For example, it may be desired to rerun the program with a new Δt in order to check the accuracy of the computation.

The print-out of TRESP1 is shown in Fig. 9-1(*b*). We shall outline the procedure for the program execution. The ***sign-on*** procedure varies with computer centers and will not be described here. When the system is READY, after a successful sign-on, the user types the following commands to execute the program:

LOAD TRESP1 *c*

RUN *c*

where *c* denotes the RETURN key. We assume that TRESP1 is on file in the system.* The first command loads TRESP1 in the working area; the command is executed when the user hits the RETURN key. When the system is READY again, the user types the command RUN and hits the RETURN key. The terminal then prints various statements until it requests for input data, using the character question mark (?). After entering the appropriate information, the user hits the RETURN key. The rest of the print-out is self-evident.

* After the sign-on procedure and when the system is ready, a program can be entered from the teletype by means of typing or a paper tape.

(1) To type-in a program, the terminal command is

ENTER FORTRAN *c*

Then the program, such as TRESP1, is typed-in. To save the program for later usage, the command is

SAVE TRESP1 *c*

TRESP1 is now on file and ready for the user.

(2) To enter a program by means of a paper tape, the commands are

ENTER FORTRAN *c*

TAPE *c*

Assume the tape of TRESP1 is available. The tape is fed to the teletype. When the tape is read, the user returns the system to the KEY mode in order to communicate with the terminal by the command

KEY *c*

The program can now be saved for later usage as before. In both cases, the program can be locked by the command

LOCK TRESP1 *c*

in order to avoid any accidental alterations of the program. Of course, the program can be unlocked for changes by the command

UNLOCK TRESP1 *c*

Let us illustrate the commands for program editing. If it is desired to change line 480 in order to have a print-out for every $10\Delta t$, the user types the command

REPLACE 480, 'I = 1,4', 'I = 1,10' *c*

where *c* denotes the RETURN key. To verify the editing, he types

LIST-N 480 *c*

The system will list the program without heading (LIST-NO-HEAD) starting from line 480. The BREAK key is used to interrupt the print-out after the desired information is listed, otherwise the system will list the rest of the program. A statement can be inserted in-between lines by typing a line number and the desired statement. Similarly, a statement is replaced by typing its line number and the replacing statement. A statement is deleted by typing its line number and striking the RETURN key.

If the existing program is not locked, the edited statements can be saved by the command SAVE. Alternatively, the edited program can be saved under the user's name by the command

SAVE *name* *c*

where *name* is any file name the user may choose. It should be added that the user should delete his program when it is no longer used, or else the file space will soon be filled with many copies of similar programs. If the file is not locked, the command

PURGE *name* *c*

can be used to delete the named program from the user's library. The computer center should be consulted for a list of the terminal commands.

9-4 PROGRAM—TRESPSUB

The program TRESPSUB uses a *sub*routine to compute the *t*ransient *resp*onse of a one-degree-of-freedom system described in Eq. (9-6). It is obtained by modifying TRESP1 in the last section to allow for arbitrary excitations.

The program is organized as described previously and is listed in Fig. 9-2(*a*). Paragraph I gives the format and input statements. Line 150 shows that extended precision, REAL*8, is specified for the variables in order to be consistent with the specification in the subroutine. The excitation $F(t)/m$ in Eq. (9-6) can be arbitrary. A terminated ramp function as shown in Fig. 9-3 is used for purpose of illustration. The number of data points NDATA is entered in line 350 and the NDATA quantized values of the excitation entered in line 410.

The subroutine $TRESP is called to perform the computation in Par. II.

TRESPSUB

```
100 "   *** ONE-DEGREE-OF-FREEDOM SYSTEMS ***
110 "       SUBROUTINE CALCULATES X AND Y FOR EVERY DT/4.
120 "       SUBROUTINE REQD:   $TRESP
130 "
140 "   *** #I. FORMAT AND INPUT. ***
150     REAL*8 DT, FORCE(100), T, WN, X, XO, Y, YO, Z
160 80  FORMAT (' TRANSIENT RESPONSE OF ONE-DEGREE-OF-FREEDOM LINEAR ' %
170     'SYSTEMS.',/,' RUNGE-KUTTA METHOD.',/,' EQUATION:',/,'      ' %
180     'D2X/DT2 + 2*Z*WN*DX/DT + WN↑2*X = FORCE(T)/M',/, %
190     '    X = DISPLACEMENT          WN = NATURAL FREQ., RAD/S',/, %
200     '    Y = VELOCITY, DX/DT    Z   = DAMPING FACTOR',/)
210 81  FORMAT (' ENTER: (1) WN = NATURAL FREQUENCY, RAD/S',/,8X, %
220     '(2) Z   = DAMPING FACTOR',/,8X,'(3) XO = INITIAL DISPLACEMENT'%
230     ,/,8X,'(4) YO = INITIAL VELOCITY',/,8X,'(5) DT = TIME INCRE' %
240     'MENT',/,8X,'(6) NDATA = NO. DATA POINTS FOR FORCE(T)/M (<= 61)')
250 82  FORMAT (' WN =',F7.4,'      Z =',F7.4,/, %
260     ' XO =',F7.4,'     YO =',F7.4,'      DT =',F7.4,'      NDATA =', I4)
270 83  FORMAT (' *** IS THIS CORRECT?    1 = YES;   2 = NO.')
280 84  FORMAT (' ENTER FORCE(I)/M FOR I=1 TO I=NDATA:')
290 85  FORMAT (6F11.4)
300 86  FORMAT (T11, 'TIME', T25, 'FORCE', T37, 'DISPLACEMENT', T54, %
310     'VELOCITY',/)
320 87  FORMAT (5F15.4)
330     WRITE (6,80)
340 10  WRITE (6,81)
350     READ (5,*) WN, Z, XO, YO, DT, NDATA
360     WRITE (6,82) WN, Z, XO, YO, DT, NDATA
370     WRITE (6,83)
380     READ (5,*) IANS
390     IF (IANS = 2) GOTO 10
400 11  WRITE (6,84)
410     READ (5,*) (FORCE(I), I=1,NDATA)
420     WRITE (6,85) (FORCE(I), I=1,NDATA)
430     WRITE (6,83)
440     READ (5,*) IANS
450     IF (IANS = 2) GOTO 11
460     WRITE (6,86)
470     T = 0
480 "
490 "   *** #II. CALL SUBROUTINE $TRESP AND OUTPUT. ***
500     DO 40 I=1,NDATA
510       WRITE (6,87) T, FORCE(I), XO, YO
520       CALL $TRESP (XO, YO, FORCE(I), DT, X, Y, Z, WN)
530       XO = X
540       YO = Y
550 40    T = T + DT
560     STOP
570     END
```

(*a*) Program listing

FIG. 9-2. *Transient response of one-degree-of-freedom systems with transient excitation—TRESPSUB.*

$TRESP is found in Fig. C-1 of App. C. The initial values are XO and YO and the values returned by the subroutine are X and Y. Note that the subroutine (1) assumes the quantized value of $F(t)/m$ is constant for a given time interval Δt, (2) performs an integration for every $\Delta t/4$, and (3) returns the values of X and Y for every Δt.

We assume TRESPSUB and $TRESP are on file in the system. To run the program when the system is READY, the user types the commands

MERGE TRESPSUB, $TRESP *c*

RUN *c*

```
MERGE TRESPSUB, $TRESP
READY

RUN

TRANSIENT RESPONSE OF ONE-DEGREE-OF-FREEDOM LINEAR SYSTEMS.
RUNGE-KUTTA METHOD.
EQUATION:
  D2X/DT2 + 2*Z*WN*DX/DT + WN↑2*X = FORCE(T)/M
  X = DISPLACEMENT         WN = NATURAL FREQ., RAD/S
  Y = VELOCITY, DX/DT      Z  = DAMPING FACTOR

ENTER: (1) WN = NATURAL FREQUENCY, RAD/S
       (2) Z  = DAMPING FACTOR
       (3) XO = INITIAL DISPLACEMENT
       (4) YO = INITIAL VELOCITY
       (5) DT = TIME INCREMENT
       (6) NDATA = NO. DATA POINTS FOR FORCE(T)/M (<= 61)
?3   .2   0   0   .1   8

WN = 3.0000      Z = 0.2000
XO = 0.0         YO = 0.0        DT = 0.1000     NDATA =    8
*** IS THIS CORRECT?   1 = YES;   2 = NO.
?1

ENTER FORCE(I)/M FOR I=1 TO I=NDATA:
?1 3 5 7 9   10 10 10

    1.0000      3.0000       5.0000       7.0000       9.0000      10.0000
   10.0000     10.0000
*** IS THIS CORRECT?   1 = YES;   2 = NO.
?1

         TIME            FORCE         DISPLACEMENT       VELOCITY

         0.0             1.0000          0.0              0.0
         0.1000          3.0000          0.0048           0.0928
         0.2000          5.0000          0.0275           0.3530
         0.3000          7.0000          0.0829           0.7397
         0.4000          9.0000          0.1814           1.2060
         0.5000         10.0000          0.3285           1.7038
         0.6000         10.0000          0.5203           2.0947
         0.7000         10.0000          0.7401           2.2650
STOP
```

(*b*) Print-out

FIG. 9-2. (*Continued*)

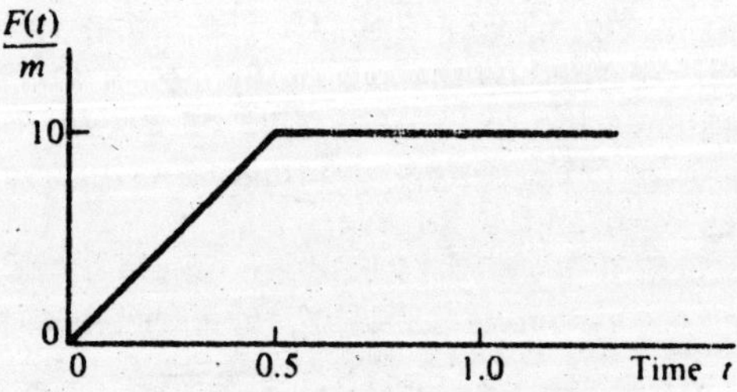

FIG. 9-3. *A terminated ramp function.*

where c denotes the RETURN key. The MERGE command creates a combined program from the programs enumerated. The print-out shown in Fig. 9-2(b) is self-explanatory.

9-5 PROGRAM—TRESPF1

The program TRESPF1 computes the *t*ransient *resp*onse of a one-degree-of-freedom system described in Eq. (9-6) and writes the output in a *f*ile. The program, as listed in Fig. 9-4, is obtained by modifying TRESP-SUB. We shall explain (1) the modifications, (2) the program execution, and (3) the procedure for plotting the computed data.

Comparing with TRESPSUB in Fig. 9-2(a), the modifications, as shown in Fig. 9-4 in the blocks, are as follows:

1. Line 150: The dummy variable (III), to be used for a file name, is added. Thus, each user of the program can name his own file and will not interfere with other users.
2. Lines 330–350: The format statements are self-evident.
3. Lines 490–540: The appropriate file name is entered. The name must consist of *four* characters, such as TSE1.
4. Lines 550–570: The CALL OPEN statement gives the file the reference number (1) and opens the file to receive the OUTPUT from the system. Assume the values of X and Y are to be plotted. Hence NVAR = 2. The number of data points NDATA and NVAR are written in the file by the WRITE statement.
5. Line 640: In executing the program, the values of X and Y are calculated for each $F(t)/m$ by the subroutine \$TRESP. They are renamed as XO and YO and are written in the file.
6. Line 690: The file is closed at the end of the program.

The execution of TRESPF1 in Fig. 9-4(b) uses the commands

FILE *name* c

MERGE TRESPF, \$TRESP c

RUN c

where c denotes the RETURN key. The FILE command creates a file for the data storage, to be used later for the plotting. The print-out, as shown in Fig. 9-4(b), is the same as for TRESPSUB in Fig. 9-2(b) except that (1) NDATA = 21, and (2) the file name TSE1 is entered. Hence the numerical values of the calculations are omitted.

```
TRESPF1

100 "    *** ONE-DEGREE-OF-FREEDOM SYSTEMS ***
110 "        SUBROUTINE CALCULATES X AND Y FOR EVERY DT/4.
120 '        SUBROUTINE REQD:  $TRESP
130 "
140 "    *** #I. FORMAT AND INPUT. ***
150      REAL*8 DT, FORCE(100), T, WN, X, X0, Y, Y0, Z, III
160 80   FORMAT (' TRANSIENT RESPONSE OF ONE-DEGREE-OF-FREEDOM LINEAR ' %
170      'SYSTEMS.',/,' RUNGE-KUTTA METHOD.',/,' EQUATION:',/, '   ' %
180      'D2X/DT2 + 2*Z*WN*DX/DT + WN↑2*X = FORCE(T)/M',/, %
190      '    X = DISPLACEMENT         WN = NATURAL FREQ., RAD/S',/, %
200      '    Y = VELOCITY, DX/DT    Z  = DAMPING FACTOR',/)
210 81   FORMAT (' ENTER: (1) WN = NATURAL FREQUENCY, RAD/S',/,8X, %
220      '(2) Z  = DAMPING FACTOR',/,8X,'(3) X0 = INITIAL DISPLACEMENT'%
230      ,/,8X,'(4) Y0 = INITIAL VELOCITY',/,8X,'(5) DT = TIME INCRE' %
240      'MENT',/,8X,'(6) NDATA = NO. DATA POINTS FOR FORCE(T)/M (<= 61)')
250 82   FORMAT (' WN =',F7.4,'     Z =',F7.4,/, %
260      ' X0 =',F7.4,'     Y0 =',F7.4,'     DT =',F7.4,'     NDATA =', I4)
270 83   FORMAT (' *** IS THIS CORRECT?    1 = YES;  2 = NO.')
280 84   FORMAT (' ENTER FORCE(I)/M FOR I=1 TO I=NDATA:')
290 85   FORMAT (6F11.4)
300 86   FORMAT (T11, 'TIME', T25, 'FORCE', T37, 'DISPLACEMENT', T54, %
310      'VELOCITY',/)
320 87   FORMAT (5F15.4)
330 88   FORMAT (' ENTER DATA FILE NAME.')
340 89   FORMAT (A4)
350 90   FORMAT (2X,A4)
360      WRITE (6,80)
370 10   WRITE (6,81)
380      READ (5,*) WN, Z, X0, Y0, DT, NDATA
390      WRITE (6,82) WN, Z, X0, Y0, DT, NDATA
400      WRITE (6,83)
410      READ (5,*) IANS
420      IF (IANS = 2) GOTO 10
430 11   WRITE (6,84)
440      READ (5,*) (FORCE(I), I=1,NDATA)
450      WRITE (6,85) (FORCE(I), I=1,NDATA)
460      WRITE (6,83)
470      READ (5,*) IANS
480      IF (IANS = 2) GOTO 11
490 12   WRITE (6,88)
500      READ (5,89) III
510      WRITE (6,90) III
520      WRITE (6,83)
530      READ (5,*) IANS
540      IF (IANS = 2) GOTO 12
550      CALL OPEN (1,III,'OUTPUT')
560      NVAR = 2
570      WRITE (1,*) NDATA, NVAR
580      WRITE (6,86)
590      T = 0
600 "
610 "    *** #II. CALL SUBROUTINE $TRESP AND OUTPUT. ***
620      DO 40 I=1,NDATA
630        WRITE (6,87) T, FORCE(I), X0, Y0
640        WRITE (1,*) X0, Y0
650        CALL $TRESP (X0, Y0, FORCE(I), DT, X, Y, Z, WN)
660        X0 = X
670        Y0 = Y
680 40     T = T + DT
690      CALL CLOSE (1)
700      STOP
710      END
```

(*a*) Program listing

Fig. 9-4. *Transient response of one-degree-of-freedom systems; Output stored in file for plotting—TRESPF1.*

```
FILE TSE1
TSE1 LIMIT = 4 UNITS

MERGE TRESPF1, $TRESP
READY

RUN

TRANSIENT RESPONSE OF ONE-DEGREE-OF-FREEDOM LINEAR SYSTEMS.
RUNGE-KUTTA METHOD.
EQUATION:
  D2X/DT2 + 2*Z*WN*DX/DT + WN↑2*X = FORCE(T)/M
  X = DISPLACEMENT          WN = NATURAL FREQ., RAD/S
  Y = VELOCITY, DX/DT       Z  = DAMPING FACTOR

ENTER: (1) WN = NATURAL FREQUENCY, RAD/S
       (2) Z  = DAMPING FACTOR
       (3) X0 = INITIAL DISPLACEMENT
       (4) Y0 = INITIAL VELOCITY
       (5) DT = TIME INCREMENT
       (6) NDATA = NO. DATA POINTS FOR FORCE(T)/M (<= 61)
?3   .2   0   0   .1   21

WN = 3.0000     Z = 0.2000
X0 = 0.0        Y0 = 0.0        DT = 0.1000     NDATA =  21
*** IS THIS CORRECT?   1 = YES;  2 = NO.
?1

ENTER FORCE(I)/M FOR I=1 TO I=NDATA:
?1 3 5 7 9    10 10 10 10 10    10 10 10 10 10    10 10 10 10 10    10

    1.0000      3.0000      5.0000      7.0000      9.0000     10.0000
   10.0000     10.0000     10.0000     10.0000     10.0000     10.0000
   10.0000     10.0000     10.0000     10.0000     10.0000     10.0000
   10.0000     10.0000     10.0000
*** IS THIS CORRECT?   1 = YES;  2 = NO.
?1

ENTER DATA FILE NAME.
?TSE1

 TSE1
*** IS THIS CORRECT?   1 = YES;  2 = NO.
?1

        TIME          FORCE        DISPLACEMENT     VELOCITY

STOP
```

(*b*) Print-out

FIG. 9-4. (*Continued*)

The program PLOTFILE for *plot*ting data from a *file* is listed in Fig. 9-5(*a*). The associated subroutine $PLOTF is listed in App. C. The procedure for using PLOTFILE consists of a sequence of commands as shown in Fig. 9-5(*b*).

FILE *name* *c*

MERGE PLOTFILE, $PLOTF *c*

RUN *c*

where *c* denotes the RETURN key. The *name*, consisting of *four* characters, is the name of the file created previously for the plotting.

```
PLOTFILE

   " *** PROGRAM TO PLOT FROM A DATA FILE. ***
   "      SUBROUTINE REQD:   $PLOTF
   "
        REAL*8  III
     80 FORMAT (' PROGRAM TO PLOT FROM A DATA FILE.',/)
     81 FORMAT (' ENTER NAME OF DATA FILE TO BE PLOTTED.')
     82 FORMAT (A4)
     83 FORMAT (2X,A4)
     84 FORMAT (' *** IS THIS CORRECT?    1 = YES;  2 = NO.')
     85 FORMAT (/,' WANT ANOTHER PLOT?    1 = YES;  2 = NO.')
        WRITE (6,80)
     10 WRITE (6,81)
        READ (5,82) III
        WRITE (6,83) III
        WRITE (6,84)
        READ (5,*) IANS
        IF (IANS = 2) GOTO 10
        CALL $PLOTF (III)
        WRITE (6,85)
        READ (5,*) IANS
        IF (IANS = 1) GOTO 10
        STOP
        END
```

(a) Program listing

FIG. 9-5. *Program to plot from a data file—PLOTFILE.*

The output of PLOTFILE is as shown in Fig. 9-5(b). Before plotting, the system requests

DO THE VALUES TO BE PLOTTED USE THE SAME SCALE?

1. When the answer is YES, the plot is as shown. The curves are normalized relative to the largest numerical value in the data set. The computed values are obtained from the print-out of the originating program. In the figure, the displacement X is denoted by the symbol (1) and the velocity Y by (2). The symbol (2) is printed if X and Y have the same integer values.
2. When the answer is NO, each curve is normalized relative to its own largest numerical value. This technique of scaling allows the curves with small numerical values to be magnified.

9-6 ONE-DEGREE-OF-FREEDOM SYSTEMS—HARMONIC RESPONSE

The impedance method for the harmonic analysis of one-degree-of-freedom systems was presented in Chap. 2 and the applications illustrated in Chap. 3. We shall consider a typical problem in order to discuss the program FRESP1.

Consider the system and its harmonic response described in Eqs. (2-49) and (2-50).

$$m\ddot{x} + c\dot{x} + kx = \bar{F}e^{j\omega t} \tag{9-9}$$

$$x = \bar{X}e^{j\omega t} \tag{9-10}$$

```
FILE TSE1
TSE1 LIMIT = 4 UNITS

MERGE PLOTFILE, $PLOTF
READY

RUN

PROGRAM TO PLOT FROM A DATA FILE.

ENTER NAME OF DATA FILE TO BE PLOTTED.
?TSE1

 TSE1
*** IS THIS CORRECT?   1 = YES;  2 = NO.
?1

DO THE VALUES TO BE PLOTTED USE THE SAME SCALE?
1 = YES;   2 = NO.
```

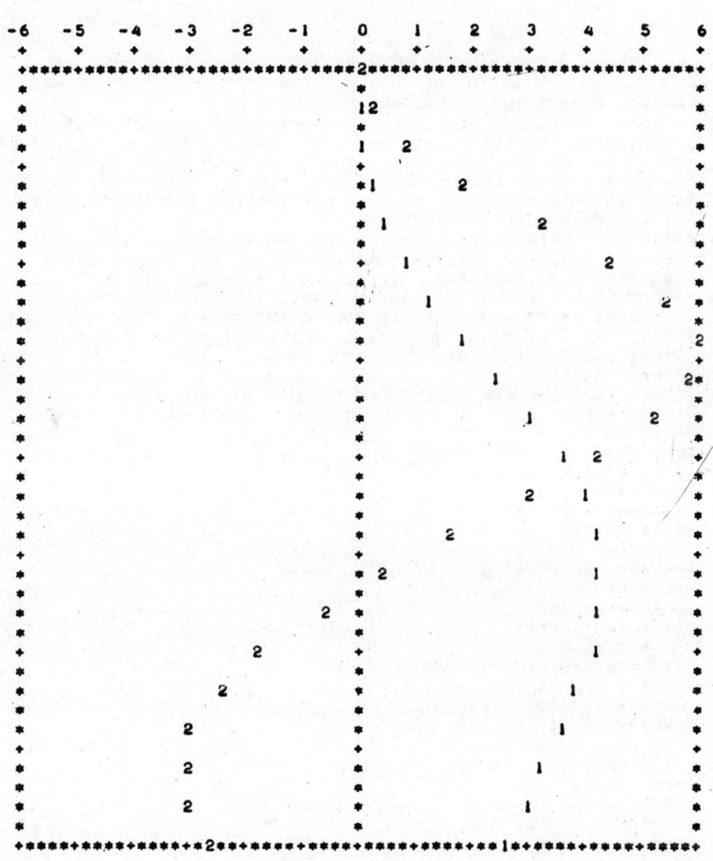

(*b*) Print-out

Fig. 9-5. (*Continued*)

Let the phase angle of the phasor $\bar{F}$ be zero, that is, $\bar{F} = F$. From Eq. (2-52), the phasor $\bar{X}$ is

$$\bar{X} = \frac{F}{k - \omega^2 m + j\omega c} = Xe^{-j\phi} \tag{9-11}$$

This can be obtained by substituting Eq. (9-10) in (9-9), factoring out the $e^{j\omega t}$ term, and simplifying. Factoring out k, defining $r = \omega/\omega_n$, and simplifying, we have

$$\frac{X}{F/k} = \frac{1}{|1 - r^2 + j2\zeta r|} \tag{9-12}$$

or

$$\frac{X}{F/k} = \frac{1}{\sqrt{(1-r^2)^2 + (2\zeta r)^2}} \tag{9-13}$$

$$\phi = \tan^{-1}\frac{2\zeta r}{1 - r^2} \tag{9-14}$$

```
FRESP1

"   *** FREQUENCY RESPONSE - ONE-DEGREE-OF-FREEDOM SYSTEMS ***
"
"   *** #I. FORMAT AND INPUT. ***
 80 FORMAT(' FREQUENCY RESPONSE - ONE-DEGREE-OF-FREEDOM SYSTEMS.',/,%
    ' MECHANICAL IMPEDANCE METHOD.',//,' EQUATIONS:',/,' AMPLITUDE' %
    ' RATIO = 1/SQRT((1-R↑2)↑2 + (2*Z*R)↑2)',/,' PHASE ANGLE = '%
    '-57.296*ATAN2(2*Z*R, (1-R↑2))',/,'   R = (EXCITATION FREQUEN' %
    'CY)/(NATURAL FREQUENCY)',/,'   Z = DAMPING FACTOR',/,'   QUOT'%
    ' = R(NEW)/R(OLD), A CONSTANT',/)
 81 FORMAT (' ENTER DATA:  (1) Z = DAMPING FACTOR',/,14X,'(2)'%
    ' RMIN = STARTING FREQUENCY RATIO',/,14X,'(3) RMAX = END'%
    'ING FREQUENCY RATIO',/,14X,'(4) QUOT = R(NEW)/R(OLD)')
 82 FORMAT (' Z =',F7.4,'    RMIN =',F7.4,'    RMAX =',F7.4,%
          QUOT =',F7.4,/,' *** IS THIS CORRECT?   1 = YES;  2 = NO.')
 84 FORMAT (5X,'FREQUENCY RATIO',3X,'AMPLITUDE RATIO',2X, %
    'PHASE ANGLE (DEG)',/)
 85 FORMAT (3E18.4)
 86 FORMAT (/,' *** RUN AGAIN?   1 = YES;  2 = NO.')
    WRITE (6,80)
 10 WRITE (6,81)
    READ (5,*) Z, RMIN, RMAX, QUOT
    WRITE (6,82) Z, RMIN, RMAX, QUOT
    READ (5,*) IANS
    IF (IANS = 2) GOTO 10
    WRITE (6,84)
"
"   *** #II. CALCULATIONS AND OUTPUT. ***
    R = RMIN
    DO 40 I=1, 100
      A = (1. - R↑2)
      B = 2.*Z*R
      IF ((A↑2+B↑2) < 1.E-20) GOTO 40
      AMP = 1./SQRT(A↑2 + B↑2)
      PHI = -57.296*ATAN2(B,A)
      WRITE (6,85) R, AMP, PHI
      IF (R > RMAX) GOTO 11
 40   R = R*QUOT
 11 WRITE (6,86)
    READ (5,*) IRUN
    IF (IRUN = 1) GOTO 10
    STOP
    END
```

(*a*) Program listing

FIG. 9-6. *Frequency response of one-degree-of-freedom systems—FRESP1.*

where ζ and ω_n are defined in Eq. (2-27). The amplitude of the harmonic response can be obtained from Eq. (9-12) or (9-13) and the phase angle ϕ, relative to the excitation, from Eq. (9-14).

The harmonic analysis of all the problems in Chap. 3 can be treated in like manner. An equivalent force F_{eq} can be substituted for F in the equation of motion, as shown in Eq. (3-22). The F_{eq} may be frequency dependent as in Eq. (3-27). The basic technique in the impedance method, however, is to substitute $j\omega$ for all the time derivatives in the equation of motion and then solve for the unknown. The terms frequency response and harmonic response are synonymous, although the former is more often used in industry.

The program FRESP1 for the *f*requency *resp*onse of one-degree-of-freedom systems is listed in Fig. 9-6(*a*). The format and input statements in Par. I are self-evident, except the term QUOT = R(NEW)/R(OLD), as in line 210.

The program is written for a nondimensional log-log plot of the magnification factor $X/(F/k)$ versus frequency ratio r, such as in Fig. 2-10.

```
LOAD FRESP1
READY

RUN

FRESP1


FREQUENCY RESPONSE - ONE-DEGREE-OF-FREEDOM SYSTEMS.
MECHANICAL IMPEDANCE METHOD.

EQUATIONS:
AMPLITUDE RATIO = 1/SQRT((1-R↑2)↑2 + (2*Z*R)↑2)
PHASE ANGLE = -57.296*ATAN2(2*Z*R, (1-R↑2))
  R = (EXCITATION FREQUENCY)/(NATURAL FREQUENCY)
  Z = DAMPING FACTOR
  QUOT = R(NEW)/R(OLD), A CONSTANT

ENTER DATA:   (1) Z = DAMPING FACTOR
              (2) RMIN = STARTING FREQUENCY RATIO
              (3) RMAX = ENDING FREQUENCY RATIO
              (4) QUOT = R(NEW)/R(OLD)
?.1   .5   3   2

Z = 0.1000     RMIN = 0.5000     RMAX = 3.0000     QUOT = 2.0000
*** IS THIS CORRECT?   1 = YES;   2 = NO.
?1

     FREQUENCY RATIO    AMPLITUDE RATIO   PHASE ANGLE (DEG)

       0.5000E+00          0.1322E+01         -0.7595E+01
       0.1000E+01          0.5000E+01         -0.9000E+02
       0.2000E+01          0.3304E+00         -0.1724E+03
       0.4000E+01          0.6657E-01         -0.1769E+03

*** RUN AGAIN?   1 = YES;   2 = NO.
?2

STOP
```

(*b*) Print-out

FIG. 9-6. (*Continued*)

In order to obtain equally spaced points in a log-scale along the r axis, we choose to increment r in terms of a quotient rather than their absolute values. For example, the length along a log-scale between the numbers 2 and 4 is the same as that between 30 and 60. The quotient of these numbers is 2:1. Hence a constant number QUOT is entered in the program to increment the frequency ratio r in order to obtain equally spaced points along the frequency-ratio r axis.

The calculations in Par. II show that Eq. (9-13) is implemented in line 420 and Eq. (9-14) in line 430. The IF statement in line 410 is to prevent an overflow when the denominator in Eq. (9-13) approaches zero.

The print-out is shown in Fig. 9-6(b). The program can be modified, as described in the last section, to have the results plotted from the terminal. Alternatively, a family of curves, as shown in Fig. 3-10, can be plotted.

9-7 N-DEGREE-OF-FREEDOM SYSTEMS—HARMONIC RESPONSE

The program FRESPN computes the *f*requency *resp*onse of n-degree-of-freedom systems. We shall first briefly review the impedance method described in Sec. 4-6.

The equations of motion of discrete systems with viscous damping from Eq. (4-3) are

$$M\{\ddot{x}\}+C\{\dot{x}\}+K\{x\}=\{F(t)\} \tag{9-15}$$

where M is the mass matrix, C the damping matrix, K the stiffness matrix, and $\{F(t)\}$ the excitation vector. When the excitation is harmonic and all the elements in $\{F(t)\}$ are of the same frequency, the vector is replaced by $\{\bar{F}\}e^{j\omega t}$. The phasor $\{\bar{F}\}$ is complex. It indicates that the elements in $\{\bar{F}\}$ need not be in phase. Applying the impedance method by substituting $j\omega$ for the time derivatives and simplifying, we obtain the equations

$$[K-\omega^2M+j\omega C]\{\bar{X}\}=\{\bar{F}\} \tag{9-16}$$

$$Z(\omega)\{\bar{X}\}=\{\bar{F}\} \tag{9-17}$$

$$\{\bar{X}\}=Z(\omega)^{-1}\{\bar{F}\} \tag{9-18}$$

which are Eqs. (4-52), (4-55), and (4-57), respectively. $Z(\omega)\triangleq[K-\omega^2M+j\omega C]$ is the impedance matrix and $\{\bar{X}\}$ the solution vector. Thus, a solution $\{\bar{X}\}$ is found for a given excitation frequency ω. The repeated application of Eq. (9-18) for a frequency range of interest yields the harmonic response of the system.

The program FRESPN for the *f*requency *resp*onse of n-degree-of-freedom systems is listed in Fig. 9-7(a). The format and input statements are listed in Par. I. Note that (1) the complex mode is specified in line 150, and (2) the input is verified after each entry, such as shown in lines 410 to 470.

```
FRESPN

100 "   *** FREQUENCY RESPONSE OF N-DEGREE-OF-FREEDOM SYSTEMS ***
110 "        SUBROUTINES REQD:   (1) $CINVS  (2) $CMPLY  (3) $CSUBN
120 "
130 "   *** #I. FORMAT AND INPUT. ***
140     REAL  AMP(10), C(10,10), K(10,10), M(10,10), PHASE(10)
150     COMPLEX  F(10), X(10), Z(10,10), ZINVS(10,10)
160 80 FORMAT (' FREQUENCY RESPONSE OF N-DEGREE-OF-FREEDOM SYSTEMS ',/,%
170     ' HARMONIC EXCITATION FORCES AT SAME FREQUENCY.',/,' MECHANICAL'%
180     ' IMPEDANCE METHOD:',/,'      SUBSTITUTE (JW) FOR THE TIME ' %
190     'DERIVATIVE IN EQUATIONS OF MOTION.',/,'     SOLVE THE RESULTANT' %
200     ' SIMULTANEOUS ALGEBRAIC EQUATIONS.',/)
210 81 FORMAT (' ENTER: N = NO. DEGREES OF FREEDOM    ** N <= 10')
220 82 FORMAT (' N =', I3,'    ** IS THIS CORRECT?   1 = YES;  2 = NO.')
230 83 FORMAT (' ENTER MATRIX-M BY ROW:-')
240 84 FORMAT (5E14.5)
250 85 FORMAT (' *** IS THIS CORRECT?    1 = YES;   2 = NO.')
260 86 FORMAT (' ENTER MATRIX-C BY ROW:-')
270 87 FORMAT (' ENTER MATRIX-K BY ROW:-')
280 88 FORMAT (' ENTER:    WMIN    WMAX    DW    COMPLEX FORCE VECTOR')
290 89 FORMAT (' WMIN =',E13.5,'     WMAX =',E13.5,'     DW =',E13.5)
300 90 FORMAT (' COMPLEX EXCITATION FORCE VECTOR:-',/,(6E12.4))
310 91 FORMAT (12X,'* * * OUTPUT * * *')
320 92 FORMAT (/,' FREQ:      ',E12.4,/,' AMP(I):   ',5E12.4,/,10X,5E12.4)
330 93 FORMAT (' PHASE(I):',5E12.4,/,10X,5E12.4)
340 94 FORMAT (/,' *** RUN AGAIN?    1 = YES;   2 = NO.')
350    WRITE (6,80)
360 10 WRITE (6,81)
370    READ (5,*) N
380    WRITE (6,82) N
390    READ (5,*) IANS
400    IF (IANS = 2) GOTO 10
410 11 WRITE (6,83)
420    READ (5,*) ((M(I,J), J=1,N), I=1,N)
430    DO 40 I=1,N
440 40   WRITE (6,84) (M(I,J), J=1,N)
450    WRITE (6,85)
460    READ (5,*) IANS
470    IF (IANS = 2) GOTO 11
480 12 WRITE (6,86)
490    READ (5,*) ((C(I,J), J=1,N), I=1,N)
500    DO 41 I=1,N
510 41   WRITE (6,84) (C(I,J), J=1,N)
520    WRITE (6,85)
530    READ (5,*) IANS
540    IF (IANS = 2) GOTO 12
550 13 WRITE (6,87)
560    READ (5,*) ((K(I,J), J=1,N), I=1,N)
570    DO 42 I=1,N
580 42   WRITE (6,84) (K(I,J), J=1,N)
590    WRITE (6,85)
600    READ (5,*) IANS
610    IF (IANS = 2) GOTO 13
620 14 WRITE (6,88)
630    READ (5,*) WMIN, WMAX, DW, (F(I), I=1,N)
640    WRITE (6,89) WMIN, WMAX, DW
650    WRITE (6,90) (F(I), I=1,N)
660    WRITE (6,85)
670    READ (5,*) IANS
680    IF (IANS = 2) GOTO 14
690    WRITE (6,91)
700 "
```

(*a*) Program listing

FIG. 9-7. *Frequency response of n-degree-of-freedom systems—FRESPN.*

```
710 "  *** #II. CALCULATIONS AND OUTPUT. ***
720     W = WMIN
730     DO 43 L=1,1000
740        DO 44 I=1,N
750        DO 44 J=1,N
760 44       Z(I,J) = CMPLX(-M(I,J)*W↑2 + K(I,J), C(I,J)*W)
770        CALL $CINVS (Z, ZINVS, N)
780        DO 45 I=1,N
790          X(I) = CMPLX(0., 0.)
800          DO 46 J=1,N
810 46         X(I) = X(I) + ZINVS(I,J)*F(J)
820          IF (ABS(REAL(X(I)))<1.E-20) X(I) = CMPLX(1.E-20,AIMAG(X(I)))
830          AMP(I) = CABS(X(I))
840 45       PHASE(I) = 57.269*ATAN2(AIMAG(X(I)), REAL(X(I)))
850        WRITE (6,92) W, (AMP(I), I=1,N)
860        WRITE (6,93) (PHASE(I), I=1,N)
870        W = W + DW
880        IF (W > WMAX) GOTO 15
890 43     CONTINUE
900 15 WRITE (6,94)
910     READ (5,*) IRUN
920     IF (IRUN = 1) GOTO 14
930     STOP
940     END
```

(*a*) Program listing (*Continued*)

FIG. 9-7. (*Continued*)

The calculation and output statements are shown in Par. II, starting from the lowest excitation frequency ω_{min} (WMIN) in line 720. The impedance matrix in Eq. (9-16) is formed in the DO 44 loop, where each element Z(I,J) is a complex number. The inverse ZINVS of the impedance $Z(\omega)$ is found by the CALL $CINVS statement in line 770. The solution vector $\{\bar{X}\}$ in Eq. (9-18) is the product of ZINVS and $\{\bar{F}\}$ as shown in line 810. AMP(I) and PHASE(I) are the magnitude and phase angle of each of the elements in $\{\bar{X}\}$. The WRITE statements for the print-out in lines 850 and 860 are self-evident.

As shown in the print-out in Fig. 9-7(*b*), the commands to execute the program are

MERGE FRESPN, $CINVS, $CMPLY, $CSUBN *c*

REPLACE ALL, 'COMPLEX*16', 'COMPLEX' *c*

RUN *c*

where *c* denotes the RETURN key. The required subroutines are $CINVS, $CMPLY, $CSUBN as stated in line 110 in Fig. 9-7(*a*). $CINVS is for the complex matrix inversion, $CMPLY for the complex matrix multiplication, and $CSUBN for the complex matrix substitution. The main program FRESPN is first merged with the subroutines to form a combined program. Since no iteration is required in the main program, it is written in single precision. The REPLACE ALL command converts the subroutines from extended to single precision. The command RUN initiates the execution.

The print-out is self-evident, except for the complex force vector $\{\bar{F}\}$. A complex number is entered as (a,b), where a and b are the real and

```
MERGE FRESPN, $CINVS, $CMPLY, $CSUBN
READY
REPLACE ALL,'COMPLEX*16','COMPLEX'
REPLACED 3
READY
RUN

FREQUENCY RESPONSE OF N-DEGREE-OF-FREEDOM SYSTEMS.
HARMONIC EXCITATION FORCES AT SAME FREQUENCY.
MECHANICAL IMPEDANCE METHOD:
  SUBSTITUTE (JW) FOR THE TIME DERIVATIVE IN EQUATIONS OF MOTION.
  SOLVE THE RESULTANT SIMULTANEOUS ALGEBRAIC EQUATIONS.

ENTER: N = NO. DEGREES OF FREEDOM     ** N <= 10
?3

N =   3    ** IS THIS CORRECT?    1 = YES;   2 = NO.
?1

ENTER MATRIX-M BY ROW:-
?3 0 0   0 1 0   0 0 2

  0.30000E+01   0.0            0.0
  0.0           0.10000E+01    0.0
  0.0           0.0            0.20000E+01
*** IS THIS CORRECT?    1 = YES;   2 = NO.
?1
ENTER MATRIX-C BY ROW:-
?10 -3 -2    -3 8 -4    -2 -4 7

  0.10000E+02  -0.30000E+01  -0.20000E+01
 -0.30000E+01   0.80000E+01  -0.40000E+01
 -0.20000E+01  -0.40000E+01   0.70000E+01
*** IS THIS CORRECT?    1 = YES;   2 = NO.
?1

ENTER MATRIX-K BY ROW:-
?140 -60 -20    -60 220 -80    -20 -80 200

  0.14000E+03  -0.60000E+02  -0.20000E+02
 -0.60000E+02   0.22000E+03  -0.80000E+02
 -0.20000E+02  -0.80000E+02   0.20000E+03
*** IS THIS CORRECT?    1 = YES;   2 = NO.
?1

ENTER:    WMIN    WMAX    DW    COMPLEX FORCE VECTOR
?2   3   1   1 0 0

WMIN =   0.20000E+01     WMAX =   0.30000E+01     DW =   0.10000E+01
COMPLEX EXCITATION FORCE VECTOR:-
 0.1000E+01  0.0          0.0          0.0          0.0          0.0
*** IS THIS CORRECT?    1 = YES;   2 = NO.
?1

                 * * * OUTPUT * * *

FREQ:        0.2000E+01
AMP(I):      0.9842E-02   0.3697E-02   0.2586E-02
PHASE(I):   -0.8516E+01  -0.5609E+01  -0.2985E+01

FREQ:        0.3000E+01
AMP(I):      0.1149E-01   0.4534E-02   0.3310E-02
PHASE(I):   -0.1480E+02  -0.1062E+02  -0.7099E+01

*** RUN AGAIN?    1 = YES;   2 = NO.
?2

STOP
```

(*b*) Print-out

FIG. 9-7. (*Continued*)

imaginary parts, respectively. The entry for $\{\bar{F}\}$ is $\{1 \quad 0 \quad 0\}$. This indicates the real parts only and implies the imaginary parts are zeros. The verification in the print-out shows that $\{\bar{F}\}$ is a complex vector, consisting of the complex numbers {(1,0) (0,0) (0,0)}.

9-8 TRANSIENT RESPONSE OF UNDAMPED DISCRETE SYSTEMS

The program TRESPUND for the *t*ransient *resp*onse of *und*amped discrete systems with arbitrary excitation employs the theory in Chaps. 4 and 6. The equations of motion from Eq. (4-35) or (6-62) are

$$M\{\ddot{q}\}+K\{q\}=\{Q(t)\} \tag{9-19}$$

where M is the mass matrix, K the stiffness matrix, $\{q\}$ the generalized coordinate, and $\{Q(t)\}$ the generalized excitation. The steps in the modal analysis are (1) to uncouple the equations of motion and to express them in terms of the principal coordinates $\{p\}$, (2) to solve each of the uncoupled equations as an independent one-degree-of-freedom system, and (3) to transform the solutions from the $\{p\}$ to the original $\{q\}$ coordinates.

Let us summarize the equations used in the program. The coordinates $\{p\}$ and $\{q\}$ are related by the transformation

$$\{q\}=[u]\{p\} \qquad \text{or} \qquad \{p\}=[u]^{-1}\{q\} \tag{9-20}$$

as shown in Eq. (4-34) or (6-23), where $[u]$ is the modal matrix. Following Eq. (4-41) or (6-63), we get

$$[u]^T M[u]\{\ddot{p}\}+[u]^T K[u]\{p\}=[u]^T\{Q(t)\} \tag{9-21}$$

or

$$[\ulcorner M \lrcorner]\{\ddot{p}\}+[\ulcorner K \lrcorner]\{p\}=\{N(t)\} \tag{9-22}$$

where $[\ulcorner M \lrcorner]$ and $[\ulcorner K \lrcorner]$ are diagonal with m_{ii} and k_{ii} as their respective diagonal elements. The excitation vector is $\{N(t)\}=[u]^T\{Q(t)\}$. Since the equations in Eq. (9-22) are uncoupled, the natural frequencies of the system are

$$\omega_i=\sqrt{k_{ii}/m_{ii}} \tag{9-23}$$

The program TRESPUND is listed in Fig. 9-8(a). Line 150 in Par. I shows that extended precision REAL*8 is specified, because an iterative method is used to find the roots of the characteristic equation in the subroutine $ROOT. Since the system is undamped, the damping factor ζ(Z) is initialized to zero in the DATA statement in line 170. The displacement and velocity are X(I) and Y(I) in the original generalized coordinates and P(I) and Q(I) in the principal coordinates. The format statements are self-evident. The excitation applied to each mass is

TRESPUND

```
"   *** TRANSIENT RESPONSE OF UNDAMPED DISCRETE SYSTEMS ***
"
"   *METHOD: (1) FIND MODAL MATRIX BY SUBROUTINE $MODL,
"            (2) SOLVE UNCOUPLED EQUATIONS BY SUBROUTINE $TRESP, THEN
"            (3) TRANSFORM SOLUTION FROM PRINCIPAL COORDINATES TO
"                ORIGINAL COORDINATES.
"   *SUBROUTINES REQD:  (1) $MODL  (2) $INVS  (3) $MPLY  (4) $SUBN
"                       (5) $ROOT  (6) $HOMO  (7) $COEFF (8) $TRESP
"
"   *** #1. FORMAT AND INPUT. ***
    REAL*8  DT, DUM(10,10), ERROR, FN(10), FORCE(10,100), K(10,10), %
    M(10,10), P(10), PO(10), Q(10), QO(10), ROOT(10), T, U(10,10),%
    UINVS(10,10), UT(10,10), WN(10), X(10), XO(10), Y(10), YO(10), %
    Z(10)
    DATA  FN, FORCE, P, PO, Q, QO, T, Z /1061*0.0/
 80 FORMAT (' TRANSIENT RESPONSE OF UNDAMPED POSITIVE DEFINITE' %
    ' SYSTEMS.',/,' EIGENVALUES DISTINCT.',/)
 81 FORMAT (' ENTER: (1) N       = ORDER OF MATRICES M AND K',/, %
    8X,'(2) NDATA = NO. DATA POINTS FOR THE QUANTIZED FORCE',/,%
    8X,'(3) DT    = TIME INCREMENT IN INTEGRATION',/, %
    8X,'(4) ERROR = ERROR SPECIFIED IN ITERATIONS',/,8X,'(5) NITER'%
    ' = MAX. ITERATIONS FOR EACH MODE')
 82 FORMAT (' N =',I3,'    NDATA =',I3,'     DT =',D12.4,/, %
    ' ERROR =',D12.4,'    NITER =',I4)
 83 FORMAT (' *** IS THIS CORRECT?   1 = YES;  2 = NO.')
 84 FORMAT (' ENTER MATRIX-M BY ROW:-')
 85 FORMAT (' ENTER MATRIX-K BY ROW:-')
 86 FORMAT (6D12.4)
 87 FORMAT (' ENTER: (1) INITIAL DISPLACEMENT XO(1) TO XO(N)',/,%
    8X,'(2) INITIAL VELOCITY YO(1) TO YO(N)')
 88 FORMAT (' ENTER FORCE(T) AT TIME T(1) TO T(NDATA) FOR EACH '%
    'OF THE N-EQUATIONS.')
 89 FORMAT (13X,'* * * OUTPUT * * *')
 90 FORMAT (/,' TIME:       ',D12.4)
 91 FORMAT (' FORCE:      ',5D12.4,/,11X,5D12.4)
 92 FORMAT (' X1 TO XN:  ',5D12.4,/,11X,5D12.4)
 93 FORMAT (' VELOCITY:  ',5D12.4,/,11X,5D12.4)
    WRITE (6,80)
 10 WRITE (6,81)
    READ (5,*) N, NDATA, DT, ERROR, NITER
    WRITE (6,82) N, NDATA, DT, ERROR, NITER
    WRITE (6,83)
    READ (5,*) IANS
    IF (IANS = 2) GOTO 10
 11 WRITE (6,84)
    READ (5,*) ((M(I,J), J=1,N), I=1,N)
    DO 40 I=1,N
 40   WRITE (6,86) (M(I,J), J=1,N)
    WRITE (6,83)
    READ (5,*) IANS
    IF (IANS = 2) GOTO 11
 12 WRITE (6,85)
    READ (5,*) ((K(I,J), J=1,N), I=1,N)
    DO 41 I=1,N
 41   WRITE (6,86) (K(I,J), J=1,N)
    WRITE (6,83)
    READ (5,*) IANS
    IF (IANS = 2) GOTO 12
```

(*a*) Program listing

FIG. 9-8. *Transient response of undamped discrete systems; transient excitation—TRESPUND.*

```
13 WRITE (6,87)
   READ (5,*) (XO(I), I=1,N), (YO(I), I=1,N)
   WRITE (6,86) (XO(I), I=1,N)
   WRITE (6,86) (YO(I), I=1,N)
   WRITE (6,83)
   READ (5,*) IANS
   IF (IANS = 2) GOTO 13
   WRITE (6,88)
   DO 42 I=1,N
14   READ (5,*) (FORCE(I,J), J=1,NDATA)
     WRITE (6,86) (FORCE(I,J), J=1,NDATA)
     WRITE (6,83)
     READ (5,*) IANS
     IF (IANS = 2) GOTO 14
42   CONTINUE
   WRITE (6,89)
"
"  *** #II. GET MODAL MATRIX TO UNCOUPLE EQUATIONS. ***
   CALL $MODL (M, K, U, ROOT, ERROR, NITER, N)
   DO 43 I=1,N
   DO 43 J=1,N
43   UT(I,J) = U(J,I)
   CALL $MPLY (UT, M, DUM, N)
   CALL $MPLY (DUM, U, M, N)
   CALL $MPLY (UT, K, DUM, N)
   CALL $MPLY (DUM, U, K, N)
   DO 44 I=1,N
44   WN(I) = DSQRT(K(I,I)/M(I,I))
"  *EXPRESS XO AND YO IN PRINCIPAL COORDINATES.
   CALL $INVS (U, UINVS, N)
   DO 45 I=1,N
   DO 45 J=1,N
     PO(I) = PO(I) + UINVS(I,J)*XO(J)
45   QO(I) = QO(I) + UINVS(I,J)*YO(J)
"
"  *** #III. CALCULATE TRANSIENT RESPONSE. ***
"  *TRANSIENT RESPONSE IN PRINCIPAL COORDINATES.
   DO 46 IT=1,NDATA
     WRITE (6,90) T
     WRITE (6,91) (FORCE (I,IT), I=1,N)
     WRITE (6,92) (XO(I), I=1,N)
     WRITE (6,93) (YO(I), I=1,N)
     DO 47 I=1,N
       FN(I) = 0
       DO 48 J=1,N
48       FN(I) = FN(I) + UT(I,J)*FORCE(J,IT)
47     FN(I) = FN(I)/M(I,I)
     DO 49 I=1,N
       CALL $TRESP(PO(I), QO(I), FN(I), DT, P(I), Q(I), Z(I),WN(I))
       PO(I) = P(I)
49     QO(I) = Q(I)
"    *TRANSIENT RESPONSE IN ORIGINAL COORDINATES.
     DO 50 I=1,N
       XO(I) = 0
       YO(I) = 0
       DO 50 J=1,N
         XO(I) = XO(I) + U(I,J)*PO(J)
50       YO(I) = YO(I) + U(I,J)*QO(J)
46   T = T + DT
    STOP
    END
```

(*a*) Program listing (*Continued*)

FIG. 9-8. (*Continued*)

```
MERGE $MODL, $INVS, $MPLY, $SUBN, $COEFF, $ROOT, $HOMO, $TRESP
READY

SAVE $TEMP
READY

MERGE TRESPUND, $TEMP
READY

RUN

TRANSIENT RESPONSE OF UNDAMPED POSITIVE DEFINITE SYSTEMS.
EIGENVALUES DISTINCT.

ENTER: (1) N     = ORDER OF MATRICES M AND K
       (2) NDATA = NO. DATA POINTS FOR THE QUANTIZED FORCE
       (3) DT    = TIME INCREMENT IN INTEGRATION
       (4) ERROR = ERROR SPECIFIED IN ITERATIONS
       (5) NITER = MAX. ITERATIONS FOR EACH MODE
?3   4   .1    1D-16   200

N =  3     NDATA =  4      DT =  0.1000D+00
ERROR =  0.1000D-15     NITER = 200
*** IS THIS CORRECT?   1 = YES;   2 = NO.
?1

ENTER MATRIX-M BY ROW:-
?2 0 0   0 1 0    0 0 2

 0.2000D+01  0.0          0.0
 0.0         0.1000D+01   0.0
 0.0         0.0          0.2000D+01
*** IS THIS CORRECT?   1 = YES;   2 = NO.
?1

ENTER MATRIX-K BY ROW:-
?4 -1 0   -1 2 -1   0 -1 4

 0.4000D+01 -0.1000D+01  0.0
-0.1000D+01  0.2000D+01 -0.1000D+01
 0.0        -0.1000D+01  0.4000D+01
*** IS THIS CORRECT?   1 = YES;   2 = NO.
?1
```

(*b*) Print-out

FIG. 9-8. (*Continued*)

quantized to become n number of data points NDATA and entered by means of the DO 42 loop in lines 430 to 460.

The steps to uncouple the equations of motion are shown in Par. II. The CALL $MODL statement calculates the modal matrix $[u]$, as described in Example 9, Chap. 4, and Fig. C-10, App. C. The statements UT(I,J) = U(J,I) in line 495 yields the transpose $[u]^T$ of the modal matrix $[u]$. The multiplications in Eq. (9-21) to find the diagonalized matrices $[\ulcorner M \lrcorner]$ and $[\ulcorner K \lrcorner]$ are computed by the CALL $MPLY statements in lines 500 to 515. The natural frequencies in Eq. (9-23) are determined in line 525. Next, the given initial conditions XO(I) and YO(I) are transformed by Eq. (9-20), $\{p\} = [u]^{-1}\{q\}$, to become PO(I) and QO(I) in lines 540 to 550.

The transient response is calculated in Par. III. The normalized force $\{N(t)\} = [u]^T\{Q(t)\}$ in Eq. (9-21) is calculated by means of the DO 48

```
ENTER: (1) INITIAL DISPLACEMENT XO(1) TO XO(N)
       (2) INITIAL VELOCITY YO(1) TO YO(N)
?0  0  0    0  0  0

 0.0          0.0          0.0
 0.0          0.0          0.0
*** IS THIS CORRECT?    1 = YES;   2 = NO.
?1

ENTER FORCE(T) AT TIME T(1) TO T(NDATA) FOR EACH OF THE N-EQUATIONS.
?1  1  1  1

 0.1000D+01   0.1000D+01   0.1000D+01   0.1000D+01
*** IS THIS CORRECT?    1 = YES;   2 = NO.
?1

?0  0  0  0

 0.0          0.0          0.0          0.0
*** IS THIS CORRECT?    1 = YES;   2 = NO.
?1

?0  0  0  0

 0.0          0.0          0.0          0.0
*** IS THIS CORRECT?    1 = YES;   2 = NO.
?1

               * * * OUTPUT * * *

TIME:          0.0
FORCE:         0.1000D+01   0.0          0.0
X1 TO XN:      0.0          0.0          0.0
VELOCITY:      0.0          0.0          0.0

TIME:          0.1000D+00
FORCE:         0.1000D+01   0.0          0.0
X1 TO XN:      0.2496D-02   0.2081D-05   0.3404D-09
VELOCITY:      0.4983D-01   0.8317D-04   0.2072D-07

TIME:          0.2000D+00
FORCE:         0.1000D+01   0.0          0.0
X1 TO XN:      0.9934D-02   0.3316D-04   0.2210D-07
VELOCITY:      0.9867D-01   0.6614D-03   0.6627D-06

TIME:          0.3000D+00
FORCE:         0.1000D+01   0.0          0.0
X1 TO XN:      0.2216D-01   0.1667D-03   0.2506D-06
VELOCITY:      0.1455D+00   0.2210D-02   0.4998D-05
STOP
```

(b) Print-out (*Continued*)

FIG. 9-8. (*Continued*)

loop. The uncoupled equations are solved by the CALL $TRESP statement in line 630. Lines 670 and 675 show that Eq. (9-20), $\{q\}=[u]\{p\}$, is used to transform the response from the $\{p\}$ to the $\{q\}$ coordinates.

The commands to execute the program, as shown in Fig. 9-8(b), are

MERGE $MODL, $INVS, $MPLY, $SUBN, $ROOT, $COEFF, $HOMO, $TRESP *c*

SAVE $TEMP *c*

MERGE TRESPUND, $TEMP *c*

RUN *c*

where c denotes the RETURN key. The first MERGE command combines eight subroutines. The SAVE command names their combination as a temporary subroutine $TEMP. The second MERGE command combines the main program TRESPUND and $TEMP. The programs are merged in two steps, because the CALL/OS system allows the merging of not more than eight programs at a time. The RUN command executes the program. The print-out is shown under the heading OUTPUT.

9-9 RAYLEIGH'S METHOD—UNDAMPED MULTIROTOR SYSTEMS

Rayleigh's method to find the fundamental frequency of multirotor systems was described in Chap. 5. Consider a gear-and-shaft system in which (1) the shaft is approximated by a simply supported beam between bearings as shown in Fig. 9-9, and (2) the gears are represented by the concentrated loads 10 kN and 5 kN, respectively. The static load due to a mass m is mg, where $g \simeq 9.81$. Hence the masses of the gears are $10{,}000/g$ kg and $5{,}000/g$ kg.

We shall first find the shaft deflection and then the fundamental frequency by Rayleigh's method. The shaft deflection is calculated from equations in strength of materials.

$$Q = \text{Shear at a point} = \sum \text{Loads to the point} \tag{9-24}$$

$$M = \text{Moment at a point} = \sum Q\,dx \tag{9-25}$$

$$S = \text{Slope at a point} = \sum (M/EI)\,dx \tag{9-26}$$

$$y = \text{Deflection at a point} = \sum S\,dx \tag{9-27}$$

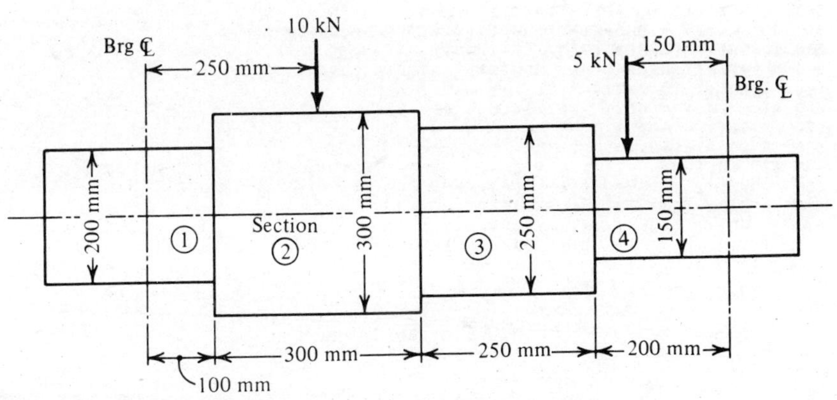

FIG. 9-9. *A gear and shaft assembly.*

RAYLEIGH

```
"   *** RAYLEIGH METHOD ***
"       FUNDAMENTAL FREQUENCY OF UNDAMPED MULTI-ROTOR SYSTEMS.
"
"   *ASSUME (1) SHAFT HAS (N) EQUALLY SPACED STATIONS BETWEEN
"               BEARINGS AND THEREFORE (N-1) EQUAL LENGTH SECTIONS,
"           (2) EXTERNAL LOADS (CW) CONCENTRATED AT MID-SECTIONS, AND
"           (3) SIMPLE TRAPEZOIDAL RULE FOR INTEGRATIONS.
"
"   *** #I. FORMAT AND INPUT. ***
    REAL  CW(100), DIAM(100), EI(100), L, M(100), Q(100), %
      SLOPE(100), W(100), Y(100), YAV(100)
    DATA  M, Q, SLOPE, W, Y, YAV, WY, WYSQ, PI /602*0.0, 3.14159265/
"
"
 80 FORMAT (' RAYLEIGH METHOD.',/,' FIND FUNDAMENTAL FREQUENCY ' %
    'OF UNDAMPED MULTI-ROTOR SYSTEMS.',/)
 81 FORMAT (' ENTER: (1) L   = SHAFT LENGTH BETWEEN BEARINGS',/, %
    8X,'(2) N   = NO. STATIONS   ** N <= 100',/,8X,'(3) DX  = LEN'%
    'GTH OF EACH SECTION   ** L = (N-1)*DX',/,8X,'(4) RHO = DENSI'%
    'TY OF MATERIAL, KG/M↑3',/,8X,'(5) E   = YOUNG''S ' %
    'MODULUS',/,8X,'(6) G   = SHEAR MODULUS')
 82 FORMAT (' L =',E13.5,T22,'N =',I4,T44,'DX =',E13.5,/, %
    ' RHO =', E13.5,T22,'E =',E13.5,T44,'G =',E13.5)
 83 FORMAT (' *** IS THIS CORRECT?   1 = YES;  2 = NO.')
 84 FORMAT (' DIAMATER OF SECTIONS (1) TO (N-1):-',/,(6E12.4))
 85 FORMAT (/,' CONCENTRATED LOADS AT EACH SECTION:-',/,(6E12.4))
 86 FORMAT (/,' TOTAL LOAD AT MID-SECTIONS:-',/,(6E12.4))
 87 FORMAT (/,' SHEAR AT STATIONS 1 TO N:-',/,(6E12.4))
 88 FORMAT (/,' MOMENTS AT STATIONS 1 TO N:-',/,(6E12.4))
 89 FORMAT (/,' SHAFT DEFLECTIONS AT STATION 1 TO N:-',/,(6E12.4))
 90 FORMAT (/,' FUNDAMENTAL FREQUENCY, HZ =', E12.4)
    WRITE (6,80)
 10 WRITE (6,81)
    READ (5,*) L, N, DX, RHO, E, G
    WRITE (6,82) L, N, DX, RHO, E, G
    WRITE (6,83)
    READ (5,*) IANS
    IF (IANS = 2) GOTO 10
    NM1 = N - 1
    WRITE (6,84) (DIAM(I), I=1,NM1)
    WRITE (6,85) (CW(I), I=1,NM1)
"
"   *** #II. CALCULATIONS AND OUTPUT. ***
"   *NOTE: (1) NO. STATIONS, 1 TO N;  (2) NO. SECTIONS, 1 TO (N-1)
"   *CALCULATE EI AND LOAD-W OF EACH SECTION.
    DO 40 I=1,NM1
      EI(I) = E*PI*(DIAM(I)↑4)/64.
 40   W(I) = RHO*PI*(DIAM(I)↑2)*DX/4. + CW(I)
    WRITE (6,86) (W(I), I=1,NM1)
"   *FIND SHEAR-Q AND MOMENT-M AT STATIONS 1 TO N.
    DO 41 I=1,NM1
 41   Q(1) = Q(1) + W(I)*(L - (I-.5)*DX)
    Q(1) = -Q(1)/L
    DO 42 I=2,N
 42   Q(I) = Q(I-1) + W(I-1)
    WRITE (6,87) (Q(I), I-1,N)
    CALL $EXT (Q, M, DX, N)
    WRITE (6,88) (M(I), I=1,N)
```

(*a*) Program listing

FIG. 9-10. *Rayleigh method; fundamental frequency of undamped multirotor systems—RAYLEIGH.*

```
680 "    *FIRST FIND UNCORRECTED SLOPE AND SHAFT DEFLECTION-Y.
690      DO 43 I=2,N
700 43     M(I) = -M(I)/EI(I-1)
710      CALL $EXT (M, SLOPE, DX, N)
720      CALL $EXT (SLOPE, Y, DX, N)
730 "    *THEN THE CORRECTED DEFLECTION-Y AND YAV AT MID-SECTION.
740      DO 44 I=2,N
750        Y(I) = Y(N)*(I-1.)*DX/L - Y(I)
760 44     YAV (I-1) = (Y(I-1) + Y(I))/2.
770      WRITE (6,89) (Y(I), I=1,N)
780 "    *CALCULATE FREQUENCY.
790      DO 45 I=1,NM1
800        WY = WY + W(I)*YAV(I)
810 45     WYSQ = WYSQ + W(I)*YAV(I)↑2
820      FREQ = SQRT(9.81*WY/WYSQ)/(2.*PI)
830      WRITE (6,90) FREQ
840      STOP
850      END
860 "
870 "    *** SUBROUTINE $EXT ***
880      SUBROUTINE $EXT (A, B, DX, N)
890      REAL  A(N), B(N)
900      DO 40 I=2,N
910 40     B(I) = B(I-1) + (A(I-1) + A(I))*DX/2.
920      RETURN
930      END
```

(*a*) Program listing (*Continued*)

FIG. 9-10. (*Continued*)

where dx = an elementary length of the beam and EI the flexural stiffness. To include the effect of shear deformation, let

$$\frac{d^2y}{dx^2}=\frac{M}{EI}+K\frac{W}{GAL} \tag{9-28}$$

where d^2y/dx^2 is the curvature, E the Young's modulus, G the shear modulus, I the moment of inertia, A the cross-sectional area, L the length of the beam, W the load, and $K=4/3$ for a solid beam of circular section. The term $KW/(GAL)$ is added to Eq. (9-26) if the deflection due to shear is to be included for the frequency calculation. If the deflection y_i and the loads W_i are known, the natural frequency ω is obtained from Eq. (5-9).

$$\omega^2=\frac{g\sum m_i y_i}{\sum m_i y_i^2} \quad \text{or} \quad \omega^2=\frac{g\sum W_i y_i}{\sum W_i y_i^2} \tag{9-29}$$

The program RAYLEIGH to solve the problem is listed in Fig. 9-10(*a*). The technique for the solution is briefly described in lines 130 to 160. In Par. I, the variables are initialized to zero in the DATA statement in line 210. Various techniques, such as creating a file, can be used to enter the diameter of the sections. We choose the simple DATA statement shown in Fig. 9-10(b) for entering the diameters in this problem. The remaining format and input statements are self-explanatory.

Comments are inserted in Par. II to relate the program to Eqs. (9-24) to (9-29). The calculations for EI(I) and the load W(I) distribution in lines 550 to 570 are self-explanatory. The shear distribution Q in Eq. (9-24) is computed as illustrated in Fig. 9-11(*a*). The CALL $EXT statement in

```
LOAD RAYLEIGH
READY

220     DATA   DIAM /2*0.2, 6*0.3, 5*0.25, 4*0.15, 83*0.0/
230     DATA   CW /4*0.0, 2*10E3, 7*0.0, 2*5E3, 85*0.0/
RUN

RAYLEIGH


RAYLEIGH METHOD.
FIND FUNDAMENTAL FREQUENCY OF UNDAMPED MULTI-ROTOR SYSTEMS.

ENTER: (1) L    = SHAFT LENGTH BETWEEN BEARINGS
       (2) N    = NO. STATIONS    ** N <= 100
       (3) DX   = LENGTH OF EACH SECTION    ** L = (N-1)*DX
       (4) RHO  = DENSITY OF MATERIAL, KG/M↑3
       (5) E    = YOUNG'S MODULUS
       (6) G    = SHEAR MODULUS
?0.85   18   0.05   8E3   200E9   77E9

L =  0.85000E+00     N =  18              DX =  0.50000E-01
RHO =  0.80000E+04   E =  0.20000E+12     G =  0.77000E+11
*** IS THIS CORRECT?    1 = YES;  2 = NO.
?1

DIAMATER OF SECTIONS (1) TO (N-1):-
 0.2000E+00  0.2000E+00  0.3000E+00  0.3000E+00  0.3000E+00  0.3000E+00
 0.3000E+00  0.3000E+00  0.2500E+00  0.2500E+00  0.2500E+00  0.2500E+00
 0.2500E+00  0.1500E+00  0.1500E+00  0.1500E+00  0.1500E+00

CONCENTRATED LOADS AT EACH SECTION:-
 0.0         0.0         0.0         0.0         0.1000E+05  0.1000E+05
 0.0         0.0         0.0         0.0         0.0         0.0
 0.0         0.5000E+04  0.5000E+04  0.0         0.0

TOTAL LOAD AT MID-SECTIONS:-
 0.1257E+02  0.1257E+02  0.2827E+02  0.2827E+02  0.1003E+05  0.1003E+05
 0.2827E+02  0.2827E+02  0.1963E+02  0.1963E+02  0.1963E+02  0.1963E+02
 0.1963E+02  0.5007E+04  0.5007E+04  0.7069E+01  0.7069E+01

SHEAR AT STATIONS 1 TO N:-
-0.1607E+05 -0.1605E+05 -0.1604E+05 -0.1601E+05 -0.1598E+05 -0.5957E+04
 0.4072E+04  0.4100E+04  0.4128E+04  0.4148E+04  0.4167E+04  0.4187E+04
 0.4207E+04  0.4226E+04  0.9233E+04  0.1424E+05  0.1425E+05  0.1425E+05

MOMENTS AT STATIONS 1 TO N:-
 0.0        -0.8030E+03 -0.1605E+04 -0.2407E+04 -0.3207E+04 -0.3755E+04
-0.3802E+04 -0.3598E+04 -0.3392E+04 -0.3185E+04 -0.2978E+04 -0.2769E+04
-0.2559E+04 -0.2348E+04 -0.2012E+04 -0.1425E+04 -0.7125E+03  0.3149E-01

SHAFT DEFLECTIONS AT STATION 1 TO N:-
 0.0         0.1347E-05  0.2567E-05  0.3608E-05  0.4522E-05  0.5337E-05
 0.6038E-05  0.6621E-05  0.7091E-05  0.7428E-05  0.7586E-05  0.7550E-05
 0.7333E-05  0.6949E-05  0.6194E-05  0.4716E-05  0.2537E-05 -0.2910E-10

FUNDAMENTAL FREQUENCY, HZ =  0.2105E+03
STOP
```

(*b*) Print-out

FIG. 9-10. (*Continued*)

line 660 sums the shear Q to yield the moment M as shown in Eq. (9-25). The simple trapezoidal rule, as shown in Fig. 9-11(*b*), is used for this summing. The summing subroutine $EXT in lines 870–930 is also used for the summing processes to find the slope S and the deflection y in Eqs. (9-26) and (9-27), respectively. The uncorrected deflection curve is illustrated in Fig. 9-11(*c*).

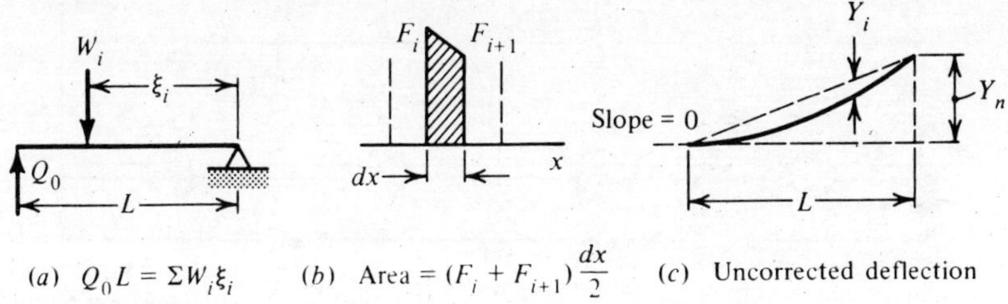

(a) $Q_0L = \Sigma W_i \xi_i$ (b) Area $= (F_i + F_{i+1})\frac{dx}{2}$ (c) Uncorrected deflection

FIG. 9-11. *Method of computation in Sec. 9-9.*

Since the shaft deflections at the bearings are assumed zero, the corrected deflection y_i is the vertical distance between the uncorrected deflection curve (solid line) and a straight line (dash line) joining the ends of the shaft. The corrected deflection is obtained from lines 730 to 760. Thus, the average deflection (YAV) at midsection and the natural frequency in Eq. (9-29) can be calculated.

The procedure for the program execution is shown in Fig. 9-10(*b*). When the system is READY, the user types

LOAD RAYLEIGH *c*
220 DATA DIAM /. . . / *c*
230 DATA CW / . . . / *c*
RUN *c*

First, the program RAYLEIGH is loaded in the working area. The shaft diameters are entered as DATA in line 220 and the concentrated loads (CW) in line 230, as shown in the print-out. The rest of the print-out is self-explanatory.

9-10 MYKLESTAD-PROHL METHOD—TRANSFER MATRIX TECHNIQUE

Myklestad-Prohl method with transfer matrix technique was illustrated in Example 8, Chap. 5. The technique is applied in this example to estimate the natural frequencies of a uniform cantilever beam.

A cantilever is approximated by a four-mass system as shown in Fig. 9-12. Hence the equivalent discrete system has only four natural frequencies. Probably the first two calculated frequencies are close to those of the continuous system. Greater accuracy can be achieved by subdividing the beam into larger number of sections.

The general theory from Eq. (5-35) is that the state vector $\{Z\}_i^R$ at the end of the ith segment is related to $\{Z\}_{i-1}^R$ at the beginning of the ith segment by the transfer matrix T_i.

$$\{Z\}_i^R = T_i\{Z\}_{i-1}^R \tag{9-30}$$

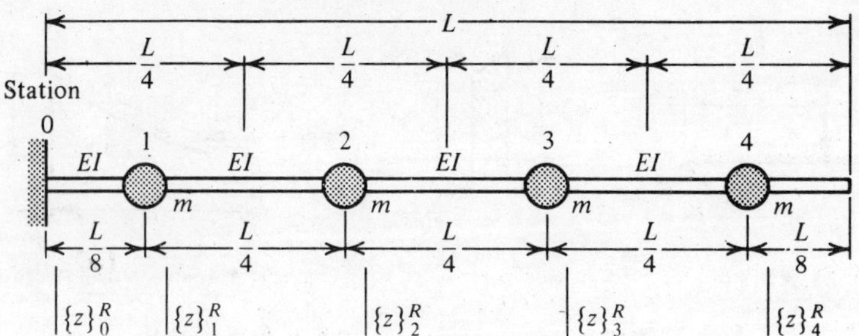

FIG. 9-12. *Uniform cantilever approximated N(=4) equal sections.*

Neglecting J, the rotational effect of the masses in Eq. (5-35), the equation above can be expressed as

$$\begin{bmatrix} Y \\ \Phi \\ M \\ V \end{bmatrix}_i^R = \begin{bmatrix} 1 & L & \dfrac{L^2}{2EI} & -\dfrac{L^3}{6EI} \\ 0 & 1 & \dfrac{L}{EI} & -\dfrac{L^2}{2EI} \\ 0 & 0 & 1 & -L \\ -\omega^2 m & -\omega^2 mL & -\omega^2 m\dfrac{L^2}{2EI} & 1+\omega^2 m\dfrac{L^3}{6EI} \end{bmatrix}_i \begin{bmatrix} Y \\ \Phi \\ M \\ V \end{bmatrix}_{i-1}^R \quad \textbf{(9-31)}$$

where Y = deflection, Φ = slope, M = moment, V = shear, and ω = frequency in rad/s. Note that L in Eq. (9-31) is the length of a segment. Applying the recurrence formula in Eq. (9-30), the state vector $\{Z\}_n^R$ and $\{Z\}_o^R$ at station o are related as

$$\{Z\}_n^R = T_n[T_{n-1} \cdots T_2 T_1]\{Z\}_o^R \quad \textbf{(9-32)}$$

Since a fixed-free beam is considered, the analysis is identical to that in Example 8, Chap. 5. The frequency is calculated from the equation

$$\Delta(\omega) = \begin{vmatrix} T_{33} & T_{34} \\ T_{43} & T_{44} \end{vmatrix} = 0 \quad \textbf{(9-33)}$$

The program TMXCANT1, as listed in Fig. 9-13(*a*), employs the transfer matrix technique to find the frequencies of a cantilever shown in Fig. 9-12. The problem is normalized such that (*EI*)/[(mass/length)(area of beam section)(total length)] = 1. The format and input statements in Par. I are self-explanatory.

The number of stations is $n = 5$. The $(n-1)$ approximate frequencies are estimated in Par. II by means of the DO 41 loop. The IF statement in line 405 stipulates the maximum number of iterations NITER. The recurrence formula in Eq. (9-30) is implemented in the DO 42 loop, using the CALL $CAL statement in line 425. For a given frequency, line 445 checks the sign of the determinant $\Delta(\omega)$ in Eq. (9-33). The frequency is incremented and the sign of $\Delta(\omega)$ is rechecked. A root occurs when there is a sign change. In other words, assuming a frequency ω_1 (W1), the steps, as shown in lines 410, 465, and 490, are

W = W1 + DW

DW(new) = 1.6*DW(old) If no sign change.

DW(new) = (W-W1)/3.2 If have sign change.

If there is no sign change, the increment $\Delta\omega$ (DW) is increased by a factor (1.6). If there is a sign change and an overestimate, $\Delta\omega$ is decreased by a factor of (3.2) and the calculation starts from ω_1 (W1), the frequency before the sign change occurs.

An approximate frequency FREQ(J) is obtained IF $(\text{DW}-\text{DEL})\leq 0$, where DEL $(=1\times 10^{-7})$ is the criterion for the approximate frequency. It is anticipated that the second root is at least twice the value of the first. Hence a larger DW in line 510 is assumed for the iteration of the second root.

The more exact frequencies can be estimated from the approximate frequencies by many methods.* Since the criterion $\text{DEL}=1\times 10^{-7}$ for the approximate frequency in line 480 is fairly stringent, a straight line interpolation is used in Par. III to find the more exact frequency. Let X(1) and X(2) in lines 535 and 540 be two frequencies about an estimated FREQ(J). The values of $\Delta(\omega)$ in Eq. (9-33), corresponding to X(1) and X(2), are found in line 570. The more exact frequency is calculated in line 580 by straight line interpolation.

The state vectors are calculated for each of the more exact frequencies in Par. IV. The method is the same as before except that the state vectors are normalized such that the shear at the fixed end is unity.

The subroutines are shown in lines 705 to 860. Thus, no merging of the programs is necessary. The print-out shown in Fig. 9-13(*c*) is self-explanatory.

9-11 MATRIX ITERATION—UNDAMPED DISCRETE SYSTEMS

The method of matrix iteration, due to Duncan and Collar, for finding the natural frequencies and modal vectors of undamped discrete systems was described in Sec. 6-9. The program ITERATE runs parallel to Example 6, Chap. 6. It first yields the lowest eigenvalue and the corresponding modal vector. Then this mode is suppressed, a new dynamic matrix calculated, and the process repeated for the next mode.

* See, for example, R. L. Ketter and S. P. Prawel, *Modern Methods of Engineering Computation*, McGraw-Hill Book Company, New York, 1969, Chap. 8.

```
TMXCANTI

"  *** APPROX. NATURAL FREQUENCIES OF UNIFORM CANTILEVER ***
"      TRANSFER MATRIX TECHNIQUE.  NEGLECT MASS MOMENT OF INERTIA-J.
"      ELEMENTS OF STATE VECTORS ARE:- LATERAL DEFLECTION, SLOPE,
"        MOMENT, AND SHEAR.
"      LET BEAM HAVE N-STATIONS.  THEREFORE, NO. SECTIONS = (N-1).
"      ASSMUE MASSES CONCENTRATED AT STATIONS.
"      METHOD:  MYKLESTAD
"
"  *** #I. FORMAT AND INPUT. ***
   REAL*8  D(4,4), DEL, DET, DW, DW1, EI(10), FREQ(10), L(10), %
     M(10), T(4,4), UNIT(4,4), W, W1, X(2), Y(2), Z(4,11)
   DATA  EI, L, M, UNIT, W1, I.D, ITER /47*0.0, 2*0/
   DATA  Z(1,1), Z(2,1), Z(4,1) /2*0.0, 1.0/
80 FORMAT (' APPROXIMATE NATURAL FREQUENCIES OF UNIFORM CANTI' %
   'LEVER.',/,' TRANSFER MATRIX TECHNIQUE.  MYKLESTAD METHOD.',/)
81 FORMAT (' ENTER: N     = NO. STATIONS   *** N <= 11 ',/,8X, %
     'NITER = MAX. NO. ITERATIONS',/,8X,'DW1    = INITIAL INCREMENT'%
     ' IN FREQUENCY ESTIMATE',/,8X,'DEL    = DW FOR APPROXIMATE ' %
     'FREQUENCY ESTIMATION')
82 FORMAT (' N =', I3,'   NITER =',I4,'   DW1 =',D11.4,'   DEL =',%
     D11.4)
83 FORMAT (' *** IS THIS CORRECT?   1 = YES;  2 = NO.')
84 FORMAT (' ENTER LENGTH OF SPAN BETWEEN STATIONS.')
85 FORMAT (6D12.4)
86 FORMAT (' ENTER FLEXURAL STIFFNESS EI OF EACH SPAN.')
87 FORMAT (' ENTER VALUES OF POINT MASS.')
88 FORMAT (/,' STATE VECTORS OF STATIONS (1) TO (N):-')
89 FORMAT (/,' (FREQ =',D11.4,'  RAD/S)')
90 FORMAT (' TOTAL ITERATIONS FOR APPROX. FREQ. EST. =',I4)
   WRITE (6,80)
10 WRITE (6,81)
   READ (5,*) N, NITER, DW1, DEL
   WRITE (6,82) N, NITER, DW1, DEL
   WRITE (6,83)
   READ (5,*) IANS
   IF (IANS = 2) GOTO 10
11 WRITE (6,84)
   READ (5,*) (L(I), I=2,N)
   WRITE (6,85) (L(I), I=2,N)
   WRITE (6,83)
   READ (5,*) IANS
   IF (IANS = 2) GOTO 11
12 WRITE (6,86)
   READ (5,*) (EI(I), I=2,N)
   WRITE (6,85) (EI(I), I=2,N)
   WRITE (6,83)
   READ (5,*) IANS
   IF (IANS = 2) GOTO 12
13 WRITE (6,87)
   READ (5,*) (M(I), I=2,N)
   WRITE (6,85) (M(I), I=2,N)
   WRITE (6,83)
   READ (5,*) IANS
   IF (IANS = 2) GOTO 13
   DO 40 I=1,4
40   UNIT(I,I) = 1.
"
```

(*a*) Program listing

FIG. 9-13. *Myklestad method with transfer matrix technique; approximate frequencies of cantilever—TMXCANTI.*

```
"    *** #II. APPROXIMATE FREQUENCIES. ***
     DW = DW1
     DO 41 J=2,N
14     ITER = ITER + 1
       IF (ITER > NITER) GOTO 17
       W = W1 + DW
       CALL $SUBN (UNIT, T, 4)
       DO 42 I=2,N
         CALL $CAL (M(I), L(I), EI(I), W, T, D)
42       CALL $SUBN (D, T, 4)
       DET = D(3,3)*D(4,4) - D(3,4)*D(4,3)
       IF (DABS(DET) < 1.D-15) GOTO 16
       SGN = DET/DABS(DET)
       K = IFIX (SGN)
       IF ((ID+K).EQ.0) GOTO 15
       ID = K
       DW = 1.6*DW
       W1 = W
       GOTO 14
15     IF ((DW-DEL).LE.0) GOTO 16
       ID = -K
       DW = (W - W1)/3.2
       GOTO 14
16     FREQ(J) = W
       ID = 0
41     DW = 2.*DW1
17   WRITE (6,90) ITER
"
"    *** #II. GET MORE EXACT FREQUENCY BY INTERPOLATION. ***
     DO 43 J=2,N
       X(1) = FREQ(J) - DEL
       X(2) = FREQ(J) + DEL
       DO 44 K=1,2
         CALL $SUBN (UNIT, T, 4)
         DO 45 I=2,N
           CALL $CAL (M(I), L(I), EI(I), X(K), T, D)
45         CALL $SUBN (D, T, 4)
44       Y(K) = D(3,3)*D(4,4) - D(3,4)*D(4,3)
       IF (DABS(Y(2)-Y(1)).LE.0) GOTO 43
       FREQ(J) = (X(1)*Y(2) - X(2)*Y(1))/(Y(2) - Y(1))
43     CONTINUE
"
"    *** #IV. CALCULATE STATE VECTORS FOR EACH FREQUENCY. ***
     WRITE (6,88)
     DO 46 J=2,N
       WRITE (6,89) FREQ(J)
       CALL $SUBN (UNIT, T, 4)
       DO 47 I=2,N
         CALL $CAL (M(I), L(I), EI(I), FREQ(J), T, D)
47       CALL $SUBN (D, T, 4)
       Z(3,1) = -D(3,4)/D(3,3)
       CALL $SUBN (UNIT,T,4)
       DO 48 I=2,N
         CALL $CAL (M(I), L(I), EI(I), FREQ(J), T, D)
         CALL $SUBN (D, T, 4)
         DO 48 K=1,4
           Z(K,I) = 0.
           DO 48 JK =1,4
48           Z(K,I) = Z(K,I) + T(K,JK)*Z(JK,1)
       DO 46 K=1,4
46       WRITE (6,85) (Z(K,I), I=1,N)
     STOP
     END
"
```

(*a*) Program listing (*Continued*)

FIG. 9-13. (*Continued*)

```
"   *** SUBROUTINE $CAL.  RECEIVES DATA AND OLD TRANSFER MATRIX. ***
"        CALCULATES NEW TRANSFER MATRIX.  RETURNS PRODUCT OF
"        NEW AND OLD TRANSFER MATRICES.
      SUBROUTINE $CAL (A, B, C, D, E, F)
      REAL*8  A, B, C, D, E(4,4), F(4,4), G(4,4)
      DATA  G /1., 4*0., 1., 4*0., 1., 5*0./
      G(1,2) =  B
      G(1,3) =  B↑2/(2.*C)
      G(1,4) = -G(1,3)*B/3
      G(2,3) =  B/C
      G(2,4) = -G(1,3)
      G(3,4) = -B
      G(4,1) = -A*D↑2
      G(4,2) =  G(4,1)*B
      G(4,3) =  G(4,1)*G(1,3)
      G(4,4) =  1 + G(1,4)*G(4,1)
      DO 40 I=1,4
      DO 40 J=1,4
        F(I,J) = 0
        DO 40 K = 1,4
40        F(I,J) = F(I,J) + G(I,K)*E(K,J)
      RETURN
      END
"
"   *** SUBROUTINE $SUBN.  REAL MATRIX SUBSTITUTION. ***
      SUBROUTINE $SUBN (X, Y, N)
      REAL*8  X(N,N), Y(N,N)
      DO 40 I=1,N
      DO 40 J=1,N
40      Y(I,J) = X(I,J)
      RETURN
      END
```

(*b*) Subroutines

Fig. 9-13. (*Continued*)

The equations of motion for the free vibration of an undamped system from Eq. (6-9) are

$$G\{\ddot{q}\}+\{q\}=0 \tag{9-34}$$

where

$$G=K^{-1}M \tag{9-35}$$

and M is the mass matrix, K the stiffness matrix, and G the dynamic matrix. Let $\{V\}_o$ be a trial modal vector to start the iteration. Following Eq. (6-55), we form a sequence of vectors

$$\begin{aligned} &\{V\}_1=G\{V\}_o, \qquad \{V\}_2=G\{V\}_1, \qquad \{V\}_3=G\{V\}_2, \\ &\quad \cdots \qquad , \qquad \{V\}_i=G\{V\}_{i-1} \end{aligned} \tag{9-36}$$

As shown in Eq. (6-57) and in Example 6, Chap. 6, a constant can be factored from $\{V\}_i$ after each iteration. For the $(s+1)$th iteration and for s sufficiently large, we have

$$\{V\}_{s+1}=\lambda_1\{V\}_s \tag{9-37}$$

where $\sqrt{\lambda_1}$ is the fundamental frequency in rad/s and $\{V\}_s$ the corresponding modal vector.

```
TMXCANT1

APPROXIMATE NATURAL FREQUENCIES OF UNIFORM CANTILEVER.
TRANSFER MATRIX TECHNIQUE.  MYKLESTAD METHOD.

ENTER: N      = NO. STATIONS    *** N <= 11
       NITER = MAX. NO. ITERATIONS
       DW1   = INITIAL INCREMENT IN FREQUENCY ESTIMATE
       DEL   = DW FOR APPROXIMATE FREQUENCY ESTIMATION
?5   400   1   1D-7

N =  5   NITER = 400   DW1 = 0.1000D+01    DEL = 0.1000D-06
*** IS THIS CORRECT?   1 = YES;  2 = NO.
?1

ENTER LENGTH OF SPAN BETWEEN STATIONS.
?.125   .25   .25   .25

 0.1250D+00  0.2500D+00  0.2500D+00  0.2500D+00
*** IS THIS CORRECT?   1 = YES;  2 = NO.
?1

ENTER FLEXURAL STIFFNESS EI OF EACH SPAN.
?1   1   1   1

 0.1000D+01  0.1000D+01  0.1000D+01  0.1000D+01
*** IS THIS CORRECT?   1 = YES;  2 = NO.
?1

ENTER VALUES OF POINT MASS.
?.25   .25   .25   .25

 0.2500D+00  0.2500D+00  0.2500D+00  0.2500D+00
*** IS THIS CORRECT?   1 = YES;  2 = NO.
?1

TOTAL ITERATIONS FOR APPROX. FREQ. EST. = 175

STATE VECTORS OF STATIONS (1) TO (N):-

(FREQ = 0.3567D+01   RAD/S)
 0.0          0.5273D-02  0.4165D-01  0.9998D-01  0.1675D+00
 0.0          0.8177D-01  0.1990D+00  0.2588D+00  0.2755D+00
 0.7167D+00  0.5917D+00  0.3459D+00  0.1332D+00 -0.1388D-16
 0.1000D+01  0.9832D+00  0.8507D+00  0.5327D+00  0.8882D-15

(FREQ = 0.2318D+02   RAD/S)
 0.0          0.1257D-02  0.5895D-02  0.4458D-02 -0.4167D-02
 0.0          0.1751D-01  0.1093D-01 -0.2284D-01 -0.4033D-01
 0.2026D+00  0.7758D-01 -0.1302D+00 -0.1400D+00 -0.2220D-15
 0.1000D+01  0.8311D+00  0.3909D-01 -0.5599D+00  0.3553D-14

(FREQ = 0.6632D+02   RAD/S)
 0.0          0.5875D-03  0.1111D-02 -0.1226D-02  0.4371D-03
 0.0          0.6796D-02 -0.6299D-02 -0.3360D-02  0.1166D-01
 0.1169D+00 -0.8128D-02 -0.9664D-01  0.1201D+00 -0.3553D-14
 0.1000D+01  0.3540D+00 -0.8671D+00  0.4806D+00  0.0

(FREQ = 0.1410D+03   RAD/S)
 0.0          0.2629D-03 -0.9211D-04  0.3937D-04 -0.8877D-05
 0.0          0.1602D-02 -0.1252D-02  0.7259D-03 -0.6524D-03
 0.7531D-01 -0.4969D-01  0.2685D-01 -0.1103D-01 -0.5684D-13
 0.1000D+01 -0.3062D+00  0.1515D+00 -0.4411D-01  0.9095D-12
STOP
```

(*c*) Print-out

FIG. 9-13. (*Continued*)

The first mode is suppressed by the constraint $p_1 = 0$, as shown in Eq. (6-58). For example, the sweeping matrix S_1 to suppress the first mode for a 3×3 matrix from Eq. (6-59) is

$$\begin{bmatrix} q_1 \\ q_2 \\ q_3 \end{bmatrix} = \begin{bmatrix} v_{11} & v_{12} & v_{13} \\ 0 & 1 & 0 \\ 0 & 0 & 1 \end{bmatrix}^{-1} \begin{bmatrix} 0 & 0 & 0 \\ 0 & 1 & 0 \\ 0 & 0 & 1 \end{bmatrix} \begin{bmatrix} q_1 \\ q_2 \\ q_3 \end{bmatrix} \tag{9-38}$$

or

$$\{q\} = S_1\{q\} \tag{9-39}$$

where the row vector $\lfloor v_{ij} \rfloor$ is obtained from the transpose $\lfloor u \rfloor_1$ of the modal vector $\{u\}_1$ as shown in Eq. (6-61).

$$\lfloor v_{11} \quad v_{12} \quad v_{13} \rfloor = (\text{constant}) \lfloor u \rfloor_1 M \tag{9-40}$$

The new dynamic matrix G_1 for the subsequent iterations is formed as indicated in Eq. (6-60).

$$G_1 = GS_1 \tag{9-41}$$

The program ITERATE is listed in Fig. 9-14(a). The format and input statements in Par. I are self-explanatory.

The iterations in Par II begins by finding the dynamic matrix $G = K^{-1}M$ in lines 365 and 370. The iterations in Eq. (9-36) are performed in the DO 45 loop, where TRY is the trial modal vector. The constant TEMP is factored out from the trial vector in line 430. The trial vector is normalized as shown in line 440. The criterion $|\mathrm{T1} - \mathrm{T2}| \leq \mathrm{ERROR}$ $(= 1 \times 10^{-16})$ is used to test the convergence of the iteration in line 465, where $\mathrm{T1} = (u_{11}^2 + u_{21}^2 + u_{31}^2)$ is the norm of the *normalized* ith trial vector and T2 is that of the normalized $(i+1)$st trial vector. Convergence is obtained when the trial vectors remain essentially unchanged in subsequent iterations. When convergence is obtained, the constant TEMP becomes λ as shown in Eq. (9-37). A natural frequency is calculated as shown in line 485.

The sweeping matrix is obtained in Par. III. Equation (9-40) can be identified in line 515. The matrix inversion and multiplications in Eq. (9-38) are performed in lines 525 to 540. The GOTO 13 statement directs the program to start again to iterate for the next mode. The output statements in Par. IV are self-explanatory and the print-out is shown in Fig. 9-14(b).

9-12 TRANSIENT RESPONSE OF DAMPED SYSTEMS

The analysis of the transient response of discrete systems with viscous damping in Sec. 6-13 is similar to that for undamped systems, except that the modal matrix is complex and it is applied to the reduced equations.

ITERATE

```
"   *** MATRIX ITERATION ***
"       SUBROUTINES REQD:   (1) $SUBN  (2) $MPLY   (3) $INVS
"
"   *** #1. FORMAT AND INPUT. ***
      REAL*8  A(10,10), DUM(10,10), ERROR, FREQ(10), G(10,10),%
        K(10,10), KINVS(10,10), M(10,10), SMATX(10,10), T1, T2, %
        TEMP, TRY(10), TRY1(10), U(10,10), V(10,10), VINVS(10,10)
      DIMENSION ITER(10)
      DATA  A, V/ 200*0.0 /
   80 FORMAT(' MATRIX ITERATION.',//,' DISCRETE UNDAMPED POSITIVE DEF'%
      'INITE SYSTEMS.',/,' FIND NATURAL FREQUENCIES AND MODAL MATRIX.'%
      ,/,' EIGENVALUES DISTINCT.',/,' METHOD:   DUNCAN & COLLAR.',/)
   81 FORMAT (' ENTER: (1) N       = NO. OF DEGREES OF FREEDOM   ** N <'%
      '= 10',/,8X,'(2) NFREQ = NO. OF FREQUENCIES AND MODES DESIRED'%
      ,/,8X,'(3) NITER = MAX. ITERATIONS FOR EACH MODE',/,8X,'(4) ' %
      'ERROR = ERROR SPECIFIED IN ITERATION',/,8X,'(5) TRY1   = '%
      'VALUES OF TRIAL VECTOR TO BEGIN ITERATION',/)
   82 FORMAT (' N =',I3,'    NFREQ =',I3,'    NITER =',I4,'    ERROR ='%
      ,D12.5,/,' TRIAL VECTOR:',/,(5D14.5))
   83 FORMAT (' *** IS THIS CORRECT?    1 = YES;   2 = NO.')
   84 FORMAT (' ENTER MASS MATRIX-M BY ROW:-')
   85 FORMAT ( 5D14.5)
   86 FORMAT (' ENTER STIFFNESS MATRIX-K BY ROW:-')
   87 FORMAT (' NATRUAL FREQUENCIES, RAD/S:-')
   88 FORMAT (/,' NO. ITERATIONS FOR EACH MODE:-',/, 10I7)
   89 FORMAT (/,' MODAL MATRIX - LISTED BY ROW:-',/)
      WRITE (6,80)
   10 WRITE (6,81)
      READ (5,*) N, NFREQ, NITER, ERROR, (TRY1(I), I=1,N)
      WRITE (6,82) N, NFREQ, NITER, ERROR, (TRY1(I), I=1,N)
      WRITE (6,83)
      READ (5,*) IANS
      IF (IANS = 2) GOTO 10
   11 WRITE (6,84)
      READ (5,*) ((M(I,J), J=1,N), I=1,N)
      DO 40 I=1,N
   40   WRITE (6,85) (M(I,J), J=1,N)
      WRITE (6,83)
      READ (5,*) IANS
      IF (IANS = 2) GOTO 11
   12 WRITE (6,86)
      READ (5,*) ((K(I,J), J=1,N), I=1,N)
      DO 41 I=1,N
   41   WRITE (6,85) (K(I,J), J=1,N)
      WRITE (6,83)
      READ (5,*) IANS
      IF (IANS = 2) GOTO 12
      DO 42 I=1,N
        A(I,I) = 1.
   42   V(I,I) = 1.
"
```

(*a*) Program listing

FIG. 9-14. *Matrix iteration by method of Duncan and Collar—ITERATE.*

(1) The equations of motion are uncoupled by means of the modal matrix. (2) The uncoupled equations are solved independently. (3) The desired solution in the given coordinates is obtained from the solutions of the uncoupled equations by means of a coordinate transformation.

The equations of motion for discrete systems with viscous damping are

$$M\{\ddot{q}\}+C\{\dot{q}\}+K\{q\}=\{Q(t)\} \tag{9-42}$$

where M is the mass matrix, C the damping matrix, K the stiffness

```
"    *** #II. ITERATIONS. ***
"    *GET DYNAMIC MATRIX-G FOR ITERATIONS.
     CALL $INVS (K, KINVS, N)
     CALL $MPLY (KINVS, M, G, N)
"    *ITERATIONS
"    *J = NFREQ, ALSO IDENTIFIES THE 1ST, 2ND, ETC. MODES
     J = 1
13   T2 = N
     DO 43 I1=1,N
43     TRY(I1) = TRY1(I1)
     DO 44 L=1,NITER
       DO 45 I=1,N
         U(I,J) = 0
         DO 45 J1=1,N
45         U(I,J) = U(I,J) + G(I,J1)*TRY(J1)
       TEMP = U(N,J)
       DO 46 I2=1,N
46       U(I2,J) = U(I2,J)/U(N,J)
       ITER(J) = L
       T1 = 0.
       DO 47 I3=1,N
47       T1 = T1 + U(I3,J)↑2
       IF (DABS(T1-T2).LE.ERROR) GOTO 14
       DO 48 I4=1,N
48       TRY(I4) = U(I4,J)
44     T2 = T1
14   FREQ(J) = 1./DSQRT(DABS(TEMP))
"
"    *** #III. GET SMATX TO ITERATE FOR HIGHER MODES. ***
     DO 49 I=1,N
       V(J,I) = 0
       DO 49 I1=1,N
49     V(J,I) = V(J,I) + U(I1,J)*M(I1,I)
     A(J,J) = 0
     CALL $INVS (V, VINVS, N)
     CALL $MPLY (VINVS, A, SMATX, N)
     CALL $MPLY (G, SMATX, DUM, N)
     CALL $SUBN (DUM, G, N)
     IF ((J-NFREQ).GE.0) GOTO 15
     J = J+1
     GOTO 13
"
"    *** #IV. OUTPUT. ***
15   WRITE (6,87)
     WRITE (6,85) (FREQ(J), J=1,NFREQ)
     WRITE (6,88) (ITER(JJ), JJ=1,NFREQ)
     WRITE (6,89)
     DO 50 I=1,N
50     WRITE (6,85) (U(I,J2), J2=1,NFREQ)
     STOP
     END
```

(*a*) Program listing (*Continued*)

FIG. 9-14. (*Continued*)

matrix, and $\{Q(t)\}$ the excitation vector. This can be reduced to a set of $2n$ simultaneous first-order equations as shown in Eqs. (6-86) and (6-87).

$$\begin{bmatrix} 0 & M \\ M & C \end{bmatrix}\begin{bmatrix} \ddot{q} \\ \dot{q} \end{bmatrix} + \begin{bmatrix} -M & 0 \\ 0 & K \end{bmatrix}\begin{bmatrix} \dot{q} \\ q \end{bmatrix} = \begin{bmatrix} 0 \\ Q(t) \end{bmatrix} \tag{9-43}$$

$$A\{\dot{y}\} + B\{y\} = \{E(t)\} \tag{9-44}$$

where the vector $\{y\} = \{\dot{q} \quad q\}$ and $\{E(t)\} = \{0 \quad Q(t)\}$ as shown in the last two equations. Introducing the transformation from Eq. (6-88), we get*

$$\{y\} = [u]\{z\} \qquad \text{or} \qquad \{z\} = [u]^{-1}\{y\} \tag{9-45}$$

* The symbol $[\Psi]$ was used to denote the modal matrix in Eq. (6-88). The symbol $[u]$ is used in this program.

```
MERGE ITERATE, $INVS, $MPLY, $SUBN
READY

RUN

MATRIX ITERATION.

DISCRETE UNDAMPED POSITIVE DEFINITE SYSTEMS.
FIND NATURAL FREQUENCIES AND MODAL MATRIX.
EIGENVALUES DISTINCT.
METHOD:   DUNCAN & COLLAR.

ENTER: (1) N       = NO. OF DEGREES OF FREEDOM   ** N <= 10
       (2) NFREQ = NO. OF FREQUENCIES AND MODES DESIRED
       (3) NITER = MAX. ITERATIONS FOR EACH MODE
       (4) ERROR = ERROR SPECIFIED IN ITERATION
       (5) TRY1   = VALUES OF TRIAL VECTOR TO BEGIN ITERATION

?3   3   200    1D-16   1 1 1

N =  3    NFREQ =  3    NITER = 200    ERROR = 0.10000D-15
TRIAL VECTOR:
  0.10000D+01    0.10000D+01    0.10000D+01
*** IS THIS CORRECT?    1 = YES;   2 = NO.
?1

ENTER MASS MATRIX-M BY ROW:-
?3 0 0   0 1 0   0 0 2

  0.30000D+01    0.0            0.0
  0.0            0.10000D+01    0.0
  0.0            0.0            0.20000D+01
*** IS THIS CORRECT?    1 = YES;   2 = NO.
?1

ENTER STIFFNESS MATRIX-K BY ROW:-
?140 -60 -20    -60 220 -80    -20 -80 200

  0.14000D+03  -0.60000D+02  -0.20000D+02
 -0.60000D+02   0.22000D+03  -0.80000D+02
 -0.20000D+02  -0.80000D+02   0.20000D+03
*** IS THIS CORRECT?    1 = YES;   2 = NO.
?1

NATRUAL FREQUENCIES, RAD/S:-
  0.57645D+01    0.93095D+01    0.15709D+02

NO. ITERATIONS FOR EACH MODE:-
    39       35        2

MODAL MATRIX - LISTED BY ROW:-

  0.21723D+01  -0.38095D+00   0.34196D+00
  0.11262D+01   0.42857D+00  -0.37548D+01
  0.10000D+01   0.10000D+01   0.10000D+01
STOP
```

(*b*) Print-out

FIG. 9-14. (*Continued*)

where $[u]$ is the modal matrix. Substituting $\{y\}=[u]\{z\}$ in Eq. (9-44) and then premultiplying by $[u]^T$, we get

$$[u]^T A[u]\{\dot{z}\}+[u]^T B[u]\{z\}=[u]^T\{E(t)\} \tag{9-46}$$

$$[\ulcorner A \lrcorner]\{\dot{z}\}+[\ulcorner B \lrcorner]\{z\}=\{N(t)\} \tag{9-47}$$

where $[\ulcorner A \lrcorner]$ and $[\ulcorner B \lrcorner]$ are diagonal and $\{N(t)\}=[u]^T\{E(t)\}$ as shown in the equations above. The uncoupled equations in Eq. (9-47) are identical to Eq. (6-89). Each can be expressed as

$$a_{ii}\dot{z}_i + b_{ii}z_i = N_i(t) \qquad \text{or} \qquad \dot{z}_i - \gamma_i z_i = \frac{1}{a_{ii}} N_i(t) \tag{9-48}$$

where

$$\gamma_i = -\frac{b_{ii}}{a_{ii}} \tag{9-49}$$

The program TRESPDAM for the *t*ransient *resp*onse of discrete systems with viscous *dam*ping is listed in Fig. 9-15(a). Since an iterative procedure is used in the subroutine $CROOT, extended precision REAL*8 and COMPLEX*16 are specified as shown in Par. I. The variables X and V in the program represent $\{q\}$ and $\{\dot{q}\}$ in Eq. (9-43); Y and Z correspond to $\{y\}$ and $\{z\}$ in Eq. (9-45). The format and input statements are self-explanatory. The excitations $\{Q(t)\}$ in Eq. (9-42) are quantized into n-number of data points NDATA. NDATA is entered in line 290 and the quantized excitation entered in line 465.

The reduced equations are uncoupled in Par. II. First, the modal matrix $[u]$ is found by the CALL $CMODL statement in line 515. The matrices A and B in Eq. (9-44) are formed by means of the DO 44 loop. The multiplications in Eq. (9-46) to obtain the uncoupled equations are performed by the CALL $CMPLY statements in lines 570 to 585. The variable BDA in line 595 denotes b_{ii}/a_{ii} in Eq. (9-49).

The calculations in Par. III involve (1) the coordinate transformations in Eq. (9-45) and (2) the solution of the first-order equation in Eq. (9-48) for each data point. The vector $\{y\}$ is formed in the DO 48 loop. The vector $\{z\}$ and the normalized force $\{N(t)\}$ in Eq. (9-47) are formed in the DO 50 loop. The uncoupled equations in Eq. (9-48) are solved by the CALL $CRKUT statement. The subroutine $CRKUT is listed in Fig. C-12, App. C. The rest of the program consists of the coordinate transformation between $\{y\}$ and $\{z\}$.

The print-out is as shown in Fig. 9-15(b). First, eight of the required subroutines are merged to give the combined program $TEMP. This temporary program is then merged with the main program TRESPDAM and the remaining subroutines. When the system is again READY, the command RUN is given for the execution. The rest of the print-out is self-evident.

9-13 SUMMARY

The sections in the chapter are arranged to conform to the sequence of the topics presented in the text. The programs, as listed in Table 9-1, are written in FORTRAN-IV for time sharing. The subroutines listed in App. C are extensively used in order to simplify the main programs. Although the programs are written to illustrate the text material, they can be

TRESPDAM

```
"   *** TRANSIENT RESPONSE OF POSITIVE DEFINITE SYSTEMS WITH ***
"       VISCOUS DAMPING.  EIGENVALUES DISTINCT.
"   *SUBROUTINES REQD: (1) $COEFF  (2) $CHOMO  (3) $CROOT  (4) $CINVS
"                      (5) $CMPLY  (6) $CSUBN  (7) $CMODL  (8) $CRKUT
"                      (9) $INVS   (10) $MPLY  (11) $SUBN
"
"   *** #1. FORMAT AND INPUT. ***
    REAL*8  DT, C(10,10), ERROR, FORCE(10,80), K(10,10), M(10,10), %
     T, V(5), X(5), ZRO
    COMPLEX*16  A(10,10), B(10,10), DUM(10,10), FN(10), BDA(10), %
     U(10,10), UINVS(10,10), UT(10,10), Y(10), Z(10), ZO(10)
    DATA  FORCE, T, ZRO /802*0.0/
    DATA  A, B, Y, ZO /220*(0., 0.)/
 80 FORMAT (' TRANSIENT RESPONSE OF POSITIVE DEFINITE SYSTEMS ' %
    'WITH VISCOUS DAMPING.',/,' EIGENVALUES DISTINCT.',/)
 81 FORMAT (' ENTER: (1) N       = ORDER OF MATRICES M, C, AND K   '%
    '** N <= 5',/,8X,'(2) NDATA = NO. DATA POINTS FOR FORCE IN EACH'%
    ' EQUATION',/,8X,'(3) DT    = TIME INCREMENT IN INTEGRATION',/,%
    8X'(4) ERROR = ERROR SPECIFIED IN ITERATION',/,8X'(5) NITER =' %
    ' MAX. ITERATIONS FOR EACH MODE')
 82 FORMAT (' N =',I3,'    NDATA =',I4,'    DT =',D12.4,/, %
    ' ERROR =',D12.4,'    NITER =',I4)
 83 FORMAT (' *** IS THIS CORRECT?   1 = YES;  2 = NO.')
 84 FORMAT (' ENTER MASS MATRIX-M BY ROW:-')
 85 FORMAT (6D12.4)
 86 FORMAT (' ENTER DAMPING MATRIX-C BY ROW:-')
 87 FORMAT (' ENTER STIFFNESS MATRIX-K BY ROW:-')
 88 FORMAT (' ENTER: (1) INITIAL DISPLACEMENT XO(1) TO XO(N)',/,8X, %
    '(2) INITIAL VELOCITY VO(1) TO VO(N)')
 89 FORMAT (' ENTER FORCE(T) FOR TIME T(1) TO T(NDATA) FOR EACH ' %
    'OF THE N EQUATIONS.')
 90 FORMAT (13X,'* * * OUTPUT * * *')
 91 FORMAT (/,' TIME:        ',D12.4)
 92 FORMAT (' FORCE:       ',5D12.4)
 93 FORMAT (' X1 TO XN:  ',5D12.4)
 94 FORMAT (' VELOCITY:  ',5D12.4)
    WRITE (6,80)
 10 WRITE (6,81)
    READ (5,*) N, NDATA, DT, ERROR, NITER
    WRITE (6,82) N, NDATA, DT, ERROR, NITER
    WRITE (6,83)
    READ (5,*) IANS
    IF (IANS = 2) GOTO 10
 11 WRITE (6,84)
    READ (5,*) ((M(I,J), J=1,N), I=1,N)
    DO 40 I=1,N
 40   WRITE (6,85) (M(I,J), J=1,N)
    WRITE (6,83)
    READ (5,*) IANS
    IF (IANS = 2) GOTO 11
 12 WRITE (6,86)
    READ (5,*) ((C(I,J), J=1,N), I=1,N)
    DO 41 I=1,N
 41   WRITE (6,85) (C(I,J), J=1,N)
    WRITE (6,83)
    READ (5,*) IANS
    IF (IANS = 2) GOTO 12
```

(*a*) Program listing

FIG. 9-15. *Transient response of discrete systems with viscous damping—TRESPDAM.*

```
13 WRITE (6,87)
   READ (5,*) ((K(I,J), J=1,N), I=1,N)
   DO 42 I=1,N
42   WRITE (6,85) (K(I,J), J=1,N)
   WRITE (6,83)
   READ (5,*) IANS
   IF (IANS = 2) GOTO 13
14 WRITE (6,88)
   READ (5,*) (X(I), I=1,N), (V(I), I=1,N)
   WRITE (6,85) (X(I), I=1,N)
   WRITE (6,85) (V(I), I=1,N)
   WRITE (6,83)
   READ (5,*) IANS
   IF (IANS = 2) GOTO 14
   WRITE (6,89)
   DO 43 I=1,N
15   READ (5,*) (FORCE(N+I,J), J=1,NDATA)
     WRITE (6,85) (FORCE(N+I,J), J=1,NDATA)
     WRITE (6,83)
     READ (5,*) IANS
     IF (IANS = 2) GOTO 15
43   CONTINUE
   NT2 = N*2
   NP1 = N + 1
"
"  *** #II. GET MODAL MATRIX-U.  UNCOUPLE THE REDUCED EQUATIONS. ***
   CALL $CMODL (M, C, K, U, ERROR, NITER, N)
   DO 44 I=1,N
   DO 44 J=1,N
     A(I+N,J) = M(I,J)
     A(I,N+J) = M(I,J)
     A(I+N,J+N) = C(I,J)
     B(I,J) = -M(I,J)
44   B(I+N,J+N) = K(I,J)
   DO 45 I=1,NT2
   DO 45 J=1,NT2
45   UT(I,J) = U(J,I)
   CALL $CMPLY (UT, A, DUM, NT2)
   CALL $CMPLY (DUM, U, A, NT2)
   CALL $CMPLY (UT, B, DUM, NT2)
   CALL $CMPLY (DUM, U, B, NT2)
   DO 46 I=1,NT2
46   BDA(I) = B(I,I)/A(I,I)
   CALL $CINVS (U, UINVS, NT2)
   WRITE (6,90)
"
"  *** #III. CALCULATE TRANSIENT RESPONSE AND OUTPUT. ***
   DO 47 IT=1,NDATA
     WRITE (6,91) T
     WRITE (6,92) (FORCE(I,IT), I=NP1,NT2)
     WRITE (6,93) (X(I), I=1,N)
     WRITE (6,94) (V(I), I=1,N)
"    *TRANSIENT RESPONSE IN Z-COORDINATES.
     DO 48 I=1,N
       Y(I) = V(I)
48     Y(N+I) = X(I)
     DO 49 I=1,NT2
       Z0(I) = DCMPLX(ZRO, ZRO)
       FN(I) = Z0(I)
       DO 50 J=1,NT2
         Z0(I) = Z0(I) + UINVS(I,J)*Y(J)
50       FN(I) = FN(I) + UT(I,J)*FORCE(J,IT)
49     FN(I) = FN(I)/A(I,I)
     CALL $CRKUT (FN, Z0, BDA, DT, Z, NT2)
```

(*a*) Program listing (*Continued*)

FIG. 9-15. (*Continued*)

```
705 "     *TRANSIENT RESPONSE IN ORIGINAL COORDINATES.
710       DO 51 I=1,NT2
715 51      ZO(I) = Z(I)
720       DO 52 I=1,NT2
725         Y(I) = DCMPLX(ZRO, ZRO)
730         DO 52 J=1,NT2
735 52        Y(I) = Y(I) + U(I,J)*ZO(J)
740       DO 53 I=1,N
745         V(I) = Y(I)
750 53      X(I) = Y(N+I)
755 47    T = T + DT
760     STOP
765     END
```

(*a*) Program listing (*Continued*)

```
MERGE $COEFF, $CHOMO, $CROOT, $CINVS, $CMPLY, $CSUBN, $CMODL, $CRKUT
READY

SAVE $TEMP
READY

MERGE TRESPDAM, $TEMP, $INVS, $MPLY, $SUBN
READY

RUN

TRANSIENT RESPONSE OF POSITIVE DEFINITE SYSTEMS WITH VISCOUS DAMPING.
EIGENVALUES DISTINCT.

ENTER: (1) N      = ORDER OF MATRICES M, C, AND K    ** N <= 5
       (2) NDATA = NO. DATA POINTS FOR FORCE IN EACH EQUATION
       (3) DT    = TIME INCREMENT IN INTEGRATION
       (4) ERROR = ERROR SPECIFIED IN ITERATION
       (5) NITER = MAX. ITERATIONS FOR EACH MODE
?3   4   .2   1D-16   200

N =  3     NDATA =    4     DT =   0.2000D+00
ERROR =   0.1000D-15     NITER = 200
*** IS THIS CORRECT?    1 = YES;   2 = NO.
?1

ENTER MASS MATRIX-M BY ROW:-
?3 0 0   0 1 0   0 0 2

 0.3000D+01  0.0           0.0
 0.0         0.1000D+01    0.0
 0.0         0.0           0.2000D+01
*** IS THIS CORRECT?    1 = YES;   2 = NO.
?1

ENTER DAMPING MATRIX-C BY ROW:-
?10 -3 -2   -3 8 -4   -2 -4 7

 0.1000D+02 -0.3000D+01 -0.2000D+01
-0.3000D+01  0.8000D+01 -0.4000D+01
-0.2000D+01 -0.4000D+01  0.7000D+01
*** IS THIS CORRECT?    1 = YES;   2 = NO.
?1
```

(*b*) Print-out

FIG. 9-15. (*Continued*)

```
ENTER STIFFNESS MATRIX-K BY ROW:-
?140 -60 -20    -60 220 -80    -20 -80 200

 0.1400D+03 -0.6000D+02 -0.2000D+02
-0.6000D+02  0.2200D+03 -0.8000D+02
-0.2000D+02 -0.8000D+02  0.2000D+03
*** IS THIS CORRECT?    1 = YES;   2 = NO.
?1

ENTER: (1) INITIAL DISPLACEMENT XO(1) TO XO(N)
       (2) INITIAL VELOCITY VO(1) TO VO(N)
?3  2  2   1  4  7

 0.3000D+01  0.2000D+01  0.2000D+01
 0.1000D+01  0.4000D+01  0.7000D+01
*** IS THIS CORRECT?    1 = YES;   2 = NO.
?1

ENTER FORCE(T) FOR TIME T(1) TO T(NDATA) FOR EACH OF THE N EQUATIONS.
?1  2  3  4

 0.1000D+01  0.2000D+01  0.3000D+01  0.4000D+01
*** IS THIS CORRECT?    1 = YES;   2 = NO.
?1

?5  6  7  8

 0.5000D+01  0.6000D+01  0.7000D+01  0.8000D+01
*** IS THIS CORRECT?    1 = YES;   2 = NO.
?1

?9  8  7  6

 0.9000D+01  0.8000D+01  0.7000D+01  0.6000D+01
*** IS THIS CORRECT?    1 = YES;   2 = NO.
?1

              * * * OUTPUT * * *

TIME:         0.0
FORCE:        0.1000D+01   0.5000D+01   0.9000D+01
X1 TO XN:     0.3000D+01   0.2000D+01   0.2000D+01
VELOCITY:     0.1000D+01   0.4000D+01   0.7000D+01

TIME:         0.2000D+00
FORCE:        0.2000D+01   0.6000D+01   0.8000D+01
X1 TO XN:     0.1843D+01   0.1110D+01   0.1277D+01
VELOCITY:    -0.1111D+02  -0.1037D+02  -0.1237D+02

TIME:         0.4000D+00
FORCE:        0.3000D+01   0.7000D+01   0.7000D+01
X1 TO XN:    -0.6618D+00  -0.8506D+00  -0.1141D+01
VELOCITY:    -0.1156D+02  -0.6651D+01  -0.7379D+01

TIME:         0.6000D+00
FORCE:        0.4000D+01   0.8000D+01   0.6000D+01
X1 TO XN:    -0.1921D+01  -0.1080D+01  -0.1092D+01
VELOCITY:     0.1775D+00   0.3826D+01   0.6349D+01
STOP
```

(*b*) Print-out (*Continued*)

FIG. 9-15. (*Continued*)

modified readily for other applications, such as by increasing the size of the arrays, by additional statements to attain greater flexibility, or by transcribing the programs for batch processes.

Each program described in the chapter is preceded by a brief description, a recapitulation of the equations employed, and the terminal commands for the program execution. With slight modifications, the programs are sufficient for the computation and the plotting of the results for the problems in Chaps. 2 and 7. Procedures for program editing and modification are briefly described in Secs. 9-3 to 9-5. A detail listing of the programs and the typical print-outs are shown in the figures.

The first four programs in Secs. 9-3 to 9-5 employ the transient response of one-degree-of-freedom systems to illustrate (1) the organization of the programs, (2) the use of subroutines, and (3) the application of the plotting subroutine.

Mechanical impedance method is used to find the frequency response of systems in Secs. 9-6 and 9-7. Essentially, the method consists of (1) substituting $j\omega$ for the time derivatives in the equations of motion, (2) using phasors to respresent the excitation and response, and (3) solving the resulting algebraic equations, which may be real or complex.

Modal analysis is applied to find the transient response of undamped discrete systems in Sec. 9-8. (1) A modal vector is found by substituting the corresponding natural frequency in the unforced system and solving the resulting algebraic equations. (2) The equations of motion are uncoupled by the modal matrix as shown in Eq. (9-22). (3) The uncoupled equations are solved as independent one-degree-of-freedom systems. The rest of the problem simply uses the modal matrix for the transformation between the generalized and the principal coordinates.

Rayleigh's method to find the fundamental frequency of a multirotor system is shown in Sec. 9-9. Myklestad method with the transfer matrix technique is illustrated in Sec. 9-10. Matrix iteration, due to Duncan and Collar, to find the natural frequencies and modal vectors of undamped systems is shown in Sec. 9-11. Each of the methods was fully described in their respective sections in the text.

Modal analysis is used to find the transient response of discrete systems with viscous damping in Sec. 9-12. The analysis is applied to the reduced equations in Eq. (9-47), which is a set of $2n$ simultaneous first-order differential equations. Aside from the reduced equations and the complex modal matrix, the procedure is essentially the same as that for undamped systems described in Sec. 9-8.

PROBLEMS

The object of the problems in this chapter is to give students additional exercises in using the digital computer for the types of programs considered in the text. It is understood that every program must be checked, preferably by means of a problem with a known solution.

9-1 Using the subrouting $INVG in Fig. C-4, write a program to find the inverse A^{-1} of a real matrix A. Verify that $A^{-1}A = I$, where I is a unit matrix.

$$A = \begin{bmatrix} 1 & 2 & 5 \\ 3 & 1 & 4 \\ 1 & 1 & 2 \end{bmatrix}$$

9-2 Repeat Prob. 9-1, using the subroutine $INVS in Fig. C-5(*a*).

9-3 Using the subroutine $CINVS in Fig. C-5(*b*), write a program to find the inverse A^{-1} of a complex matrix A. Verify that $A^{-1}A = I$, where I is a unit matrix.

$$A = \begin{bmatrix} 5+j1 & 2+j2 & 3+j3 \\ 7+j3 & 5+j2 & 6+j1 \\ 5+j6 & 2+j5 & 3+j4 \end{bmatrix}$$

9-4 Using the subroutine $CINVS in Fig. C-5(*b*), write a program to solve the simultaneous equations $AX = B$.

$$\begin{bmatrix} 1+j1 & 3-j2 & 5+j1 \\ 6-j2 & 2+j1 & 6+j5 \\ 2+j7 & 1-j6 & 3+j7 \end{bmatrix} \begin{bmatrix} x_1 \\ x_2 \\ x_3 \end{bmatrix} = \begin{bmatrix} -5-j6 \\ -3+j4 \\ -9+j8 \end{bmatrix}$$

Use the solution vector X to verify that $AX = B$.

9-5 (**a**) Write a subroutine for the inversion of a complex matrix, using the Gaussian elimination method with pivot elements.

(**b**) Write a program to check the subroutine. Use the data in Prob. 9-3 for the numerical check. *Hint:* See subroutine $INVG in Fig. C-4.

9-6 Using the subroutine $COEFF in Fig. C-6, find the coefficients of the characteristic equation $|\lambda I - H| = 0$.

$$H = \begin{bmatrix} 1 & 2 & 5 \\ 3 & 1 & 4 \\ 1 & 1 & 2 \end{bmatrix}$$

9-7 Given the matrices M and K of an undamped system, write a program to use the subroutine $COEFF in Fig. C-6 to find the coefficients of the characteristic equation $|\lambda I - H| = 0$, where $H = M^{-1}K$.

$$M = \begin{bmatrix} 3 & 0 & 0 \\ 0 & 1 & 0 \\ 0 & 0 & 2 \end{bmatrix} \qquad K = \begin{bmatrix} 140 & -60 & -20 \\ -60 & 220 & -80 \\ -20 & -80 & 200 \end{bmatrix}$$

9-8 Using the Faddeev-Leverrier method illustrated in Figs. C-5 and C-6, write a program to find (**a**) the inverse H^{-1} of a matrix H, and (**b**) the coefficients of the characteristic equation $|\lambda I - H| = 0$.

$$H = \begin{bmatrix} 21 & 4 & 2 \\ 12 & 12 & 6 \\ 3 & 3 & 10 \end{bmatrix}$$

9-9 Write a program to factorize a polynomial $f(x)$. Assume

$$\begin{aligned} f(x) &= c_0 + c_1 x + c_2 x^2 + c_3 x^3 + \cdots c_n x^n \\ &= (1 + a_1 x)(1 + a_2 x) \cdots (1 + a_n x) \end{aligned}$$

where the a's are real. If $f(x) = -24 + 26x - 9x^2 + x^3$, find (**a**) the values of a's, and (**b**) the number of iterations to obtain each factor, when the error in the iteration is of the order of 10^{-16}. *Hint:* See subroutine $ROOT in Fig. C-7.

9-10 Repeat Prob. 9-9 by means of a subroutine. Note that the subroutine can be obtained by extracting the appropriate statements from the program in Prob. 9-9.

9-11 Write a program to find the quadratic factors of a polynomial $f(x)$. Assume

$$f(x) = c_0 + c_1 x + c_2 x^2 + c_3 x^3 + c_4 x^4 + \cdots + c_n x^n$$
$$= (1 + a_1 x + a_2 x^2)(\cdots)\cdots(\cdots)$$

where the a's are real. If $f(x) = 1 + 8x + 50x^2 + 172x^3 + 445x^4 + 668x^5 + 576x^6$, find (**a**) the values of a's, and (**b**) the number of iterations to obtain each factor, when the error in the iteration is of the order of 10^{-16}. *Hint:* See subroutine $CROOT in Fig. C-8.

9-12 Repeat Prob. 9-11 by means of a subroutine. Note that the subroutine can be obtained by extracting the appropriate statements from the program in Prob. 9-11.

9-13 Repeat Prob. 9-11 and find the roots of each quadratic factor.

$$f(x) = 1 + 19x + 132x^2 + 496x^3 + 1096x^4 + 1440x^5 + 960x^6$$

9-14 Given the matrices M and K of an undamped system $M\{\ddot{q}\} + K\{q\} = \{0\}$, write a program to use the subroutine $ROOT in Fig. C-7 to find the eigenvalues of the system.

$$M = \begin{bmatrix} 2 & 0 & 0 \\ 0 & 1 & 0 \\ 0 & 0 & 2 \end{bmatrix} \qquad K = \begin{bmatrix} 4 & -1 & 0 \\ -1 & 2 & -1 \\ 0 & -1 & 4 \end{bmatrix}$$

9-15 Using the subroutine $CROOT, write a program to find the complex eigenvalues of a system with viscous damping.

$$M\{\ddot{q}\} + C\{\dot{q}\} + K\{q\} = \{0\}$$

$$M = \begin{bmatrix} 3 & 0 & 0 \\ 0 & 1 & 0 \\ 0 & 0 & 2 \end{bmatrix} \qquad C = \begin{bmatrix} 10 & -3 & -2 \\ -3 & 8 & -4 \\ -2 & -4 & 7 \end{bmatrix} \qquad K = \begin{bmatrix} 140 & -60 & -20 \\ -60 & 220 & -80 \\ -20 & -40 & 200 \end{bmatrix}$$

9-16 Given the coefficient matrix A of a real homogeneous algebraic equation $AX = 0$, use the subroutine $HOMO in Fig. C-9($a$) to find the solution vector X. Verify that $AX = 0$.

$$A = \begin{bmatrix} 1 & 2 & 3 & 1 \\ 2 & 5 & 4 & 7 \\ 1 & 1 & 3 & -2 \\ 4 & 2 & 1 & -3 \end{bmatrix}$$

9-17 Using the subroutine $CHOMO in Fig. C-9($b$), write a program to solve the complex homogeneous algebraic equation $AX = 0$. Use your own values for the complex coefficient matrix A. Verify that $AX = 0$.

9-18 Using the subroutine $CRKUT in Fig. C-12, write a program to solve a set of N first-order differential equations, where $N \leq 10$. Use your own values for the constant coefficients of the differential equations.

RECOMMENDED GENERAL REFERENCES

The following short list of references is recommended for readers who would like additional discussions on the topics presented in the text. There are many good books in each area. The list is purposely abbreviated to several books for each subject. It is primarily intended as a general reference.

General Reference

ANDERSON, R. A., *Fundamentals of Vibrations*, Macmillan Co., New York, 1967.

CRANDAL, S. H., *Engineering Analysis*, 4th ed., McGraw-Hill Book Co., New York, 1956.

DEN HARTOG, J. P., *Mechanical Vibrations*, 4th ed., McGraw-Hill Book Co., New York, 1956.

HARRIS, C. M., and C. E. CREDE, (eds.), *Shock and Vibration Handbook*, McGraw-Hill Book Co., New York, 1961.

KELLY, R. D., and G. RICHMAN, *Principles and Techniques of Shock Data Analysis*, Shock and Vibration Information Center, U. S. Department of Defense, SVM-5, 1971.

MEIROVITCH, L., *Analytical Methods in Vibrations*, Macmillan Co., New York, 1967.

TIMOSHENKO, S., D. H. YOUNG, and W. WEAVER, JR., *Vibration Problems in Engineering, 4th ed.*, John Wiley & Sons, New York, 1967.

TONG, K. N., *Theory of Mechanical Vibration*, John Wiley & Sons, New York, 1960.

VERNON, J. B., *Linear Vibration Theory*, John Wiley & Sons, New York, 1967.

Nonlinear Vibration

ANDRONOW, A., and S. CHAIKEN, *Theory of Oscillations*, Moscow, 1937, English translation by S. Lefschitz, Princeton University Press, Princeton, N.J., 1949.

CUNNINGHAM, W. J., *Introduction to Nonlinear Analysis*, McGraw-Hill Book Co., New York, 1958.

MINORISKY, N., *Nonlinear Oscillations*, D. Van Nostrand Co., Princeton, N.J., 1962.

STOKER, J. J., *Nonlinear Vibrations*, Interscience Publishers, Inc., New York, 1950.

Applied Mathematics and Computation

CRESS, P., P. DIRKSEN, and J. W. GRAHAM, *FORTRAN IV with WAT-FOR and WATFIV*, Prentice-Hall, Inc., Englewood Cliffs, N.J., 1970.

FADDEEVA, V. N., *Computational Methods of Linear Algebra*, Translated by C. D. Benster, Dover Publications, Inc., New York, 1959.

FRAZER, R. A., W. J. DUNCAN, and A. R. COLLAR, *Elementary Matrices*, Cambridge University Press, London, 1957.

HAMMING, R. W., *Numerical Methods for Scientists and Engineers*, 2d ed., McGraw-Hill Book Co., New York, 1973.

HILDEBRAND, F. B., *Methods of Applied Mathematics*, Prentice Hall, Inc., Englewood Cliffs, N.J., 1965.

IBM CALL/360-OS, GH20-0710-2, *FORTRAN Language Reference Manual*, White Plains, N.Y., 1971.

VON KARMAN, T., and M. A. BIOT, *Mathematical Methods in Engineering*, McGraw-Hill Book Co., New York, 1940.

SCARBOROUGH, J. B., *Numerical Mathematical Analysis*, 4th ed. Johns Hopkins Press, Baltimore, Md., 1958.

SHAPIRO, W., G. C. HORNER, and W. D. PILKEY, *Shock and Vibration Computer Programs*, Shock and Vibration Information Center, Naval Research Laboratory, SVM-10, 1975.

WYLIE, C. R., JR., *Advanced Engineering Mathematics*, 4th ed., McGraw-Hill Book Co., New York, 1975.

Appendix A

Elements of Matrix Algebra

We shall briefly review matrix techniques useful for the discussions in Chaps. 4 and 6 and illustrate with examples.

A-1 DEFINITIONS

A *matrix* is a rectangular array of numbers, real or complex. An array with m rows and n columns enclosed in brackets is called an m-by-n matrix. The symbols commonly used to denote a matrix are

$$\mathrm{A} = [\mathrm{A}] = [a] = [a_{ij}] = \begin{bmatrix} a_{11} & a_{12} & a_{13} & \cdots & a_{1n} \\ a_{21} & a_{22} & a_{23} & \cdots & a_{2n} \\ \cdots & \cdots & \cdots & \cdots & \cdots \\ a_{m1} & a_{m2} & a_{m3} & \cdots & a_{mn} \end{bmatrix}$$

Each *element* a_{ij} in a matrix has two subscripts. The first subscript indicates that the element is in the ith row and the second subscript the jth column of the array. A 2×3 matrix is shown in Fig. A-1(a). The element in the second row and first column is -5.

If $m = n$, the $n \times n$ array is called a *square matrix* of *order* n. A 3×3 square matrix is shown in Fig. A-1(b). The *principal diagonal* of a square matrix consists of all the a_{ii} terms. For example, the principal diagonal of the square matrix in Fig. A-1(b) consists of 3, 9, and 5. Figure A-1(c) shows a 4×4 *symmetric matrix* in which the elements are symmetrical with respect to the principal diagonal, that is, $a_{ij} = a_{ji}$. A square matrix which has zero for all the off diagonal elements is called a *diagonal matrix.* A 4×4 diagonal matrix is shown in Fig. A-1(d). A diagonal matrix with unity for all the diagonal elements is called a *unit matrix*, or *identity matrix*, denoted by the symbol I. A 3×3 unit matrix is illustrated in Fig. A-1(e). A *null matrix* has all its elements equal to zero. The unit matrix and the null matrix correspond to unity and zero, respectively, in ordinary algebra. A unit matrix, however, must be square while a null matrix can be square, rectangular, or just a row or a column. Figure A-1(*f*) shows a 2×3 null matrix.

An $m\times1$ matrix is called a *column matrix* or a *column vector.* A $1\times n$ matrix is called a row matrix or a *row vector.* We shall use the symbol { } to denote a column vector and ⌊ ⌋ to denote a row vector as illustrated in Figs. A-1(g) and (*h*), respectively. A 1×1 matrix is called a *scalar.*

If the determinant of a square matrix is zero, the matrix is termed *singular*, otherwise, the matrix is *nonsingular.* All rectangular matrices are singular. Note that a determinant is denoted by the symbol $|A|$ and a matrix by $[A]$. A determinant has a specific value when expanded according to the rules. A matrix is merely an array and cannot be expanded to yield a value.

The *transpose* A^T of an $m\times n$ matrix A is the $n\times m$ matrix obtained

$$A=\begin{bmatrix}2 & 0 & 1\\ -5 & 3 & 7\end{bmatrix}$$

(*a*) A 2×3 matrix

$$B=\begin{bmatrix}3 & 2 & 1\\ 4 & 9 & 11\\ 6 & 7 & 5\end{bmatrix}$$

(*b*) A 3×3 square matrix

$$C=\begin{bmatrix}3 & 2 & 1 & 0\\ 2 & 9 & 6 & 8\\ 1 & 6 & 11 & 7\\ 0 & 8 & 7 & 5\end{bmatrix}$$

(*c*) A 4×4 symmetric matrix

$$D=\begin{bmatrix}3 & 0 & 0 & 0\\ 0 & 9 & 0 & 0\\ 0 & 0 & 11 & 0\\ 0 & 0 & 0 & 5\end{bmatrix}$$

(*d*) A 4×4 diagonal matrix

$$I=[I]=\begin{bmatrix}1 & 0 & 0\\ 0 & 1 & 0\\ 0 & 0 & 1\end{bmatrix}$$

(*e*) A 3×3 unit matrix

$$0=[0]=\begin{bmatrix}0 & 0 & 0\\ 0 & 0 & 0\end{bmatrix}$$

(*f*) A 2×3 null matrix

$$X=\{x\}=\{x_1, x_2\cdots x_m\}$$

$$X=\begin{bmatrix}x_1\\ x_2\\ \cdots\\ x_m\end{bmatrix}$$

(g) An $m\times1$ column vector

$$Y=\lfloor y\rfloor$$

$$Y=[y_1, y_2\cdots y_n]$$

(*h*) A $1\times n$ row vector

$$A^T=\begin{bmatrix}2 & -5\\ 0 & 3\\ 1 & 7\end{bmatrix}$$

(*i*) Transpose of A in Part *a*

$$\text{Adj } B=\begin{bmatrix}-32 & -3 & 13\\ 46 & 9 & -29\\ -26 & -9 & 19\end{bmatrix}$$

(*j*) Adj B of B in Part *b*

FIG. A-1. *Examples of Matrices.*

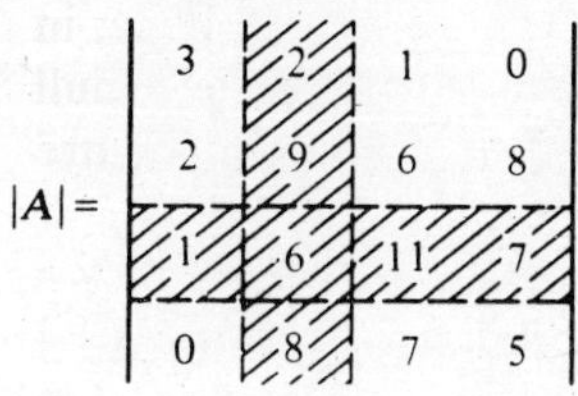

$$A_{32} = (-1)^{3+2} \begin{vmatrix} 3 & 1 & 0 \\ 2 & 6 & 8 \\ 0 & 7 & 5 \end{vmatrix}$$

$$= -(-88) = 88$$

(*a*) A 4th-order determinant (*b*) Cofactor of a_{32}

FIG. A-2. *Cofactor* A_{ij} *of* a_{ij} *in* $|A|$.

by interchanging the rows and columns of A. Figure A-1(*i*) illustrates the A^T. Evidently, the transpose of a column vector is a row vector and vice versa.

The *cofactor* A_{ij} of the element a_{ij} in a square matrix A is the cofactor of a_{ij} in $|A|$. Consider a 4th-order determinant shown in Fig. A-2(*a*). If the elements in the ith row and the jth column are deleted from the determinant, the resultant 3rd-order determinant is a *minor* of a_{ij}. The minor multiplied by $(-1)^{i+j}$ is the cofactor A_{ij} of a_{ij} in $|A|$. The *adjoint* adj A of a square matrix A is formed by replacing every element in A by its cofactor and transposing. The adjoint of matrix B is shown in Fig. A-1(*j*).

If a square matrix A is nonsingular, it possesses an *inverse* (reciprocal) A^{-1}, which has the property

$$A^{-1}A = AA^{-1} = I \tag{A-1}$$

where I is a unit matrix. The inverse is defined by

$$A^{-1} = \frac{\text{adj } A}{|A|} \tag{A-2}$$

A-2 BASIC MATRIX OPERATIONS

Two matrices $A = [a_{ij}]$ and $B = [b_{ij}]$ are *equal* if and only if $a_{ij} = b_{ij}$ for every i and j.

The *sum*, or difference, of two matrices A and B having the same number of rows and columns is defined by

$$\underset{m \times n}{[A]} + \underset{m \times n}{[B]} = \underset{m \times n}{[C]}$$

or

$$[a_{ij}] + [b_{ij}] = [c_{ij}] = [(a_{ij} + b_{ij})]$$

The operation is commutative, that is,

$$A + B = B + A$$

and associative, that is,

$$(A+B)+C=A+(B+C)$$

The *product of a scalar* k and matrix A is given as

$$B=kA \qquad \text{or} \qquad [b_{ij}]=[ka_{ij}]$$

Each element in B equals k times the corresponding element in A.

The *product* of two matrices $A=[a_{ij}]$ and $B=[b_{ij}]$ is defined as

$$\underset{m\times p}{[A]}\ \underset{p\times n}{[B]} = \underset{m\times n}{[C]} \tag{A-3}$$

where

$$[C]=[c_{ij}]=\left[\sum_p a_{ip}b_{pj}\right] \tag{A-4}$$

Note that (1) the matrices A and B must be *conformable*, that is, the multiplication is defined only if the number of columns in A is equal to the number of rows in B as shown in Eq. (A-3), and (2) each element c_{ij} in matrix C is the sum of the product of all the elements in the ith row of A and the corresponding elements in the jth column of B as indicated in Eq. (A-4). This is called the *row-on-column multiplication.*

Matrix multiplication is in general not commutative.

$$AB \neq BA$$

For example, consider the product of matrices A and B below.

$$A=\begin{bmatrix}0 & 1\\ 0 & 1\end{bmatrix} \qquad B=\begin{bmatrix}1 & 1\\ 0 & 0\end{bmatrix} \qquad AB=\begin{bmatrix}0 & 0\\ 0 & 0\end{bmatrix} \qquad BA=\begin{bmatrix}0 & 2\\ 0 & 0\end{bmatrix}$$

Note that if $AB=0$, it does not imply $A=0$, or $B=0$, or $BA=0$, or $A=B=0$. If the matrices are conformable, the multiplication is associative, that is,

$$(AB)C=A(BC)$$

and distributive, that is,

$$(A+B)C=AC+BC, \qquad (A+B)(A+B)=AA+AB+BA+BB$$

Note that AB is the *premultiplication* of B by A, but BA is the *postmultiplication* of B by A. The product AB is not necessarily equal to BA.

The *transpose of a product* matrix is the product of the transposes of the separate matrices in reverse order.

$$(AB)^T=B^TA^T \qquad (ABC)^T=C^TB^TA^T \tag{A-5}$$

The *inverse of a product* matrix is the product of the inverses of the separate matrices in reverse order.

$$(ABC)^{-1}=C^{-1}B^{-1}A^{-1} \tag{A-6}$$

The *derivative* of a matrix $[a_{ij}]$, when it exists, is defined as

$$\frac{d}{dt}[a_{ij}] = \left[\frac{d}{dt} a_{ij}\right] \tag{A-7}$$

Each element of the derivative matrix is the derivative of the corresponding element in the original matrix. The *integral* of a matrix $[a_{ij}]$ is defined as

$$\int [a_{ij}]\,dt = \left[\int a_{ij}\,dt\right] \tag{A-8}$$

Hence the integral exists only if the integral of every element of the original matrix exists.

Example 1. Product of matrices A and B, Eq. (A-3)

$$AB = \begin{bmatrix} 1 & 2 & 5 \\ 3 & 1 & 4 \\ 1 & 1 & 2 \end{bmatrix}\begin{bmatrix} 1 & 2 \\ 4 & 1 \\ 6 & 3 \end{bmatrix} = \begin{bmatrix} 1+8+30 & 2+2+15 \\ 3+4+24 & 6+1+12 \\ 1+4+12 & 2+1+6 \end{bmatrix} = \begin{bmatrix} 39 & 19 \\ 31 & 19 \\ 17 & 9 \end{bmatrix}$$

Example 2. Inverse of a matrix A, Eq. (A-2)

$$A = \begin{bmatrix} 1 & 2 & 5 \\ 3 & 1 & 4 \\ 1 & 1 & 2 \end{bmatrix} \qquad |A| = 4$$

$$[A_{ij}] = \begin{bmatrix} -2 & -2 & 2 \\ 1 & -3 & 1 \\ 3 & 11 & -5 \end{bmatrix} \qquad \text{adj } A = \begin{bmatrix} -2 & 1 & 3 \\ -2 & -3 & 11 \\ 2 & 1 & -5 \end{bmatrix}$$

$$A^{-1} = \frac{\text{adj } A}{|A|} = \frac{1}{4}\begin{bmatrix} -2 & 1 & 3 \\ -2 & -3 & 11 \\ 2 & 1 & -5 \end{bmatrix}$$

Let us check A^{-1} using Eq. (A-1).

$$AA^{-1} = \begin{bmatrix} 1 & 2 & 5 \\ 3 & 1 & 4 \\ 1 & 1 & 2 \end{bmatrix}\begin{bmatrix} -2 & 1 & 3 \\ -2 & -3 & 11 \\ 2 & 1 & -5 \end{bmatrix}\frac{1}{4}$$

$$= \frac{1}{4}\begin{bmatrix} -2-4+10 & 1-6+5 & 3+22-25 \\ -6-2+8 & 3-3+4 & 9+11-20 \\ -2-2+4 & 1-3+2 & 3+11-10 \end{bmatrix} = \begin{bmatrix} 1 & 0 & 0 \\ 0 & 1 & 0 \\ 0 & 0 & 1 \end{bmatrix}$$

Example 3

Let us illustrate A^{-1} in Eq. (A-2) by relating A^{-1} to the solution of simultaneous algebraic equations. Consider

$$\begin{aligned} a_{11}x_1 + a_{12}x_2 &= b_1 \\ a_{21}x_1 + a_{22}x_2 &= b_2 \end{aligned} \tag{A-9}$$

This can be expressed in matrix notations as

$$\begin{bmatrix} a_{11} & a_{12} \\ a_{21} & a_{22} \end{bmatrix}\begin{bmatrix} x_1 \\ x_2 \end{bmatrix} = \begin{bmatrix} b_1 \\ b_2 \end{bmatrix} \quad \text{or} \quad A\{X\} = \{B\} \tag{A-10}$$

where A is the *coefficient matrix* and $\{X\}$ and $\{B\}$ are vectors.

$$A = [a_{ij}] = \begin{bmatrix} a_{11} & a_{12} \\ a_{21} & a_{22} \end{bmatrix} \qquad \{X\} = \begin{bmatrix} x_1 \\ x_2 \end{bmatrix} \qquad \{B\} = \begin{bmatrix} b_1 \\ b_2 \end{bmatrix}$$

The unknown $\{X\}$ is the *solution vector.* If A is nonsingular, $\{X\}$ can be found by *premultiplying* Eq. (A-10) by A^{-1}. Since $A^{-1}A = I$, the solution vector is

$$\{X\} = A^{-1}\{B\} \tag{A-11}$$

Alternatively, the unknowns (x_1, x_2) in Eq. (A-9) can be found by Cramer's rule.

$$x_1 = \frac{\begin{vmatrix} b_1 & a_{12} \\ b_2 & a_{22} \end{vmatrix}}{\begin{vmatrix} a_{11} & a_{12} \\ a_{21} & a_{22} \end{vmatrix}} \quad \text{and} \quad x_2 = \frac{\begin{vmatrix} a_{11} & b_1 \\ a_{21} & b_2 \end{vmatrix}}{\begin{vmatrix} a_{11} & a_{12} \\ a_{21} & a_{22} \end{vmatrix}}$$

The equations above can be written in matrix notations as

$$\begin{bmatrix} x_1 \\ x_2 \end{bmatrix} = \frac{\begin{bmatrix} a_{22} & -a_{12} \\ -a_{21} & a_{11} \end{bmatrix}\begin{bmatrix} b_1 \\ b_2 \end{bmatrix}}{|A|} = \frac{\text{adj } A}{|A|}\begin{bmatrix} b_1 \\ b_2 \end{bmatrix} = A^{-1}\{B\} \tag{A-12}$$

where

$$|A| = \begin{vmatrix} a_{11} & a_{12} \\ a_{21} & a_{22} \end{vmatrix} \quad \text{and} \quad \text{adj } A = \begin{bmatrix} a_{22} & -a_{12} \\ -a_{21} & a_{11} \end{bmatrix}$$

Equation (A-12) is identical to (A-11). Hence, if A^{-1} exists, Eq. (A-11) can be used to solve any set of n nonhomogeneous equations in Eq. (A-10).

A-3 MATRIX INVERSION BY METHOD OF ELIMINATION

The inversion of a large matrix by Eq. (A-2) is rather inefficient, because a large number of determinants must be calculated to find the cofactors. The *method of elimination* is closely related to the Gauss-Jordan reduction method. It can be programmed readily in a digital computer.

The operations in the elimination method are (1) the interchange of two rows of a matrix, (2) the multiplication of a row by a nonzero constant, and (3) the adding to a row the multiple of another row of the matrix. The procedure is to perform identical operations on the corresponding rows of a matrix A and a unit matrix I until A is reduced to a unit matrix. Let us illustrate this with an example.

Example 4. Inversion by elimination

Given matrices A and I of the same order.

	$[A]$				$[I]$	
1	2	5	|	1	0	0
3	1	4	|	0	1	0
1	1	2	|	0	0	1

Multiply row 1 by −3 and add to row 2; multiply row 1 by −1 and add to row 3:

1	2	5	|	1	0	0
0	−5	−11	|	−3	1	0
0	−1	−3	|	−1	0	1

Multiply row 3 by 2 and add to row 1; multiply row 3 by −6 and add to row 2:

1	0	−1	|	−1	0	2
0	1	7	|	3	1	−6
0	−1	−3	|	−1	0	1

Add row 2 to row 3:

1	0	−1	|	−1	0	2
0	1	7	|	3	1	−6
0	0	4	|	2	1	−5

Divide row 3 by 4:

1	0	−1	|	−1	0	2
0	1	7	|	3	1	−6
0	0	1	|	2/4	1/4	−5/4

Add row 3 to row 1; multiply row 3 by −7 and add to row 2:

1	0	0	|	−2/4	1/4	3/4
0	1	0	|	−2/4	−3/4	11/4
0	0	1	|	2/4	1/4	−5/4

The array on the right side is the inverse A^{-1}. It is identical to that illustrated in Example 2.

A-4 MODAL MATRIX

Eigenvalues, eigenvectors. and modal matrix will be discussed in this section. It will be shown in the text that (1) an eigenvalue corresponds to a natural frequency, (2) an eigenvector (modal vector) represents a principal mode of vibration, and (3) a modal matrix describes all the natural modes of vibration of a system.

Consider a set of homogeneous equations

$$[\lambda I - A]\{X\} = \{0\} \tag{A-13}$$

where λ is a scalar parameter, A a square matrix of order n, I a unit matrix of the same order, and $\{X\}$ the solution vector. $[\lambda I - A]$ is a square matrix of order n and is called the *characteristic matrix* of A. A necessary and sufficient condition that Eq. (A-13) has a solution other than $\{X\} = \{0\}$ is that $[\lambda I - A]$ be singular, that is,

$$\Delta(\lambda) = |\lambda I - A| = 0 \tag{A-14}$$

This is called the *characteristic equation* of A. The roots, $\lambda_1, \lambda_2, \ldots, \lambda_n$, of the equations are the *eigenvalues* of A. The λ's are variously called the latent roots, characteristic values, or characteristic numbers. We shall assume that all the eigenvalues are distinct.

If Eq. (A-14) gives n distinct values of λ, correspondingly, Eq. (A-13) has n solutions. If $\lambda = \lambda_s$, the solution vector is

$$\{X(\lambda_s)\} \triangleq \{X\}_s = \{x_{1s} \; x_{2s} \; \cdots \; x_{ns}\} \tag{A-15}$$

Two subscripts are assigned to each element of the solution vector $\{X\}_s$. The first subscript refers to the position of the element in the column vector and the second subscript the particular value of s. $\{X(\lambda_s)\}$ is called an *eigenvector* (modal column or *modal vector*). A *modal matrix* is formed from a collection of all the eigenvectors $\{X\}_s$. Thus,

$$[x_{ij}] = [\{X\}_1 \quad \{X\}_2 \quad \cdots \quad \{X\}_n] = \begin{bmatrix} x_{11} & \cdots & x_{1n} \\ \cdots & \cdots & \cdots \\ x_{n1} & \cdots & x_{nn} \end{bmatrix} \tag{A-16}$$

An eigenvector can be multiplied by an arbitrary constant. For $\lambda = \lambda_s$, Eq. (A-13) can be expressed as

$$\lambda_s\{X(\lambda_s)\} = A\{X(\lambda_s)\} \tag{A-17}$$

The equation is also satisfied if $\{X(\lambda_s)\}$ is multiplied by a constant c_s.

$$\lambda_s c_s\{X(\lambda_s)\} = Ac_s\{X(\lambda_s)\} \qquad \text{or} \qquad \lambda_s\{c_s X(\lambda_s)\} = A\{c_s X(\lambda_s)\}$$

that is, each element in $\{X(\lambda_s)\}$ is multiplied by the constant c_s. Let us use $\{u_{is}\}$ to denote $c_s\{X(\lambda_s)\}$. Thus a more general form of Eq. (A-17) is

$$\lambda_s\{u_{is}\} = A\{u_{is}\} \tag{A-18}$$

where $\{u_{is}\} = c_s\{X(\lambda_s)\}$ is a modal vector corresponding to λ_s. Note that the relative values of the elements in the vector remain unchanged. The eigenvector $\{X(\lambda_s)\}$ is *normalized* by the constant c_s.

Considering all the λ's, the modal matrix in Eq. (A-16) can be expressed in terms of $\{u_{is}\}$ as

$$[u] = [u_{ij}] = \begin{bmatrix} u_{11} & \cdots & u_{1n} \\ \cdots & \cdots & \cdots \\ u_{n1} & \cdots & u_{nn} \end{bmatrix} \tag{A-19}$$

Thus, considering all the λ's, Eq. (A-18) becomes

$$[u]\Lambda = A[u] \tag{A-20}$$

where Λ is a diagonal matrix with λ_s as its diagonal elements. For example, if A is a 3×3 matrix, we get

$$\begin{bmatrix} u_{11} & u_{12} & u_{13} \\ u_{21} & u_{22} & u_{23} \\ u_{31} & u_{32} & u_{33} \end{bmatrix} \begin{bmatrix} \lambda_1 & 0 & 0 \\ 0 & \lambda_2 & 0 \\ 0 & 0 & \lambda_3 \end{bmatrix} = \begin{bmatrix} a_{11} & a_{12} & a_{13} \\ a_{21} & a_{22} & a_{23} \\ a_{31} & a_{32} & a_{33} \end{bmatrix} \begin{bmatrix} u_{11} & u_{12} & u_{13} \\ u_{21} & u_{22} & u_{23} \\ u_{31} & u_{32} & u_{33} \end{bmatrix}.$$

Premultiplying Eq. (A-20) by $[u]^{-1}$ gives

$$\Lambda = [u]^{-1}A[u] \tag{A-21}$$

Hence matrix A is *diagonalized* by its modal matrix.

A-5 EVALUATION OF MODAL MATRIX

We shall present two methods to evaluate the modal matrix. An obvious method is to solve the set of homogeneous equations in Eq. (A-13). Another method is to define a lambda matrix and then obtain the modal vector from the adjoint of the lambda matrix. The methods will be illustrated in the examples to follow.

Let us define a lambda matrix $[f(\lambda)] \triangleq [\lambda I - A]$. From Eq. (A-13) we have

$$[f(\lambda)]\{X\} = \{0\} \tag{A-22}$$

Hence the characteristic equation $|f(\lambda)| = |\lambda I - A| = 0$ is identical to Eq. (A-14). From Eq. (A-2) the inverse of $[f(\lambda)]$ is

$$[f(\lambda)]^{-1} = \frac{\text{adj}\,[f(\lambda)]}{\Delta(\lambda)} \triangleq \frac{[F(\lambda)]}{\Delta(\lambda)} \tag{A-23}$$

where $\Delta(\lambda)$ is the *characteristic function* $|\lambda I - A|$ and the adjoint $[F(\lambda)]$ is also a function of λ.

Premultiplying Eq. (A-23) by $[f(\lambda)]$ and rearranging, we get

$$[f(\lambda)][F(\lambda)] = \Delta(\lambda)I$$

Since $\Delta(\lambda_s) = 0$ for $\lambda = \lambda_s$, the equation above yields

$$[f(\lambda_s)][F(\lambda_s)] = [0] \quad \textbf{(A-24)}$$

Note that, for each value of λ_s, Eq. (A-22) also gives

$$[f(\lambda_s)]\{u_{is}\} = \{0\} \quad \textbf{(A-25)}$$

where $\{u_{is}\}$ is an eigenvector. Comparing Eqs. (A-24) and (A-25) we deduce that every column of $[F(\lambda_s)]$ must be proportional to $\{u_{is}\}$. Furthermore, every column of $(F(\lambda_s)]$ must be proportional to one another. Hence any nonzero column of $[F(\lambda_s)]$ can be taken as an eigenvector appropriate to λ_s.

In summary, the procedure for finding the modal matrix of A is (1) to calculate the eigenvalues by Eq. (A-14), (2) to form a lambda matrix $[f(\lambda)]$ and its adjoint $[F(\lambda)]$, (3) to find $[F(\lambda_s)]$ for each value of λ_s, (4) to select any nonzero column of $[F(\lambda_s)]$ as an eigenvector, and (5) to form the modal matrix from the selected eigenvectors.

Example 5. Modal Matrix from Solution of Algebraic Equations

For a given matrix A, we use Eq. (A-13) to form a set of homogeneous algebraic equations $[\lambda I - A]\{X\} = \{0\}$, that is,

$$A = \begin{bmatrix} 2 & -1 & 1 \\ -8 & 3 & 7 \\ -8 & -1 & 11 \end{bmatrix}$$

$$\begin{bmatrix} \lambda-2 & 1 & -1 \\ 8 & \lambda-3 & -7 \\ 8 & 1 & \lambda-11 \end{bmatrix}\begin{bmatrix} x_1 \\ x_2 \\ x_3 \end{bmatrix} = \begin{bmatrix} 0 \\ 0 \\ 0 \end{bmatrix}$$

The characteristic equation from Eq. (A-14) is

$$\Delta(\lambda) = \begin{vmatrix} \lambda-2 & 1 & -1 \\ 8 & \lambda-3 & -7 \\ 8 & 1 & \lambda-11 \end{vmatrix} = (\lambda-2)(\lambda-4)(\lambda-10) = 0$$

Thus, the eigenvalues are 2, 4, and 10. Form a set of homogeneous equations for each value of λ and solve for $\{X\}$.

For $\lambda = \lambda_1 = 2$, we have

$$\begin{bmatrix} 0 & 1 & 1 \\ 8 & -1 & -7 \\ 8 & 1 & -9 \end{bmatrix}\begin{bmatrix} x_1 \\ x_2 \\ x_3 \end{bmatrix} = \begin{bmatrix} 0 \\ 0 \\ 0 \end{bmatrix}$$

It can be shown that the solutions are

$$\{x_{11} \quad x_{21} \quad x_{31}\} = c_1\{1 \quad 1 \quad 1\}$$

Similarly, if $\lambda = \lambda_2 = 4$, we obtain

$$\{x_{12} \quad x_{22} \quad x_{32}\} = c_2\{1 \quad -1 \quad 1\}$$

If $\lambda = \lambda_3 = 10$, we get

$$\{x_{13} \quad x_{23} \quad x_{33}\} = c_3\{0 \quad 1 \quad 1\}$$

The modal matrix may be selected as

$$[u] = [u_{ij}] = \begin{bmatrix} 1 & 1 & 0 \\ 1 & -1 & 1 \\ 1 & 1 & 1 \end{bmatrix}$$

Example 6. Modal matrix from the adjoint matrix

Let us repeat Example 5 using the lambda matrix $[f(\lambda)]$.

$$[f(\lambda)] = \begin{bmatrix} \lambda-2 & 1 & -1 \\ 8 & \lambda-3 & -7 \\ 8 & 1 & \lambda-11 \end{bmatrix}$$

The adjoint $[F(\lambda)]$ of $[f(\lambda)]$ is

$$[F(\lambda)] = \begin{bmatrix} (\lambda-3)(\lambda-11)+7 & -(\lambda-11)-1 & (\lambda-3)-7 \\ -8(\lambda-11)-56 & (\lambda-2)(\lambda-11)+8 & 7(\lambda-2)-8 \\ -8(\lambda-3)+8 & -(\lambda-2)+8 & (\lambda-2)(\lambda-3)-8 \end{bmatrix}$$

From Example 5, the eigenvalues are 2, 4, and 10.

For $\lambda = \lambda_1 = 2$, we obtain

$$[F(2)] = \begin{bmatrix} 16 & 8 & -8 \\ 16 & 8 & -8 \\ 16 & 8 & -8 \end{bmatrix}$$

The eigenvector can be selected proportional to any nonzero column as

$$\{u_{i1}\} = \{1 \quad 1 \quad 1\}$$

For $\lambda = 4$ and 10, respectively, we get

$$[F(4)] = \begin{bmatrix} 0 & 6 & -6 \\ 0 & -6 & 6 \\ 0 & 6 & -6 \end{bmatrix} \quad \text{and} \quad [F(10)] = \begin{bmatrix} 0 & 0 & 0 \\ -48 & 0 & 48 \\ -48 & 0 & 48 \end{bmatrix}$$

The corresponding eigenvectors are

$$\{u_{i2}\} = \{1 \quad -1 \quad 1\} \quad \text{and} \quad \{u_{i3}\} = \{0 \quad 1 \quad 1\}$$

The modal matrix is formed from the eigenvectors as

$$[u] = [u_{ij}] = \begin{bmatrix} 1 & 1 & 0 \\ 1 & -1 & 1 \\ 1 & 1 & 1 \end{bmatrix}$$

Appendix B
Lagrange's Equations

We shall derive Lagrange's equations from the concepts of virtual displacement and d'Alembert's principle rather than Hamilton's principle.

B-1 VIRTUAL DISPLACEMENT

The method of virtual displacement is a static analysis. A *virtual displacement* is an infinitesimal arbitrary change in the coordinate compatible with the geometric constraints of the system.

The *principle of virtual work*, also known as the principle of virtual displacements, may be stated as follows: If a system with workless constraint is in equilibrium, the total virtual work done by the applied forces on all the virtual displacements must equal to zero. The reactive forces at the constraints are not considered, because the constraints are frictionless and the reactive forces must be perpendicular to the virtual displacement. For example, if a particle is constrained to move on a frictionless surface, the virtual displacement satisfying the constraint must be tangential to the surface at the point of contact. In the absence of friction, the reactive force at the contact must be perpendicular to the surface. Hence, for workless constraints, the contribution of the reactive force to the virtual work must be nil. It may be restated that virtual work requires (1) the reaction be frictionless, (2) the virtual displacements be compatible with the geometric constraints, and (3) the system be at equilibrium.

To illustrate the principle, consider a rigid bar pivoting about a frictionless pin C as shown in Fig. B-1. The bar is constrained to rotate about the longitudinal axis of the pin. The applied forces X_1 and X_2 are of constant magnitude and the bar is in equilibrium. The corresponding displacements are δx_1 and δx_2. The work done by the applied forces is

$$\delta W_v = X_1 \delta x_1 - X_2 \delta x_2$$

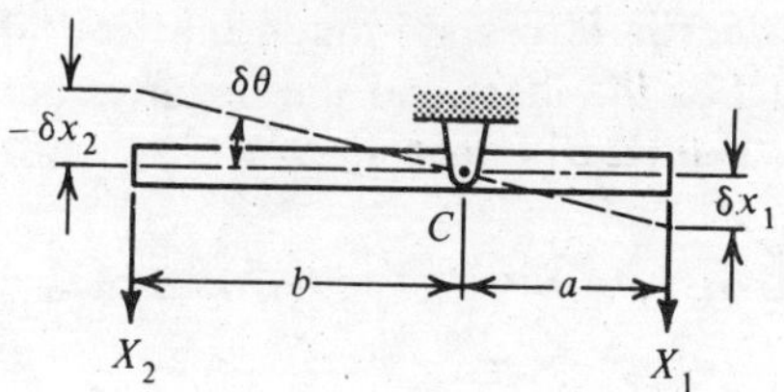

FIG. B-1. *Rigid bar hinged at C.*

The work done by the reactive force is zero and need not be considered. The displacements δx_1 and δx_2 are not independent. The geometric constraint requires that

$$b\delta x_1 = a\delta x_2$$

Substituting this in the previous equation gives

$$\delta W_v = (X_1 - X_2 b/a)\delta x_1$$

Since $\delta W_v = 0$ and δx_1 arbitrary, we deduce that

$$X_1 a - X_2 b = 0$$

Evidently, the same result can be obtained by using the relation $\delta x_1 = a\delta\theta$ and $\delta x_2 = b\delta\theta$ with the first equation.

Now consider a general system with p mass points. Let there be r geometric constraints of the form

$$f_j(x_1, y_1, z_1, \ldots, x_p, y_p, z_p) = 0 \qquad \textbf{(B-1)}$$

where $j = 1, 2, \ldots, r$ and (x, y, z) are Cartesian coordinates. The virtual displacements δx_i, δy_i, δz_i are arbitrary but not independent. They must satisfy the r relations

$$\sum_{i=1}^{p} \left(\frac{\partial f_j}{\partial x_i} \delta x_i + \frac{\partial f_j}{\partial y_i} \delta y_i + \frac{\partial f_i}{\partial z_i} \delta z_i \right) = 0 \qquad \textbf{(B-2)}$$

Let the components of the forces applied to the particle m_i be X_i, Y_i, and Z_i. Neglecting the work done by the reactions and applying the principle of virtual work, we have

$$\delta W_v = \sum_{i=1}^{p} (X_i \delta x_i + Y_i \delta y_i + Z_i \delta z_i) = 0 \qquad \textbf{(B-3)}$$

Equations (B-2) and (B-3) can be solved simultaneously to obtain the desired equations.

B-2 D'ALEMBERT'S PRINCIPLE

D'Alembert's principle enables the discussion in the previous section to be extended to dynamic systems. The principle may be stated as follows: Every state of motion may be considered at any instant as a state of equilibrium if the inertia forces are taken into consideration. The inertia force is the product of the mass and the negative acceleration. Including the inertia forces, Eq. (B-3) becomes

$$\delta W_v = \sum_{i=1}^{p} [(X_i - m_i\ddot{x}_i)\delta x_i + (Y_i - m_i\ddot{y}_i)\delta y_i + (Z_i - m_i\ddot{z}_i)\delta z_i] = 0 \quad \textbf{(B-4)}$$

The simultaneous solution of this equation and the equations of constraint in Eq. (B-2) leads to the equation of motion of the system.

B-3 LAGRANGE'S EQUATIONS

The Lagrange method expresses the energies of the system in terms of the generalized coordinates and thereby obtains the equations of motion directly. We shall derive the Lagrange equations for a single mass particle and then extend the theory to cover a system of particles. The proof can be generalized readily to a system of rigid bodies.

Consider a mass particle m described by the Cartesian coordinates (x,y,z). The forces applied in the corresponding directions are (X,Y,Z). From Eq. (B-4), we get

$$\begin{aligned} \delta W_v &= (X\delta x + Y\delta y + Z\delta z) - (m\ddot{x}\delta x + m\ddot{y}\delta y + m\ddot{z}\delta z) \\ &= \delta W_1 - \delta W_2 = 0 \end{aligned} \quad \textbf{(B-5)}$$

where δW_1 and δW_2 are as defined in the equation above. The coordinates (x,y,z) and the generalized coordinates $(q_1,q_2, \ldots ,q_n)$ are related as

$$\begin{aligned} x &= x(q_1,q_2, \ldots ,q_n) \\ y &= y(q_1,q_2, \ldots ,q_n) \\ z &= z(q_1,q_2, \ldots ,q_n) \end{aligned} \quad \textbf{(B-6)}$$

For infinitesimal virtual changes in the coordinates, the functional relations of the coordinates are

$$\delta x = \sum_{i=1}^{n} \frac{\partial x}{\partial q_i}\delta q_i \qquad \delta y = \sum_{i=1}^{n} \frac{\partial y}{\partial q_i}\delta q_i \qquad \text{and} \qquad \delta z = \sum_{i=1}^{n} \frac{\partial z}{\partial q_i}\delta q_i \quad \textbf{(B-7)}$$

Since the q's are independent, their variations can be examined one at a time. For $\delta q_1 \neq 0$ and $\delta q_2 = \delta q_3 = \cdots = \delta q_n = 0$, we have

$$\delta x = \frac{\partial x}{\partial q_1}\delta q_1 \qquad \delta y = \frac{\partial y}{\partial q_1}\delta q_1 \qquad \text{and} \qquad \delta z = \frac{\partial z}{\partial q_1}\delta q_1 \quad \textbf{(B-8)}$$

Let us examine the term δW_1 in Eq. (B-5). Applying Eq. (B-8) yields

$$\delta W_1 = \left(X\frac{\partial x}{\partial q_1} + Y\frac{\partial y}{\partial q_1} + Z\frac{\partial z}{\partial q_1}\right)\delta q_1 = Q_1\delta q_1 \quad \textbf{(B-9)}$$

Hence δW_1 defines the virtual work associated with the virtual displacement δq_1 due to the applied forces (X, Y, Z). Consequently, a generalized force Q_1 is defined as the force associated with δq_1.

Similarly, by applying Eq. (B-8), the δW_2 term in Eq. (B-5) becomes

$$\delta W_2 = \left(m\ddot{x}\frac{\partial x}{\partial q_1} + m\ddot{y}\frac{\partial y}{\partial q_1} + m\ddot{z}\frac{\partial z}{\partial q_1} \right)\delta q_1 \qquad \textbf{(B-10)}$$

This is the virtual work associated with the virtual displacement δq_1 due to the inertia forces. We shall simplify Eq. (B-10) by introducing (1) a kinetic energy function, and (2) the functional relations of the coordinates.

The kinetic energy function in the $(x, y, z$ coordinates is

$$T = \tfrac{1}{2}(m\dot{x}^2 + m\dot{y}^2 + m\dot{z}^2) \qquad \textbf{(B-11)}$$

Considering only the x-coordinate for the moment, we have

$$\frac{\partial T}{\partial \dot{x}} = m\dot{x} \quad \text{and} \quad \frac{d}{dt}\left(\frac{\partial T}{\partial \dot{x}}\right) = m\ddot{x} \qquad \textbf{(B-12)}$$

Substituting Eq. (B-12) in (B-10) and including the y and z coordinates, the δW_2 term becomes

$$\delta W_2 = \left[\frac{d}{dt}\left(\frac{\partial T}{\partial \dot{x}}\right)\frac{\partial x}{\partial q_1} + \frac{d}{dt}\left(\frac{\partial T}{\partial \dot{y}}\right)\frac{\partial y}{\partial q_1} + \frac{d}{dt}\left(\frac{\partial T}{\partial \dot{z}}\right)\frac{\partial z}{\partial q_1}\right]\delta q_1 \qquad \textbf{(B-13)}$$

Let us develop the functional relations for the coordinates to further reduce the equation above. Since Eq. (B-6) is linear, we get

$$\dot{x} = \sum_{i=1}^{n} \frac{\partial x}{\partial q_i}\dot{q}_i \quad \text{or} \quad \frac{\partial \dot{x}}{\partial \dot{q}_i} = \frac{\partial x}{\partial q_i} \qquad \textbf{(B-14)}$$

Consider the relations

$$\frac{d}{dt}\left(\frac{\partial T}{\partial \dot{x}}\frac{\partial x}{\partial q_i}\right) = \frac{d}{dt}\left(\frac{\partial T}{\partial \dot{x}}\right)\frac{\partial x}{\partial q_i} + \frac{\partial T}{\partial \dot{x}}\frac{\partial \dot{x}}{\partial q_i}$$

or

$$\frac{d}{dt}\left(\frac{\partial T}{\partial \dot{x}}\right)\frac{\partial x}{\partial q_i} = \frac{d}{dt}\left(\frac{\partial T}{\partial \dot{x}}\frac{\partial x}{\partial q_i}\right) - \frac{\partial T}{\partial \dot{x}}\frac{\partial \dot{x}}{\partial q_i} \qquad \textbf{(B-15)}$$

Noting that $\partial\dot{x}/\partial\dot{q}_i = \partial x/\partial q_i$ in Eq. (B-14), the equation above yields

$$\frac{d}{dt}\left(\frac{\partial T}{\partial \dot{x}}\right)\frac{\partial x}{\partial q_i} = \frac{d}{dt}\left(\frac{\partial T}{\partial \dot{x}}\frac{\partial \dot{x}}{\partial \dot{q}_i}\right) - \frac{\partial T}{\partial \dot{x}}\frac{\partial \dot{x}}{\partial q_i} \qquad \textbf{(B-16)}$$

To relate T to the generalized coordinates q_i, we assume that T is a function of $\dot{q}_i$ and q_i. Thus,

$$\frac{\partial T}{\partial \dot{q}_i} = \frac{\partial T}{\partial \dot{x}}\frac{\partial \dot{x}}{\partial \dot{q}_i} \quad \text{and} \quad \frac{\partial T}{\partial q_i} = \frac{\partial T}{\partial \dot{x}}\frac{\partial \dot{x}}{\partial q_i} \qquad \textbf{(B-17)}$$

Substituting this in Eq. (B-16) yields

$$\frac{d}{dt}\left(\frac{\partial T}{\partial \dot{x}}\right)\frac{\partial x}{\partial q_i}=\frac{d}{dt}\left(\frac{\partial T}{\partial \dot{q}_i}\right)-\frac{\partial T}{\partial q_i} \qquad \textbf{(B-18)}$$

Including the y and z coordinates but considering only δq_1, the term δW_2 in Eq. (B-13) becomes

$$\delta W_2=\left[\frac{d}{dt}\left(\frac{\partial T}{\partial \dot{q}_1}\right)-\frac{\partial T}{\partial q_1}\right]\delta q_1 \qquad \textbf{(B-19)}$$

An equation of motion is obtained by substituting Eqs. (B-9) and (B-19) in (B-5). Since δq_1 is arbitrary, we have

$$\frac{d}{dt}\left(\frac{\partial T}{\partial \dot{q}_1}\right)-\frac{\partial T}{\partial q_1}=Q_1 \qquad \textbf{(B-20)}$$

Similar expressions hold for the variations $\delta q_2, \ldots, \delta q_n$, each taken separately. Considering $i=1,2,\ldots,n$, we have the *Lagrange's equations* for an n-degree-of-freedom system in the generalized coordinate q_i.

$$\frac{d}{dt}\left(\frac{\partial T}{\partial \dot{q}_i}\right)-\frac{\partial T}{\partial q_i}=Q_i \qquad \textbf{(B-21)}$$

B-4 GENERALIZED FORCES

The generalized force Q_i in Lagrange's equations is defined in Eq. (B-9). It includes all the forces external to the mass particles. This includes (1) the spring force, (2) the damping force, and (3) the applied force.

The spring force is due to the change in the potential energy U of the spring. Let

$$U=U(q_1, q_2, \ldots, q_n)$$

Expanding this about the stable equilibrium position of the system gives

$$U=U_o+\sum_{i=1}^{n}\left(\frac{\partial U}{\partial q_i}\right)_o q_i+\frac{1}{2}\sum_{i=1}^{n}\sum_{j=1}^{n}\left(\frac{\partial^2 U}{\partial q_i \partial q_j}\right)_o q_i q_j+\cdots \qquad \textbf{(B-22)}$$

where the subscript (o) denotes the values evaluated at the equilibrium position. U_o can be defined equal to zero, that is, the potential energy is measured with respect to U_o as datum. Since U is minimum at U_o, its first derivative must vanish. Thus, if the third and higher order terms are neglected, Eq. (B-22) becomes

$$U=\tfrac{1}{2}\sum_{i=1}^{n}\sum_{j=1}^{n}\left(\frac{\partial^2 U}{\partial q_i \partial q_j}\right)_o q_i q_j \qquad \textbf{(B-23)}$$

For small variations, the second partial derivative term may be assumed constant. Denoting its value by an equivalent spring constant k_{ij}, we obtain

$$U = \frac{1}{2}\sum_{i=1}^{n}\sum_{j=1}^{n} k_{ij}q_iq_j \tag{B-24}$$

where

$$k_{ij} = k_{ji} \tag{B-25}$$

The spring force associated with the coordinate q_j is

$$Q_j = -\frac{\partial U}{\partial q_j} = -\sum_{i=1}^{n} k_{ij}q_i \tag{B-26}$$

By analogy with the potential energy function U in Eq. (B-23), a *dissipation function* D may be defined as

$$D = \frac{1}{2}\sum_{i=1}^{n}\sum_{j=1}^{n} c_{ij}\dot{q}_i\dot{q}_j \tag{B-27}$$

where $c_{ij} = c_{ji}$. Hence the damping force Q_j associated with the velocity $\dot{q}_j$ is

$$Q_j = -\frac{\partial D}{\partial \dot{q}_j} = -\sum_{i=1}^{n} c_{ij}\dot{q}_i \tag{B-28}$$

Note that the dissipation function defined in Eq. (B-27) is equal to half of the rate of energy dissipation in a damper.

Combining the spring force, the damping force, and the applied force Q_i, Lagrange's equations become

$$\frac{d}{dt}\left(\frac{\partial T}{\partial \dot{q}_i}\right) - \frac{\partial T}{\partial q_i} + \frac{\partial D}{\partial \dot{q}_i} + \frac{\partial U}{\partial q_i} = Q_i \tag{B-29}$$

Comparing with Eq. (B-21), the generalized force in the equation above is separated into the damping force, the spring force, and the externally applied force. For the free vibration of a conservative system, we get

$$\frac{d}{dt}\left(\frac{\partial T}{\partial \dot{q}_i}\right) - \frac{\partial T}{\partial q_i} + \frac{\partial U}{\partial q_i} = 0 \tag{B-30}$$

Since U is a function of the coordinates only, this equation can be further simplified by introducing the Lagrangian function $L = (T - U)$. Hence Eq. (B-30) becomes

$$\frac{d}{dt}\left(\frac{\partial L}{\partial \dot{q}_i}\right) - \frac{\partial L}{\partial q_i} = 0 \tag{B-31}$$

Appendix C
Subroutines

C-1 INTRODUCTION

Subroutines are used extensively to simplify the programs in Chap. 9. The subroutines of this appendix, as listed in Table C-1, are in extended precision,† REAL*8 or COMPLEX*16. Their names begin with the character dollar ($) to denote a subroutine. If the second character is the letter (C), the subroutine is in the complex mode; the only exception is $COEFF. The remaining characters identify the name of the subroutine.

Note that the subroutines in the real and the complex modes are almost alike if the same technique is used to solve the problem. The essential difference is in the specification of the real or the complex mode. The subroutines are listed in the figures without line numbers by the command

LIST-TEXT $\quad c$

where c is the RETURN to execute the command. The subroutine $PLOTF for plotting is not a vibration problem and therefore will not be discussed.

C-2 $TRESP—TRANSIENT RESPONSE OF ONE-DEGREE-OF-FREEDOM SYSTEMS. METHOD: 4TH-ORDER RUNGE-KUTTA.

The subroutine $TRESP, as listed in Fig. C-1, finds the *transient response* of one-degree-of-freedom systems. It is obtained from TRESP1 by extracting lines 460 through 610 in Fig. 9-1(a). The initial conditions are XO and YO. The excitation F is assumed constant for the time interval Δt. The system parameters are the damping factor ζ(Z) and the

† It was shown in Sec. 9-6 that the extended precision specification can be changed to single precision by the REPLACE command after the subroutines are merged with the main program.

TABLE C-1 *Subroutines*

SUBROUTINE REAL*8	COMPLEX*16	TITLE AND DESCRIPTION
1. $TRESP		Transient response of one-degree-of-freedom linear systems. Method: 4th-order Runge–Kutta. Assume excitation = constant for Δt. Integrate for every $\Delta t/4$. Return X and Y for given Δt.
2. $SUBN	$CSUBN	Matrix substitution. Given matrix A of order n, find $B = A$.
3. $MPLY	$CMPLY	Matrix multiplication. Given matrices A and B of order n, find $C = A * B$
4. $INVG		Matrix inversion. Method: Gaussian elimination with pivoting elements.
5. $INVS	$CINVS	Matrix inversion. Method: Faddeev–Leverrier.
6. $COEFF		Coefficients of characteristic equation. Method: Faddeev–Leverrier. Given matrix H of order n, find C's in $\|\lambda I - H\| = C_0 + C_1\lambda + \cdots + C_n\lambda^n$
7. $ROOT	$CROOT	Roots of characteristic equation. Method: Iterative. Given matrix H of order n, find roots of equation $\|\lambda I - H\| = 0$
8. $HOMO	$CHOMO	Solution of homogeneous algebraic equations. Given coefficient matrix A of order n, form $AX = 0$, and find solution vector X.
9. $MODL	$CMODL	Modal matrix of discrete systems. Given matrices M, C, and K, find roots of characteristic equation and modal matrix.
10.	$CRKUT	To solve n-equations: $\dot{z} + Cz = F$ Method: 4th-order Runge–Kutta. Integrate for every $\Delta t/4$. Return Z for given Δt.
11. $PLOTF		Plotting subroutine to plot data from a file.

```
$TRESP

"   *** TRENSIENT RESPONSE OF ONE-DEGREE-OF-FREEDOM SYSTEMS ***
"       ASSUME FORCE F = CONSTANT FOR GIVEN DT.
"       CALCULATES VALUES OF X & Y FOR EVERY DT/4.  RETURNS X & Y
"       RUNGE-KUTTA METHOD.
"
      SUBROUTINE $TRESP (X0, Y0, F, DT, X, Y, Z, WN)
      REAL*8  K1, K2, K3, K4, L1, L2, L3, L4, %
        A, B, D, DT, DX, DY, F, WN, X, X0, Y, Y0, Z
      A = WN↑2
      B = 2*Z*WN
      D = DT/4
      X = X0
      Y = Y0
      DO 40 I=1,4
        K1 = D*Y
        L1 = D*(-B*Y - A*X + F)
        K2 = D*(Y + L1/2.)
        L2 = D*(-B*(Y+L1/2.) - A*(X+K1/2.) + F)
        K3 = D*(Y + L2/2.)
        L3 = D*(-B*(Y+L2/2.) - A*(X+K2/2.) + F)
        K4 = D*(Y + L3)
        L4 = D*(-B*(Y+L3) - A*(X+K3) + F)
        DX = (K1 + 2.*K2 + 2.*K3 + K4)/6.
        DY = (L1 + 2.*L2 + 2.*L3 + L4)/6.
        X = X + DX
40      Y = Y + DY
      RETURN
      END
```

FIG. C-1. *Transient response of one-degree-of-freedom systems.*

natural frequency ω_n (WN). Since D = DT/4, the subroutine performs a computation for every $\Delta t/4$ and returns the values X and Y for the given Δt.

C-3 $SUBN AND $CSUBN—MATRIX SUBSTITUTION.

The subroutine $SUBN for real matrix *sub*stitutio*n* is listed in Fig. C-2(*a*). Matrix *A* is the input and the subroutine returns matrix *B*. For

```
$SUBN

"   *** REAL MATRIX SUBSTITUTION:  GETS  B = A ***
"
      SUBROUTINE $SUBN (A, B, N)
      REAL*8  A(10,10), B(10,10)
      DO 40 I=1,N
      DO 40 J=1,N
40      B(I,J) = A(I,J)
      RETURN
      END
```

(*a*) Real

```
$CSUBN

"   *** COMPLEX MATRIX SUBSTITUTION:  GETS  B = A ***
"
      SUBROUTINE $CSUBN (A, B, N)
      COMPLEX*16  A(10,10), B(10,10)
      DO 40 I=1,N
      DO 40 J=1,N
40      B(I,J) = A(I,J)
      RETURN
      END
```

(*b*) Complex

FIG. C-2. *Matrix substitution B = A; returns matrix B*

the real matrices A and B of order n, the substitution is

$$B = A \qquad \text{or} \qquad b_{ij} = a_{ij} \tag{C-1}$$

The subroutine $CSUBN for complex matrix substitution is listed in Fig. C-2(b). Except for the complex mode, it is identical to $SUBN.

C-4 $MPLY AND $CMPLY—MATRIX MULTIPLICATION.

The subroutine $MPLY for real matrix multiplication is listed in Fig. C-3(a). Matrices A and B are the input and the subroutine returns matrix C. For the real matrices A and B of order n, the operation is

$$C = A * B \qquad \text{or} \qquad c_{ij} = \sum a_{ik} b_{kj} \tag{C-2}$$

Except for the complex mode, $CMPLY as listed in Fig. C-3(b) is identical to $MPLY.

```
$MPLY

"    *** REAL MATRIX MULTIPLICATION:   CALCULATES  C = A*B ***
"
     SUBROUTINE $MPLY (A, B, C, N)
     REAL*8   A(10,10), B(10,10), C(10,10)
     DO 40 I=1,N
     DO 40 J=1,N
       C(I,J) = 0
       DO 40 K=1,N
40       C(I,J) = C(I,J) + A(I,K)*B(K,J)
     RETURN
     END
```

(a) Real

```
$CMPLY

"    *** COMPLEX MATRIX MULTIPLICATION:   CALCULATES  C = A*B ***
"
     SUBROUTINE $CMPLY (A, B, C, N)
     COMPLEX*16  A(10,10), B(10,10), C(10,10)
     DO 40 I=1,N
     DO 40 J=1,N
       C(I,J) = 0
       DO 40 K=1,N
40       C(I,J) = C(I,J) + A(I,K)*B(K,J)
     RETURN
     END
```

(b) Complex

FIG. C-3. *Matrix multiplication $C = A * B$; returns matrix C.*

C-5 $INVG—REAL MATRIX INVERSION. METHOD: GAUSSIAN ELIMINATION WITH PIVOT ELEMENTS.

The subroutine $INVG for the *inv*ersion of real matrices by the *G*aussian method is listed in Fig. C-4. The input is the matrix H of order N and the output is the inverse of H (HINVS).

```
$INVG

"  *** REAL MATRIX INVERSION:   FINDS INVERSE (HINVS) OF MATRIX H ***
"      METHOD: GAUSSIAN ELMINIATION WITH PIVOT ELEMENTS.
"
     SUBROUTINE $INVG (H, HINVS, N)
     REAL*8  A, B, D(10,20), H(10,10), HINVS(10,10), QUOT, TEMP, %
       UNIT(10,10)
     DATA  UNIT /100*0.0/
     NT2 = N*2
     NP1 = N + 1
     NM1 = N - 1
     DO 40 I=1,N
40     UNIT(I,I) = 1.
     DO 41 I=1,N                          ]
     DO 41 J=1,N                          ] Form nx2n
       D(I,J) = H(I,J)                    ] matrix.
41     D(I,N+J) = UNIT(I,J)               ]
     DO 42 J=1,NM1
       A = DABS(D(J,J))                                       |
       JP1 = J + 1                                            |
       DO 43 I=JP1,N                                          | Search for pivot
         B = DABS(D(I,J))                                     | elements.
         IF (A-B) 10, 43, 43                                  |
10       DO 44 K=J,NT2                    ]                   |
           TEMP = D(I,K)                  ] Interchange       |
           D(I,K) = D(J,K)                ] rows.             |
44         D(J,K) = TEMP                  ]                   |
43       A = B                                                |
       DO 42 I=JP1,N                                          ]
         QUOT = D(I,J)/D(J,J)                                 ] Eliminate    Have upper
         DO 42 K=JP1,NT2                                      ] elements.    triangular
42         D(I,K) = D(I,K) - QUOT*D(J,K)                      ]              matrix.
     K = N                                                                   Begin back
11   I = K - 1                                                               solution.
12   QUOT = D(I,K)/D(K,K)                                     ]
     DO 45 J=NP1,NT2                                          ] Eliminate
45     D(I,J) = D(I,J) - QUOT*D(K,J)                          ] elements.
     IF (I = 1) GOTO 13                                       ]
     I = I - 1
     GOTO 12
13   IF (K = 2) GOTO 14
     K = K - 1                                              —— Have diagonal
     GOTO 11                                                   matrix.
14   DO 46 I=1,N
     DO 46 J=1,N
46     HINVS(I,J) = D(I,N+J)/D(I,I)                         —— Get inverse.
     RETURN
     END
```

FIG. C-4. *Real matrix inversion; returns matrix*—HINVS.

The Gaussian method for matrix inversion eliminates the off diagonal elements of a matrix by substitution, in the manner for solving algebraic equations. Given a matrix H and a unit matrix I of order n, identical elementary row operations, as described in Sec. A-3, are performed on H and I. When H becomes a unit matrix, the resultant matrix in I is the inverse of H.

Let us describe the logic in the program with a numerical example instead of algebraic equations. The steps can be traced by comparing the remarks in the example and those in the program.

Example C-1

Consider the 3×3 matrices H and I below. The procedure is (1) to eliminate the elements below the principal diagonal to get an upper triangular matrix

H			*I*				*Remarks*
1	2	5	1	0	0		
1	1	2	0	1	0		Form $n \times 2n$ matrix.
3	1	4	0	0	1		
3	1	4	0	0	1 ↰		Search for pivot elements in col 1.
1	1	2	0	1	0		Pivot element is (3). Interchange rows 1 and 3
1	2	5	1	0	0 ↲		by DO 44 loop.
3	1	4	0	0	1		
x	2/3	2/3	0	1	−1/3	QUOT = 1/3	Eliminate elements (*x*).
x	5/3	11/3	1	0	−1/3		
3	1	4	0	0	1		Search for pivot elements in col 2.
0	5/3	11/3	1	0	−1/3 ↰		Pivot element is (5/3). Interchange rows 2 and 3
0	2/3	2/3	0	1	−1/3 ↲		by DO 44 loop.
3	1	4	0	0	1		
0	5/3	11/3	1	0	−1/3	QUOT = 2/5	Eliminate element (*x*) Have upper triangular
0	*x*	−4/5	−2/5	1	−1/5		matrix in *H*.
3	1	*x*	−2	5	0	QUOT = −5/1	Begin back-solution.
0	5/3	*x*	−5/6	55/12	−5/4	= −55/12	Eliminate elements (*x*) by DO 45 loop.
0	0	−4/5	−2/5	1	−1/5		
3	*x*	0	−3/2	9/4	3/4	QUOT = 3/5	Eliminate element (*x*).
0	5/3	0	−5/6	55/12	−5/4		Have diagonal matrix in *H*.
0	0	−4/5	−2/5	1	−1/5		
1	0	0	−2/4	3/4	1/4		Divide by diagonal elements to get unit
0	1	0	−2/4	11/4	−3/4		matrix in *H*. Matrix on right is
0	0	1	2/4	−5/4	1/4		inverse of *H*.

in *H*, then (2) to eliminate the elements above the diagonal by "back solution." A *pivot element* is the number with the largest absolute value in the column for forming the upper triangular matrix. This will be illustrated in the example. Since the quotient (QUOT) for the elimination process has the pivot element in its denominator, better accuracy is obtained with pivot elements.*

C-6 $INVS AND $CINVS—MATRIX INVERSION. METHOD: FADDEEV-LEVERRIER.

The subroutines $INVS and $CINVS for the *inversion* of real and complex matrices are listed in Fig. C-5. The subroutine receives a matrix

```
$INVS

"  *** REAL MATRIX INVERSION:  FINDS INVERSE (HINVS) OF MATRIX H ***
"      SUBROUTINES REQD: (1) $SUBN  (2) $MPLY
"
      SUBROUTINE $INVS (H, HINVS, N)
      REAL*8  H(10,10), HINVS(10,10), A(10,10), B(10,10), SUM
      CALL $SUBN (H, A, N)
      NM1 = N - 1
      DO 40 I=1,NM1
        SUM = 0
        DO 41 K=1,N
41        SUM = SUM + A(K,K)
        SUM = SUM/I
        DO 42 J=1,N
42        A(J,J) = A(J,J) - SUM
        IF (I = NM1) CALL $SUBN (A, HINVS, N)
        CALL $MPLY (H, A, B, N)
40      CALL $SUBN (B, A, N)
      DO 43 I=1,N
      DO 43 J=1,N
43      HINVS(I,J) = HINVS(I,J)/A(1,1)
      RETURN
      END
```

(*a*) Real

```
$CINVS

"  *** COMPLEX MATRIX INVERSION:  FINDS INVERSE (HINVS) OF H. ***
"      SUBROUTINES REQD:  (1) $CMPLY  (2) $CSUBN
"
      SUBROUTINE $CINVS (H, HINVS, N)
      COMPLEX*16  H(10,10), HINVS(10,10), A(10,10), B(10,10), SUM
      CALL $CSUBN (H, A, N)
      NM1 = N - 1
      DO 40 I=1,NM1
        SUM = 0
        DO 41 K=1,N
41        SUM = SUM + A(K,K)
        SUM = SUM/I
        DO 42 J=1,N
42        A(J,J) = A(J,J) - SUM
        IF (I = NM1) CALL $CSUBN (A, HINVS, N)
        CALL $CMPLY (H, A, B, N)
40      CALL $CSUBN (B, A, N)
      DO 43 I=1,N
      DO 43 J=1,N
43      HINVS(I,J) = HINVS(I,J)/A(1,1)
      RETURN
      END
```

(*b*) Complex

FIG. C-5. *Matrix inversion; returns matrix*—HINVS.

* Hamming, op. cit. p. 362.

H of order N and returns the inverse of H (HINVS). Except for the real and the complex modes, the two subroutines are identical. Let us trace the steps in $INVS with a numerical example.

Example C-2

Consider the matrix H in Example C-1. The steps in the program $INVS to find H^{-1} are as follows:

(1) The CALL $SUBN statement yields

$$\mathrm{A} = \mathrm{H} = \begin{bmatrix} 1 & 2 & 5 \\ 1 & 1 & 2 \\ 3 & 1 & 4 \end{bmatrix}$$

(2) For I = 1 in the DO 40 loop, we get

$$\mathrm{SUM} = (1 + 1 + 4) = 6$$

$$\mathrm{SUM} = \mathrm{SUM/I} = 6/1 = 6$$

(a) The DO 42 loop gives

$$\mathrm{A} = \begin{bmatrix} 1-6 & 2 & 5 \\ 1 & 1-6 & 2 \\ 3 & 1 & 4-6 \end{bmatrix} = \begin{bmatrix} -5 & 2 & 5 \\ 1 & -5 & 2 \\ 3 & 1 & -2 \end{bmatrix}$$

(b) The CALL $MPLY and CALL $SUBN statements yield

$$\mathrm{B} = \mathrm{H} * \mathrm{A} = \begin{bmatrix} 1 & 2 & 5 \\ 1 & 1 & 2 \\ 3 & 1 & 4 \end{bmatrix} \begin{bmatrix} -5 & 2 & 5 \\ 1 & -5 & 2 \\ 3 & 1 & -2 \end{bmatrix}$$

$$\mathrm{A} = \mathrm{B} = \begin{bmatrix} 12 & -3 & -1 \\ 2 & -1 & 3 \\ -2 & 5 & 9 \end{bmatrix}$$

(3) For I = 2 in the DO 40 loop, we get from the matrix A above

$$\mathrm{SUM} = (12 - 1 + 9) = 20$$

$$\mathrm{SUM} = \mathrm{SUM/I} = 20/2 = 10$$

(a) The DO 42 loop gives

$$\mathrm{A} = \begin{bmatrix} 12-10 & -3 & -1 \\ 2 & -1-10 & 3 \\ -2 & 5 & 9-10 \end{bmatrix} = \begin{bmatrix} 2 & -3 & -1 \\ 2 & -11 & 3 \\ -2 & 5 & -1 \end{bmatrix}$$

$$= \text{HINVS (temporary), since } \mathrm{I} = \mathrm{N} - 1$$

(b) The CALL $MPLY and CALL $SUBN statements yield

$$B = H*A = \begin{bmatrix} 1 & 2 & 5 \\ 1 & 1 & 2 \\ 3 & 1 & 4 \end{bmatrix} \begin{bmatrix} 2 & -3 & -1 \\ 2 & -11 & 3 \\ -2 & 5 & -1 \end{bmatrix}$$

$$A = B = \begin{bmatrix} -4 & 0 & 0 \\ 0 & -4 & 0 \\ 0 & 0 & -4 \end{bmatrix}$$

(4) The DO 43 loop gives

HINVS = HINVS(temporary)/A(1, 1)

$$= -\frac{1}{4} \begin{bmatrix} 2 & -3 & -1 \\ 2 & -11 & 3 \\ -2 & 5 & -1 \end{bmatrix} = \begin{bmatrix} -2/4 & 3/4 & 1/4 \\ -2/4 & 11/4 & -3/4 \\ 2/4 & -5/4 & 1/4 \end{bmatrix}$$

C-7 $COEFF—COEFFICIENTS OF CHARACTERISTIC EQUATION. METHOD: FADDEEV-LEVERRIER.

The subroutine $COEFF to find the *coefficients* of a characteristic equation is listed in Fig. C-6. For a real matrix H of order n, the program yields the values of C's in the polynomial

$$|\lambda I - H| = C_0 + C_1\lambda + C_2\lambda^2 + \cdots + C_n\lambda^n \qquad \textbf{(C-3)}$$

Note that (1) the values of C's in Eq. (C-3) are obtained from the statements C(NP2 − I) = −SUM/I and C(N+4−I) = C(N+4−I)/C(2),

```
$COEFF

"     *** TO FIND COEFFICIENTS OF CHARACTERISTIC EQUATION ***
"         GIVEN REAL MATRIX-H, PROGRAM YIELDS COEFFICIENTS-C'S.
"         SUBROUTINES REQD:   (1) $SUBN,    (2) $MPLY
"
      SUBROUTINE $COEFF (H, C, N)
      REAL*8   A(10,10), B(10,10), C(12), H(10,10), SUM
      CALL $SUBN (H, A, N)
      NP2 = N + 2
      C(1) = 0
      C(NP2) = 1
      DO 40 I=1,N
        SUM = 0
        DO 41 K=1,N
41        SUM = SUM + A(K,K)
        C(NP2-I) = - SUM/I
        DO 42 J=1,N
42        A(J,J) = A(J,J) + C(NP2-I)
        CALL $MPLY (H, A, B, N)
40      CALL $SUBN (B, A, N)
      DO 43 I=2,NP2
43      C(N+4-I) = C(N+4-I)/C(2)
      RETURN
      END
```

FIG. C-6. *Find C's in* $|\lambda I - H| = C_0 + C_1\lambda + C_2\lambda^2 + \cdots + C_n\lambda^n$.

and (2) the rest of the listings in $COEFF is essentially the same as that in $INVS in Fig. C-5(*a*). Hence the two subroutines use the same logic. Using the same matrix H in Example C-2, the reader may wish to verify that the steps to evaluate the C's in $COEFF run parallel to those in $INVS. The corresponding characteristic equation is

$$|\lambda I - H| = 1.0 - 2.5\lambda - 1.5\lambda^2 + 0.25\lambda^3$$

Note also the dimension of the array C is (12) while that of H is (10,10). If H is of order n, there are $(n+1)$ values of C's in Eq. (C-3). We assume the characteristic equation is of the form

$$|\lambda I - H| = C_1 + C_2\lambda^0 + C_3\lambda^1 + C_4\lambda^2 + \cdots + C_{n+2}\lambda^n$$

where $C_1 = 0$ and $C_2 = 1$. The subscripts are adjusted in order to accommodate the subscripts in the subroutines $ROOT and $CROOT. Note that the subscripts (−1) and (0) are not allowed in FORTRAN.

C-8 $ROOT—REAL ROOTS OF CHARACTERISTIC EQUATION. METHOD: ITERATIVE.

The subroutine $ROOT finds the roots of the characteristic equation in Eq. (C-3). For the case of real roots, let

$$\begin{aligned} C_0 + C_1\lambda + C_2\lambda^2 + C_3\lambda^3 + \cdots + C_n\lambda^n \\ = (1 + A\lambda)(B_0 + B_1\lambda + B_2\lambda^2 + \cdots + B_n\lambda^n) \\ = 0 \end{aligned} \quad \text{(C-4)}$$

where $C_0 = B_0 = 1$ and $(1 + A\lambda)$ is the first factor. The procedure can be repeated to obtain all the factors of the polynomial. Hence the roots are

$$\lambda_i = -1/A_i \quad \text{(C-5)}$$

To find the first factor, we equate the coefficients of λ^i in Eq. (C-4) and obtain

$$B_i = C_i - AB_{i-1} \qquad \text{for } i = 1 \text{ to } n \quad \text{(C-6)}$$

Since B_n must be identically zero, the constraint equation is

$$B_n = 0 \quad \text{(C-7)}$$

Differentiating Eq. (C-6) with respect to A gives

$$\frac{dB_i}{dA} = -A\frac{dB_{i-1}}{dA} - B_{i-1} \qquad \text{for } i = 1 \text{ to } n \quad \text{(C-8)}$$

The recurrence formulas for the iterations are Eqs. (C-6) and (C-8).

```
$ROOT

"   *** TO FIND THE REAL ROOTS OF CHARACTERISTIC EQUATION OF ***
"       POSITIVE DEFINITE UNDAMPED SYSTEMS.   ROOTS DISTINCT.
"       GIVEN COEFFICIENT MATRIX-H, PROGRAM YIELDS EIGENVALUES.
"       SUBROUTINES REQD:   (1) $COEFF   (2) $MPLY   (3) $SUBN
"
      SUBROUTINE $ROOT (H, ROOT, ERROR, NITER, N)
      REAL*8  A, B(11), C(12), DA, DBDA(11), ERROR, H(10,10), ROOT(10)
      DATA  B(1), DBDA(1)/1.0, 0.0/
      CALL $COEFF (H, C, N)
      I = N + 1
   10 A = 0
      IMI = I - 1
      DO 40 LI=1,NITER
        DO 41 J=2,I
          B(J) = C(J+1) - A*B(J-1)
   41     DBDA(J) = -A*DBDA(J-1) - B(J-1)
        DA = -B(I)/DBDA(I)
        A = A + DA
        IF (DABS(DA) - ERROR) 11, 11, 40
   40   CONTINUE
   11 DO 42 JI=1,IMI
   42   C(JI+1) = B(JI)
      ROOT(I-1) = -1/A
      IF (I = 2) GOTO 12
      I = I - 1
      GOTO 10
   12 RETURN
      END
```

FIG. C-7. *Real roots of* $|\lambda I - H| = 0$; *returns* λ*'s.*

Assume a trial value for A to start the iteration. If the trial value does not satisfy the constraint $B_n = 0$, let

$$A(\text{new}) = A(\text{old}) + \Delta A \tag{C-9}$$

$$B_n(\text{new}) = B_n(\text{old}) + \Delta B_n \tag{C-10}$$

$$\Delta B_n = \frac{dB_n}{dA}\Delta A \tag{C-11}$$

where ΔA and ΔB_n are both unknown, although (dB_n/dA) can be estimated from Eq. (C-8). For the second iteration, assume from Eq. (C-10) that

$$B_n(\text{new}) \simeq 0 \qquad \text{or} \qquad \Delta B_n \simeq -B_n(\text{old}) \tag{C-12}$$

Thus, ΔA can be estimated from Eq. (C-11)

$$\Delta A \simeq -B_n(\text{old})/(dB_n/dA) \tag{C-13}$$

Referring to the subroutine $ROOT listed in Fig. C-7,

1. The CALL $COEEF statement gives the C's in Eq. (C-3) and (C-4). We iterate for the first factor by assuming A = 0, as shown in statement (10).
2. For the DO 40 loop, NITER limits the number of iterations for each factor.

3. Equations (C-6) and (C-8) can be identified in the DO 41 loop, to be followed by Eq. (C-13). Note that the subscript of C is adjusted as described in the last section.
4. After obtaining the first factor, the order of the polynomial is reduced to $(n-1)$. The B's are renamed as C's in the DO 42 loop.
5. The GOTO 10 statement directs the program to the starting point to iterate for additional factors.

C-9 $CROOT—COMPLEX ROOTS OF CHARACTERISTIC EQUATION. METHOD: ITERATIVE.

The subroutine $CROOT finds the roots of the characteristic equation in Eq. (C-3) for the case of complex roots. Let

$$\begin{aligned} C_0 + C_1\lambda + C_2\lambda^2 + \cdots + C_{n-1}\lambda^{n-1} + C_n\lambda^n \\ = (1 + A_1\lambda + A_2\lambda^2)(B_0 + B_1\lambda + B_2\lambda^2 \cdots \\ \cdots + B_{n-2}\lambda^{n-2} + B_{n-1}\lambda^{n-1} + B_n\lambda^n) \\ = 0 \end{aligned} \tag{C-14}$$

where $C_0 = B_0 = 1$ and the first factor is $(1 + A_1\lambda + A_2\lambda^2)$. Equating the coefficients of λ^i in Eq. (C-14) gives

$$B_i = C_i - A_1B_{i-1} - A_2B_{i-2} \qquad \text{for } i = 1 \text{ to } n \tag{C-15}$$

where $B_{-1} = 0$. This may be compared with Eq. (C-6). Similar to Eq. (C-7), the constraint equations are

$$B_n = B_{n-1} = 0 \tag{C-16}$$

Differentiating Eq. (C-15), we get

$$\begin{aligned} \frac{\partial B_i}{\partial A_1} &= -A_1\frac{\partial B_{i-1}}{\partial A_1} - A_2\frac{\partial B_{i-2}}{\partial A_1} - B_{i-1} \\ \frac{\partial B_i}{\partial A_2} &= -A_1\frac{\partial B_{i-1}}{\partial A_2} - A_2\frac{\partial B_{i-2}}{\partial A_2} - B_{i-2} \end{aligned} \tag{C-17}$$

The equations above may be compared with Eq. (C-8).

Assume trial values for A_1 and A_2 to start the iteration. If the trial values do not satisfy the constraints in Eq. (C-16), let

$$\begin{aligned} A_1(\text{new}) &= A_1(\text{old}) + \Delta A_1 \\ A_2(\text{new}) &= A_2(\text{old}) + \Delta A_2 \end{aligned} \tag{C-18}$$

$$\begin{aligned} \Delta B_{n-1} &= \frac{\partial B_{n-1}}{\partial A_1}\Delta A_1 + \frac{\partial B_{n-1}}{\partial A_2}\Delta A_2 \\ \Delta B_n &= \frac{\partial B_n}{\partial A_1}\Delta A_1 + \frac{\partial B_n}{\partial A_2}\Delta A_2 \end{aligned} \tag{C-19}$$

The equations above may be compared with Eqs. (C-9) and (C-11). Similar to Eq. (C-12), we assume

$$\begin{aligned} B_{n-1}(\text{new}) \simeq 0 \quad &\text{or} \quad \Delta B_{n-1} \simeq -B_{n-1}(\text{old}) \\ B_n(\text{new}) \simeq 0 \quad &\text{or} \quad \Delta B_n \simeq -B_n(\text{old}) \end{aligned} \tag{C-20}$$

Thus, substituting Eq. (C-20) in (C-19) and solving the equations simultaneously, ΔA_1 and ΔA_2 can be estimated. They can be substituted into Eq. (C-18) to repeat the iteration.

```
$CROOT

"   *** TO FIND THE COMPLEX ROOTS OF CHARACTERISTIC EQUATION OF ***
"       POSITIVE DEFINITE SYSTEMS WITH VISCOUS DAMPING.
"       ROOTS DISTINCT.
"       GIVEN COEFFICIENT MATRIX-H, PROGRAM YIELDS COMPLEX ROOTS.
"       SUBROUTINES REQD:   (1) $COEFF  (2) $MPLY  (3) $SUBN
"
      SUBROUTINE $CROOT (H, ROOT, ERROR, NITER, N)
      REAL*8  A(2,5), A1, A2, B(12), C(12), DA1, DA2, %
        DBDA1(12), DBDA2(12), DET, ERROR, H(10,10), P, Q, Z
      COMPLEX*16  ROOT(2,5)
      DATA  B, DBDA1, DBDA2, Z/ 36*0.0, 0.0 /
      B(2) = 1
      CALL $COEFF (H, C, N)
      I = N + 2
   10 L = I/2 - 1
      A1 = 0
      A2 = 0
      DO 40 IT=1,NITER
        DO 41 J=3,I
          B(J) = C(J) - A1*B(J-1) - A2*B(J-2)
          DBDA1(J) = -A1*DBDA1(J-1) - A2*DBDA1(J-2) - B(J-1)
   41     DBDA2(J) = -A1*DBDA2(J-1) - A2*DBDA2(J-2) - B(J-2)
        DET = DBDA1(I-1)*DBDA2(I) - DBDA1(I)*DBDA2(I-1)
        DA1 = (-B(I-1)*DBDA2(I) + B(I)*DBDA2(I-1))/DET
        DA2 = (+B(I-1)*DBDA1(I) - B(I)*DBDA1(I-1))/DET
        A1 = A1 + DA1
        A2 = A2 + DA2
        IF ((DABS(DA1)-ERROR).GE.0) GOTO 40
        IF ((DABS(DA2)-ERROR).GE.0) GOTO 40
        DO 42 I1=4,I
   42     C(I1-2) = B(I1-2)
        A(1,L) = A1
        A(2,L) = A2
        IF (I = 4) GOTO 11
        I = I-2
        GOTO 10
   40   CONTINUE
   11 ND2 = N/2
      DO 43 J=1,ND2
        P = A(1,J)↑2 - 4*A(2,J)
        Q = DSQRT(DABS(P))
        IF (P.GE.0) GOTO 12
        ROOT(1,J) = DCMPLX(-A(1,J), Q)
        ROOT(2,J) = DCMPLX(-A(1,J),-Q)
        GOTO 13
   12   ROOT(1,J) = DCMPLX(-A(1,J)-Q,Z)
        ROOT(2,J) = DCMPLX(-A(1,J)+Q,Z)
   13   ROOT(1,J) = ROOT(1,J)/(2*A(2,J))
   43   ROOT(2,J) = ROOT(2,J)/(2*A(2,J))
      RETURN
      END
```

FIG. C-8. *Complex roots of* $|\lambda I - H| = 0$; *returns* λ*'s.*

Referring to the subroutine $CROOT in Fig. C-8,

1. The CALL $COEFF statement gives the C's in Eq. (C-3) and (C-14). We let A1 = A2 = 0 to start the iteration.
2. For the DO 40 loop, NITER limits the number of iterations for each quadratic factor.
3. Equations (C-15) and (C-17) can be identified readily in the DO 41 loop. Then ΔA_1 (DA1) and ΔA_2 (DA2) are obtained from the simultaneous solution of Eq. (C-19). The iteration is repeated.
4. After obtaining the first quadratic factor, the order of the polynomial is reduced to $(n-2)$. The B's are renamed as C's in the DO 42 loop.
5. The GOTO 10 statement directs the program to iterate for the subsequent quadratic factors.
6. The rest of the program finds the complex roots from the quadratic factors.

C-10 $HOMO AND $CHOMO—SOLUTION OF HOMOGENEOUS ALGEBRAIC EQUATIONS.

The subroutine $HOMO, as listed in Fig. C-9(a), finds the solution of a set of real-*homo*geneous-algebraic equations. Let us illustrate the technique with a numerical example.

Example C-3

Consider the homogeneous equations

$$AX = \begin{bmatrix} 1 & 2 & 3 & 1 \\ 2 & 5 & 4 & 7 \\ 1 & 1 & 3 & -2 \\ 4 & 2 & 1 & -3 \end{bmatrix} \begin{bmatrix} x_1 \\ x_2 \\ x_3 \\ x_4 \end{bmatrix} = \{0\}$$

where A is the coefficient matrix and X the solution vector.* Since only the relative values of x's can be determined, let $x_4 = 1$, arbitrarily. The first three equations† above can be rearranged to give

$$BZ = \begin{bmatrix} 1 & 2 & 3 \\ 2 & 5 & 4 \\ 1 & 1 & 3 \end{bmatrix} \begin{bmatrix} z_1 \\ z_2 \\ z_3 \end{bmatrix} = - \begin{bmatrix} 1 \\ 7 \\ -2 \end{bmatrix} = \begin{bmatrix} y_1 \\ y_2 \\ y_3 \end{bmatrix} \triangleq Y$$

where $Z = \{z\} = \{x_1/x_4 \quad x_2/x_4 \quad x_3/x_4\} = \{x_1 \quad x_2 \quad x_3\}$. Thus,

$$Z = B^{-1}Y = \begin{bmatrix} 1 & 2 & 3 \\ 2 & 5 & 4 \\ 1 & 1 & 3 \end{bmatrix}^{-1} \begin{bmatrix} 1 \\ 7 \\ -2 \end{bmatrix} = \begin{bmatrix} 2 \\ -3 \\ 1 \end{bmatrix} = \begin{bmatrix} x_1 \\ x_2 \\ x_3 \end{bmatrix}$$

Recalling $x_4 = 1$, the solution is $X = \{2 \quad -3 \quad 1 \quad 1\}$.

* A nontrivial solution for X can be found if the determinant of the matrix A is zero.
† Any $(n-1)$ of the (n) simultaneous equations can be used to find the solution

```
$HOMO

"  *** SOLUTION OF REAL ALGEBRAIC HOMOGENEOUS EQUATIONS ***
"      SUBROUTINES REQD:   (1) $INVS   (2) $MPLY   (3) $SUBN
"
      SUBROUTINE $HOMO (A, X, N)
      REAL*8   A(10,10), B(10,10), BINVS(10,10), X(10), Y(10)
      X(N) = 1
      NM1 = N - 1
      DO 40 I=1,NM1
        Y(I) = -A(I,N)
        DO 40 J=1,NM1
40        B(I,J) = A(I,J)
      IF (NM1 = 1) BINVS(1,1) = 1./B(1,1)
      IF (NM1 = 1) GOTO 10
      CALL $INVS (B, BINVS, NM1)
10    DO 41 I=1,NM1
        X(I) = 0
        DO 41 J=1,NM1
41        X(I) = X(I) + BINVS(I,J)*Y(J)
      RETURN
      END
```

(*a*) Real

```
$CHOMO

"  *** SOLUTION OF COMPLEX ALGEBRAIC HOMOGENEOUS EQUATIONS ***
"      SUBROUTINES REQD:   (1) $CINVS   (2) $CMPLY   (3) $CSUBN
"
      SUBROUTINE $CHOMO (A, X, N)
      COMPLEX*16   A(10,10), B(10,10), BINVS(10,10), X(10), Y(10)
      REAL*8   U, Z
      DATA  U, Z /1.0, 0.0/
      X(N) = DCMPLX(U, Z)
      NM1 = N - 1
      DO 40 I=1,NM1
        Y(I) = -A(I,N)
        DO 40 J=1,NM1
40        B(I,J) = A(I,J)
      IF (NM1 = 1) BINVS(1,1) = 1./B(1,1)
      IF (NM1 = 1) GOTO 10
      CALL $CINVS (B, BINVS, NM1)
10    DO 41 I=1,NM1
        X(I) = DCMPLX(Z, Z)
        DO 41 J=1,NM1
41        X(I) = X(I) + BINVS(I,J)*Y(J)
      RETURN
      END
```

(*b*) Complex

FIG. C-9. *Solution of homogeneous algebraic equations*

Comparing the subroutine $HOMO and Example C-3, for which $n = 4$, first we define X(N) = 1, that is, $x_4 = 1$. Then matrix B and vector Y are formed in the DO 40 loop. The CALL $INVS statement gives B^{-1} (BINVS). Then the product of B^{-1} and Y is obtained in the DO 41 loop. The solution vector X follows.

Except for the complex mode, the subroutine $CHOMO, listed in Fig. C-9(*b*), is the same program.

C-11 $MODL—MODAL MATRIX OF UNDAMPED DISCRETE SYSTEMS.

The subroutine $MODL finds the *modal* matrix of positive-definite-undamped-discrete systems with distinct roots. We shall follow the presentation in Example 9, Chap. 4, to derive the modal matrix.

The equations of motion and the frequency equation from Eqs. (4-36) and (4-38) are

$$M\{\ddot{q}\}+K\{q\}=\{0\} \tag{C-21}$$

$$\Delta(\omega)=|K-\omega^2 M|=0 \tag{C-22}$$

Alternatively, premultiplying Eq. (C-21) by M^{-1} and defining $H=M^{-1}K$, Eq. (C-22) can be expressed as

$$\Delta(\lambda)=|\lambda I-H|=0 \tag{C-23}$$

where $\lambda=\omega^2$. Thus, the subroutine $COEFF can be used to find the λ's and therefore the natural frequencies ω's

At a principal mode, the entire system executes synchronous harmonic notion at a natural frequency ω. Hence

$$\{\ddot{q}\}=\{-\omega^2 q\}=-\omega^2\{q\} \tag{C-24}$$

The displacement vector $\{q\}$ at a principal mode is also a modal vector. Substituting Eq. (C-24) into (C-21) yields

$$[-\omega^2 M+K]\{q\}=\{0\} \tag{C-25}$$

The equation above is identical to Eq. (4-37). A modal vector is obtained from the solution of the homogeneous equations as described in the last section. Considering all modes of vibration of the system, the modal matrix is formed from a combination of all the modal vectors as shown in Eq. (4-23).

The equations above can be traced readily in $MODL, as listed in Fig. C-10.

```
$MODL

"   *** CALCULATION OF THE MODAL MATRIX OF ***
"       UNDAMPED POSITIVE DEFINITE DISCRETE SYSTEMS.
"       SUBROUTINES REQD:  (1) $INVS  (2) $MPLY  (3) $ROOT  (4) $HOMO
"
      SUBROUTINE $MODL (M, K, U, ROOT, ERROR, NITER, N)
      REAL*8  DUM(10,10), ERROR, H(10,10), K(10,10), M(10,10), %
        MINVS(10,10), ROOT(10), U(10,10), X(10)
      CALL $INVS (M, MINVS, N)
      CALL $MPLY (MINVS, K, H, N)
      CALL $ROOT (H, ROOT, ERROR, NITER, N)
      DO 40 L=1,N
        DO 41 I=1,N
        DO 41 J=1,N
41        DUM(I,J) = - M(I,J)*ROOT(L) + K(I,J)
        CALL $HOMO (DUM, X, N)
        DO 40 I1=1,N
40        U(I1,L) = X(I1)
      RETURN
      END
```

FIG. C-10. *Modal matrix of undamped discrete systems; roots distinct.*

1. The CALL \$INVS and CALL \$MPLY statements give $H = M^{-1}K$. The CALL \$ROOT statement finds the roots λ_r of Eq. (C-23), for $r = 1$ to n.

2. For a given frequency ω, the coefficient matrix $[-\omega^2 M + K]$ is obtained by means of the DO 41 loop, where ROOT $= -\omega^2$.

3. The CALL \$HOMO statement solves Eq. (C-25) to yield a modal vector.

4. The modal matrix is formed in the last DO loop.

C-12 \$CMODL—MODAL MATRIX OF DISCRETE SYSTEMS WITH VISCOUS DAMPING.

The subroutine \$CMODL finds the modal matrix of positive-definite-discrete systems with viscous damping, assuming the eigenvalues are complex and distinct. Except for the reduced equations, as described in Sec. 6-12, the technique for this program runs parallel to that of undamped systems.

The equations of motion and the corresponding reduced equations from Eqs. (6-73), (6-75) and (6-77) are

$$M\{\ddot{q}\} + C\{\dot{q}\} + K\{q\} = \{0\} \tag{C-26}$$

$$\begin{bmatrix} 0 & M \\ M & C \end{bmatrix} \begin{bmatrix} \ddot{q} \\ \dot{q} \end{bmatrix} + \begin{bmatrix} -M & 0 \\ 0 & K \end{bmatrix} \begin{bmatrix} \dot{q} \\ q \end{bmatrix} = \begin{bmatrix} 0 \\ 0 \end{bmatrix} \tag{C-27}$$

$$\{\dot{y}\} - H\{y\} = \{0\} \tag{C-28}$$

where $\{y\} = \{\dot{q} \quad q\}$,

$$-H = \begin{bmatrix} -M^{-1}C & -M^{-1}K \\ I & 0 \end{bmatrix} \triangleq \begin{bmatrix} -H1 & -H2 \\ I & 0 \end{bmatrix} \tag{C-29}$$

$I =$ unit matrix of order n, $H1 = M^{-1}C$, and $H2 = M^{-1}K$. The characteristic equation from Eq. (C-28) or (6-80) is

$$\Delta(\gamma) = |\gamma I - H| = 0 \tag{C-30}$$

where γ is a complex root.

The solution of Eq. (C-28) is of the form $\{y\} = \{\Psi\}e^{\gamma t}$, where $\{\Psi\}$ is a modal vector. Thus, $\{\dot{y}\} = \gamma\{y\}$. Since $\{y\} = \{\dot{q} \quad q\}$, we deduce that $\{\ddot{q} \quad \dot{q}\} = \gamma\{\dot{q} \quad q\}$, that is, $\{\dot{q}\} = \gamma\{q\}$ and $\{\ddot{q}\} = \gamma^2\{q\}$. Substituting these in Eq. (C-26) yields

$$[M\gamma^2 + C\gamma + K]\{q\} = \{0\} \tag{C-31}$$

Note that this may be compared with Eq. (C-25) for undamped systems in the last section. In other words, except for the reduced equations and complex numbers, the same technique is applicable to both programs.

```
$CMODL

"  *** CALCULATION OF THE MODAL MATRIX OF ***
"      POSITIVE DEFINITE SYSTEMS WITH VISCOUS DAMPING.
"      SUBROUTINES REQD:  (1) $COEFF  (2) $CHOMO  (3) $CROOT
"                         (4) $CINVS  (5) $CMPLY  (6) $CSUBN
"                         (7) $INVS   (8) $MPLY   (9) $SUBN
"
      SUBROUTINE $CMODL (M, C, K, U, ERROR, NITER, N)
      REAL*8  C(10,10), ERROR, H(10,10), H1(10,10), H2(10,10),    %
        K(10,10), M(10,10), MINVS(10,10), UNIT(10,10), ZERO(10,10)
      COMPLEX*16  DUM(10,10), ROOT(2,5), U(10,10), X(10)
      DATA  UNIT, ZERO /200*0.0/
      NT2 = N*2
      DO 40 I=1,N
40      UNIT(I,I) = 1.
      CALL $INVS (M, MINVS, N)
      CALL $MPLY (MINVS, C, H1, N)
      CALL $MPLY (MINVS, K, H2, N)
      DO 41 I=1,N
      DO 41 J=1,N
        H(I,J) = -H1(I,J)
        H(I,N+J) = -H2(I,J)
        H(N+I,J) =  UNIT(I,J)
41      H(N+I,N+J) = ZERO(I,J)
      CALL $CROOT (H, ROOT, ERROR, NITER, NT2)
      DO 42 JJ=1,N
      DO 42 II=1,2
        DO 43 I=1,N
        DO 43 J=1,N
43        DUM(I,J) = M(I,J)*ROOT(II,JJ)**2 + C(I,J)*ROOT(II,JJ) + K(I,J)
        CALL $CHOMO (DUM, X, N)
        JI = 2*(JJ-1) + II
        DO 42 L=1,N
          U(N+L,JI) = X(L)
42        U(L,JI) = U(N+L,JI)*ROOT(II,JJ)
      RETURN
      END
```

FIG. C-11. *Modal matrix of discrete systems with viscous damping; roots distinct.*

Let us compare the equations above with the statements in $CMODL listed in Fig. C-11.

1. the CALL $INVS statement gives M^{-1} (MINVS). The two CALL $MPLY statements give $M^{-1}C = H1$ and $M^{-1}K = H2$ in Eq. (C-29).
2. The matrix H is formed in the DO 41 loop. The rest of the program is almost identical to $MODL.
3. The CALL $CROOT statement yields the complex roots γ_i in Eq. (C-30).
4. The coefficient matrix $[M\gamma^2 + C\gamma + K]$ in Eq. (C-31) is obtained by means of the DO 43 loop, to be followed by the CALL $CHOMO statement for finding the complex modal vectors.
5. The modal matrix is determined by means of the last DO loop in the program.

C-13 $CRKUT—SOLUTION OF COMPLEX 1ST-ORDER DIFFERENTIAL EQUATION. METHOD: 4TH-ORDER RUNGE–KUTTA.

The subroutine $CRKUT finds the solution of a first-order differential equation

$$\frac{dz}{dt} + Cz = F \tag{C-32}$$

or

$$\Delta z = \Delta t(-Cz + F) \tag{C-33}$$

where C and F are constants. For the initial condition z_1, applying the fourth-order Runge–Kutta method gives

$$\begin{aligned} K_1 &= \Delta t(-C \times z_1 + F \\ K_2 &= \Delta t[-C \times (z_1 + K_1/2) + F] \\ K_3 &= \Delta t[-C \times (z_1 + K_2/2) + F] \\ K_4 &= \Delta t[-C \times (z_1 + K_3) + F] \end{aligned} \tag{C-34}$$

$$\Delta z = (K_1 + 2K_2 + 2K_3 + K_4)/6 \tag{C-35}$$

$$z = z_1 + \Delta z \tag{C-36}$$

Let us compare the equations above with the statements in $CRKUT listed in Fig. C-12.

1. Since A = DT/4, the subroutine computes a value of Z for every DT/4 and returns Z for the given DT.
2. *The DO* 40 *loop accounts for the number of equations N to be solved by the subroutine.*
3. The Eqs. (C-34) to (C-36) can be identified readily in the DO 41 loop.

```
$CRKUT

"   *** SOLUTION OF COMPLEX 1ST-ORDER DIFFERENTIAL EQUATIONS ***
"       RUNGE-KUTTA METHOD.  ASSUME CONSTANT FORCE.
"       SUBROUTINE CALCULATES VARIABLE-Z FOR EVERY DT/4.
"
      SUBROUTINE $CRKUT (F, Z1, C, DT, Z, N)
      REAL*8  A, DT
      COMPLEX*16  K1, K2, K3, K4, C(10), F(10), Z(10), Z1(10)
      A = DT/4.
      DO 40 I=1,N
        DO 41 J=1,4
          K1 = A*(-C(I)*Z1(I)  + F(I))
          K2 = A*(-C(I)*(Z1(I)+K1/2.)  + F(I))
          K3 = A*(-C(I)*(Z1(I)+K2/2.)  + F(I))
          K4 = A*(-C(I)*(Z1(I)+K3)  + F(I))
41        Z1(I) = Z1(I) + (K1 + 2.*K2 + 2.*K3 + K4)/6.
40      Z(I) = Z1(I)
      RETURN
      END
```

FIG. C-12. *Solution of* 1*st-order differential equations.*

C-14 $PLOTF—TO PLOT DATA FROM A FILE.

The subroutine $PLOTF listed in Fig. C-13 pots data stored in a file. It assumes that (1) the number of variables NVAR to be plotted is ≤9, and (2) the number of equally spaced data points NDATA≤61. The variables are plotted in the form of a family of curves.

```
$PLOTF

"   *** SUBROUTINE TO PLOT FROM A DATA FILE. ***
"
      SUBROUTINE $PLOTF (III)
      REAL*8  III
      DIMENSION  Y(9,61),PAGE(61,121),YMINI(9),YMAXI(9),IY(9,61),PT(9)
      DATA  BLANK,DOT,PLUS/' ','*','+'/
      DATA  PT/'1','2','3','4','5','6','7','8','9'/
80    FORMAT (5X,61A1)
81    FORMAT (4X,'-6    -5    -4    -3    -2    -1    0    1',%
      '    2    3    4    5    6')
82    FORMAT (5X,'0    1    2    3    4    5    6    7    8',%
      '    9    10    11    12 ')
83    FORMAT (5X,'+',12(4X,'+'))
84    FORMAT (' DO THE VALUES TO BE PLOTTED USE THE SAME SCALE?',/,%
      ' 1 = YES;    2 = NO.')
      CALL OPEN (1,III,'INPUT')
      READ(1,*) NDPT,N
      DO 40 I=1,NDPT
40      READ (1,*) (Y(K,I), K=1,N)
      CALL CLOSE (1)
      NSIZE = (NDPT*2) - 1
      DO 41 I=1,61
      DO 41 J=1,NSIZE
41      PAGE(I,J) = BLANK
"     FIND THE MAX AND MIN VALUES IN Y1,Y2 .... YN.
      DO 42 I=1,N
        CALL $MINAX (Y, YMIN, YMAX, I, NDPT)
        YMINI(I) = YMIN
42      YMAXI(I) = YMAX
      DO 43 I=1,N
        IF (YMINI(I)) 10,43,43
43      CONTINUE
      IKG = 1
      NGPT = 60
      GOTO 11
"     NEGATIVE VALUES OF Y1,Y2 ..... YN.
10    IKG = 31
      NGPT = 30
11    CONTINUE
      DO 44 I=1,N
44      IF (ABS(YMINI(I)) > ABS(YMAXI(I))) YMAXI(I) = ABS(YMINI(I))
"     YMAX CONTAINS ALL THE MAX VALUES ONLY.
      DO 45 I=1,61
        PAGE(I,1) = DOT
45      PAGE(I,NSIZE) = DOT
      DO 46 I=1,NSIZE
        PAGE(1,I) = DOT
46      PAGE(61,I) = DOT
      DO 47 I=1,61,5
        PAGE(I,1) = PLUS
47      PAGE(I,NSIZE) = PLUS
      DO 48 I=1,NSIZE,5
        PAGE(1,I) = PLUS
48      PAGE(61,I) = PLUS
      DO 49 I=1,N
49      YMAXI(I) = YMAXI(I)/NGPT
"     NOW YMAX HAS ALL THE SCALES.
```

FIG. C-13. *Plotting subroutine.*

```
      WRITE (6,84)
      READ (5,*) IDIFF
      IF (IDIFF = 2) GOTO 12
      TEM = 0.
      DO 50 I=1,N
50      IF (TEM < YMAXI(I)) TEM = YMAXI(I)
      DO 51 I=1,N
51      YMAXI(I) = TEM
12    CONTINUE
      IF (IKG - 31) 14,13,14
13    DO 52 L2=2,NSIZE
52      PAGE(31,L2) = PAGE(1,L2)
14    CONTINUE
      DO 53 J=1,NDPT
      DO 53 I=1,N
53      IY(I,J) = Y(I,J)/YMAXI(I) + .1
"     STORE THE INTEGER VALUES IN THE ARRAY PAGE.
      DO 54 J=1,NDPT
      DO 54 K=1,N
54      IY(K,J) = IY(K,J) + IKG
      J2=1
      DO 56 J=1,NDPT
        DO 55 K=1,N
55      PAGE(IY(K,J),J2)=PT(K)
56      J2=J2+2
      IF (IKG = 31) WRITE (6,81)
      IF (IKG = 1) WRITE (6,82)
      WRITE(6,83)
      WRITE (6,80) ((PAGE(J,I), J=1,61), I=1,NSIZE)
      RETURN
      END
"
      SUBROUTINE $MINAX (X, TEMIN, TEMAX, K, NDPT)
      DIMENSION  X(9,61)
      TEMIN = 0.
      TEMAX = 0.
      DO 40 I=1,NDPT
        IF (TEMIN > X(K,I)) TEMIN = X(K,I)
40      IF (TEMAX < X(K,I)) TEMAX = X(K,I)
      RETURN
      END
```

FIG. C-13 (*continued*).

Appendix D

Linear Ordinary Differential Equations with Constant Coefficients

We shall briefly review the solution of linear ordinary differential equations with constant coefficients by the "classical" method. This type of equation is used extensively in the text.

D-1 DEFINITIONS

A *differential equation* is an equation relating the variables and their derivatives or differentials. A typical linear ordinary differential equation with constant coefficients is of the form

$$m\frac{d^2x}{dt^2}+c\frac{dx}{dt}+kx=f(t) \tag{D-1}$$

where x, the unknown, is the *dependent variable*, t the *independent variable*, m, c, and k are the *constant coefficients*, and $f(t)$ a known function of t. The equation relates the dependent variable x and its derivatives with the independent variable t. If x represents a displacement and t time, then $m(d^2x/dt^2)$ is the inertia force, $c(dx/dt)$ the viscous damping force, kx the spring force, and $f(t)$ the excitation force applied to the system. For convenience, the symbol ($\dot{}$) is used to denote a time derivative, that is, $\dot{x}=dx/dt$ and $\ddot{x}=d^2x/dt^2$.

The *order* of the differential equation is the order of the highest derivative; Eq. (D-1) is of second order.

Since there is only one independent variable in Eq. (D-1), it is an *ordinary differential equation*. When the unknown is a function of two or more independent variables, partial derivatives are used to describe the functional relation of the variables, resulting in a partial differential equation.

A differential equation is *linear* if none of the terms in the equation involving the dependent variable and its derivatives (1) appear as products or (2) are raised to a power different from unity. For example, the equation

$$t^2\ddot{x} + t\dot{x} + (t^2 - n^2)x = 0$$

which has the $t^2\ddot{x}$, $t\dot{x}$, and t^2x terms, is linear. The unknown x and its derivatives do not appear as products and each is raised to a power of unity. The coefficients of x and its derivatives, however, are variables. This is a linear ordinary differential equation with variable coefficients. No formulas are available for the general solution of such equations of order greater than one. In contrast, Eq. (D-1) is a linear ordinary differential equation with constant coefficients, which can be solved by a "standard" procedure.

D-2 SOLUTION

The *solution* of a differential equation is a functional relation of the dependent and independent variables and no derivatives are involved in the relation. For example,

$$x = C_1 \sin t \tag{D-2}$$

is a solution of the differential equation

$$\ddot{x} + x = 0 \tag{D-3}$$

The solution in Eq. (D-2) relates the dependent variable x and the independent variable t. In other words, a plot of x versus t can be obtained from Eq. (D-2).

To verify that Eq. (D-2) is a solution of (D-3), we differentiate Eq. (D-2) twice with respect to t. Substituting in Eq. (D-3) gives

$$-C_1 \sin t + C_1 \sin t = 0$$

and Eq. (D-3) is satisfied. The value of C_1 is arbitrary and C_1 is called a *constant of integration.* Similarly, it can be shown that $x = C_2 \cos t$ is also a solution of Eq. (D-3). Hence a more general solution of Eq. (D-3) is

$$x = C_1 \sin t + C_2 \cos t$$

Note that the solution of the second-order differential equation in Eq. (D-3) has two constants of integration, C_1 and C_2.

To generalize the observation, consider an nth order equation

$$a_n \frac{d^n x}{dt^n} + a_{n-1}\frac{d^{n-1}x}{dt^{n-1}} + \cdots + a_1 \frac{dx}{dt} + a_0 x = f(t) \tag{D-4}$$

where the a's are constants. This is a linear ordinary differential equation with constant coefficients. For brevity, we shall call it an nth-order differential equation. It is stated without proof that the *general solution* of an nth-order differential equation has n arbitrary constants of integration. Solutions obtained from the general solution by giving particular values to these constants are called *particular solutions.*

It is convenient to introduce the D and L *linear differential operators* for this discussion. The *D operator* is defined by the operations

$$Dx = \frac{d}{dt}x = \frac{dx}{dt}, \qquad D^2x = \frac{d}{dt}\frac{d}{dt}x = \frac{d^2x}{dt^2}, \ldots, \qquad D^n x = \frac{d^n x}{dt^n}$$

The symbol $D = d/dt$ is an operator. It has no physical interpretation unless it is applied to a differentiable function $x(t)$. When it is used in a linear ordinary differential equation with constant coefficients, it can be manipulated as an algebraic quantity. With the D operator, Eq. (D-4) can be expressed as

$$(a_n D^n + a_{n-1}D^{n-1} + \cdots + a_2 D^2 + a_1 D + a_0)x = f(t) \qquad \textbf{(D-5)}$$

or more precisely as

$$Lx = f(t) \qquad \textbf{(D-6)}$$

where the L operator is defined as

$$L = (a_n D^n + a_{n-1}D^{n-1} + \cdots + a_1 D + a_0) \qquad \textbf{(D-7)}$$

Since derivatives have the property

$$\frac{d^p}{dt^p}(C_1x_1 + C_2x_2) = C_1\frac{d^p}{dt^p}x_1 + C_2\frac{d^p}{dt^p}x_2$$

The *L operator* is linear in the respect that

$$L(C_1x_1 + C_2x_2) = C_1Lx_1 + C_2Lx_2 \qquad \textbf{(D-8)}$$

Hence the *superposition principle* applies to the L operator.

Equation (D-4) or (D-6) is called an nth-order *nonhomogeneous* differential equation. The solution that satisfies the nonhomogeneous equation is called the *particular integral* $x_p(t)$, that is,

$$Lx_p = f(t) \qquad \textbf{(D-9)}$$

The corresponding nth-order *homogeneous* differential equation is obtained if $f(t) = 0$ in Eq. (D-6), that is,

$$Lx = 0 \qquad \textbf{(D-10)}$$

The solution of Eq. (D-10) is called the *complementary function* $x_c(t)$. Let x_{c1} be a solution of Eq. (D-10), that is, $Lx_{c1} = 0$. Similarly, let x_{c2},

$x_{c3}, \ldots, x_{cn}$ be the solutions of Eq. (D-10). Let $x_c(t)$ be a linear combination of the individual solutions.

$$x_c = C_1 x_{c1} + C_2 x_{c2} + C_3 x_{c3} + \cdots + C_n x_{cn}) \tag{D-11}$$

where the C's are constants. From the linear property of the L operator, we have

$$\begin{aligned} Lx_c &= L(C_1 x_{c1} + C_2 x_{c2} + \cdots + C_n x_{cn}) \\ &= C_1 L x_{c1} + C_2 L x_{c2} + \cdots + C_n L x_{cn} = 0 \end{aligned} \tag{D-12}$$

Hence a linear combination of the solutions of the homogeneous equation $Lx = 0$ is also a solution of the equation. It is stated without proof that an nth-order homogeneous differential equation has n linearly independent* solutions and its general solution is a linear combination of the n independent solutions.

The *general solution* $x(t)$ of an nth-order nonhomogeneous differential equation $Lx = f(t)$ in Eq. (D-6) consists of the complementary function $x_c(t)$ and the particular integral $x_p(t)$, that is,

$$x = x_c + x_p \tag{D-13}$$

To verify the statement, substituting this in Eq. (D-6) gives

$$Lx = L(x_c + x_p) = 0 + f(t) \tag{D-14}$$

Hence Eq. (D-6) is satisfied and the general solution is of the form shown in Eq. (D-13). The next step is to evaluate $x_c(t)$ and $x_p(t)$.

D-3 COMPLEMENTARY FUNCTION

The complementary function is the solution of the homogeneous differential equation

$$a_n \frac{d^n x}{dt^n} + a_{n-1} \frac{d^{n-1}}{dt^{n-1}} + \cdots + a_1 \frac{dx}{dt} + a_0 x = 0 \tag{D-15}$$

Let $x_c = e^{st}$ be a solution, where s is a constant. Substituting x_c in Eq. (D-15) and factoring out the e^{st} term, we get

$$a_n s^n + a_{n-1} s^{n-1} + \cdots + a_1 s + a_0 = 0 \tag{D-16}$$

This is called the auxiliary or the *characteristic equation.* It is the condition that must be satisfied in order that $x_c = e^{st}$ be a solution. By the fundamental theorem of algebra, Eq. (D-16) has n number of roots, $s_1, s_2, \ldots, s_n$. If the n roots are distinct, the general solution of Eq. (D-15) is

$$x_c = C_1 e^{s_1 t} + C_2 e^{s_2 t} + \cdots + C_n e^{s_n t} \tag{D-17}$$

* For a discussion of linear dependence, see, for example, F. B. Hildebrand, *Advanced Calculus for Engineers,* Prentice-Hall, Inc., Englewood Cliffs, N.J., 1949, p. 3.

Thus, it is seen that the roots of the characteristic equation establish the behavior of the system as shown in Eq. (D-17).

The roots of Eq. (D-16) can be distinct or repeating. For completeness, we shall discuss the cases for which the roots are real, imaginary, or complex. For simplicity, we confine the discussion to the second-order differential equation, which is used extensively in Chaps. 2 and 3.

Case 1. Roots Real and Distinct

Consider the second-order differential equation

$$m\ddot{x} + c\dot{x} + kx = 0 \tag{D-18}$$

The corresponding characteristic equation is

$$ms^2 + cs + k = 0 \tag{D-19}$$

The roots $s_{1,2}$ of Eq. (D-19) and the general solution of (D-18) are

$$s_{1,2} = \frac{1}{2m}(-c \pm \sqrt{c^2 - 4mk}) \tag{D-20}$$

$$x_c = C_1 e^{s_1 t} + C_2 e^{s_2 t} \tag{D-21}$$

Note that $c^2 > 4mk$ in order that the roots be real. Since $\sqrt{c^2 - 4mk} < c$, the roots are real and negative. Thus, Eq. (D-21) indicates that the solution $x_c(t)$ decreases exponentially with increasing time.

Case 2. Roots Real and Repeating.

If the roots are repeating, then $s_1 = s_2$ and the constants C_1 and C_2 in Eq. (D-21) can be combined to give a new constant. Hence, only one solution of the form e^{st} is obtained. For equal roots from Eq. (D-20), we have $c^2 = 4mk$ or

$$s = s_1 = s_2 = -c/2m \tag{D-22}$$

Dividing Eq. (D-18) by m and using the D operator, we get

$$(D^2 - 2sD + s^2)x = 0 \quad \text{or} \quad (D - s)(D - s)x = 0 \tag{D-23}$$

where $s^2 = c^2/(2m)^2 = 4mk/(2m)^2 = k/m$. Defining

$$(D - s)x = y \tag{D-24}$$

Equation (D-23) can be written as

$$(D - s)y = 0 \tag{D-25}$$

Clearly, $y = e^{st}$ is a solution of this equation. Substituting $y = e^{st}$ in Eq. (D-24) gives

$$(D - s)x = e^{st} \tag{D-26}$$

It can be shown by the method of undetermined coefficients (see Sec. D-4) that the solution of Eq. (D-26) is of the form

$$x = te^{st} \tag{D-27}$$

This is the second solution of Eq. (D-23). The general solution is a linear combination of the two solutions, that is,

$$x_c = (C_1 + C_2 t)e^{st} \tag{D-28}$$

Case 3. Roots Complex.

The roots of the characteristic equation $s_{1,2}$ in Eq. (D-20) are complex if $c^2 < 4mk$. Let the complex conjugate roots be

$$s_{1,2} = -\alpha \pm j\beta \tag{D-29}$$

where $j = \sqrt{-1}$, $\alpha = c/2m$, $\beta = \sqrt{4mk - c^2}/2m$, and α and β are real and positive. For distinct roots, the solution from Eq. (D-17) is

$$\begin{aligned} x_c &= C_1 e^{(-\alpha + j\beta)t} + C_2 e^{(-\alpha - j\beta)t} \\ &= e^{-\alpha t}(C_1 e^{+j\beta t} + C_2 e^{-j\beta t}) \end{aligned} \tag{D-30}$$

Applying Euler's formula $e^{\pm j\theta} = \cos\theta \pm j \sin\theta$, this equation becomes

$$\begin{aligned} x_c &= e^{-\alpha t}[(C_1 + C_2)\cos\beta t + j(C_1 - C_2)\sin\beta t] \\ &= e^{-\alpha t}(A\cos\beta t + B\sin\beta t) \end{aligned} \tag{D-31}$$

Since $x_c(t)$ is real, the constants A and B above must be real quantities. This implies that the constants $C_{1,2}$ in Eq. (D-30) must be complex conjugates.

Case 4. Roots Imaginary.

The roots of the characteristic equation $s_{1,2}$ in Eq. (D-20) are imaginary if $c = 0$. thus, $s_{1,2} = \pm j\sqrt{k/m} = \pm j\omega_n$. Since the roots are distinct, the solution is of the form

$$x_c = C_1 e^{+j\omega_n t} + C_2 e^{-j\omega_n t}$$

Comparing with Eq. (D-30), it is evident that this is a special case of complex roots when the real part α of the complex roots is zero. Hence, the solution can be expressed as

$$x_c = A\cos\omega_n t + B\sin\omega_n t \tag{D-32}$$

D-4 PARTICULAR INTEGRAL

The particular integral $x_p(t)$ is a solution of the nonhomogeneous equation as shown in Eq. (D-4) or (D-6). It satisfies the equation

$$Lx_p = f(t) \tag{D-33}$$

By definition, $x_p(t)$ does not contain any arbitrary constants. Thus, $x_p(t)$ does not contain the complementary function $x_c(t)$. If $x_c(t)$ is known, $x_p(t)$ can be found by the method of variation of parameters,* which in general is laborious.

For many practical problems the function $f(t)$ in Eq. (D-33) is a sine or cosine, a polynomial, or a product or a linear combination of such functions. For these functions, the *method of undetermined coefficients* give a simple procedure to solve for $x_p(t)$. The method is applicable, if Eq. (D-33) is a linear ordinary differential equation with constant coefficients and $f(t)$ consists of functions, possessing a finite number of linearly independent derivatives.

If $f(t)$ consists of a number of terms, such as

$$Lx = f(t) = f_1(t) + f_2(t)$$

the particular integral is $x_p(t) = x_{p1}(t) + x_{p2}(t)$, where $Lx_{p1} = f_1(t)$ and $Lx_{p2} = f_2(t)$. Since the L operator is linear, we have

$$Lx_p = L(x_{p1} + x_{p2}) = f_1(t) + f_2(t) \qquad \textbf{(D-34)}$$

The method of undetermined coefficients can be stated in Rules 1 and 2 to follow. It is assumed that $x_c(t)$, the solution of $Lx = 0$, has been found.

Rule 1. *The trial function to evaluate $x_p(t)$ in Eq. (D-33) is a linear combination of $f(t)$ and all its independent derivatives.*

Example 1

Find the particular integral of the equation

$$\ddot{x} + \omega_n^2 x = F \sin \omega t \qquad \textbf{(D-35)}$$

Solution:

Since the successive derivatives of $\sin \omega t$ consist of $\sin \omega t$ and $\cos \omega t$ only, the trial function is of the form

$$x_p = A \sin \omega t + B \cos \omega t \qquad \textbf{(D-36)}$$

The coefficients A and B are as yet undetermined. Substituting Eq. (D-36) in (D-35) and collecting the sine and cosine terms gives

$$(\omega_n^2 - \omega^2) A \sin \omega t + (\omega_n^2 - \omega^2) B \cos \omega t = F \sin \omega t$$

Hence, the values of A and B are

$$A = F/(\omega_n^2 - \omega^2) \qquad \text{and} \qquad B = 0$$

Rule 2. *When the trial function $x_p(t)$ of Rule 1 has $f(t)$ or its independent derivatives proportional to a term in $x_c(t)$, a new trial function is substituted.*

* *Ibid.*, p. 26.

The new trial function is the product of the initial trial function and the lowest integral power of t, such that none of the terms in $x_p(t)$ is proportional to the terms in $x_c(t)$.

Example 2.

Find the particular integral of the equation

$$\ddot{x} + \omega_n^2 x = F \sin \omega_n t \tag{D-37}$$

Solution:

It can be shown that the complementary function of the equation is $x_c = C_1 \sin \omega_n t + C_2 \cos \omega_n t$. By Rule 1, the trial function for $x_p(t)$ is $x_p = A \sin \omega_n t + B \cos \omega_n t$, which is proportional to the terms in $x_c(t)$. Hence, the new trial function by Rule 2 is

$$x_p = (A \sin \omega_n t + B \cos \omega_n t)t \tag{D-38}$$

Substituting Eq. (D-38) in (D-37) and simplifying, we get

$$2\omega_n A \cos \omega_n t - 2\omega_n B \sin \omega_n t - \omega_n^2 At \sin \omega_n t - \omega_n^2 Bt \cos \omega_n t$$
$$+ \omega_n^2 At \sin \omega_n t + \omega_n^2 Bt \cos \omega_n t = F \sin \omega_n t$$

Further simplifying and equating the coefficients of the sine and cosine terms, we obtain

$$A = 0 \qquad \text{and} \qquad B = -F/2\omega_n$$

Thus, the trial function in Eq. (D-38) yields $x_p(t) = \dfrac{-Ft}{2\omega_n} \cos \omega t$ as prescribed by Rule 2.

Example 3.

Find the general solution of the equation

$$\ddot{x} - \dot{x} = 3t^2 - 4t + 5 + 2e^t + \sin t$$

Solution:

Writing the equation in operator form gives

$$Lx = D(D-1)x = f_1(t) + f_2(t) + f_3(t) \tag{D-39}$$

where

$f_1(t) = 3t^2 - 4t + 5$, a polynomial

$f_2(t) = 2e^t$, an exponential

$f_3(t) = \sin t$, a sine function

The corresponding homogeneous equation and the characteristic equation are

$$Lx = D(D-1)x = 0 \tag{D-40}$$

$$s(s-1) = 0 \tag{D-41}$$

The roots of Eq. (D-41) are $s = 0$ and $s = 1$. The complementary function is

$$x_c = C_1 + C_2 e^t \qquad \textbf{(D-42)}$$

By Rule 1, $x_{p1}(t)$ due to $f_1(t)$ is a linear combination of $f_1(t)$ and all of its independent derivatives. Thus, the trial function is

$$x_{p1} = At^2 + Bt + C$$

Note that C_1 in $x_c(t)$ in Eq. (D-42) is proportional to C in the equation above. By Rule 2, the new trial function is

$$x_{p1} = (At^2 + Bt + C)t$$

Substituting the new trial function $x_{p1}(t)$ in $Lx = f_1(t)$ gives

$$6At + 2B - 3At^2 - 2Bt - C = 3t^2 - 4t + 5$$

Thus, $-3A = 3$, $6A - 2B = -4$, and $2B - C = 5$, giving $A = -1$, $B = -1$, and $C = -7$. Hence $x_{p1}(t)$ becomes

$$x_{p1} = -t^3 - t^2 - 7t \qquad \textbf{(D-43)}$$

Similarly, by Rule 1, $x_{p2}(t)$ due to $f_2(t)$ is Ee^t. Since this is proportional to $C_2 e^t$ in Eq. (D-42), the new trial function, by Rule 2, is Ete^t. Substituting this in $Lx = f_2(t)$ gives

$$x_{p2} = 2te^t \qquad \textbf{(D-44)}$$

By Rule 1, the $x_{p3}(t)$ due to $f_3(t)$ is of the form

$$x_{p3} = F \sin t + G \cos t$$

Substituting x_{p3} in $Lx = f_3(t)$ gives

$$-F \sin t - G \cos t - F \cos t + G \sin t = \sin t$$

Thus, the coefficients are $F = -\frac{1}{2}$ and $G = \frac{1}{2}$ and x_{p3} is

$$x_{p3} = -\tfrac{1}{2} \sin t + \tfrac{1}{2} \cos t \qquad \textbf{(D-45)}$$

The *general solution* of the given differential equation is the combination of Eqs. (D-42) through (D-45).

$$\begin{aligned} x &= x_c + x_{p1} + x_{p2} + x_{p3} \\ &= (C_1 + C_2 e^t) - (t^2 + t + 7)t + 2te^t - \tfrac{1}{2}(\sin t - \cos t) \end{aligned} \qquad \textbf{(D-46)}$$

where C_1 and C_2 are constants of integration, which are generally determined by the initial conditions of the problem.

D-5 PARTICULAR SOLUTION

The general solution $x(t)$ of a differential equation is the sum of the complementary function $x_c(t)$ and the particular integral $x_p(t)$ as shown in Eq. (D-13), that is, $x = x_c + x_p$. Note that $x_c(t)$ has as many arbitrary constants, or constants of integration, as the order of the differential equation as indicated in Eq. (D-11). For example, the general solution of the second-order differential equation in Example 3 above has two arbitrary constants as shown in Eq. (D-46). Solutions obtained from the general solution by giving particular values to these constants are called *particular solutions.* We shall use Example 3 to illustrate the particular solution due to initial conditions.

Example 4.

Find the particular solution of Example 3 for the initial conditions $x(0) = 1$ and $\dot{x}(0) = 0$.

Solution:

From Eq. (D-46), we have

$$x = (C_1 + C_2 e^t) - (t^2 + t + 7)t + 2te^t - \tfrac{1}{2}(\sin t - \cos t)$$

$$\dot{x} = (0 + C_2 e^t) - (3t^2 + 2t + 7) + 2(te^t + e^t) - \tfrac{1}{2}(\cos t + \sin t)$$

Substituting the initial conditions, we get

$$x(0) = 1 = (C_1 + C_2) - 0 + 0 - \tfrac{1}{2}(0 - 1) \qquad \text{or} \qquad C_1 + C_2 = \tfrac{1}{2}$$

$$\dot{x}(0) = 0 = (0 + C_2) - (0 + 0 + 7) + 2(0 + 1) - \tfrac{1}{2}(1 + 0) \qquad \text{or} \qquad C_2 = 5\tfrac{1}{2}$$

Hence the particular solution is

$$x = -(5 - 5\tfrac{1}{2}e^t) - (t^2 + t + 7)t + 2te^t - \tfrac{1}{2}(\sin t - \cos t)$$

It is important to note that the arbitrary constants are evaluated by applying the initial conditions to the general solution of the given differential equation. Although these constants occur only in the complementary function $x_c(t)$, it is the entire system that must satisfy the given initial conditions, or whatever constraint applied to the system.

D-6 SIMULTANEOUS DIFFERENTIAL EQUATIONS

The method described in the previous sections can be used to solve simultaneous linear ordinary differential equations with constant coefficients, such as described in Chap. 4. The procedure is to (1) substitute the D operator for the time derivatives in the equations, (2) treat the resulting equations as algebraic equations and solve for each of the dependent variables, and then (3) solve each of the equations from part 2 by the method described in the previous sections. The procedure is relatively simple, but the algebra can be laborious. We shall briefly describe the method.

Consider Eq. (4-4) from Chap. 4.

$$\begin{bmatrix} m_{11} & m_{12} \\ m_{21} & m_{22} \end{bmatrix} \begin{bmatrix} \ddot{x}_1 \\ \ddot{x}_2 \end{bmatrix} + \begin{bmatrix} c_{11} & c_{12} \\ c_{21} & c_{22} \end{bmatrix} \begin{bmatrix} \dot{x}_1 \\ \dot{x}_2 \end{bmatrix} + \begin{bmatrix} k_{11} & k_{12} \\ k_{21} & k_{22} \end{bmatrix} \begin{bmatrix} x_1 \\ x_2 \end{bmatrix} = \begin{bmatrix} f_1(t) \\ f_2(t) \end{bmatrix}$$

Using the D operator, the equations can be expressed as

$$(m_{11}D^2 + c_{11}D + k_{11})x_1 + (m_{12}D^2 + c_{12}D + k_{12})x_2 = f_1(t)$$
$$(m_{21}D^2 + c_{21}D + k_{21})x_1 + (m_{22}D^2 + c_{22}D + k_{22})x_2 = f_2(t)$$

Using the L operator, the equations can be written concisely as

$$\begin{aligned} L_{11}x_1 + L_{12}x_2 &= f_1(t) \\ L_{21}x_1 + L_{22}x_2 &= f_2(t) \end{aligned} \qquad \textbf{(D-47)}$$

The operator L_{ij} can be readily identified by comparing the last two sets of equations. For example, $L_{12} = (m_{12}D^2 + c_{12}D + k_{12})$. Since $m_{12} = m_{21}$, $c_{12} = c_{21}$, and $k_{12} = k_{21}$, we have $L_{12} = L_{21}$.

The dependent variables $x_1(t)$ and $x_2(t)$ in Eq. (D-47) are obtained from the simultaneous solution of the equations. Using Cramer's rule, we get

$$\begin{aligned} (L_{11}L_{22} - L_{12}L_{21})x_1 &= L_{22}f_1(t) - L_{12}f_2(t) \\ (L_{11}L_{22} - L_{12}L_{21})x_2 &= L_{11}f_2(t) - L_{21}f_1(t) \end{aligned} \qquad \textbf{(D-48)}$$

Each of the equations in Eq. (D-48) is a linear ordinary differential equation with constant coefficients. Hence the method described previously can be applied directly, although the algebra can be very laborious.

INDEX